MATRIX METHODS OF STRUCTURAL ANALYSIS

Matrix Methods of Structural Analysis

M. B. Kanchi
Professor of Civil Engineering,
Roorkee University, India

A HALSTED PRESS BOOK

John Wiley & Sons
New York Chichester Brisbane Toronto

Published in the U.S.A., Canada and Latin America
by Halsted Press, a Division of John Wiley & Sons,
Inc, New York

Library of Congress Cataloging in Publication Data

Kanchi, M B
Matrix methods of structural analysis
"A Halsted Press book"
1. Structures, Theory of matrix methods
I Title
TA 642.K36 624.1′71 80-18442
ISBN 0-470-26945-6

Printed in India at India Printers, Delhi

To
Gitanjali, Anuradha
and Kanaklata

Preface

With the increasing use of digital computers in industry and education considerable changes have taken place over the last two decades in the teaching of the courses of structural mechanics. Many texts have been continually reviewed so as to include matrix method and computer applications. Students now have more background of matrix algebra and computer programming during the early years of their undergraduate studies. They have been able to accept the use of matrix notations and computer aspects during the study of the courses of structural analysis. However it is more or less an accepted practice nowadays to introduce matrix notations as and when required during the teaching of the basic courses and reserve the matrix methods of analysis and computing details as a separate course to be given during the final year. Such a course is invariably compulsory for students of civil engineering and optional to those of mechanical and aeronautical disciplines. This is recognized as a more effective practice wherein the matrix notations and computing aspects do not obscure the very important requirement of having to put across to the student the basic concepts of analysis during the initial courses. Also, such a course can conveniently find place in the post-graduate curriculum of structural engineering where for some reasons it cannot be offerred as elective at the final year of undergraduation. This is common till today in many institutions. One finds it assuring that more and more spacializations are being introduced in the undergraduate program of civil engineering with possible majors in structurces, water resources etc.

This book has been written as a text keeping in view the above features of engineering curricula. Obviously it has not been written for any particular course of any institution. Rather the subject matter has been treated in a way that it is complete in all its necessary aspects, so that it becomes a useful piece by itself, usable by students for one or more than one of their courses of study, by serving engineers in industry, who might not have studied this course during their undergraduate program, and also specially by the young teacher who is preparing to teach this subject. I believe that there are many aspects of the book which are characteristically different and are of special appeal and use to many in the academic profession.

The book has essentially two objectives: the first, to identify and expose the principles and approaches of structural analysis common to all structures in general, such as requirements of equilibrium, kinematics, member response laws, virtual work principles, approaches of force and displacement methods, etc. The second is the presentation of the analysis procedures so as to enable one to translate them into computer algorithms such as reducing force and displacement methods of mere algebraic steps involving matrices and vectors, matters related to input of data, considerations of the limitations as well as versatilities of the computer etc. To this end I have tried to bring under one cover the treatment of these two essential scopes, namely, exposition of the general principles and approaches of analysis and presentation of essential aspects of computer applications. Each of these is indeed complementary to the other, and when they are combined together a completeness can be achieved. One of the noticeable impact of the digital computer is the emergence of ready-made program packages, requiring on the part of the user only the learning of 'user manual'. To such program users the first aspect of having to gain the theoretical insight into the program packages is not required. The book is however addressed to those programmers who are expected to develop programs on their own, or are able to modify them to meet new requirements. These programmers have necessarily to combine within themselves the physical understanding of the structural manipulations and the skill of computer programming. Though this is a continuous process, University is the right place to start.

The book displays a few important features which are largely my preferences. The first preference is the choice of dealing the subject matter under the headings such as equilibrium, kinematics, member stiffness etc. rather than the structure-oriented headings such as pinjointed frames, rigid jointed frames etc. Such a treatment would give the student the confidence of being able to identify the principles and issues underlying all structures in general. The second is the preference given to virtual work techniques over energy extremum ones, in view of the appeal and the applicability to linear as well as nonlinear problems involving either line or finite elements.

The third preference is to adopt a treatment which would be the same for analysis of structural assemblies consisting of either beam (line) elements or finite (area and volume) elements. A chapter on finite elements has been added at the end of the book taking the case of only plane elasticity, which is well within the grasp of the final year undergraduate student. The essential objective has been to demonstrate that the procedures of analysis need not distinguish between the types of elements involved in a structure. This helps in recognizing common features of analysis and in handling practical situations where it is common to find structural elements like edge-beams, plates and shells assembled together as in buildings, bridges, aircraft etc. Ideas and derivations have been shown to be identical and at many places the matrix forms of the equations are also shown to be identical. The method of element analysis based on an assumed deformation pattern is introduced while treating beam element itself, so that its use in the chapter on finite element comes without surprise. It is not unusual to find in many texts on finite elements its principles illustrated through the examples of line elements.

Now as to computer aspect: the extent to which a book should include computing details is rather debatable. Clearly, it is the duty of the teacher to bridge the gap between the equations he puts down on the board and computer programming which his student knows. Bridging of this gap is often neglected and it is not unusual to find perplexed student wondering as to how to start, even when he has some one else's program listing on his table. A lot of time of student can be saved if some hints are given in implementing the equations of structural analysis, so that they do not remain the tricks of the trade. Comments have been introduced on computing aspects as and where required in the text. Two sample programs have been appended with adequate comments so that they may serve as models to help the student to draw his own programs.

As another aid to the student are the illustrative examples to clarify to him the details behind the matrix equations. Anyone aspiring to master programming structural algorithms will seldom succeed unless he solves at least one example or writes the scalar details of the matrix equations. Many times one has to have a hand-computed problem to test the program. Examples have been discussed keeping this objective in view rather than demonstrating the solution of the structure involved in the example.

I am deeply indebted to the numerous books and papers; in particular the work of B.G. Neal, R.K. Livesley and O.C. Zienkiewicz. Also to the student community to whom this book is addressed and who are the motivation behind the present effort. Special thanks to one of my students, Vinod Kumar Agarwal for his contribution in drawing the computer programs given in the book.

M. B. KANCHI

Tucson, Arizona
June 30, 1981

Contents

Chapter 1

Review

A broad review of the scope, the basic principles, and the approach of structural analysis will be given in this chapter with a view to orient the student reader to take up the study of structural analysis with generality employing matrix notations and computer oriented formulations. This text presupposes that the student reader has undergone the basic courses of structural analysis. It is likely that some of them may have been exposed to the use of matrix notations during the basic courses themselves, such as when dealing with analysis of pinjointed frameworks, member stiffness and flexibility studies, use of slope-deflection equations, etc. The review however is undertaken here to bring out the general features of analysis common to all structures. This is very essential in the study of the subject leading to computer oriented formulations. Also, the chapter serves to show the approach adopted in the book.

1.1 STRUCTURAL ANALYSIS

a. STRUCTURES

A structure is an assembly of a number of small units generally called members or elements. A building frame—both from the point of view of construction and analysis—consists of beams, columns, floors, wall panels and foundation blocks. Similarly the components of a bridge are longitudinal girders, cross girders, decking slab, braces, etc. We generally call such components as *members* or *elements* and their assembly as a *structure*. A structure while serving its functional purpose is subjected to forces, and the essential task of the designer is to proportion the different elements and build a stiff enough structure consistent with strength and economy.

A load resisting body may be conveniently classified, at least from the point of view of analysis as (a) Framed structures; (b) Surface structures;

and (c) Solids: A *framed structure* is a network of beam (line) elements. A beam element is a unidimensional element with its cross-sectional dimensions very small compared to its length. In most of the framed structures straight beam elements are invariably used. They also form from the point of view of analysis, a simplest class of structural elements. *Surface structures* are built up of two dimensional elements such as thin slabs, membranes and thin shells. The thickness is very small compared to the lateral dimensions. Massive structures consist of *solid elements.* We can classify a dam, or a rock mass, or a thick container like the modern prestressed concrete pressure vessel into this class. A solid element is truely three dimensional from the point of view of analysis.

In a practical structure different kinds of elements invariably occur together. Thus in a building unidimensional (line) elements such as beams and columns interact with two dimensional (plane) elements, namely floors and wall panels. Similar situation occurs in a bridge with edge beams and deck slab elements. In the modern context where digital computers are invariably used, the analysis procedure does not distinguish between such elements but treats whole of the structure as an assembly of structural elements of different types. The *finite element technique* which is now an established procedure is concerned with the problem of providing a numerical solution to the analysis of surface structures and solids. These are imagined to be subdivided into a number of surface or solid elements.

b. Requirements of Structural Analysis

The essential requirement of a structure is that it should resist the loads without undue deformations. All materials develop internal resistance and in doing so undergo deformations. The first requirement is that the structural member is strong enough to develop internal forces of resistance at any location equal to the resultant action of the loads at that location. This is the *requirement of equilibrium* between the internal forces of resistance and the external loads and must always hold good both for static and dynamic behaviour of any structure. Since the material is not ideally rigid deformations (distortions) of the members occur, such as axial straining, bending, twisting, etc. The second aspect of structural analysis is to study the load-deformation characteristics of the structural member. Deformations are therefore related to the loads through stiffness (or flexibility) parameters of the member which depend on the strength of the material and the shape of the member. These can be determined by direct experiment or constructing theories based partly on observed behaviour and on simplifying assumptions with a view to evolving analytical expressions applicable under a given set of conditions. This study of *stiffness properties* of structural members is the second aspect with which structural analysis is invariably concerned. The *third* aspect with which structural analysis is concerned is the one of obtaining displacements, namely deflections and rotations at certain locations, such as at the centre of a bridge or the top of a building. The deformations of structural members, which are in general different for different members, do not occur freely since the members are connected to one another to form the structure which as a whole displaces itself according to the constraints exerted by the connectivity of the members and the supports. For example, the deflection of a node point of a pinconnected truss work can only be determined by considering simultaneously the elongations of all members and the constraints exerted by the members which are connected to one another. This is the problem of the familiar Williot-Mohr construction. Similarly the calculation of the change in the sag of a suspended cable due to extensions or to temperature change, or the calculation of the amount by which the axial load applied at the top of a column descends due to the bending of the column, are examples which are purely matters of geometry. This is referred to as *kinematics* which is the property of the structural form. Kinematics relates the deformations of structural members to the displacements of certain chosen locations.

In the analysis of any structure the above three aspects namely of equilibrium, stiffness properties of members, and kinematics of the structure are always to be considered. We shall briefly discuss the general features of these three aspects in the following sections.

Equilibrium Requirements

The requirement of equilibrium between the loads and internal forces is an absolutely necessary condition to be satisfied in any structure under any kind of equilibrium, static, dynamic or neutral. The conditions of statics are to be satisfied between the forces acting on the structure as a whole in space or between those acting on any portion, or any element of the structure. The nature of equilibrium may be either stable, unstable or neutral. In case of a structure under vibration the conditions of equilibrium are changing with respect to time, but at any given instant the internal forces are in equilibrium with the inertia and external forces.

Statically determinate structures are the ones in which the internal forces are directly determinable from the conditions of statics applied to different parts of the structure. If the displacements are small the changes in the geometrical configuration of the structure due to loads can be ignored. This leads to equations of statics which will be linear in load terms enabling the use of superposition principle. Determinate structures form an important class of structures possessing certain advantages and are widely used in practice. They also occupy an important position in structural analysis since statically indeterminate structures are reduced to equivalent determinate structures as an intermediate stage in the process of analysis. (Recall the force method.)

The consideration of equilibrium between the internal forces and the loads in a *redundant* structure is rather involved on account of the important fact that as far as the requirement of statics alone is concerned, there is no unique internal force distribution for a given set of loads. In other words more than one distributions are possible. Additional conditions of the behaviour of the structure are to be supplemented. The distribution determined by the elastic theory forms one of the possible distributions consistent with the conditions of elastic behaviour of members. The internal force distribution would change with the changes in the stiffness of structural elements.

Redundant structures also have the feature that the members can have forces without any apparent external loads such loads are temperature, creep, lack of fit of members and yielding of supports. There would be similar self-equilibrating internal forces when some of the members have residual deformations such as when loaded beyond the yield limit and then unloaded.

It is pertinent to emphasize here the significance of statics in structural analysis relative to that of other requirements, specially from the practical viewpoint of the safety of a structure. The criteria of deformations are equally important specially from the point of view of functional efficiency of the structure, such as limitations on deflections, crack widths, etc. The present day *Limit State Design* codes reflect very much the importance of permissible deformations. The modern computing facilities permit consideration of complex material and structural properties in the analysis. The complexities of analytical expressions and mathematical manipulations should be carefully understood without allowing them to obscure the importance of the basic requirement of statics. The flexibility of the structure as a result of plasticity and creep permits redistribution of internal forces and is a great factor of safety against the errors introduced on account of the assumptions regarding material behaviour and simplified structural analysis. But if the conditions of statics are not ensured there is no other aspect of structural behaviour which can come to the rescue, except the partial factor of safety assumed for the working loads.

Kinematical Requirements

Invariably all materials deform under stress. Points get displaced to neighbouring points after the development of the stress. The displacements of the points in general are different but vary continuously from point to point. Therefore there are *relative displacements* between the ends of a small segment and also angular changes between two such segments. These relative displacements and angular changes are collectively called as *deformations*. The elongation of a longitudinal fibre of a bar in tension or rotation of one cross-section relative to the adjacent one of a beam in bending, are examples of the deformations of structural members. The deformations also in general vary from point to point. The rate of change of deformation is ***strain***. Thus in the case of bar in tension the elongation per unit length of the fibre is the tensile strain and in bending the angular deformation per unit length is the curvature. These in general are different from point to point along the axis of the bar such as when it is held at top and being strained by its own weight, or while it is loaded transversely.

Kinematics is the term applied to the requirement of certain purely geometrical relationships between deformations of structual members and displacements of certain convenient points on the structure. In a pinjointed frame the deformation of any given member is the relative elongation (or shortening) and can be obtained in terms of the displacements of the joint to which the member is connected. In general, like statical relationships kinematic relations too are linear. Strictly speaking kinematic relationships are nonlinear. This is rendered linear by the common assumption that the deformations and the displacements are both small. This is the essential assumption of linear structural analysis, consistent with neglecting the effect of displacements in writing equations of equilibrium.

Different types of kinematic considerations are made in structural analysis depending on the restrictive conditions or approximations. For example, in pinconnected frameworks bending moments in the members are absent by virtue of pin connections between the members, and the problem of kinematics is one of obtaining only the member elongations in terms of joint displacements; while if the members are monolithic with one another at their junctions, in addition to elongations, rotational deformations are also required to be evaluated in terms of the joint displacements.

Another kinematic assumption in the analysis of certain class of frames, predominantly in bending, is the one of *inextensional* bending. Here the axial deformations are neglected, though in the context of modern automated computations such assumptions are not necessary.

Another important type of kinematics is one of *collapse* of structure, namely the study of deformations and displacements of structure which behaves as a mechanism at the instant of collapse. For frames in bending the collapse occurs when at sufficient number of sections, moments have reached the full moment capacity of the sections and the small segments near about such sections behave as hinges, causing the structure to collapse as a mechanism.

Consideration of kinematics is invariably necessary besides statics. It has to be necessarily considered for determining the internal forces in a redundant frame in which they are not unique but depend on the deformations. In a determinate frame it would be required only if the displacements are required.

It is relatively easy to visualize the kinematics of a structure than the internal force distributions. Thus it is easier to incorporate approximations more confidently in the kinematical behaviour of structure than in the internal force distributions. The *Rayleigh-Ritz* method and methods of weighted residuals (Galerkin or Least Square methods) are powerful tools of structural analysis based on assumed displaced configuration of the structure. The derivation of the stiffness of finite elements used in the analysis of surface and solid structures is based on this principle and forms one of the most powerful tools of solid and structual analysis.

Stiffness Requirements

The relationship between the deformation of a member and the corresponding force can be expressed in the form of either *flexibility* or *stiffness* of the member. For small deformations these are constants depending only on the geometrical properties of the bar and the material property. The stiffness or flexibility relations can be derived either by direct experiment or deduced from the known material properties and suitable assumptions regarding stress and strain distributions, such as using the assumptions of simple theory of bending. The stiffness of a material or of a structural member such as a beam is constant in the initial stages of the load followed by a decreasing stiffness characterized by elastoplastic behaviour.

In any structure the calculated internal force and the corresponding deformation should be such that they satisfy the stiffness (or flexibility) relations. This is the characteristic relation between the resisting force and deformation of a member and is always to be satisfied.

In the analysis of determinate frames the internal forces are determinable without the consideration of the stiffness properties of the members. However in a redundant frame the internal force distributions directly depend on the stiffness characteristics of the members and change with the changes in the stiffness of the members. Since the members lose their stiffness after reaching their load carrying capacities (yield limits) the stiffness of the structure as a whole, which is in direct proportion to the stiffnesses of the members, goes on decreasing. The distribution of internal forces changes with changes in stiffness. Thus, as briefly stated earlier, there are more than one possible distributions of internal forces in a redundant frame depending on the stiffnesses of members.

In the elastic range the stiffnesses are constants depending only on the area of section, length, etc. They do not depend on the deformation level. If one assumes perfectly plastic behaviour continuing deformation takes place at constant load with zero stiffness. In other words it is also independent of the deformation level. Stiffness being independent of the deformation level is the reason behind the relative ease of elastic and collapse load analysis of structures compared to elastoplastic analysis or analysis involving large displacements wherein stiffness depends on the deformation level.

c. Methods of Structural Analysis

So far we had an overview of the general requirements of structures, namely of equilibrium, member stiffness characteristics and kinematics. Any method of structural analysis must therefore necessarily ensure that (a) the conditions of equilibrium between the internal and external forces are satisfied; (b) the internal forces of the members and the deformations of the member are consistent with their stiffness properties; and (c) the deformation of members obey the kinematical characteristics and restraints of the structure. Statically determinate and indeterminate structures pose different problems to the analyst. In a determinate structure the above three aspects are separate and are handled independently. Laws of statics give the internal forces whatever may be the stiffness properties of the members. When the internal forces are thus determined each element can be isolated with the forces acting on it and can be analyzed separately for the resulting deformations. The displacements of the structure are then computed from these known member deformations and the kinematical conditions of the structural assembly. However, in a redundant structure all the three aspects are coupled and are required to be handled simultaneously.

There are two basic ways in which these can be considered. Accordingly two types of methods of analysis have been in practice: (i) *Force methods*, and (ii) *Displacement methods*. These methods have been used in variety of forms as applied to different types of structures. These methods differ in the way the above three basic requirements are combined. In force methods certain force parameters are the unknowns in terms of which the structual problem is formulated. In displacement methods certain displacement parameters achieve the same objective.

There is no apparent restriction on the part of the analyst to choose either forces alone or displacements alone as unknown parameters. A *mixed* approach is also possible. Obviously the convenience of such an approach is limited to certain type of problems in which they prove efficient. Force and displacement methods by far are the dual methods of structural analysis. We will consider the details of these methods in subsequent sections.

d. Analysis Approach

Any approach of analysis, be it a force or displacement approach, breaks up the total effort into a number of steps. *The analysis proceeds from the part to the whole.* This subdivision is dependent upon the methods adopted but the general principle remains the same.

As stated earlier a given structure is an assembly of members (elements). The analysis procedures essentially consist of two steps: (i) Member (element) analysis; and (ii) Structural (global) analysis. The first step is the study of the member (element) itself as an independent body. Properties (such as stiffness or flexibility) which are required in the second stage are determined. The second step then consists of the analysis of the assembly of such members. In the first step certain displacement or force parameters are taken as unknowns and the member analysis is formulated in terms of these unknowns. The second step determines the set of all such unknown parameters. This step involves the knowledge of structural connectivity and from the computational viewpoint is a major issue. A set of discrete unknowns has to be solved. The present day digital computers have almost removed the restriction on the total number of unknowns to be determined.

Consider the frame shown in Fig. 1.1 (a), which we will now suppose, for purposes of illustration is a plane frame loaded in its own plane. The frame is considered as an assembly of four members interconnected at nodes (joints) B, C and D, and supported at nodes A and E. Properties of each of the members are determined in terms of length, section properties and loads. Figure 1.1 (b) shows the free body of the members BC and Figure 1.1 (c) shows its deformed shape. Member analysis considers the three essential aspects—equilibrium, kinematics and stiffness (flexibility) characteristics—as applied to this member in terms of certain force, or displacement unknowns. Thus if the displacement approach is followed the six member end displacements—u_B, v_B and θ_B at B and u_c, v_c and θ_c at C—are treated as unknowns, to be determined by the second stage analysis, namely structural (global) analysis. If a force approach is used, then any three of six member end forces (say P_c, Q_c and M_c or M_B, M_c and the axial force) are treated as unknowns. Step (ii) consists of the determination of the independent unknown parameters introduced in analyzing the members. Again, the three aspects—equilibrium, kinematics and stiffness (or flexibility) aspects—are applied to the assembly of members (structure) as a whole, and the unknown parameters determined. A 'back substitution' of these into the member analysis determines the forces and displacements within the member at any point.

The above approach is naturally applicable to skeletal structures like building frames or lamella type structures consisting of unidimensional (line) members (elements). It is also applicable, in theory approximately, but almost correctly to surface structures and solids or a structure involving solid elements. The generalization to the latter class of problems is recent and is known as the *finite element technique.* This is now a well established technique which has proved its versatality and power. An introduction to this is given later in this chapter (Sec. 1.7).

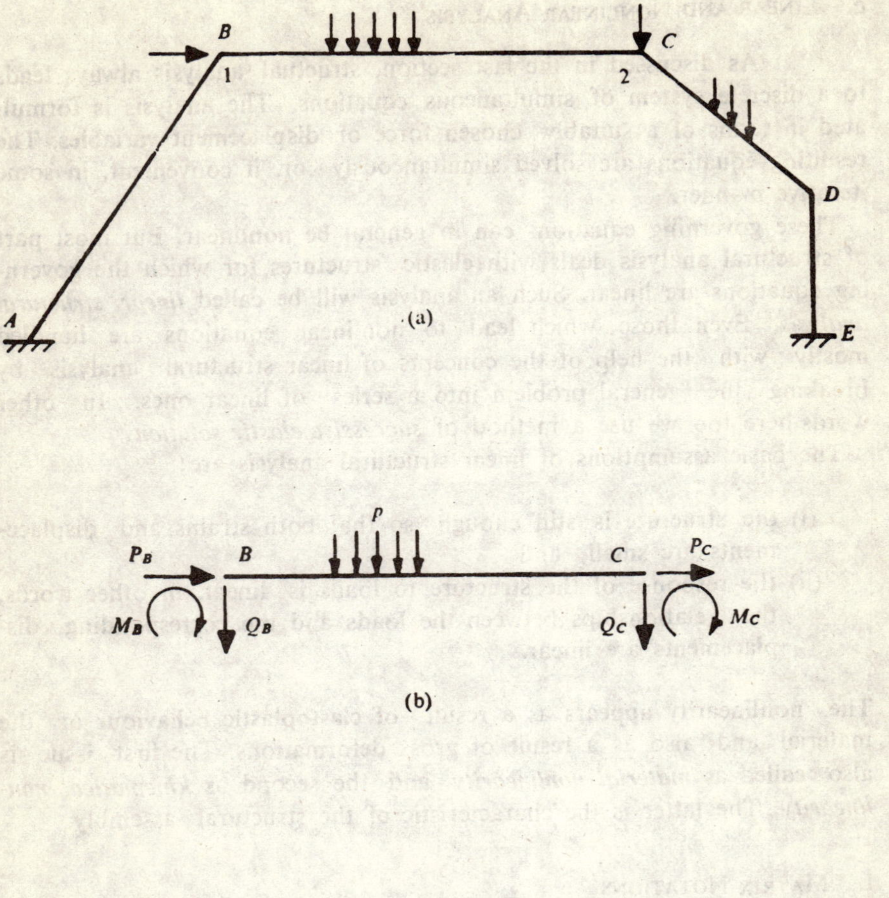

Figure 1.1

e. Linear and Nonlinear Analysis

As discussed in the last section, structual analysis always leads to a discrete system of simultaneous equations. The analysis is formulated in terms of a suitably chosen force or displacement variables. The resulting equations are solved simultaneously or, if convenient, in some iterative manner.

These governing equations can in general be nonlinear. But most part of structural analysis deals with elastic structures for which the governing equations are linear. Such an analysis will be called *linear structural analysis*. Even those which lead to nonlinear equations are handled mostly with the help of the concepts of linear structural analysis by breaking the general problem into a series of linear ones. In other words here too we use a method of *successive elastic solutions*.

The basic assumptions of linear structural analysis are:

(i) the structure is stiff enough so that both strains and displacements are small, and

(ii) the response of the structure to loads is linear, in other words, the relationships between the loads and the corresponding displacements are linear.

The nonlinearity appears as a result of elastoplastic behaviour of the material and also as a result of gross deformations. The first issue is also called as *material nonlinearity* and the second as *kinematical nonlinearity*. The latter is the characteristic of the structural assembly.

f. Matrix Notations

Methods of structural analysis gradually developed with the types of structures built. Naturally the methods of analysis were largely oriented towards handling a given structure such as a pinjointed truss a continuous beam, or a bridge, of certain geometry and under a set of loads. Special methods came to be developed to handle such problems. With the development of the digital computer, matrix notations and algebra came to be used in structural analysis with consequent change in the organization of calculations. In the modern context where the concepts of matrix algebra are studied at secondary school level education, and courses of computer programming and numerical analysis are given side by side the structural mechanics courses, matrix notations are being introduced from the very beginning in structural mechanics courses. As such the barrier between the socalled classical and matrix methods is fast disappearing. It is presumed here that the reader has the basic knowledge of matrix notations and algebra.

The matrix methods of structural analysis have two chief purposes to serve :

(a) to provide a compact and efficient notation to treat the principles and methods of structural analysis in generality with least restriction on the type of structure; and

(b) to provide a notation and organisation of the steps of structural analysis adaptable for use with a digital computer.

The first purpose appears to be academic; indeed it is so but with special advantages. It affords a language in which general principles, such as equilibrium, kinematics, flexibility and stiffness, etc., can be discussed without reference to any specific topological data of a structure and in terms of general parameters defining topology of a given class of structure. The use of matrix notations reveals certain correspondence between the structure of the equations and the actual structure. Such correspondence in form of equations and the structural network can be rewarding and many times serve as an useful check and save labour. Reflection of the reciprocal relationship in the symmetry of the stiffness and flexibility matrices, contragredience between statical and kinematical relations, are some of the examples.

The second purpose has great practical utility. With the present-day large-size computing machines having sophisticated peripherals, the organization of the calculations is of immense importance. Matrix notations and algebra are used with advantage in the programming languages of the computers. Thus the matrix methods of structural analysis lend themselves to be directly programmed. This is the main and the true purpose of the method. The organization of the calculation procedure using a digital computer is generally different from the one using hand computations. The organization in matrix methods is suited for use with a computer and hence of no special advantage, and instead become cumbersome, when used for hand computations. Keeping this in view one ought to view matrix methods in the true perspective. Likewise, the use of the methods suited for hand computations, such as moment distribution or similar iterative procedures for a digital computer proves cumbersome and inefficient. The detailed discussion on the type of organization required for solving problems on a digital computer will be discussed later. Briefly the type of organization required is as follows.

In using a computer, the process of calculation is required to be given in the form of a *program.* A program is a series of instructions stating the nature of action such as reading a numerical value, performing arithmetical calculations, taking logical decisions, performing a set of calculations again and again with different numerical values, etc. This program is written to work upon certain numerical figures which form the *data* to be provided. Thus the program, which is in an algebraic language characteristic to the computer, and the data which is set of numbers (or even alphabetic data), are both to be provided as an *input* to the computer in the form of punched cards or tapes. The computer then executes the series of instructions of the program on the data and prints out the results. This is in a nutshell the organization of the work.

It is therefore noticed that, the series of instructions detailing the process of calculations is required to be prepared assuming that the data would be supplied later. It is equivalent to, when we think of an elementary example, developing a formula in terms of certain formal parameters and get it evaluated for different numerical values of the parameters. But the problems are rarely reducible to evaluation of formulae. They normally require a series of complex instructions calling for addressing various intermediate steps of calculations and decisions to be taken depending on the results of the calculations. However, in essence it is equivalent to developing a series of mathematical operations involving formal parameters to be substituted later by actual numerical values given as data.

Matrix methods afford the structural analysis of a class of problems to be represented as a series of algebraic operations on matrices representing force and displacement groups, and thus permit programming in terms of these symbols. These are to be substituted later by numerical values. Thus by changing the numerical data, solutions to structures of different geometry and loading are readily obtained. Therefore we develop general programs for a given class of structures such as pinjointed frames, plane frames, etc, without any specific geometry of the structure or loading in view, but with adequate parameters introduced to define the geometry of the structure and the loading. The power of such computations from the practical viewpoint of design office is immense. General purpose programs once written out and tested (which itself may involve hundreds of manhours) together with a powerful computer form a readily available source of computational service. Several practical programs are commercially made available for users.

1.2 EQUILIBRIUM

We will now review the different aspects of equilibrium (statical) considerations encountered in analysis. Equilibrium considerations are applied in variety of forms and in a variety of situations. These are expressed either between stresses, forces (stress-resultants) or some generalized stresses or forces. Equilibrium (statical) relations are written either for a differential element, a member, or a structure or may be a part of a structure (substructure).

The basic assumption in linear structural analysis is the one of small displacements. Thus the equilibrium relations are written in the original (undeformed) geometry of the body. Thus an equilibrium relation is a *linear* relation between stresses or forces, involving length and angular parameters. When the restriction of the assumption of small displacement regime is removed, the displacements appear in equilibrium relations. Since these would be unknown and depend on the forces themselves the relationships are *nonlinear*. An elementary example is the well known beam-column situation wherein the interaction of axial load and the transverse displacement are considered.* Another example is the case of a freely suspended cable where its large displacements are to be taken into account while writing the equilibrium relations.

For beam elements the statical relations are written in terms of forces namely bending moment, shearing force and axial force. These are the resultants of the stresses acting on the crosssection of the beam. Figure 1.2(a) shows the free body of a straight beam of length L. The forces P_1, Q_1 and M_1 at end 1 and forces P_2, Q_2 and M_2 at end 2 will be called together as a set of *member end forces*. These are supposed to be determined from the structural analysis of the assembly of the members. These are the forces exerted by the group of adjoining members. Since these are

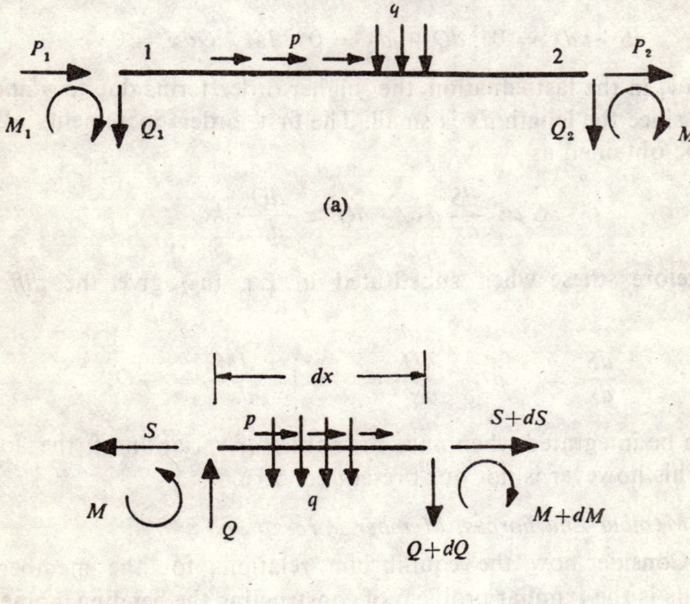

(b)

Figure 1.2

*See Example 5.2, Chapter 5.

unknown they are generally taken to be positive according to a common reference system, such as clockwise moments reckoned positive, forces from left to right or downward as positive, etc., admitting the interpretation that if the analysis gives negative signs for these, then their directions are understood to be opposite to the positive directions set out. These are according to the so-called *statical* conventions. Fig. 1.2(b) shows the equilibrium of a small element of the beam. The internal forces are according to the *deformation* convention namely axial tensile forces taken positive, hogging (or sagging) moments taken positive, etc. At certain stages in analysis, a need arises to relate the forces taken according to this and the statical sign conventions. Let ends 1 and 2 correspond to $x = 0$ and $x = L$ respectively in Fig. 1.2(a). Then we have the following relations between the forces reckoned according to the two systems of conventions:

$$P_1 = -S(0) \; ; P_2 = S(L) \; ; Q_1 = -Q(0) \; ; Q_2 = Q(L)$$
$$M_1 = -M(0) \; ; M_2 = M(L) \qquad (a_1)$$

These follow by noting that, say, P_1 is compressive, M_1 is sagging, etc., and are opposite to the ones used in deformation convention.

The equations of equilibrium for the differential element shown in Fig. 1.2(b) are

$$dS+pdx = 0 \; ; dQ+qdx = 0 \; ; dM+Qdx = 0. \qquad (a_2)$$

Recall how, in the last equation the higher order terms due to p and q are neglected since the length dx is small. The first order increments dS, dQ, etc. can be obtained as

$$dS \approx \frac{dS}{dx}\,dx \; ; \quad dQ \approx \frac{dQ}{dx}\,dx, \qquad (a_3)$$

etc. Therefore these when substituted in Eq. (a_2) give the *differential* equations

$$\frac{dS}{dx} = -p \; ; \quad \frac{dQ}{dx} = -q \; ; \quad \frac{dM}{dx} = -Q. \qquad (1.1)$$

These can be integrated when once the nature of variation of the loads is known. This however is not our present concern.

Member End Forces, Member Forces

Consider now the equilibrium relations for the member as a whole. This is the familiar problem of constructing the bending moment and shearing force diagrams for beams. Consider to begin with the definitions of certain groups of forces. The member end forces shown in Fig. 1.2(a) are not all independent but satisfy three statical relations between themselves. Thus any three of them are independent. The independent set of three forces will be called the set of *member forces*, just to differentiate the two sets. The set of forces $\{P_2, M_1, M_2\}$ is one of the possible sets and can be called, *member forces* of a *beam model*. The axial force at end 2 and the two end moments are taken as independent quantities. The other three forces P_1, Q_1 and Q_2 are determinable from the three conditions of equilibrium applied to the beam. Alternatively we can consider a *cantilever model* in which, all the three forces at one of the ends, usually end 2, are treated as independent. Thus, in the absence of lateral loads, the forces at end 1 are determined as $P_1 = -P_2$, $Q_1 = -Q_2$, etc. As we will see later these definitions have many advantages. The *member forces* are the independent unknowns for a member. Thus in a structure such as of Fig. 1.1(a) the total number of unknown member forces is 12. In a pinjointed truss the total unknown member forces is equal to the number of members, one for each member, namely the internal axial force S.

Equilibrium Relations for a Structure

Conditions of equilibrium between member forces (all or certain selected forces) and loads can always be written for a structure. This is possible regardless as to whether the structure is statically indeterminate or determinate. The converse form of statical relations, namely solution of internal forces in terms of known loads is possible only for a statically determinate structure.

The most general approach is to write the conditions of equilibrium at the joints (nodes) at which the members meet. As discussed in the previous paragraph, the unknowns for each member are its member forces, namely the independent set of forces which can be arbitrarily prescribed. Since the members meet at the joint the forces at the ends of the member should balance amongst themselves or must equilibrate the external loads if any which may be applied to the joint. The plane frame shown in Fig. 1.1(a) consists of four members meeting at five joints. Three equations of equilibrium at each of the five joints are required to be satisfied. The total number of unknown member forces is 12 and since it is a plane frame we have six independent unknown reactive components. Thus the total force unknowns are 18. The total available conditions of equilibrium are 15 (three at each of the five joints). Therefore we have a statically indeterminate situation wherein the number of unknown forces are more than the available conditions of equilibrium. Force and displacement methods of structural analysis provide the required additional conditions to make the available equations equal in number to the total number of unknowns.

Equations of equilibrium are also written for a *part of a structure* consisting of a number of members. Such equations are often termed 'overall equations of equilibrium' or 'sway equations' which generally arise in the analysis of frames in which axial deformations are ignored. In general, the number and nature of the equations of equilibrium are controlled by the number and nature of the displacement variables introduced in the analysis. *The most general approach of obtaining the required equations of equilibrium is to use the virtual work principle*. This approach is explained in a later chapter after the ideas of the virtual work are reviewed. Recall at this stage the use of virtual work in obtaining equations of equilibrium in the collapse load analysis of portal frames using a chosen collapse mechanism as the virtual displacement pattern. Recall also the use of Muller-Breslau principle for the construction of influence lines.

1.3 KINEMATICS

Kinematic relations are based on physical judgement and approximations. Unlike the internal stresses, the strains and displacements are physical descriptions which are readily observed and measured. *Kinematic approximations* are therefore more common in structural anylysis. The assumption of plane transverse section of a beam remaining plane after deformation is an approximation based on observation. Similarly ignoring of the axial defomations of members of certain frames predominantly in bending, ignoring of shearing deformations in beams are some of the examples. The concept of plastic hinge is another good example. Analysis of bodies and structure based on an *assumed strain distribution* is one of the powerful technique of analysis and is widely employed in continuum and structural analysis. Virtual work and energy minimization techniques have been very well applied to problems based on an assumed strain distribution.

Kinematical considerations proceed from the *part to the whole*. A small differential element is considered and the resulting deformations studied. Then follows the study of the kinematical behaviour of the member. The member elongation and end rotations are determined by *integrating* the deformations taking place in the number of differential elements into which the member is assumed to be subdivided. The summing up of the effect is achieved either in a closed form where possible and in a numerical way, in general. This study is then followed by that of the *overall kinematics* of the structure consisting of several members. In the elastic range the overall problem is based on the assumption that full continuity exists between the member ends meeting at a node. When elastoplastic behaviour is in question appropriate assumptions are made such as development of plastic hinges. In collapse load analysis the kinematics of the *collapse of a structure* as a mechanism is also investigated.

Linear Kinematics

Like statical relations kinematic relations too are linear, when once the assumption of small displacements is made. Consider the expression for axial strain ϵ_x of a small segment dx shown in Fig. 1.3 displacing in a plane. The displacements u and v vary smoothly over the body. Thus they are continuous functions of x and y. Thus in distance dx alone we

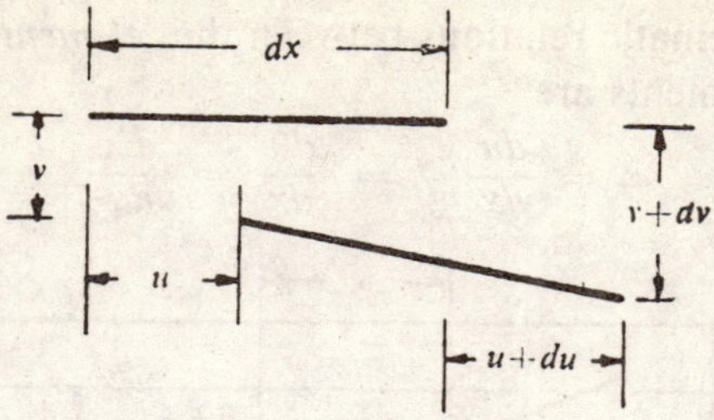

Figure 1.3

have
$$du = \frac{\partial u}{\partial x}\, dx \text{ and } dv = \frac{\partial v}{\partial x} dx.$$

If the axial strain is ϵ_x the deformed length is $dx+\epsilon_x dx = dx\,(1+\epsilon_x)$. This is given by

$$\left\{dx(1+\epsilon_x)\right\}^2 = (dx+du)^2+dv^2$$

This gives

$$\epsilon_x+\frac{1}{2}\epsilon_x^2 = \frac{\partial u}{\partial x}+\frac{1}{2}\left\{\left(\frac{\partial u}{\partial x}\right)^2+\left(\frac{\partial v}{\partial x}\right)^2\right\} \tag{a$_1$}$$

In linear structural analysis the nonlinear terms are neglected which are terms of higher order. We have therefore the linear strain displacement relation

$$\epsilon_x = \frac{\partial u}{\partial x} \tag{a$_2$}$$

This is an elementary demonstration, but the principle remains applicable to more general cases. Considerations of higher order terms would be necessary only when stability effects or flexible structures are considered.

It should be noted that when the kinematic relations are nonlinear the corresponding static relations too are nonlinear. This will become clear when equations of equilibrium are derived by applying the principle of virtual work. In other words the static and kinematic relations are both required to be either linear or nonlinear.

Deformations of Beam Element

The elementary theory of the bending and twisting of beams is based on certain basic assumptions for the variation of strains within the cross-section. These are expressed in terms of certain quantities defined at the axis of the beam. According to the simple theory of stretching and bending, the axial strain, the curvature and the shearing strain completely define the behaviour of the crosssection. The shearing deformation is small and is generally neglected. We will at this stage neglect it since only the basic features of kinematics are being considered.

The displaced shape of the beam axis is characterized by the two components u and v, the axial and the transverse components respectively (Fig. 1.4 a) The kinematic relations between the *element strains* (Fig. 1.4 b, c) and the displacements are

$$\epsilon_x = \frac{du}{dx};\ \chi = \frac{d\theta}{dx} = \frac{d^2v}{dx^2} \tag{a_3}$$

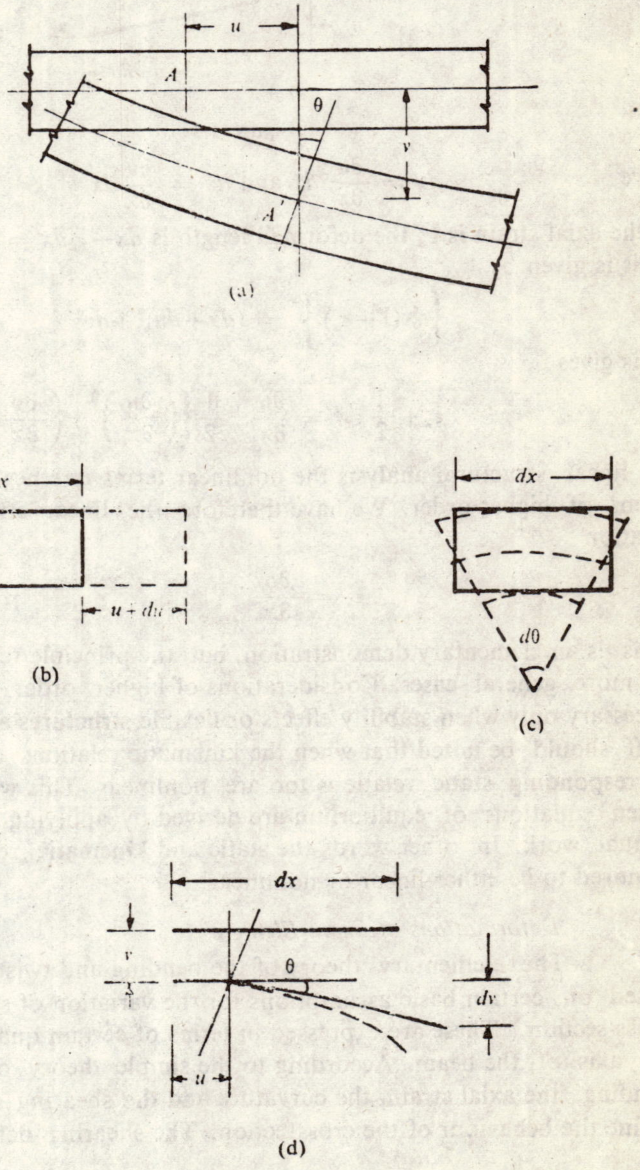

Figure 1.4

Here, since shearing deformation is neglected, the rotation θ of the cross-section is equal to (dv/dx), which is the rotation of the beam axis* (see Fig. 1.4 d). The *deformations* of the element are du and $d\theta$, respectively equal to $\epsilon_x d_x$ and χdx, the *axial elongation* and the *relative angular rotation.*

Member Deformations

The deformed shape of the beam is determined by summing up the element deformations du and $d\theta$. This summation depends on the end conditions of the member.

The general case is as shown in Fig. 1.5 (a). The member end displacements are as a result of the cumulative effect of the deformations occurring within the beam, and a rigid body movement as a result of the translation of joints. Therefore, we have to separate the two effects before we compute the cumulative effect of deformations of the beam elements. There are two possible ways: first is a *beam model* idea in which the rigid body motion is considered to consist of (a) a rotation $(v_2-v_1)/L$ and (b) a translation characterized by u_1 and v_1. The member deformations are obtained by *subtracting* the rigid body effects:

$$\varphi_1 = \theta_1 - \frac{v_2-v_1}{L}$$

$$\varphi_2 = \theta_2 - \frac{v_2-v_1}{L} \tag{1.2}$$

$$e = u_2-u_1$$

Here φ_1 and φ_2 are rotations measured from the displaced chord. We will call these as *chord rotations.* The quantity e is the axial elongation of the member, obtained simply as u_2-u_1. Here the assumption of small displacements is made. Therefore also the approximate term (v_2-v_1) L for the rigid body displacement of the beam. These deformations are shown in Fig. 1.5(b).

Alternatively the second approach is to have the *cantilever model* idea, in which the rigid body movement is considered to consist of (a) a translation characterized by u_1 and v_1; and (b) a rotation θ_1. Thus the member deformations are those relative to end 1 being fixed. These are

$$e = u_2-u_1$$

$$v = v_2-v_1-L\theta_1 \tag{1.3}$$

$$\varphi = \theta_2-\theta_1\,.$$

*See eq.(a_4), Sec. 4.1, Chapter 4 for the general case.

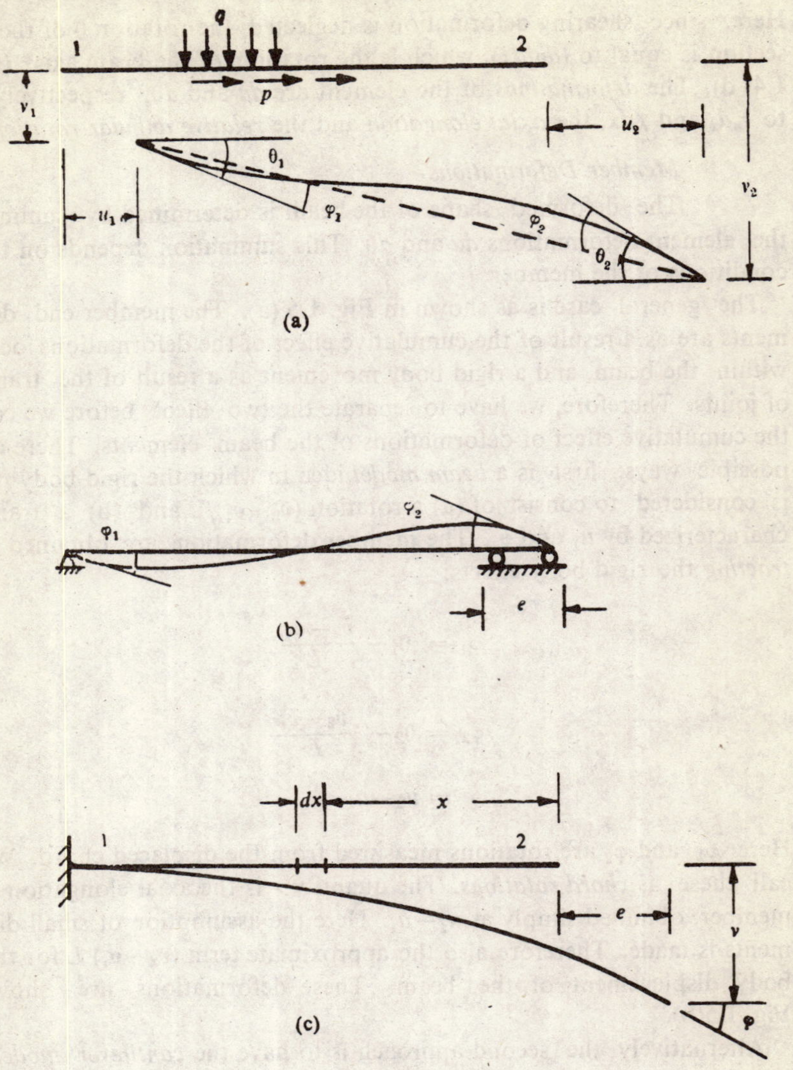

Figure 1.5

These are obtained by subtracting from u_2, v_2 and θ_2 the components due to rigid body translations u_1 and v_1 and a rotation θ_1. Note that due to $-\theta_1$ we have $-L\theta_1$ occurring at end 2 in a transverse direction.

These two models correspond exactly to the member forces which were defined as a set of independent forces for a member. The deformations $\{e, \varphi_1, \varphi_2\}$ or $\{e, v, \varphi\}$ correspond to standard end conditions and are determined by summing the deformations occuring within the beam.

Consider now the methods to determine these beam deformations. These are (a) Direct consideration of the kinematics or (b) Application of virtual work principle.

Consider for the present the former approach. The cantilever model is relatively easier. Let du and $d\theta$ be the deformations of an element distant x from end 2. Due to these the values of beam displacements at end 2 are obtained by noting that the portion x of the beam acts as a rigid body; Fig. 1.5 (c). Therefore

$$de = du \; ; \; dv = x d\theta \; ; \; d\phi = d\theta \tag{a_4}$$

The total deformations are obtained by summing over the entire beam :

$$e = \int_0^L du \; ; \; v = \int_0^L x d\theta \; ; \; \varphi = \int_0^L d\theta \tag{1.4}$$

The derivation of the expression for the deformations φ_1, φ_2 and e of the beam model (Eq. 1.2, Fig. 1.5 b) can also be obtained from there by noting that these can be obtained by superimposing a rigid body rotation equal to $(-v/L)$. The rotation of end 1 will then be $(-v/L)$ and that at end 2 will be $\varphi - v/L$

Therefore we have

$$\varphi_1 = -\frac{v}{L} = -\int_0^L \frac{x}{L} d\theta$$

$$\varphi_2 = \varphi - \frac{v}{L} = \int_0^L \left(1 - \frac{x}{L}\right) d\theta \tag{1.5}$$

The axial elongation however remains the same.

Note that the deformations du and $d\theta$ occurring at an element of the beam may be due to the *applied loads* or due to *temperature, shrinkage,* or *creep.* When we consider the flexibility relations of the differential element we can express these in terms of the forces. As this stage recall the well known elastic relation $d\theta = Mdx/EI$ and observe that the second and the third relations of Eq. (1.4) then state the *area-moment theorems* and that the considerations made in deriving Eq. (1.5) are those involved in deriving *conjugate beam analogy* (See Ex. 1.9).

The geometrical considerations of the type discussed above are *avoided* when *virtual work* principles are employed. We will be discussing this in a later chapter.

Kinematics of a Structure

Relations of the type (1.2) or (1.3) define the member deformations in terms of member end displacements. The problem for a structure as a whole is to relate the *member end displacements* to the *displacements of the joints* or to certain displacement variables defined for the structure. For *elastic analysis* where full continuity exists at joints the problem is rather trivial. We have to simply state the name of the joint to which a particular member end is connected. Thus for the frame of Fig. 1.1(a) we simply have, say at B,

$$u_{BA} = u_{BC} = u_B;\ v_{BA} = v_{BC} = v_B;\ \theta_{BA} = \theta_{BC} = \theta_B \qquad (a_5)$$

with similar statements at other joints.

The above relations require modifications when full continuity between members does not exist as when plastic hinge formations occur.

Many times we have to consider degenerate cases of structural behaviour. The inextensional bending of frames is one such example. These are departures from the general case and will be considered at appropriate places in the book. The kinematics of plastic collapse of a frame as a mechanism due to the formation of hinges is another example.

1.4 FLEXIBILITY AND STIFFNESS

We will discuss the flexibility and stiffness characteristics of a differential element, then of a member, and finally of structure. This is in line with the review followed in the two previous sections. We will be restricting ourselves to elastic range wherein these do not depend on the level of force or deformation. The material constant (stiffness) is defined by E, the Young's Modulus. The Poisson ratio is implicitly zero since only beam elements are in question. Later in Chapter 9 an introduction to the plane elasticity case will be given.

Flexibility and Stiffness of a Differential Element

The flexibility relationship between the axial force S and the axial elongation du, and between the moment M and the relative angular deformation $d\theta$ between two crosssections dx apart (Fig. 1.4 a, b), are the well known relations

$$du = \frac{dx}{AE} S\ ;\ d\theta = \frac{dx}{EI} M \qquad (a_1)$$

Here dx/AE and dx/EI are the axial and flexural flexibilities. The inverses, EI/dx and AE/dx are respectively the axial and flexural stiffness of the element.

Flexibility and Stiffness of a Member

These are developed between a chosen set of forces and deformations for a member and are defined in a variety of ways depending on the choice of the force and deformation systems. Consider the simply supported beam shown in Fig. 1.6(a) with end moments M_1 and M_2 producing chord rotations φ_1 and φ_2. The basic flexibility relations are

$$\begin{aligned} \varphi_1 &= f_{11}\, M_1 + f_{12}\, M_2 \\ \varphi_2 &= f_{21}\, M_1 + f_{22}\, M_2 \end{aligned} \qquad (a_2)$$

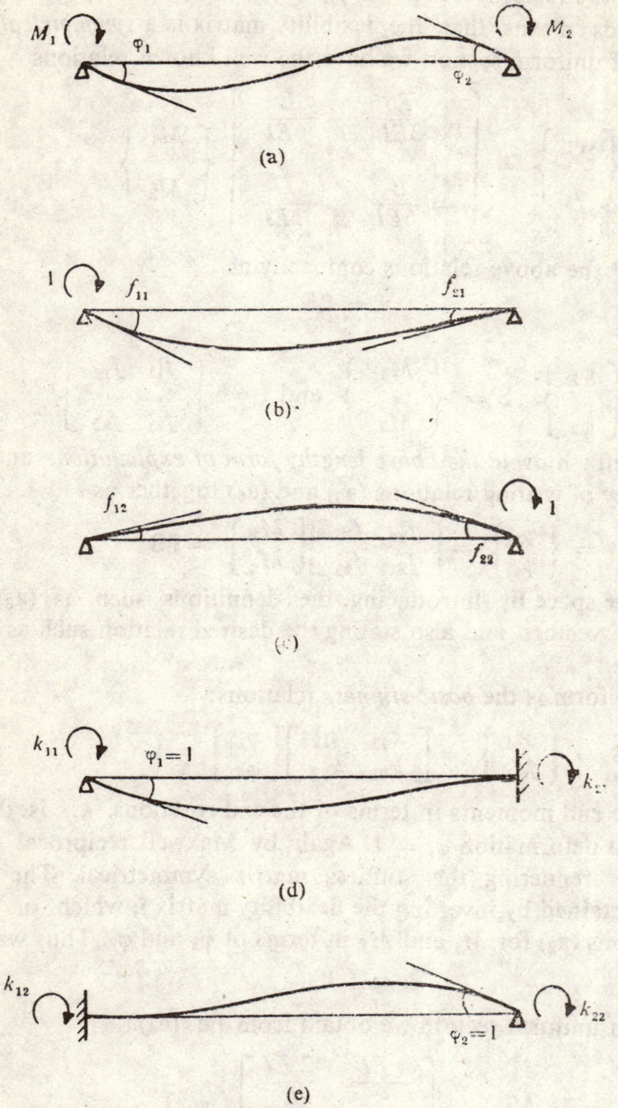

Figure 1.6

where f_{ij} is the flexibility coefficient. These state the principle of superposition of the effect of M_1 and M_2 as shown in Fig. 1.6(b) and (c). The above relations are better expressed in matrix form as follows:

$$\begin{Bmatrix} \varphi_1 \\ \varphi_2 \end{Bmatrix} = \begin{bmatrix} f_{11} & f_{12} \\ f_{21} & f_{22} \end{bmatrix} \begin{Bmatrix} M_1 \\ M_2 \end{Bmatrix} \tag{a$_3$}$$

Here the square matrix is the flexibility relating the vector of rotations to that of moments. f_{ij} denotes the rotation φ_i due to moment $M_j = 1$. Thus on account of the reciprocal theorem $f_{ji} = f_{ij}$. Thus we have $f_{12} = f_{21}$. This in other words means that the flexibility matrix is a *symmetrical* matrix. For a beam of uniform section we have the well known relations

$$\begin{Bmatrix} \varphi_1 \\ \varphi_2 \end{Bmatrix} = \begin{bmatrix} \frac{L}{3EI} & -\frac{L}{6EI} \\ -\frac{L}{6EI} & \frac{L}{3EI} \end{bmatrix} \begin{Bmatrix} M_1 \\ M_2 \end{Bmatrix} \tag{a$_4$}$$

We will write* the above relations compactly as

$$\mathbf{v} = \mathbf{f}\,\mathbf{S} \tag{a$_5$}$$

where

$$\mathbf{v} = \begin{Bmatrix} \varphi_1 \\ \varphi_2 \end{Bmatrix}; \mathbf{S} = \begin{Bmatrix} M_1 \\ M_2 \end{Bmatrix} \text{ and } \mathbf{f} = \begin{bmatrix} f_{11} & f_{12} \\ f_{21} & f_{22} \end{bmatrix} \tag{a$_6$}$$

We will henceforth *avoid the above lengthy form of explanations* and adopt a compact way of writing relations (a$_5$) and (a$_6$) together as

$$\mathbf{v} = \begin{Bmatrix} \varphi_1 \\ \varphi_2 \end{Bmatrix} = \begin{bmatrix} f_{11} & f_{12} \\ f_{21} & f_{22} \end{bmatrix} \begin{Bmatrix} M_1 \\ M_2 \end{Bmatrix} = \mathbf{f}\,\mathbf{S} \tag{1.6}$$

These conserve space by introducing the definitions such as (a$_6$) of the matrices and vectors, and also stating the desired relation such as (a$_5$) between them.

The inverse form is the *basic stiffness* relations:

$$\mathbf{S} = \begin{Bmatrix} M_1 \\ M_2 \end{Bmatrix} = \begin{bmatrix} k_{11} & k_{12} \\ k_{21} & k_{22} \end{bmatrix} \begin{Bmatrix} \varphi_1 \\ \varphi_2 \end{Bmatrix} = \mathbf{k}\,\mathbf{v} \tag{1.7}$$

These give the end moments in terms of the end rotations. k_{ij} is the force M_i due to unit deformation $\varphi_j = 1$. Again by Maxwell reciprocal theorem $k_{ji} = k_{ij}$ thus rendering the stiffness matrix symmetrical. The stiffness matrix $\mathbf{k}$ is obtained by inverting the flexibility matrix $\mathbf{f}$, which in effect is solving relations (a$_2$) for M_1 and M_2 in terms of φ_1 and φ_2. Thus we have

$$\mathbf{k} = \mathbf{f}^{-1} \tag{1.8}$$

For a beam of uniform section we obtain from Eq. (a$_4$).

$$\mathbf{S} = \begin{Bmatrix} M_1 \\ M_2 \end{Bmatrix} = \begin{bmatrix} \frac{4\,EI}{L} & \frac{2\,EI}{L} \\ \frac{2\,EI}{L} & \frac{4\,EI}{L} \end{bmatrix} \begin{Bmatrix} \varphi_1 \\ \varphi_2 \end{Bmatrix} = \mathbf{k}\,\mathbf{v} \tag{a$_7$}$$

* Boldface letters will be used for matrices and vectors. Also curly brackets for column vectors.

These are the simpler versions of the familiar 'slope-deflection' equations. The interpretations of k_{ij} are as shown in Fig. 1.6(d) and (e). The general approach is to determine the flexibility matrix and obtain stiffness as its inverse though special methods may be adopted to obtain the stiffness matrix *directly*.

General Stiffness

The matrix of stiffness coefficients which relates the set of member end forces to the set of member end displacements is a more general definition of stiffness of a member, Fig. 1.7.

$$\mathbf{P} = \begin{Bmatrix} Q_1 \\ M_1 \\ Q_2 \\ M_2 \end{Bmatrix} = \begin{bmatrix} k_{11} & k_{12} & k_{13} & k_{14} \\ k_{21} & k_{22} & k_{23} & k_{24} \\ k_{31} & k_{32} & k_{33} & k_{34} \\ k_{41} & k_{42} & k_{43} & k_{44} \end{bmatrix} \begin{Bmatrix} v_1 \\ \theta_1 \\ v_2 \\ \theta_2 \end{Bmatrix} = \mathbf{K\,u} \quad (1.9)$$

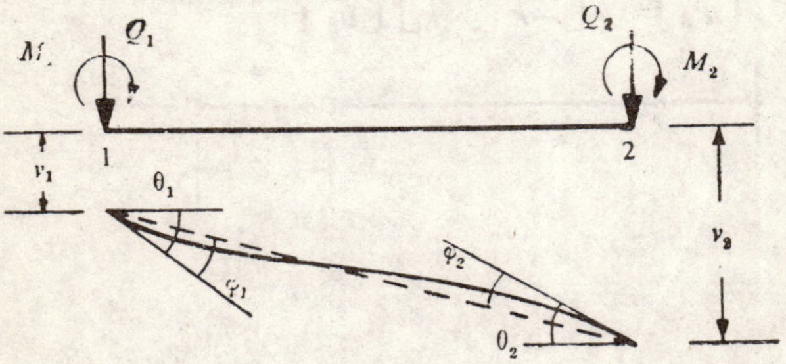

Figure 1.7

For a beam of uniform section these are

$$\mathbf{P} = \begin{Bmatrix} Q_1 \\ M_1 \\ Q_2 \\ M_2 \end{Bmatrix} = \begin{bmatrix} \frac{12\,EI}{L^3} & \frac{6\,EI}{L^2} & -\frac{12\,EI}{L^3} & \frac{6\,EI}{L^2} \\ \frac{6\,EI}{L^2} & \frac{4\,EI}{L} & -\frac{6\,EI}{L^2} & \frac{2\,EI}{L} \\ -\frac{12\,EI}{L^3} & -\frac{6\,EI}{L^2} & \frac{12\,EI}{L^3} & -\frac{6\,EI}{L^2} \\ \frac{6\,EI}{L^2} & \frac{2\,EI}{L} & -\frac{6\,EI}{L^2} & \frac{4\,EI}{L} \end{bmatrix} \begin{Bmatrix} v_1 \\ \theta_1 \\ v_2 \\ \theta_2 \end{Bmatrix} = \mathbf{K\,u} \quad (a_8)$$

It may strike the reader that these are derivable by some transformation of the basic stiffness relation (1.7). The member end displacements **u** and the deformation **v** are related by Eq. (1.2) and so are the member forces **S** to the member end forces **P** through equilibrium relations. We will discuss this issue later, limiting the attention for the present to the different definitions used in structural analysis.

Equation (a_8) are equivalent to the so-called 'slope-deflection' equations. Each *column* of the matrix denotes the set of forces due to unit value of the corresponding displacement. Thus the first column represents the set of forces and moments due to v_1 alone, the second represents those due to θ_1 alone and so on. Each *row* states the superposition effect; thus the first row gives the expression for Q_1 due to all the displacements, the second for M_1 and so on.

The above definitions are general. They can be applied to any type of structural member and for a chosen set of forces and displacements. Thus, for a pin ended plane frame member we have the following basic flexibility and stiffness relations (Fig. 1.8a):

$$e \equiv \mathbf{v} = \mathbf{f}\,\mathbf{S} = \left(\frac{L}{AE}\right) S;\ S \equiv \mathbf{S} = \mathbf{k}\,\mathbf{v} = \frac{AE}{L}\,e \qquad (a_9)$$

The general stiffness relations (Fig. 1.8b) are

$$\mathbf{P} = \begin{Bmatrix} P_1 \\ P_2 \end{Bmatrix} = \begin{bmatrix} k & -k \\ -k & k \end{bmatrix} \begin{Bmatrix} u_1 \\ u_2 \end{Bmatrix} = \mathbf{K}\,\mathbf{u} \qquad (a_{10})$$

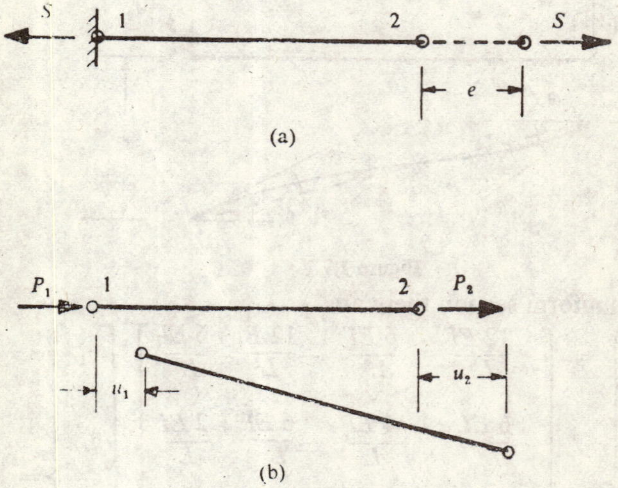

Figure 1.8

This is a simple case to understand as to how these are derived by noting that $e = u_2 - u_1$ and $P_2 = -P_1 = S$.

A point to be noted is that the converse definition 'general flexibility' does not exist. The relations $\mathbf{P} = \mathbf{K}\,\mathbf{u}$ cannot be inverted to yield the member end displacements in terms of member end forces. The general stiffness matrix is singular and can be easily verified to be so from the example of Eq. (a_{10}). The physical reasoning is that the member end displacements involve an unknown rigid body movement. The basic flexibility matrix(Eq. a_5) can always be obtained as the inverse of basic stifiness matrix (Eq. 1.7) since no rigid body movements are possible. Obviously since determination of flexibility means determination of deformations due to a given set of forces the member has got to be properly supported so that it is prevented from moving as a rigid body.

Flexibility and Stiffness of a Structure

The flexibility or stiffness of a structure is obtained by summing those of its constituent members in an appropriate sense. Qualitatively we see that the stiffness of a structure is a 'sum' of the stiffnesses of its members, the sum being defined in a particular manner. This depends on the *connectivity* of the members, in other words, on the way the members are connected to one another.

The stiffness relation for a frame such as shown in Fig. 1.1(a) will be of the form

$$\mathbf{K}\,\mathbf{U} = \mathbf{P} \tag{1.10}$$

Here **P** is the vector of all *joint loads* and **U** the vector of *joint displacements*. Matrix **K** is the *stiffness matrix of the structure*. The elements of this matrix would be obtained by summing the stiffness coefficients of the members according to the connectivity of the members. A solution gives the displacements in terms of the loads in the form

$$\mathbf{U} = \mathbf{K}^{-1}\,\mathbf{P} = \mathbf{F}\,\mathbf{P} \tag{1.11}$$

It is supposed that the connectivity of the members and provision of adequate supports is properly made so that rigid body movements of the structure or any part of it are avoided. $\mathbf{K}^{-1}$ then exists and represents the flexibility of the structure.

We will discuss later at length the methods of assembling the stiffness matrix of a structure. This process is reduced to a mere mechanical process easily adopted to computer formulations.

1.5 EQUIVALENT LOADS

Analysis of a structure is generally handled in two convenient stages. The effects of distributed lateral loads acting on the members are handled in the initial member analysis stage (see Sec. 1.1d). The second stage, which is analyzing the structural assembly (structure), is reduced to analyzing for equivalent loads applied only at the nodes (joints). Consider the frame shown in Fig. 1.9(a). The member analysis stage assumes that all the possible displacements are prevented.* Thus the structural *continuity is prevented* and therefore each member can be *analyzed separately* with zero end displacements. The forces required to restrain each member are then computed. Such forces are termed as *fixing end forces*. The total joint (nodal) loads which are obtained by summing the fixing end forces of the

*The frame (b) is said to be in a 'kinematically determinate' condition, or is recognized as a basic 'kinemetic system'.

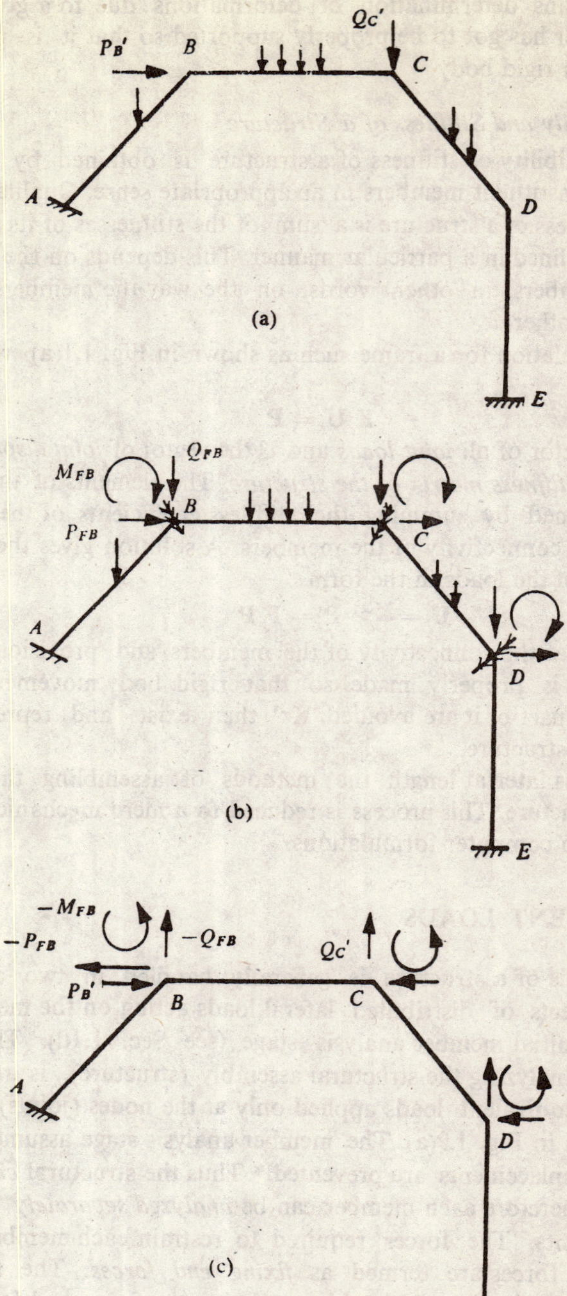

Figure 1.9

members meeting there represent the forces required to restrain the joint. Since these do not exist, forces *equal and opposite* to these forces, are to be superimposed. These are added to the loads which may be directly applied to the joints. Thus the second analysis is the analysis of the structure with loads applied only at the joints as shown in Fig. 1.9(c). Such joint loads are the *equivalent joint loads*. The displacements of the joints occurring in this case are the ones occurring in the actual structure, (Fig. 1.9 a) since the displacements at the member analysis stage are all zero.

Analysis of the member with its end displacements completely prevented (kinematically determinate stage) is a process which can be standardized. The analysis is expected to provide displacements and internal forces at any point on the beam axis as continuous functions of length coordinate. Similarly the structural analysis (second stage) of the frame under the equivalent loads is expected to provide member end displacements and member end forces. With these variables known, the displacements and the internal forces at any point on the beam axis are determined, and superimposed on those due to the internal loads alone with end displacement prevented.

Prescribed Deformations

Temperature, shrinkage, creep, and lack of fit of members, are also to be included under the term 'load'. *Equivalent loads* can be computed for a member for these effects also. The free strains due to temperature, shrinkage or creep etc. are prevented in the member analysis stage. Thus we analyze a member for these effects assuming that the member end displacements are fully prevented and determine the fixing end forces for the member due to these causes. Forces opposite to these are the equivalent loads on the joint of the structure to which the member is connected.

Complete Member Flexibitity and Stiffness Relations

We therefore write the general member stiffness relations as

$$\mathbf{P} = \mathbf{K}\,\mathbf{u} + \mathbf{P}_F \tag{1.12}$$

Here $\mathbf{P}_F$ denotes the set of fixed end forces. It represents the set of member end forces in a condition when $\mathbf{u} \equiv \mathbf{0}$. The forces $\mathbf{K}\,\mathbf{u}$ denote the member end forces due to the end displcements $\mathbf{u}$ alone.

Likewise the flexibility relations including the effects of lateral loads are obtained in the form

$$\mathbf{v} = \mathbf{f}\,\mathbf{S} + \mathbf{v}^0 \tag{1.13}$$

Here $\mathbf{v}^0$ are the deformations due to lateral loads alone ($\mathbf{S} \equiv \mathbf{0}$) and $\mathbf{f}\,\mathbf{S}$ those due to the end forces alone. Such relations are used in writing compatibility conditions between the deformations.

The derivation of relation (1.12) proceeds generally, though not necessarily from the basic flexibility relation (1.13). The latter is 'inverted' to give

$$\mathbf{S} = \mathbf{f}^{-1}(\mathbf{v}-\mathbf{v}^0) = \mathbf{k}(\mathbf{v}-\mathbf{v}^0) \tag{1.14}$$

and by transformations of **S** and **v** to **P** and **u** we arrive at relation (1.12). We will however relegate the details to later parts of the book.

1.6 FORCE AND DISPLACEMENT METHODS

Force and displacement methods are the two dual methods of analysis both of which combine the three requirements of equilibrium, kinematics and flexibility (or stiffness) characteristics, either for the study of a member or a structure as a whole. These are frequently called *compatibility method* and *equilibrium method* respectively. Also the alternative names, *flexibility method* and *stiffness method* are equally popular. The names force and displacement methods are based on the feature that certain force or displacement variables finally determine the solution. When forces are used as variables equilibrium is satisfied first and the requirements of kinematics (compatibility) are used as conditions to determine the unknown force variables. When displacements are chosen as variables the kinematical requirements are satisfied first and the requirements of equilibrium form the necessary equations to determine the governing displacement variables. Hence the alternative names. The other pair of alternative names is on account of the nature of coefficients of the matrix stating either compatibility or equilibrium condition for the structure. In force (compatibility) method flexibility coefficients appear in the formulation while as in the displacement (equilibrium) method stiffness coefficients appear.

There are two general approaches open to establish the equations of equilibrium or compatibility. These are (a) methods based on virtual work, and (b) methods based on energy extremum principles. It can be shown that they are alternative formulations and that they lead to identical results. The methods based on virtual work are most versatile and are here, preferred to those based on the latter principle. Some publications and texts [1, 2, 3] have clarified the confusion in the concepts and nomenclatures of the various energy theorems of structural analysis. The virtual work concept forms a basic one to which extremum principles can be related [1].

Both force and displacement methods lead to a set of simultaneous equations to be solved. The number of unknowns in the force method is generally smaller. However there is some arbitrariness in the force approach in the sense that the analyst has a certain freedom to choose the set of force unknowns. The displacement approach is however definite in its steps and thus becomes easier from the point of view of developing general purpose programs. Though certain publications do exist giving methods of automatic selection of redundants, it is still beset with disadvantages.

1.7 FINITE ELEMENTS

The beam type element (member) which can also be curved in space, is a unidimensional structural element. All quantities, such as forces and deformations vary only along the length of the member. They are thus functions of a single coordinate, *s*, the distance measured along the centre line of the member. In continuum problems such as a plate or a solid the physical quantities of interest namely, stresses and strains, vary from point to point in a continuous fashion. They are thus continuous functions of two or three coordinates respectively. The body such as a plate or a solid can also be imagined to be an assembly of two or three dimensional elements, such as triangular plates or tetrahedral elements. This is the extension of the structural idea. Alternatively, this can also be interpreted mathematically as a means of discretization of a continuum problem.

The chief difficulty of the so called 'classical' or 'closed form' solutions of continuum problems is that they are possible only under restrictive conditions of loads and boundary conditions. Therefore numerical solutions have been developed from time to time. Such solutions are evidently approximate in theory but tolerably accurate from a practical point of view. The analysis is formulated in terms of a discrete number of displacement, strain or stress parameters which are solved simultaneously as in structural problems.

Finite difference technique was the first numerical approach put forward and used successfully for many practical problems. This technique is a numerical technique of solving the governing partial differential equations of continuum problems. Other powerful numerical procedures were the Rayleigh-Ritz [22] and Galerkin [23] methods which are based on assumed variations of strains (or stresses). Such assumed variations are taken over the whole body (domain). This largely restricted the scope of the principle. The *finite element* technique, which originally started as an extension of the structural idea to plate and thin walled structures, greatly enhanced the scope of Ritz and Galerkin methods by admitting the idea of assuming strain distributions only in a small element (subdomain) of the body. This technique is now undoubtedly the most powerful and versatile technique of providing practical solutions to almost all problems of solid mechanics and many branches of engineering and science. Its power is really the power of the digital computer which implements the technique. We will seek here an introduction of this technique more as a generalization of the basic structural idea to two and three dimensional cases.

Consider the case of the analysis of a plane stress problem as an example for illustration. Fig. 1.10(a) shows a deep beam. The stresses and displacements vary continuously and an analysis procedure is expected to determine these at any point (x, y). The body is imagined to be sub-divided arbitrarily

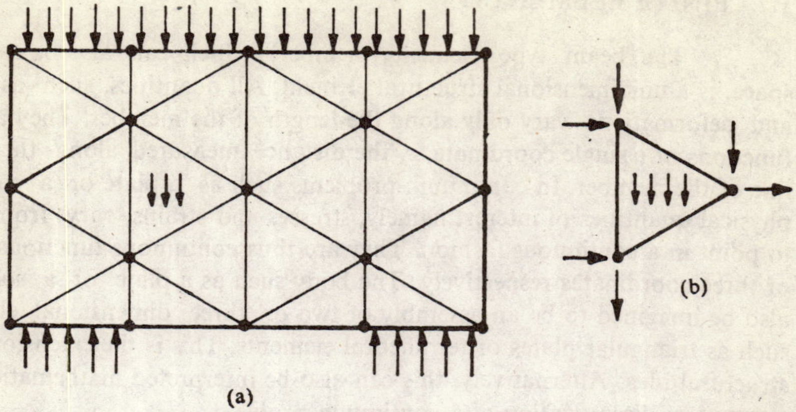

Figure 1.10

into a number of elements of convenient shapes. It is here imagined to be triangular elements.* Alternatively we can choose quadrilateral or curved elements. The structural idea consists in obtaining a set of simultaneous equations involving either displacement or stress variables defined at the nodes. The displacement approach is more convenient and is almost exclusively used in practice. Presently we have 17 nodes with two components of the displacement, u, v at each node. Thus 34 unknown displacements, describe the solution. The analysis procedure arrives at a set of equations.

$$\mathbf{K}\,\mathbf{U} = \mathbf{P} \tag{1.15}$$

where **U** will be a (34×1) vector of displacements, **P** a (34×1) vector of equivalent loads, and **K** a (34×34) coefficient matrix, having the meaning of an equivalent stiffness matrix. The approach used is similar to the one used in the displacement method as applied to frames with beam elements. The deep beam is considered as an assembly of 20 triangular structural elements. Figure 1.10(b) shows the free body of an element. The chief concept behind the structural idea is the one of replacing the boundary stresses acting on the edges of the element by *equivalent nodal forces* acting at nodes, as shown in Fig. 1.10(b). These nodal forces are statically equivalent to the boundary stresses. Thus for an element we derive force-displacement (stiffness) relations in the form

$$\mathbf{P}^{(e)} = \mathbf{K}^{(e)}\ \mathbf{U}^{(e)} + \mathbf{P}_F{}^{(e)} \tag{1.16}$$

Here $\mathbf{U}^{(e)}$ is the (6×1) vector denoting the nodal displacements, $\mathbf{P}^{(e)}$, the (6×1) vector of equivalent nodal forces, and $\mathbf{K}^{(e)}$ a (6×6) element stiffness matrix. $\mathbf{P}_F{}^{(e)}$ is a (6×1) vector of nodol forces due to element loads, in a kinematically determinate condition, namely when $\mathbf{U}^{(e)} \equiv \mathbf{O}$ (fixed end stage). Such relationships are derived by assuming strain (or stress) distri-

*See Fig. 9.6, Chap. 9

butions within the element, such as constant or linearly varying ones. Relations (1.16) are a generalization of the usual stiffness relations derived for beams.

The next step is the *global assembly* process. At a node we have, first, the statical requirement, namely that the sum of the nodal forces of the different elements meeting at the node is required to vanish. Second, we have the continuity requirement that the displacements of the nodes of the elements meeting of the node must be the same. These conditions, which are similar to the ones used in frames, then lead to Eq. (1.15). On solution we obtain the nodal displacements which also are the displacements of the nodes of the elements. The strains and stresses at any point within an element are obtained in terms of element node displacements and functions assumed in deriving the stiffness relations (1.16). We thus see that the set of nodal displacements determine the stresses and strains as continuous functions defined locally within each element in terms of element node displacements.

The problem of interconnection between the elements deserves special consideration. The elements are required to satisfy certain *interelement* requirements along the common edges. Since the sub-divisions are imaginary *full continuity* of displacements should be ensured between common edges. In simple terms this means that after deformation there should not be any 'gaps' or discontinuities across the common edges of the elements. This puts a severe demand on the element analysis. Another requirement is also of the *equality* (transmissibility) *of stresses* along such common edges.

The above discretization which started as a structural idea has received considerable attention by mathematicians too. Thus we have a mathematical interpretation. The determination of the displacements within the entire domain is achieved in terms of a discrete number of displacement values at certain points in the domain. These are the displacement values at the 17 points arbitrarily chosen for the beam in question. These points define sub-domains (finite elements). Presently these points are joined by straight lines resulting into 20 triangular sub-domains or finite regions. The displacements are defined within a sub-domain as continuous functions of x and y in terms of the displacement values of nodes constituting the sub-domain. These are defined in a local reference frame local to each sub-domain. Thus it is a numerical solution procedure in which the function to be determined is expressed piecewise within a small domain in terms of a finite (discrete) number of values at the nodes defining the sub-domain. These are then determined simultaneously by solving a set of equations.

Since surface and solid elements do not permit exact analysis satisfying equilibrium kinematics and material response laws, certain approximations are invariably required. Such approximations are not required in the case of line (beam) elements where exact solutions are possible.* Considerable attention has been paid by engineers and mathematicians alike in studying

*Recall that here too the analysis is not free of kinematic assumptions such as plane transverse sections remaining plane, shearing strains being negligible etc. which are made *a priori*.

this idealization of continuum problems for the last two decades. It is now a standard method of analysis of continuum problems.

The generalization of the idea to bodies considered as assemblies of three-dimensional elements, at least conceptually, is obvious. The resulting set of equations is generally large, but the present day computer facilities have permitted handling of sets involving thousands of unknowns. Computers are also used to draw pictorial variations of physical quantities of interest such as stress or deformation profiles instead of printing out pages of numbers.

An introduction to this technique has been given later in Chapter 9 of this book. The purpose of giving this introduction is twofold: firstly, it serves to broaden the scope of structural analysis of frames consisting of beam elements. The common principles of solid mechanics are identified. Secondly there is the practical need of developing a unified approach so as to help develop practical programs to analyse structures which actually consist of beam, plate and solid elements. The treatment of beam elements adopted in the book is similar to the one followed in handling two and three dimensional elements so that Chapter 9 comes as no surprise.

1.8 REFERENCES

Considerable published literature exists on the subject matter covered in this book. However only those quoted in the text are appended at the end of the book. Readers interested in a source for reference may consult Reference 4. Computerized sources of information too are available in this field. Some information in this regard can be found in Reference 5.

EXERCISES

1.1 Consider the simple theory of bending. Identify the steps which take into consideration the requirements of equilibrium, kinematics, and material characteristics.

1.2 Repeat Ex. 1.1 by considering the case of simple theory of torsion.

1.3 A pin ended member of length L undergoes only transverse displacements v_1 and v_2 at its ends 1 and 2. Calculate the axial strain and show that it is of a higher order magnitude.

1.4 Consider the derivation of Euler buckling load for a column. Show that we have to necessarily write equilibrium in a deformed configuration. Show that the distance through which the axial load descends in a buckling mode is $\frac{1}{2}\int\left(\frac{dv}{dx}\right)^2 dx$.

1.5 Explain the need for the two prevalent sign conventions, the statical and the deformation conventions; in other words, identify the instances where each is required to be followed.

1.6 Write Eq. (1.1) to handle the case of a concentrated load applied.

1.7 Identify the kinematic assumptions made in (a) beam in bending, (b) beam in torsion, (c) triangulated frameworks, (d) portal and building frames, (e) be haviour of a group of rivets subjected to a moment, (f) reinforced concrete beam in bending, and (h) plastic hinge idealization.

1.8 Consider the first equation of Eq. (1.4) and the case of a bar held at top and strained under its self weight. Determine the elongation of the bar. Use Eq.(a_1), Sec. 1.4.

1.9 Show how the last two equation of Eq. (1.4) can be interpreted as moment area theorems when we use $d\theta = Mdx/EI$ and that Eq. (1.5) can be used to derive the conjugate beam analogy.

1.10 Consider Eq. (1.4). Take the case of a cantilever subjected to temperature gradient varying from temperature t_1 at top to t_2 at bottom, and uniformly at all sections along the span. Determine the displacement and the rotation at free end.

1.11 Repeat Ex 1.10 for a simply supported beam and using Eq. (1.5).

1.12 Repeat Ex 1.9, 1.10 and 1.11 using virtual work method (Unit load method).

1.13 Derive Eq. (a_4) of Sec 1.4 using Unit load method, and subsequently, Eq. (a_7). Derive the latter *directly* solving the fourth order differential equation for the beam. If we do not take the help of the differential equation, is there any other way of directly obtaining them? What if the beam is of variable section?

1.14 Derive Eq. (a_8) from Eq. (a_7), of Sec. 1.4 by relating member end forces to the member forces and the deformations to displacements. Express these in terms of $\Delta = (v_1 - v_2)$ as generally adopted. Why do we generally adopt this form?

1.15 Using Eq. (a_{10}), Sec 1.4, solve the pin jointed frame of Ex. 2.12 (Chapter 2).

1.16 Taking the case of uniform load q_0 derive v^0 of relations (1.13). By inversion derive Eq. (1.14) and note that these are the familiar slope deflection equations.

1.17 Recall your experience that pinjointed frameworks are generally solved by force method. Do you see any reason for its preference to the displacement method? Likewise, building frames are generally handled by the slope deflection (or its equivalent, moment distribution) method. Identify the reasons for the preferences.

1.18 Solve the frame of Ex 2.12 by the force method and compare with the displacement method followed in Ex (1.15).

1.19 A bridge truss can be considered as an idealization of a beam. Identify how the moment and the shearing force are resisted in each case. If we increase the number of members considerably, can the solution of member forces form an approximation to the stress distribution in a homogeneous beam? In other words we can develop an equivalent 'frame work analogy' for a beam? In such a case can we consider that in Fig 1.10(a) the beam is an equivalent framework interconnected only at nodes?

1.20 Identify the assumptions which reduce the analysis of a prismatic beam from a three dimensional problem to a unidimensional one.

1.21 Take an example of a simply supported beam or a cantilever and indicate the superiority of the Unit load method over the Castigliano's method in determining displacemens.

Chapter 2

Equilibrium and Determinate Structures

We will consider in this chapter equilibrium equations as applied to a member or a structure, and also some force transformation relations. This will be followed by the analysis of determinate structures wherein consideration of statical relations alone are sufficient. Analysis of a determinate structure is also important from the viewpoint of the force method wherein a redundant structure is reduced to a determinate structure by relaxing certain continuities and later re-establishing them in terms of certain force parameters.

Recall that a statical relation, in the regime of small displacement theory, is always linear. Thus these are expressible in the forms,

$$\mathbf{C}\,\mathbf{S} = \mathbf{P}$$

or conversely

$$\mathbf{S} = \mathbf{B}\,\mathbf{P}$$

Here **S** will be the vector of internal member forces and **P** the vector of external loads. The first set relations states the conditions of equilibrium while the second states the solution of internal forces in terms of external forces. These are obtained by solutions in the form $\mathbf{S} = \mathbf{C}^{-1}\mathbf{P}$. Thus **B** is formally $\mathbf{C}^{-1}$. We will be demonstrating the details of the above relationships.

2.1 MEMBER EQUILIBRIUM RELATIONS

Recall the introduction of the terms 'member forces' and 'member end forces', in Sec. 1.2. Figure 2.1 shows the free body of a member. It is assumed, to begin with, that lateral loads are absent.

The *member end force* vectors will be denoted as*

$$\overline{\mathbf{P}}_1 = \left\{\overline{P_1}, \overline{Q_1}, \overline{M_1}\right\}\ ;\ \overline{\mathbf{P}}_2 = \left\{\overline{P_2}, \overline{Q_2}, \overline{M_2}\right\} \qquad (a_1)$$

The *member force* vector is a set of three forces which can be considered as independent, since not all the six can be independent but related to one another by the three equations of overall equilibrium. The axial tension S in the beam is constant throughout and the member end forces are related to this simply as

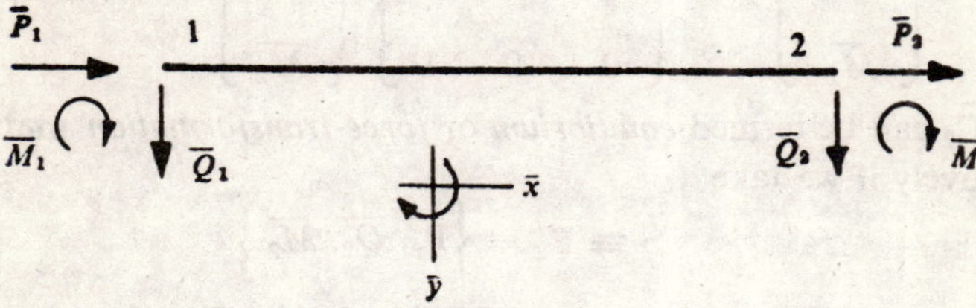

Figure 2.1

$$\overline{P_2} = -\overline{P_1} = S \qquad (a_2)$$

If we take

$$\mathbf{S} = \left\{S, \overline{M_1}, \overline{M_2}\right\} \qquad (2.1)$$

We have then in view a beam model (Fig. 2.2a) in which the shear forces are given as

$$\overline{Q_1} = -\overline{Q_2} = \frac{\overline{M_1}+\overline{M_2}}{L} \qquad (a_3)$$

Thus we relate $\overline{\mathbf{P}_1}$ and $\overline{\mathbf{P}_2}$ to $\mathbf{S}$ as follows:

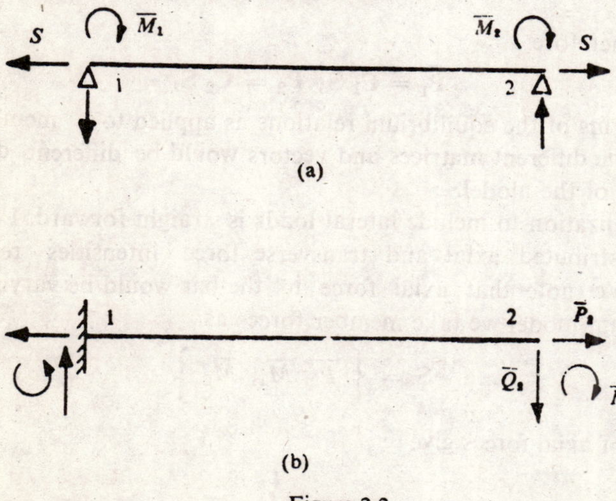

Figure 2.2

*The significance of bars on top will become clear soon when resolution of forces would be considered.

$$\overline{\mathbf{P}}_1 = \begin{Bmatrix} \overline{P}_1 \\ \overline{Q}_1 \\ \overline{M}_1 \end{Bmatrix} = \begin{bmatrix} -1 & 0 & 0 \\ 0 & \frac{1}{L} & \frac{1}{L} \\ 0 & 1 & 0 \end{bmatrix} \begin{Bmatrix} S \\ \overline{M}_1 \\ \overline{M}_2 \end{Bmatrix} = \overline{\mathbf{C}}_1\mathbf{S}$$

$$\overline{\mathbf{P}}_2 = \begin{Bmatrix} \overline{P}_2 \\ \overline{Q}_2 \\ \overline{M}_2 \end{Bmatrix} = \begin{bmatrix} 1 & 0 & 0 \\ 0 & -\frac{1}{L} & -\frac{1}{L} \\ 0 & 0 & 1 \end{bmatrix} \begin{Bmatrix} S \\ \overline{M}_1 \\ \overline{M}_2 \end{Bmatrix} = \overline{\mathbf{C}}_2\mathbf{S} \tag{2.2}$$

$\overline{\mathbf{C}}_1$ and $\overline{\mathbf{C}}_2$ can be termed *equilibrium* or *force-transformation* matrices. Alternatively if we take

$$\mathbf{S} \equiv \overline{\mathbf{P}}_2 = \left\{\overline{P}_2, \overline{Q}_2, \overline{M}_2\right\} \tag{2.3}$$

we have in view a *cantilever model* (Fig. 2.2b). The forces at end 1, obtainable from statics, are

$$\overline{\mathbf{P}}_1 = \begin{Bmatrix} \overline{P}_1 \\ \overline{Q}_1 \\ \overline{M}_1 \end{Bmatrix} = \begin{bmatrix} -1 & 0 & 0 \\ 0 & -1 & 0 \\ 0 & -L & -1 \end{bmatrix} \begin{Bmatrix} \overline{P}_2 \\ \overline{Q}_2 \\ \overline{M}_2 \end{Bmatrix} = \overline{\mathbf{C}}_1\,\mathbf{S} \tag{2.4}$$

Note that

$$\overline{\mathbf{P}}_2 = \mathbf{S} = \mathbf{I}\,\mathbf{S} = \overline{\mathbf{C}}_2\,\mathbf{S} \tag{2.5}$$

Thus we have an advantage in this choice that $\overline{\mathbf{C}}_2$ is a (3×3) identity matrix.

We have therefore

$$\overline{\mathbf{P}}_1 = \overline{\mathbf{C}}_1\,\mathbf{S};\ \overline{\mathbf{P}}_2 = \overline{\mathbf{C}}_2\,\mathbf{S}, \tag{2.6}$$

as general forms of the equilibrium relations as applied to a member. The elements of the different matrices and vectors would be different depending on the choice of the model.

The generalization to include lateral loads is straight forward. Let p and q be the distributed axial and transverse force intensities respectively (Fig. 2.3). We note that axial force in the bar would be varying. If we adopt the beam model we take member forces as

$$\mathbf{S} = \left\{ \overline{P}_2, \overline{M}_1, \overline{M}_2 \right\} \tag{a_4}$$

Equilibrium of axial forces give

$$\overline{P}_1 = -\overline{P}_2 - \int_0^L p\,dx = -\overline{P}_2 + \overline{P}_1^0 \tag{a_5}$$

The shear forces would be

$$\bar{Q}_1 = \frac{\bar{M}_1+\bar{M}_2}{L}+\bar{Q}_1^0, \quad \bar{Q}_2 = -\frac{\bar{M}_1+\bar{M}_2}{L}+\bar{Q}_2^0 \qquad (a_6)$$

where $\bar{Q}_1^0$ and $\bar{Q}_2^0$ would be the shears due to q, in the beam considered as a simply supported beam. Thus we have

$$\bar{\mathbf{P}}_1 = \begin{Bmatrix} \bar{P}_1 \\ \bar{Q}_1 \\ \bar{M}_1 \end{Bmatrix} = \begin{bmatrix} -1 & 0 & 0 \\ 0 & \frac{1}{L} & \frac{1}{L} \\ 0 & 1 & 0 \end{bmatrix} \begin{Bmatrix} \bar{P}_2 \\ \bar{M}_1 \\ \bar{M}_2 \end{Bmatrix} + \begin{Bmatrix} \bar{P}_1^0 \\ \bar{Q}_1^0 \\ 0 \end{Bmatrix}$$

$$= \bar{\mathbf{C}}_1 \mathbf{S}+\bar{\mathbf{P}}_1^0 \qquad (a_7)$$

$$\bar{\mathbf{P}}_2 = \begin{Bmatrix} \bar{P}_2 \\ \bar{Q}_2 \\ \bar{M}_2 \end{Bmatrix} = \begin{bmatrix} 1 & 0 & 0 \\ 0 & -\frac{1}{L} & -\frac{1}{L} \\ 0 & 0 & 1 \end{bmatrix} \begin{Bmatrix} \bar{P}_2 \\ \bar{M}_1 \\ \bar{M}_2 \end{Bmatrix} + \begin{Bmatrix} 0 \\ \bar{Q}_2^0 \\ 0 \end{Bmatrix}$$

$$= \bar{\mathbf{C}}_2 \mathbf{S}+\bar{\mathbf{P}}_2^0 \qquad (a_8)$$

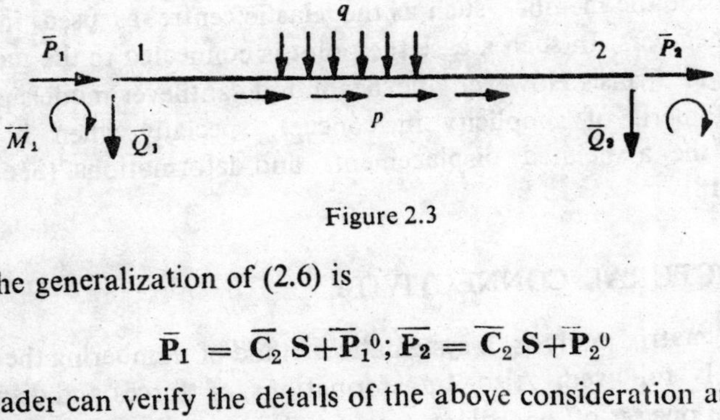

Figure 2.3

Thus the generalization of (2.6) is

$$\bar{\mathbf{P}}_1 = \bar{\mathbf{C}}_1 \mathbf{S}+\bar{\mathbf{P}}_1^0; \; \bar{\mathbf{P}}_2 = \bar{\mathbf{C}}_2 \mathbf{S}+\bar{\mathbf{P}}_2^0 \qquad (2.7)$$

The reader can verify the details of the above consideration as applied to the cantilever model.

Sometimes we combine the expressions for $\bar{\mathbf{P}}_1$ and $\bar{\mathbf{P}}_2$ and write

$$\underset{(6\times 1)}{\bar{\mathbf{P}}} = \begin{Bmatrix} \bar{\mathbf{P}}_1 \\ \bar{\mathbf{P}}_2 \end{Bmatrix} = \underset{(6\times 3)}{\begin{bmatrix} \bar{\mathbf{C}}_1 \\ \bar{\mathbf{C}}_2 \end{bmatrix}} \underset{(3\times 1)}{\{\mathbf{S}\}} = \bar{\mathbf{C}}\,\mathbf{S} \qquad (2.8)$$

where $\bar{\mathbf{P}}$, $\bar{\mathbf{C}}$ and $\mathbf{S}$ are augmented matrices in terms of submatrices. Extensive use will be made of such compact notations in this book. The beginner should always write them out in full and grasp the details. Presently these would read, for the beam model,

$$\overline{\mathbf{P}}=\left\{\begin{array}{c}\overline{P}_1\\ \overline{Q}_1\\ \overline{M}_1\\ \hdashline \overline{P}_2\\ \overline{Q}_2\\ \overline{M}_2\end{array}\right\}=\left[\begin{array}{ccc}-1 & 0 & 0\\ 0 & \frac{1}{L} & \frac{1}{L}\\ 0 & 1 & 0\\ \hdashline 1 & 0 & 0\\ 0 & -\frac{1}{L} & -\frac{1}{L}\\ 0 & 0 & 1\end{array}\right]\left\{\begin{array}{c}S\\ \overline{M}_1\\ \overline{M}_2\end{array}\right\}=\overline{\mathbf{C}}\mathbf{S} \qquad (a_9)$$

Identification of the two sets of forces (members forces **S** and member end forces $\overline{\mathbf{P}}_1$ and $\overline{\mathbf{P}}_2$) will be found useful at many stages in analysis. This will become evident when one considers the problem of solving for internal forces and also when deriving member flexibility.

Generalized Member Forces

The beam and cantilever models adopted so far are by no means the only possible models [6]. We can choose for instance, the internal forces at any point within the beam as the member forces. Alternatively, since there is no particular preference to the choice of the point, such a point can be even outside the member, such as the 'elastic centre' as used in analysis of curved members. In such a case the point is connected to the member end through rigid arms. However the beam and cantilever models possess the remarkable merit of simplicity in concept, specially when it comes to visualizing the associated displacements and deformations. (See Exercises 2.17 and 2.18).

2.2 STRUCTURAL CONNECTIVITY

In matrix analysis, a systematic method of numbering the members and joints is followed. Also transformations of forces and displacement vectors from one set of coordinates to another are effected by reducing the process to one of matrix algebra. In general two sets of coordinate axes are identified. Denoting of member forces, and analysis of member, such as finding stiffness and flexibility are made in a set of axes local to the member. Such a set of axes would be called *member axes*. For the structural analysis a common set of axes for the structural assembly is defined. This will be referred to as *structure axes* or, alternatively, as *global axes*. The loads as well as the displacements of the joints are measured in this set of axes.

While considering the equilibrium or kinematics of the structure the member forces and deformations, which are measured in the respective local axes of the members, are required to be resolved along the global axes. Such resolutions are accomplished by premultiplication by certain transformation matrices.

Figure 2.4a shows a portal frame. The member axes are denoted by $(\bar{x}, \bar{y})$. The positive direction of $\bar{x}$ is always taken proceeding from end 1 towards end 2. The $\bar{y}$ axis is then chosen according to the right handed convention. The inclination of the axes $(\bar{x}, \bar{y})$ of a typical member is denoted by the angle

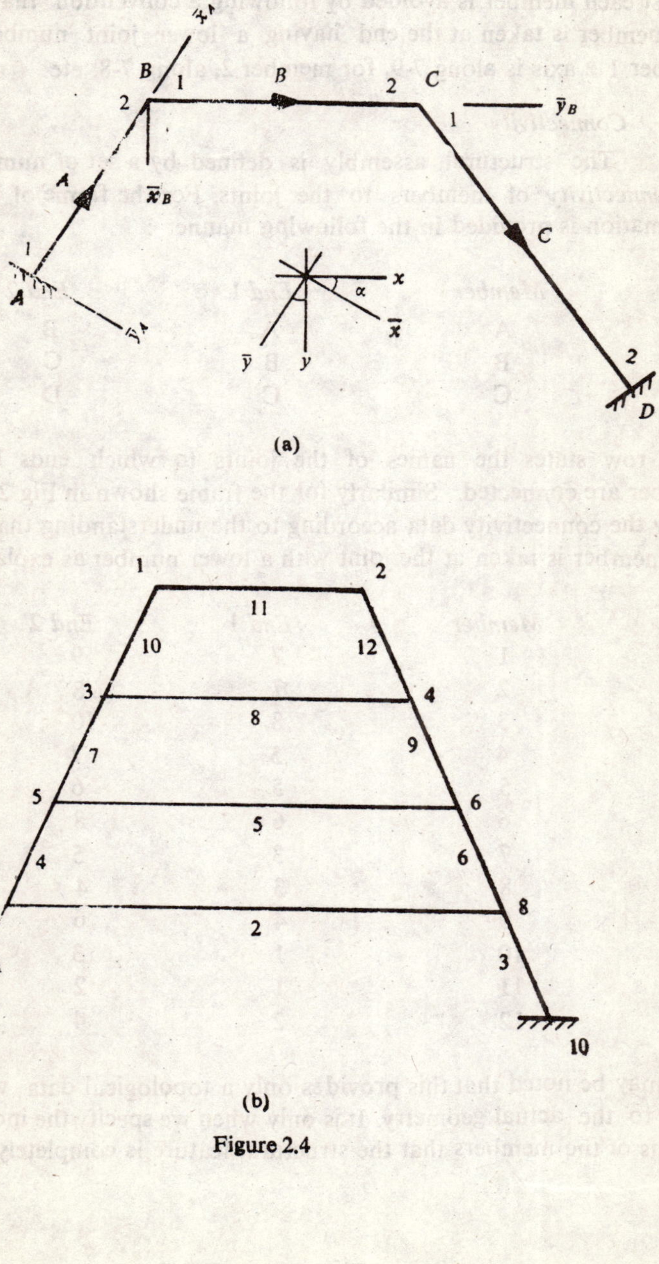

Figure 2.4

as shown. The positive direction of $\bar{x}$ axis of the member is occasionally shown by an arrow just as the flow of the electric current in a network. If this is adopted then writing of 1 and 2 to denote the ends are *not* needed.

For numbering of members or joints either letters or numbers can be used. For illustration of principles or small size examples, use of letters as in Fig. 2.4(a) is adequate. However in general, numbers are used. Figure 2.4(b) shows an example. The placing of the arrow or writing end 1 and 2 against each member is avoided by following a convention that the end 1 of a member is taken at the end having a lower joint number. Thus for member 1 $\bar{x}$ axis is along 7-9, for member 2, along 7-8, etc.

Connectivity

The structural assembly is defined by a set of numbers defining the *connectivity* of members to the joints. For the frame of Fig 2.4(a) the information is provided in the following manner :

Member	*End* 1	*End* 2	
A	A	B	
B	B	C	(a_1)
C	C	D	

Each row states the names of the joints to which ends 1 and 2 of the member are connected. Similarly for the frame shown in Fig 2.4(b) we have below the connectivity data according to the understanding that the end 1 of a member is taken at the joint with a lower number as explained above:

Member	*End* 1	*End* 2	
1	7	9	
2	7	8	
3	8	10	
4	5	7	
5	5	6	
6	6	8	(a_2)
7	3	5	
8	3	4	
9	4	6	
10	1	3	
11	1	2	
12	2	4	

It may be noted that this provides only a topological data without reference to the actual geometry. It is only when we specify the inclinations and lengths of the members that the structural feature is completely defined.

2.3 COORDINATE TRANSFORMATION

In Fig. 2.5 ($\bar{x}$, $\bar{y}$) axes are inclined at an angle α to the (x, y) axes. The components of a vector OP, which can be a force or a displacement vcctor, can be expressd in either of the systems. We can now derive the transformation relationships between the components. From the geometry of of Fig 2.5, we have

$$x = \bar{x} \cos \alpha - \bar{y} \sin \alpha ; \qquad y = \bar{x} \sin \alpha + \bar{y} \cos \alpha \tag{a$_1$}$$

These in matrix form are

$$\mathbf{X} = \begin{Bmatrix} x \\ y \end{Bmatrix} = \begin{bmatrix} \cos\alpha & -\sin\alpha \\ \sin\alpha & \cos\alpha \end{bmatrix} \begin{Bmatrix} \bar{x} \\ \bar{y} \end{Bmatrix} = \mathbf{T}_0 \, \overline{\mathbf{X}} \tag{2.9}$$

Matrix $\mathbf{T}_0$ appearing in the above equation is called *coordinate transformation* matrix. Matrix such as this has the property that its determinant is unity. This results in a very useful (orthogonality) property that

$$\mathbf{T}_0^{-1} = \mathbf{T}_0^{t} \tag{2.10}$$

Thus solving for $\overline{\mathbf{X}}$ from Eq. (2.9) we have

$$\overline{\mathbf{X}} = \mathbf{T}_0^{-1} \, \mathbf{X} = \mathbf{T}_0^{t} \, \mathbf{X} \tag{2.11}$$

which, when written out in full reads

$$\mathbf{X} = \begin{Bmatrix} \bar{x} \\ \bar{y} \end{Bmatrix} = \begin{bmatrix} \cos\alpha & \sin\alpha \\ -\sin\alpha & \cos\alpha \end{bmatrix} \begin{Bmatrix} x \\ y \end{Bmatrix} = \mathbf{T}_0^{t} \, \mathbf{X} \tag{a$_2$}$$

giving

$$\bar{x} = x \cos \alpha + y \sin \alpha ; \quad \bar{y} = -x \sin \alpha + y \cos \alpha$$

These relationships can be directly verified from the geometry of Fig. 2.5.

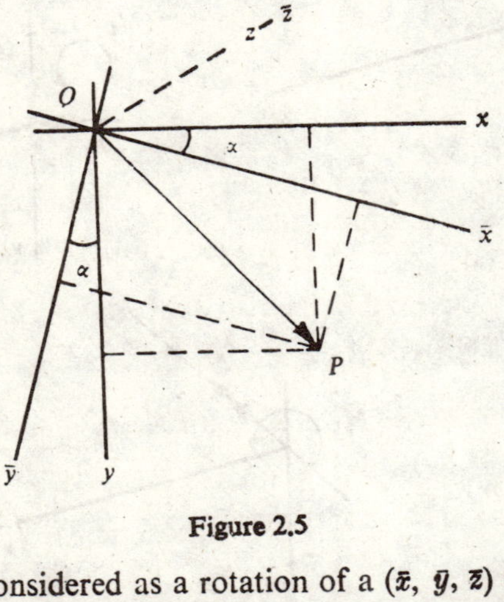

Figure 2.5

Angle α can be considered as a rotation of a ($\bar{x}$, $\bar{y}$, $\bar{z}$) system about z axis of a (x, y, z) system. Thus z and $\bar{z}$ are coincident. If the vector OP has some projection along $\bar{z}$ axis it is the same along z axis. or vice versa. Thus we can write

$$\mathbf{X} = \begin{Bmatrix} x \\ y \\ z \end{Bmatrix} = \begin{bmatrix} \cos\alpha & -\sin\alpha & 0 \\ \sin\alpha & \cos\alpha & 0 \\ 0 & 0 & 1 \end{bmatrix} \begin{Bmatrix} \bar{x} \\ \bar{y} \\ \bar{z} \end{Bmatrix} = \mathbf{T}\,\overline{\mathbf{X}} \qquad (2.12)$$

Relation (2.10) holds for this transformation matrix too.

As an application, consider the transformations required for force vectors at end 1 of a member as shown in Figs. 2.6 (a) and (b). In Fig. 2.6(a) the components are along the member axes ($\bar{x}$, $\bar{y}$) and in Fig 2.6 (b) along the global axes (x, y). By actually resolving we find that

$$\mathbf{P}_1 = \begin{Bmatrix} P_1 \\ Q_1 \\ M_1 \end{Bmatrix} = \begin{bmatrix} \cos\alpha & -\sin\alpha & 0 \\ \sin\alpha & \cos\alpha & 0 \\ 0 & 0 & 1 \end{bmatrix} \begin{Bmatrix} \overline{P_1} \\ \overline{Q_1} \\ \overline{M_1} \end{Bmatrix} = \mathbf{T}\,\overline{\mathbf{P}_1} \qquad (a_3)$$

The last row simply states, $M_1 = \overline{M}_1$. We can see that the moment when represented according to the right hand corkscrew rule (Fig. 2.6 (c), becomes a vector projecting along the positive z axis, and since the rotation of axes is about this axis itself the vector remains unchanged. To distinguish from forces, the moments are denoted by *double-headed* vectors. This will be necessary when dealing with three dimensional set of forces.

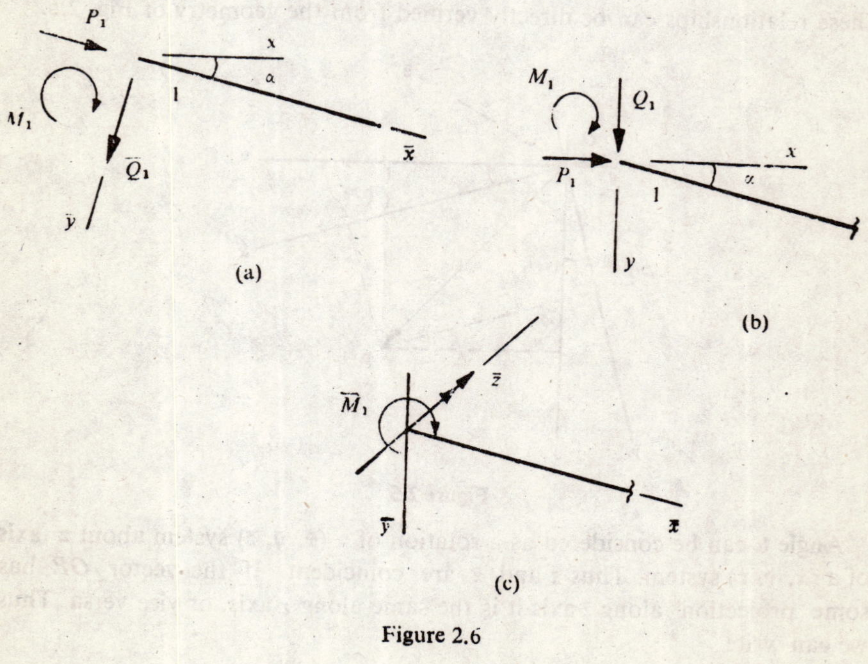

Figure 2.6

Coordinate transformation in three dimensions will be a generalization of the above ideas and will be taken up later. (See Sec. 8.5 c).

2.4 EQUILIBRIUM OF STRUCTURES

We will now consider equilibrium equations as applied to a structural framework.

Equations of equilibrium relate the member forces to the applied loads. The convenient form of member end forces to be considered is the one shown in Fig. 2.7(a). In this only three are independent. It is usual therefore to consider the forces shown in Fig. 2.7(b) as an independent set of member forces consisting of an axial tension S and end moments $\overline{M}_1$ and $\overline{M}_2$. We first relate the two member force systems and write the desired equation of equilibrium of a joint or a group of members.

Let the member end forces of Fig. 2.7(a) be,

$$\mathbf{P} = \left\{\mathbf{P}_1 \vdots \mathbf{P}_2\right\} \equiv \left\{P_1, Q_1, M_1 \vdots P_2, Q_2, M_2\right\} \qquad (a_1)$$

and let the member forces, Fig. 2.7(b), be

$$\mathbf{S} = \left\{S, \overline{M}_1, \overline{M}_2\right\} \qquad (a_2)$$

The relationship between these two can be directly written by resolving the forces of Fig 2.7(a) along the member and equate to those of Fig. 2.7(b). However we will do this using matrix operations. Figures 2.1 and 2.2 show the force systems S and $\overline{\mathbf{P}}$. Here $\overline{\mathbf{P}}$ is the vector of member end forces expressed in member coordinate system. The relationship between the two was considered in Sec. 2.1 and is given by Eq. (2.6); to rewrite

$$\overline{\mathbf{P}_1} = \overline{\mathbf{C}_1}\,\mathbf{S};\ \overline{\mathbf{P}_2} = \overline{\mathbf{C}_2}\,\mathbf{S} \qquad (a_3)$$

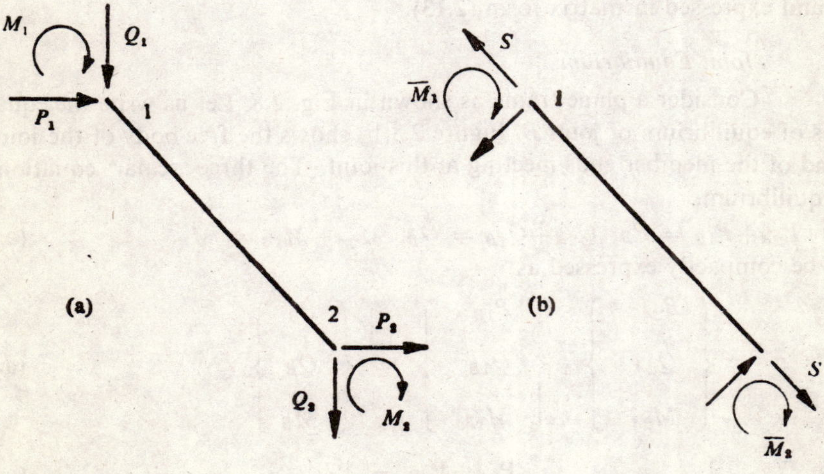

Figure 2.7

The next step is to use coordinate transformation for $\mathbf{P}_1$ and $\mathbf{P}_2$ in terms of $\overline{\mathbf{P}}_1$ and $\overline{\mathbf{P}}_2$ respectively. With the coordinate transformation matrix, **T**, as given in Eq. (2.12) we have

$$\mathbf{P}_1 = \mathbf{T}\,\overline{\mathbf{P}}_1;\ \ \mathbf{P}_2 = \mathbf{T}\,\overline{\mathbf{P}}_2 \tag{a$_4$}$$

Thus using Eq. (a_3) we write

$$\mathbf{P}_1 = \mathbf{T}\,\overline{\mathbf{C}}_1\,\mathbf{S} = \mathbf{C}_1\,\mathbf{S} \tag{a$_5$}$$

$$\mathbf{P}_2 = \mathbf{T}\,\overline{\mathbf{C}}_2\,\mathbf{S} = \mathbf{C}_2\,\mathbf{S}$$

The matrix products $\mathbf{T}\,\overline{\mathbf{C}}_1$ and $\mathbf{T}\,\overline{\mathbf{C}}_2$ can be performed easily and obtain finally the desired relationship:

$$\mathbf{P} = \begin{Bmatrix} \mathbf{P}_1 \\ \mathbf{P}_2 \end{Bmatrix} = \begin{bmatrix} \mathbf{C}_1 \\ \mathbf{C}_2 \end{bmatrix} \{\mathbf{S}\} \tag{2.13}$$

where

$$\mathbf{C}_1 = \begin{bmatrix} -c & -s/L & -s/L \\ -s & c/L & c/L \\ 0 & 1 & 0 \end{bmatrix};\ \ \mathbf{C}_2 = \begin{bmatrix} c & s/L & s/L \\ s & -c/L & -c/L \\ 0 & 0 & 1 \end{bmatrix} \tag{2.14}$$

with $c = \cos\alpha$, $s = \sin\alpha$.

The relation (2.13), which is equivalent to

$$\mathbf{P}_1 = \mathbf{C}_1\,\mathbf{S},\ \mathbf{P}_2 = \mathbf{C}_2\,\mathbf{S}, \tag{2.15}$$

simply state resolutions such as

$$P_1 = -S\cos\alpha - \frac{\overline{M}_1 + \overline{M}_2}{L}\sin\alpha$$

$$Q_1 = -S\sin\alpha + \frac{\overline{M}_1 + \overline{M}_2}{L}\cos\alpha$$

$$M_1 = \overline{M}_1 \text{ etc.}$$

Such relations could have been directly written down by inspection of Fig. 2.7 and expressed in matrix form (2.13).

Joint Equilibrium

Consider a plane frame as shown in Fig. 2.8. Let us write the equations of equilibrium of joint B. Figure 2.8(b) shows the free body of the joint B and of the member ends meeting at this joint. The three scalar equations of equilibrium,

$$P_{2A} + P_{1B} = P_B,\ Q_{2A} + Q_{1B} = Q_B,\ M_{2A} + M_{1B} = M_B, \tag{a$_6$}$$

can be compactly expressed as

$$\begin{Bmatrix} P_{2A} \\ Q_{2A} \\ M_{2A} \end{Bmatrix} + \begin{Bmatrix} P_{1B} \\ Q_{1B} \\ M_{1B} \end{Bmatrix} = \begin{Bmatrix} P_B \\ Q_B \\ M_B \end{Bmatrix} \tag{a$_7$}$$

or

$$\mathbf{P}_{2A} + \mathbf{P}_{1B} = \mathbf{P}_B \tag{a$_8$}$$

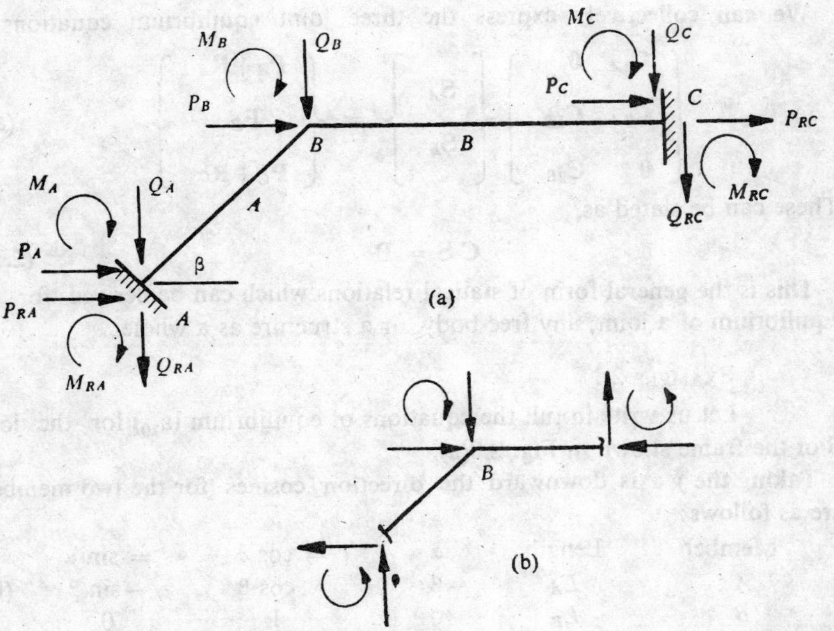

Figure 2.8

Equilibrium of Structure

Let us write the overall equilibrium equation for the structure as a whole. In terms of vectors of forces,

$$\mathbf{P}_A+\mathbf{P}_B+\mathbf{P}_C+\mathbf{R}_A+\mathbf{R}_C = 0 \tag{a_9}$$

where $\mathbf{R}_A$ and $\mathbf{R}_C$ are unknown reactive forces at A and C. Observe that these denote three scalar equations. For a space frame, the above form is the same, but representing six scalar equations.

Equilibrium in terms of member forces

Equations of equilibrium of type (a_8) are next expressed in terms of member force vectors $\mathbf{S}_A$ and $\mathbf{S}_B$ of members A and B respectively, using relation (2.15). We have

$$\mathbf{P}_{2A} = \mathbf{C}_{2A}\,\mathbf{S}_A,\ \mathbf{P}_{1B} = \mathbf{C}_{1B}\,\mathbf{S}_B$$

with $\mathbf{C}_{2A}$ and $\mathbf{C}_{1B}$ given by matrices of Eq. (2.14) as applied to member A and B respectively. Thus Eq. (a_8) can be written as

$$[\mathbf{C}_{2A}\quad \mathbf{C}_{1B}]\begin{Bmatrix}\mathbf{S}_A\\ \mathbf{S}_B\end{Bmatrix} = \mathbf{P}_B \tag{a_{10}}$$

Let us now write the equations of equilibribm at support A:

$$\mathbf{P}_{1A} = \mathbf{P}_A+\mathbf{R}_A,\ \text{or}\ \mathbf{C}_{1A}\,\mathbf{S}_A = \mathbf{P}_A+\mathbf{R}_A \tag{a_{11}}$$

Similarly at C:

$$\mathbf{P}_{2B} = \mathbf{P}_C+\mathbf{R}_C,\ \text{or}\ \mathbf{C}_{2B}\,\mathbf{S}_B = \mathbf{P}_C+\mathbf{R}_C \tag{a_{12}}$$

We can collectively express the three joint equilibrium equations as

$$\begin{bmatrix} \mathbf{C}_{1A} & \mathbf{0} \\ \mathbf{C}_{2A} & \mathbf{C}_{1B} \\ \mathbf{0} & \mathbf{C}_{2B} \end{bmatrix} \begin{Bmatrix} \mathbf{S}_A \\ \mathbf{S}_B \end{Bmatrix} = \begin{Bmatrix} \mathbf{P}_A+\mathbf{R}_A \\ \mathbf{P}_B \\ \mathbf{P}_C+\mathbf{R}_C \end{Bmatrix} \qquad (a_{13})$$

These can be stated as,

$$\mathbf{C}\,\mathbf{S} = \mathbf{P} \qquad (2.16)$$

This is the general form of statical relations which can be derived for the equilibrium of a joint, any free-body, or a structure as a whole.

EXAMPLE 2.1

Let us write in full the equations of equilibrium (a_{10}) for the joint B of the frame shown in Fig. 2.8(a).

Taking the y axis downward the direction cosines for the two members are as follows:

Member	Length	α	$c = \cos\alpha$	$s = \sin\alpha$
A	L_A	$-\beta$	$\cos\beta$	$-\sin\beta$
B	L_B	0	1	0

(b_1)

Using the above values in Eq. (2.14) for $\mathbf{C}_1$ and $\mathbf{C}_2$, Eq. (a_{10}) becomes

$$\left[\begin{array}{ccc|ccc} \cos\beta & -\dfrac{\sin\beta}{L_A} & -\dfrac{\sin\beta}{L_A} & -1 & 0 & 0 \\ -\sin\beta & -\dfrac{\cos\beta}{L_A} & -\dfrac{\cos\beta}{L_A} & 0 & \dfrac{1}{L_B} & \dfrac{1}{L_B} \\ 0 & 0 & 1 & 0 & 1 & 0 \end{array}\right] \begin{Bmatrix} S_A \\ M_{1A} \\ M_{2A} \\ \hline S_B \\ M_{1B} \\ M_{2B} \end{Bmatrix} = \begin{Bmatrix} P_B \\ Q_B \\ M_B \end{Bmatrix} \qquad (b_2)$$

Each of the scalar equations can be verified by proceeding with scalar quantities. Observe that the last row states the moment equilibrium which is readily verified. The first and the second rows likewise state equilibirum of forces along x and y directions.

2.5 SOLUTION OF EQUILIBRIUM EQUATIONS

The conditions of equilibrium are in general of the the form (2.16). The inverse form gives the internal forces in terms of external loads. Thus in general,

$$\mathbf{S} = \mathbf{B}\,\mathbf{P} \qquad (2.17)$$

Therefore, we have from Eq. (2.16),

$$\mathbf{B} = \mathbf{C}^{-1} \tag{a$_1$}$$

Herein it has been assumed first, that the matrix **C** is square, and second that it is not singular. It may be noted that merely writing the conditions of equilibrium in the form (2.16) does not in any way ensure that the matrix **C** is square or nonsingular. It is necessary to consider these aspects before attempting to invert Eq. (2.16). The reactive forces at supports are not known and therefore have to be considered as additional unknowns. If the structure is determinate the unknown internal forces and the reactive forces are just equal to the available equilibrium equations and are determined from statics alone. Hence the matrix **C** will be square and can then be inverted. In a redundant structure the matrix **C** is rectangular with less number of rows meaning thereby lesser available equations for the total unknowns.

Consider Eq. (a_{13}) of Sec. 2.4 written for the frame of Fig. 2.8, which is fixed at *A* and *C*. In Eq. (a_{13}) the terms $\mathbf{R}_A$ and $\mathbf{R}_C$ are to be brought to the left hand side together with $\mathbf{S}_A$ and $\mathbf{S}_B$. Therefore we can write,

$$\underset{(9\times 12)}{\left[\begin{array}{cc|cc} \mathbf{C}_{1A} & \mathbf{0} & -\mathbf{I} & \mathbf{0} \\ \mathbf{C}_{2A} & \mathbf{C}_{1B} & \mathbf{0} & \mathbf{0} \\ \mathbf{0} & \mathbf{C}_{2B} & \mathbf{0} & -\mathbf{I} \\ \multicolumn{2}{c}{(9\times 6)} & \multicolumn{2}{c}{(9\times 6)} \end{array}\right]} \underset{(12\times 1)}{\begin{Bmatrix} \mathbf{S}_A \\ \mathbf{S}_B \\ \mathbf{R}_A \\ \mathbf{R}_C \end{Bmatrix}} = \underset{(9\times 1)}{\begin{Bmatrix} \mathbf{P}_A \\ \mathbf{P}_B \\ \mathbf{P}_C \end{Bmatrix}} \tag{a$_2$}$$

where the unit submatrices are of size (3×3). These represent nine equations in twelve unknowns leading to third degree indeterminacy as expected. This modified set of equilibrium can be written as

$$\mathbf{C}'\,\mathbf{S}' = [\mathbf{C} \quad \mathbf{C}_R] \begin{Bmatrix} \mathbf{S} \\ \mathbf{R} \end{Bmatrix} = \mathbf{P} \tag{2.18}$$

Let us suppose that the frame shown in Fig. 2.8(a) is modified to render it determinate. Let $\mathbf{R}_C \equiv \mathbf{0}$, which means that the frame is a determinate cantilever from support *A*. In this case we write the equations as follows:

$$\underset{(9\times 9)}{\begin{bmatrix} \mathbf{C}_{1A} & 0 & -\mathbf{I} \\ \mathbf{C}_{2A} & \mathbf{C}_{1B} & 0 \\ 0 & \mathbf{C}_{2B} & 0 \end{bmatrix}} \underset{(9\times 1)}{\begin{Bmatrix} \mathbf{S}_A \\ \mathbf{S}_B \\ \mathbf{R}_A \end{Bmatrix}} = \underset{(9\times 1)}{\begin{Bmatrix} \mathbf{P}_A \\ \mathbf{P}_B \\ \mathbf{P}_C \end{Bmatrix}} \tag{a$_3$}$$

It is seen that the **C** matrix is square and can be inverted to give the internal forces (and the reactive forces at *A*) in terms the loads in the form (2.17).

2.6 DETERMINATE PINJOINTED FRAMES

The case of the member of a pinjointed framework is a degenerate case of a rigid jointed frame member with end moments omitted. Thus for a pinjointed frame the internal force in a member is an axial force S, assumed to be a tensile force (Fig. 2.9). The member end forces for the two dimensional case are shown in the figure. The relationships between these two forces are,

$$\mathbf{P}_1 = \begin{Bmatrix} P_1 \\ Q_1 \end{Bmatrix} = \begin{Bmatrix} -\cos\alpha \\ -\sin\alpha \end{Bmatrix} S = \begin{Bmatrix} -c \\ -s \end{Bmatrix} S = \mathbf{C}_1\,\mathbf{S}$$

$$\mathbf{P}_2 = \begin{Bmatrix} P_2 \\ Q_2 \end{Bmatrix} = \begin{Bmatrix} \cos\alpha \\ \sin\alpha \end{Bmatrix} S = \begin{Bmatrix} c \\ s \end{Bmatrix} S = \mathbf{C}_2\,\mathbf{S} \qquad (2.19)$$

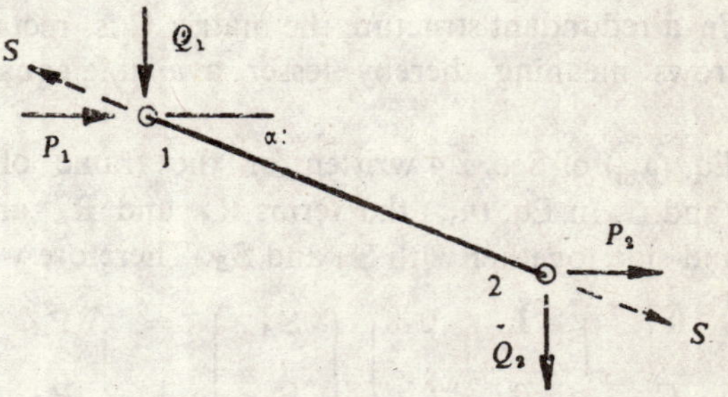

Figure 2.9

The same can be deduced from $\mathbf{C}_1$ and $\mathbf{C}_2$ matrices of Eq. (2.14), by considering the first two rows and the first column only.

Similarly for the space frame member shown in Fig. 2.10, we have

$$\mathbf{P}_1 = \begin{Bmatrix} P_{x1} \\ P_{y1} \\ P_{z1} \end{Bmatrix} = \begin{Bmatrix} -l \\ -m \\ -n \end{Bmatrix} S = \mathbf{C}_1\,\mathbf{S}$$

$$\mathbf{P}_2 = \begin{Bmatrix} P_{x2} \\ P_{y2} \\ P_{z2} \end{Bmatrix} = \begin{Bmatrix} l \\ m \\ n \end{Bmatrix} S = \mathbf{C}_2\,\mathbf{S} \qquad (2.20)$$

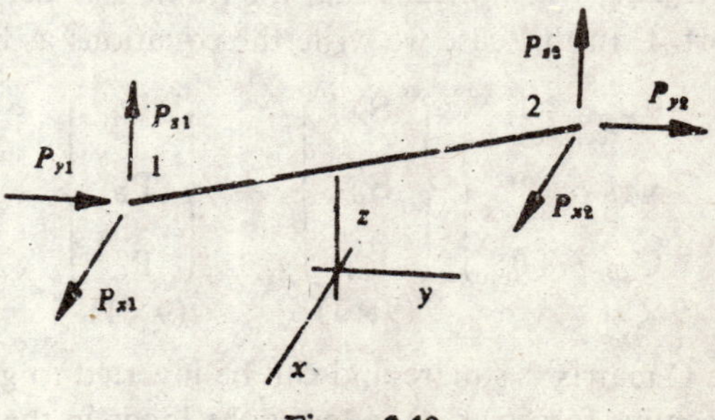

Figure 2.10

where l, m and n are the direction cosines of $\bar{x}$ axis of the member in global system. These are conveniently defined in terms of projections L_x, L_y and L_z of the member along the global axes;

$$l = \frac{x_2 - x_1}{L} = \frac{L_x}{L}\ ;\ m = \frac{y_2 - y_1}{L} = \frac{L_y}{L}\ ;\ n = \frac{z_2 - z_1}{L} = \frac{L_z}{L} \tag{2.21}$$

In the above simple expressions, Eqs. (2.19) and (2.20), the member equilibrium requirements $\mathbf{P}_1 + \mathbf{P}_2 = \mathbf{0}$ are readily verified.

In a pinjointed frame the total number of unknowns is equal to the number of bars plus three reactive components, (or six in the case of space frame). The available joint equilibrium equations are just sufficient to evaluate the unknowns.

EXAMPLE 2.2

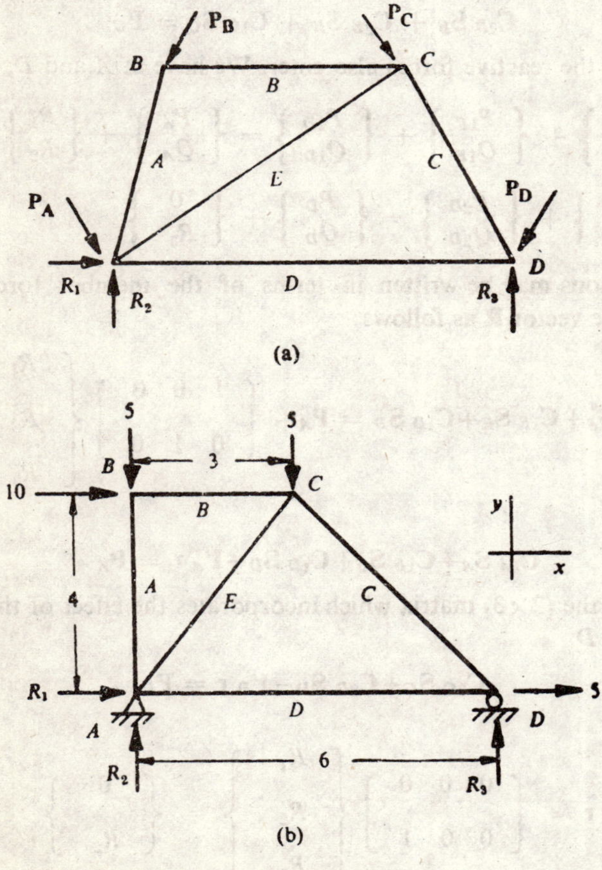

Example 2.2

We will demonstrate the application to a simple pinjointed plane frame. The frame shown in (b) is a particular case of the general one shown at (a). The number of unknown bar forces is five. These together with the three unknown reactions constitute eight unknown quantities. Two equations of equilibrium, at each of the four joints are the required number of equations to solve for these unknowns.

For this frame, the internal force, load and reactive force vectors are

$$\underset{(5\times 1)}{\mathbf{S}} = \{ S_A, S_B, S_C, S_D, S_E \}$$

$$\underset{(3\times 1)}{\mathbf{r}} = \{ R_1, R_2\ R_3 \}$$

$$\underset{(8\times 1)}{\mathbf{P}} = \{ \mathbf{P}_A, \mathbf{P}_B, \mathbf{P}_C, \mathbf{P}_D \}$$

The equations of equilibrium at B and C are:

$$\mathbf{C}_{2A}\,\mathbf{S}_A + \mathbf{C}_{1B}\,\mathbf{S}_B = \mathbf{P}_B$$

$$\mathbf{C}_{2B}\,\mathbf{S}_B + \mathbf{C}_{2E}\,\mathbf{S}_E + \mathbf{C}_{1C}\,\mathbf{S}_C = \mathbf{P}_C$$

At A and D the reactive forces also enter. We have at A and D,

$$\begin{Bmatrix} P_{1A} \\ Q_{1A} \end{Bmatrix} + \begin{Bmatrix} P_{1E} \\ Q_{1E} \end{Bmatrix} + \begin{Bmatrix} P_{1D} \\ Q_{1D} \end{Bmatrix} = \begin{Bmatrix} P_A \\ Q_A \end{Bmatrix} + \begin{Bmatrix} R_1 \\ R_2 \end{Bmatrix} \tag{a}$$

$$\begin{Bmatrix} P_{2C} \\ Q_{2C} \end{Bmatrix} + \begin{Bmatrix} P_{2D} \\ Q_{2D} \end{Bmatrix} = \begin{Bmatrix} P_D \\ Q_D \end{Bmatrix} + \begin{Bmatrix} 0 \\ R_3 \end{Bmatrix} \tag{b}$$

These equations may be written in terms of the member forces and the reactive force vector $\mathbf{R}$ as follows:

$$\mathbf{C}_{1A}\,\mathbf{S}_A + \mathbf{C}_{1E}\,\mathbf{S}_E + \mathbf{C}_{1D}\,\mathbf{S}_D = \mathbf{P}_A + \begin{bmatrix} 1 & 0 & 0 \\ 0 & 1 & 0 \end{bmatrix} \begin{Bmatrix} R_1 \\ R_2 \\ R_3 \end{Bmatrix} \tag{c}$$

or

$$\mathbf{C}_{1A}\,\mathbf{S}_A + \mathbf{C}_{1E}\,\mathbf{S}_E + \mathbf{C}_{1D}\,\mathbf{S}_D - \mathbf{I'}_A\,\mathbf{r} = \mathbf{P}_A \tag{d}$$

where $\mathbf{I'}_A$ is the (2×3) matrix which incorporates the effect of the reactions. Similarly at D

$$\mathbf{C}_{2C}\,\mathbf{S}_C + \mathbf{C}_{2D}\,\mathbf{S}_D - \mathbf{I'}_D\,\mathbf{r} = \mathbf{P}_D \tag{e}$$

where

$$\mathbf{I'}_D\,\mathbf{r} = \begin{bmatrix} 0 & 0 & 0 \\ 0 & 0 & 1 \end{bmatrix} \begin{Bmatrix} R_1 \\ R_2 \\ R_3 \end{Bmatrix} = \begin{Bmatrix} 0 \\ R_2 \end{Bmatrix} \tag{f}$$

Thus the equations of equilibrium are:

$$\underset{8\times 8}{\left[\begin{array}{ccccc|c} \mathbf{C}_{1A} & \cdot & \cdot & \mathbf{C}_{1D} & \mathbf{C}_{1E} & -\mathbf{I}'_A \\ \mathbf{C}_{2A} & \mathbf{C}_{1B} & \cdot & \cdot & \cdot & \cdot \\ \cdot & \mathbf{C}_{2B} & \mathbf{C}_{1C} & \cdot & \mathbf{C}_{2E} & \cdot \\ \cdot & \cdot & \mathbf{C}_{2C} & \mathbf{C}_{2D} & \cdot & -\mathbf{I}'_D \end{array}\right]} \underset{(8\times 1)}{\left\{\begin{array}{c} \mathbf{S}_A \\ \mathbf{S}_B \\ \mathbf{S}_C \\ \mathbf{S}_D \\ \mathbf{S}_E \\ \mathbf{r} \end{array}\right\}} = \underset{(8\times 1)}{\left\{\begin{array}{c} \mathbf{P}_A \\ \mathbf{P}_B \\ \mathbf{P}_C \\ \mathbf{P}_D \end{array}\right\}} \qquad \text{(g)}$$

This is of the form Eq. (2.18):

$$\mathbf{C}'\,\mathbf{S}' = \left[\underset{(8\times 5)}{\mathbf{C}} \;\middle|\; \underset{(8\times 3)}{\mathbf{C}_r}\right]\left\{\begin{array}{c} \mathbf{S} \\ \hline \mathbf{r} \end{array}\right\} = \underset{(8\times 1)}{\left\{\mathbf{P}\right\}} \qquad \text{(h)}$$

Let us write the equations in scalar form in full. From Eq. (2.19) we note that for any member $\mathbf{C}_1 = \{-C, -S\}$ and $\mathbf{C}_2 = \{C, S\}$ where c and s stand for cos α and sin α respectively. Consider the particular case shown in (b). The cos α and sin α terms for each of the members are as follows:

Member	L_x	L_y	L	cos α	sin α
A	0	4	4	0	1.0
B	3	0	3	1.0	0
C	3	−4	5	0.6	−0.8
D	6	0	6	1.0	0
E	3	4	5	0.6	0.8

(i)

The joint loads are

$$\mathbf{P}_A = \{0, 0\}\,;\; \mathbf{P}_B = \{10, -5\}\,;\; \mathbf{P}_C = \{0, -5\}\,;\; \mathbf{P}_D = \{5, 0\} \qquad \text{(j)}$$

Thus the equilibrium equation (g) are as follows:

$$\begin{array}{c} \\ A \\ \\ B \\ \\ C \\ \\ D \end{array} \left[\begin{array}{ccccc|ccc} A & B & C & D & E & R_1 & R_2 & R_3 \\ \hline 0 & & & -1 & -0.6 & -1 & 0 & 0 \\ -1 & & & 0 & -0.8 & 0 & -1 & 0 \\ \hline 0 & -1 & & & & & & \\ 1 & 0 & & & & & & \\ \hline & 1 & -0.6 & & 0.6 & & & \\ & 0 & 0.8 & & 0.8 & & & \\ \hline & & 0.6 & 1 & & 0 & 0 & 0 \\ & & -0.8 & 0 & & 0 & 0 & -1 \end{array}\right] \left\{\begin{array}{c} S_A \\ S_B \\ S_C \\ S_D \\ S_E \\ \hline R_1 \\ R_2 \\ R_3 \end{array}\right\} = \left\{\begin{array}{c} P_A \\ Q_A \\ \hline P_B \\ Q_B \\ \hline P_C \\ Q_C \\ \hline P_D \\ Q_D \end{array}\right\} \qquad \text{(k)}$$

Here the blank spaces are all zero. The inversion of the matrix leads to the solution in terms of **P**. In this particular case the inversion can be performed by hand. This leads to the solution as follows:

$$
\begin{Bmatrix} S_A \\ S_B \\ S_C \\ S_D \\ S_E \\ \hline R_1 \\ R_2 \\ R_3 \end{Bmatrix}
\begin{bmatrix}
\multicolumn{2}{c|}{A} & \multicolumn{2}{c|}{B} & \multicolumn{2}{c|}{C} & \multicolumn{2}{c}{D} \\
 & & & 1 & & & & \\
 & & -1 & & & & & \\
 & & -0.833 & & -0.833 & 0.625 & & \\
 & & 0.5 & & 0.5 & -0.375 & 1 & \\
 & & 0.833 & & 0.833 & 0.625 & & \\
\hline
-1 & & -1 & & -1 & & -1 & \\
 & -1 & -0.667 & -1 & -0.667 & -0.5 & & \\
 & & 0.667 & & 0.667 & -0.5 & & -1
\end{bmatrix}
\begin{Bmatrix} P_A \\ Q_A \\ \hline P_B \\ Q_B \\ \hline P_C \\ Q_C \\ \hline P_D \\ Q_D \end{Bmatrix}
\tag{l}
$$

The solution for the actual loads (Eq. j), is obtained by premultiplying it by the above matrix.

It is observed that there will be no bar forces due to P_A, Q_A and Q_D as these will be absorbed by the reactions. We can write the bar forces by considering the effective loads. Thus we can write

$$
\mathbf{S} = \begin{Bmatrix} S_A \\ S_B \\ S_C \\ S_D \\ S_E \end{Bmatrix} =
\begin{bmatrix}
0 & 1 & 0 & 0 & 0 \\
-1 & 0 & 0 & 0 & 0 \\
-0.833 & 0 & -0.833 & 0.625 & 0 \\
0.5 & 0 & 0.5 & -0.375 & 1 \\
0.833 & 0 & 0.833 & 0.625 & 0
\end{bmatrix}
\begin{Bmatrix} P_B \\ Q_B \\ P_C \\ Q_C \\ P_D \end{Bmatrix} = \mathbf{B\,P'}
\tag{m}
$$

Substituting the numerical values for the joint loads (j),

$$
\mathbf{S} = \{-5.0,\ -10.0,\ -11.458,\ 11.875,\ 5.205\}
\tag{n}
$$

These are easily verified.

2.7 MEMBERWISE ASSEMBLY

We will now consider *memberwise* assembly of the coefficients of the equilibrium matrix. This has relevance in programming where the data of the members such as the connectivity, direction cosines, length, area, etc., are stored memberwise in the form of an array on a tape or disc. The principle of the assembly is based on considering the contribution of each member towards the equilibrium equations of joints. The contributions occur only at the two joints at which the member is connected. The required space for the equilibrium matrix is first set out. This is governed by the number of joints, namely $(2n+3)\times(2n+3)$ or $(3n+6)\times(3n+6)$, where n is the number of joints for plane and space frames respectively.

Let J and S be the joints at which ends 1 and 2 of member R are connected (Fig. 2.11). The end forces $\mathbf{P}_{1R} = \mathbf{C}_{1R}\,\mathbf{S}_R$ and $\mathbf{P}_{2R} = \mathbf{C}_{2R}\,\mathbf{S}_R$ contribute towards equilibrium equations at joints J and S respectively. This means that $\mathbf{C}_{1R}$ and $\mathbf{C}_{2R}$ matrices will respectively find place in the Jth and S^{th} row of Rth column, as shown in (2.22).

$$
\begin{array}{c} \\ \\ J - \\ Q - \\ S - \\ \\ \end{array}
\left[\begin{array}{ccc} & R & \\ & \vdots & \vdots \\ \cdots & \mathbf{C}_{1R} & \cdots \\ \cdots & \cdots & \mathbf{I}'_Q \\ \cdots & \mathbf{C}_{2R} & \cdots \\ & \vdots & \vdots \end{array}\right]
\begin{Bmatrix} \mathbf{S}_A \\ \mathbf{S}_B \\ \vdots \\ \mathbf{S}_R \\ \vdots \\ \mathbf{r} \end{Bmatrix}
=
\begin{Bmatrix} \mathbf{P}_A \\ \mathbf{P}_B \\ \vdots \\ \mathbf{P}_J \\ \vdots \\ \mathbf{P}_Q \\ \vdots \\ \mathbf{P}_S \\ \vdots \end{Bmatrix}
\qquad (2.22)
$$

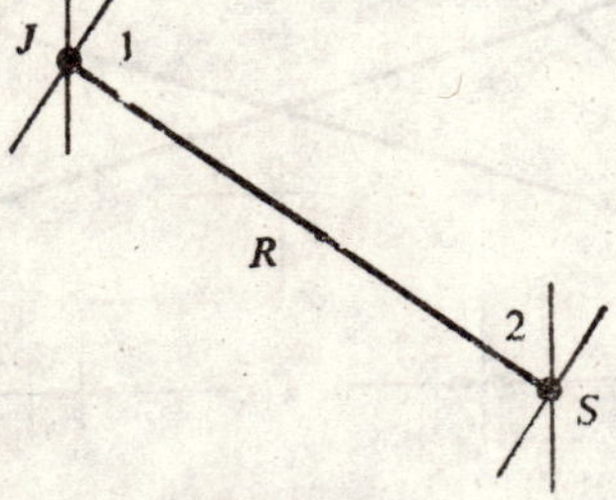

Figure 2.11

Thus the contribution of each member can be posted memberwise in the equilibrium matrix.

After the contributions of all the members are incorporated the additional columns corresponding to the reactions are introduced. The data of reactive forces consist of a *list* of the names of joints at which reactives forces are provided, and at each such joint (which is in fact a support) the elements of matrix $\mathbf{I}'$, (2×3 in case of plane frame, and 3×6 in case of space frame), are to be provided. Such matrices were illustrated in Exam.2.2. If reactive forces act at joint Q, the elements of $\mathbf{I}'_Q$ with a *minus sign* are to be placed in the rows corresponding to the equilibrium of joint Q, as shown in Eq. (2.22). We will now explain the above with an example.

EXAMPLE 2.3

Consider the solution of the pinjointed plane frame which belongs to the class of complex frames. For such frames Hennberg method [8] has to be used which consists in replacing one of the bars by a fictitious bar so that the method of solution of joints becomes tractable. The solution is obtained by superposition of solutions such that the force in the fictitious bar becomes zero. However the structure being determinate there are as many equilibrium equations as the number of unknowns. The present matrix approach is in effect the formulation of all the equations of equilibrium and their *simultaneous* solution.

The first step will be to prepare the member connectivity data, followed by the geometrical data giving projections L_x and L_y. This is shown in tabular form in Table a, where the last two columns give the calculated values of direction cosines.

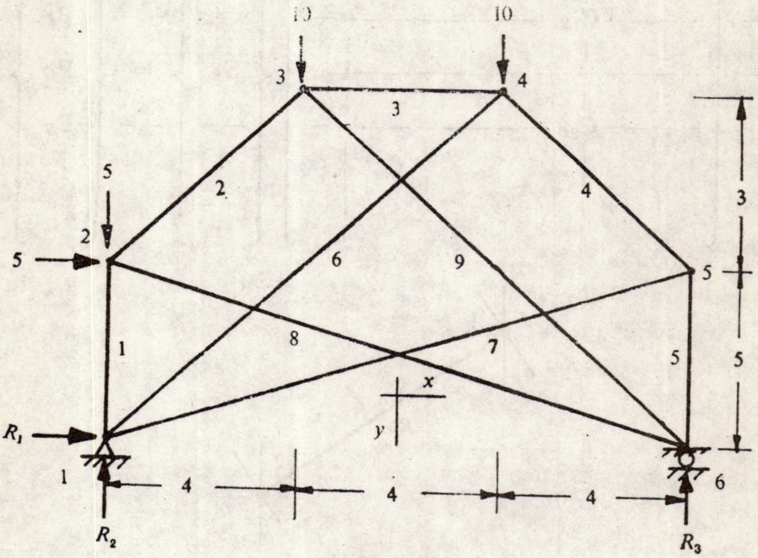

Example 2.3

TABLE a

Member	End 1	End 2	L_x	L_y	L	c	s
1	1	2	0	−5	5	0	−1
2	2	3	4	−3	5	0.8	−0.6
3	3	4	4	0	4	1	0
4	4	5	4	3	5	0.8	0.6
5	5	6	0	5	5	0	1
6	1	4	8	−8	11.313	0.707	−0.707
7	1	5	12	−5	13	0.9231	−0.3846
8	2	6	12	5	13	0.9231	0.3846
9	3	6	8	8	11.313	0.707	0.707

The next step is to have the set of joints with loads against each:

Joint	*P*	*Q*
2	5	5
3	0	10
4	0	10
5	0	0

To incorporate reactive forces, we find out the elements of I' matrices to be determined at joints 1 and 6. Refer again to Eq.(d) of Exam.2.2 in which the elements of I' matrix were determind. Presently too, we have at 1 and 3,

$$\mathbf{C_{1,1}\,S_1+C_{1,6}\,S_6+C_{1,7}\,S_7-I'_1\,r = P_1} \tag{c}$$

$$\mathbf{C_{2,8}\,S_8+C_{2,9}\,S_9+C_{2,5}\,S_5-I'_6\,r = P_6} \tag{d}$$

where,

$$\mathbf{I'_1} = \begin{bmatrix} 1 & 0 & 0 \\ 0 & 1 & 0 \end{bmatrix}; \quad \mathbf{I'_6} = \begin{bmatrix} 0 & 0 & 0 \\ 0 & 0 & 1 \end{bmatrix}$$

$$\begin{array}{c|cccccccccc}
 & 1 & 2 & 3 & 4 & 5 & 6 & 7 & 8 & 9 & r \\
1 & \mathbf{C_{1,1}} & \cdot & \cdot & \cdot & \cdot & \mathbf{C_{1,6}} & \mathbf{C_{1,7}} & \cdot & \cdot & \mathbf{-I'_1} \\
2 & \mathbf{C_{2,2}} & \mathbf{C_{1,2}} & \cdot & \cdot & \cdot & \cdot & \cdot & \mathbf{C_{1,8}} & \cdot & \cdot \\
3 & \cdot & \mathbf{C_{2,3}} & \mathbf{C_{1,3}} & \cdot & \cdot & \cdot & \cdot & \cdot & \mathbf{C_{1,9}} & \cdot \\
4 & \cdot & \cdot & \mathbf{C_{2,3}} & \mathbf{C_{1,4}} & \cdot & \mathbf{C_{2,6}} & \cdot & \cdot & \cdot & \cdot \\
5 & \cdot & \cdot & \cdot & \mathbf{C_{2,4}} & \mathbf{C_{1,5}} & \cdot & \mathbf{C_{2,7}} & \cdot & \cdot & \cdot \\
6 & \cdot & \cdot & \cdot & \cdot & \mathbf{C_{2,5}} & \cdot & \cdot & \mathbf{C_{2,8}} & \mathbf{C_{2,9}} & \mathbf{-I'_6}
\end{array}
\begin{Bmatrix} \mathbf{S_1} \\ \mathbf{S_2} \\ \mathbf{S_3} \\ \mathbf{S_4} \\ \mathbf{S_5} \\ \mathbf{S_6} \\ \mathbf{S_7} \\ \mathbf{S_8} \\ \mathbf{S_9} \\ \mathbf{r} \end{Bmatrix} = \begin{Bmatrix} \mathbf{P_1} \\ \mathbf{P_2} \\ \mathbf{P_3} \\ \mathbf{P_4} \\ \mathbf{P_5} \\ \mathbf{P_6} \end{Bmatrix} \tag{e}$$

Consider now assembling the required equations **(Eq. e) memberwise** in terms of the C_1 and C_2 at each member. Consider number 1; its ends 1 and 2 are at joints 1 and 2 as seen from the first two columns of the table. $C_{1,1}$ and $C_{2,1}$ are written in the first column at first and second rows, respectively. The next number has near and far end connections at 2 and 3. Thus $C_{1,2}$, and $C_{2,2}$ find place in column number 2 (member number 2) at row numbers 2 and 3 (near and far end joint numbers) respectively. Thus the process is reduced to a machanical process to be performed with the help of connectivity data.

TABLE b (Example 2.3)

S_1	S_2	S_3	S_4	S_5	S_6	S_7	S_8	S_9	r_1	r_2	r_3	$=P$
0					−0.707	−0.9231			−1	0	0	0
1					0.707	0.3846			0	−1	0	0
0	−0.8						−0.9231					5
−1	0.6						−0.3846					5
	0.8	−1						−0.707				0
	−0.6	0						−0.707				10
		1	−0.8		0.707							0
		0	−0.6		−0.707							10
			0.8	0		0.9231						0
			0.6	−1		−0.3846						0
				0			0.9231	0.707	0	0	0	0
				1			0.3846	0.707	0	0	−1	0

The $-I'_1$ and $-I'_6$ matrices are similarly placed in the column corresponding to the reactions and rows corresponding to 1 and 6 respectively.

The actual matrix is as shown in Table (b) ; note that for any member $\mathbf{C}_1 = \{-c, s-\}$; $\mathbf{C}_2 = \{c, s\}$. The elements are as given in Table a above. In Table b the column vector **S** is shown as a *row* on top so that matching of elements which multiply with each other is easier for our eyes.

By solving the equations we obtain the forces and reactions are as follows :

$$\mathbf{S} = \{-12.9169, -10.7140, -4.9983, -10.7132, -9.9991$$
$$-5.052, 9.2852, 3.8683, -5.0500\} \text{ kN} \qquad \text{(f)}$$

$$\mathbf{r} = \{-5.0, -12.9167, -12.0833\} \text{ kN} \qquad \text{(g)}$$

2.8 PLANE GRIDS

Transversely loaded grid frame works (loaded perpendicular to the plane) form an important class of frameworks. Fig. 2.12(a) shows a plan of such a grid work. In Fig. 2.12(b) the free body of a member is shown with end loads denoted in member coordinates. At each end there is a flexural moment, a torsional moment and a shear to balance the end moments.

Thus the load vectors are

$$\bar{\mathbf{P}} = \left\{ \mathbf{P}_1, \mathbf{P}_2 \right\} = \left\{ \overline{M}x_1, \overline{M}y_1, \overline{Q}_1 \middle| \overline{M}x_2, \overline{M}y_2, \overline{Q}_2 \right\} \qquad (2.23)$$

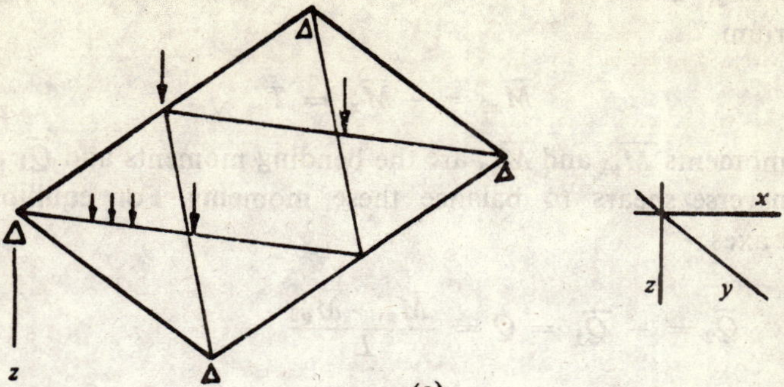

(a)

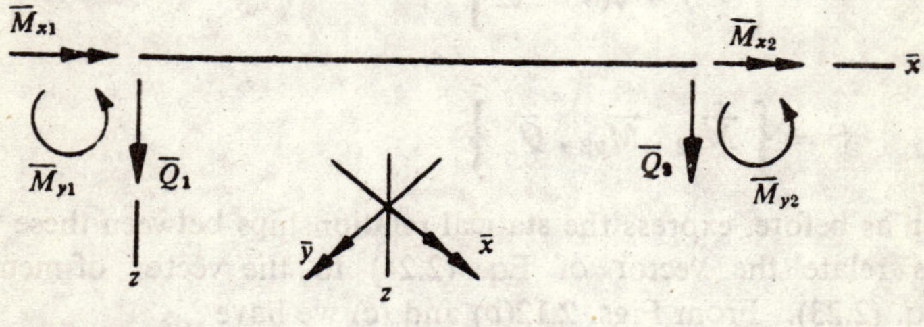

(b)

(c)

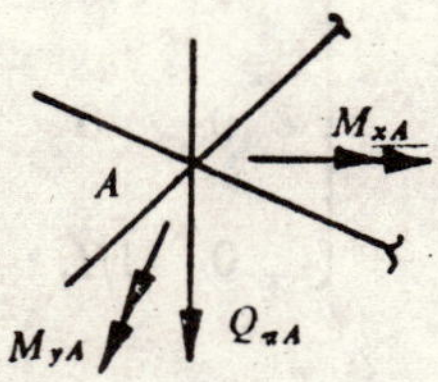

(d)

Figure 2.12

Here $\overline{M}_{x1}$ and $\overline{M}_{x2}$ are the torsional end moments requiring, for equilibrium,

$$\overline{M}_{x2} = -\overline{M}_{x1} = T \tag{a$_1$}$$

The moments $\overline{M}_{y1}$ and $\overline{M}_{y2}$ are the bending moments and $\overline{Q}_1$ and $\overline{Q}_2$ are the transverse shears to balance these moments. For equilibrium about $\bar{y}$ and $\bar{z}$ axes

$$\overline{Q}_2 = -\overline{Q}_1 = Q = \frac{\overline{M}_{y1}+\overline{M}_{y2}}{L} \tag{a$_2$}$$

Thus the independent set of member forces is either

$$\mathbf{S} = \left\{ T\ ,\ \overline{M}_{y1}\ ,\ \overline{M}_{y2} \right\} \tag{2.24}$$

or

$$\mathbf{S} = \left\{ \overline{M}_{x2}\ ,\ \overline{M}_{y2}\ ,\ \overline{Q}_2 \right\} \tag{2.25}$$

We can as before, express the statical relationships between these vectors.

Let us relate the vector of Eq. (2.24) to the vector of member end forces, Eq. (2.23). From Figs. 2.12(b) and (c) we have

$$\overline{M}_{x1} = -T\ ;\ \overline{M}_{x2} = T\ ;\ \overline{Q}_1 = -\overline{Q}_2 = -\frac{\overline{M}_{y1}+\overline{M}_{y2}}{L} \tag{a$_3$}$$

Therefore we obtain

$$\overline{\mathbf{P}} = \begin{Bmatrix} \overline{\mathbf{P}}_1 \\ \\ \overline{\mathbf{P}}_2 \end{Bmatrix} = \begin{Bmatrix} \overline{M}_{x1} \\ \overline{M}_{y1} \\ \overline{Q}_1 \\ \hline \overline{M}_{x2} \\ \overline{M}_{y2} \\ \overline{Q}_2 \end{Bmatrix} = \begin{bmatrix} -1 & 0 & 0 \\ 0 & 1 & 0 \\ 0 & -1/L & -1/L \\ \hline 1 & 0 & 0 \\ 0 & 0 & 1 \\ 0 & 1/L & 1/L \end{bmatrix} \begin{Bmatrix} T \\ \overline{M}_{y1} \\ \overline{M}_{y2} \end{Bmatrix}$$

$$= \overline{\mathbf{C}}\,\mathbf{S} \tag{2.26}$$

This is similar in form to Eq.(2.8) derived for a plane frame member. Similar equations of equilibrium between end forces $\overline{\mathbf{P}}_1$ and $\overline{\mathbf{P}}_2$ can be obtained as

$$\left\{ \overline{P}_1 \right\} = \begin{Bmatrix} \overline{M}_{x1} \\ \overline{M}_{y1} \\ \overline{Q}_1 \end{Bmatrix} = \begin{bmatrix} 1 & 0 & 0 \\ 0 & 1 & -L \\ 0 & 0 & -1 \end{bmatrix} \begin{Bmatrix} \overline{M}_{x2} \\ \overline{M}_{y2} \\ \overline{Q}_2 \end{Bmatrix} = \overline{\mathbf{C}}_1 S \tag{2.27}$$

Coordinate Transformation

Figure 2.12(a) shows the member axes ($\bar{x}$, $\bar{y}$) inclined to the global axes (x, y). The coordinate transformation matrix **T** remains the same as in Eq. (2.12). The member end forces can be transformed as in a plane frame loaded in its own plane. Observe, that here the moments get resolved similar to the (direct) forces of a plane frame loaded in its own plane. The shear force however remains unaltered.

The joint load vector (Fig 2.12 d) consists of a (3×1) vector

$$\mathbf{P} = \{M_x \quad M_y \quad Q\} \tag{2.28}$$

The loads are thus seen to consist of two applied moments measured in the global axes and a force acting transversely.

2.9 CONCLUDING REMARKS

The method discussed so far for determining the internal forces in a determinate frame leads to *unsymmetric* **C** matrix. This has a serious disadvantage from the point of view of large size problems especially when the displacement method offers a system of equations with symmetric coefficient matrix applicable even to redundant structures. In fact the displacement method expresses the equations of equilibrium in terms of displacement variables. We will show later (Sec. 8.2) as to how the displacement method transforms the unsymmetric **C** matrix to a symmetric one. In this light we can regard this as a transformation of the variable **S** to another variable so that the resulting coefficient matrix is symmetric.

The subject matter discussed in this chapter has relevance in the force method where an indeterminate frame is reduced to a determinate one by treating the extra internal force unknowns as loads (See Exercise 2.12).

EXERCISES

2.1 Express the forces $\mathbf{S}_x = \{ S_x, M_x\, Q_x \}$ at any point x in the beam in terms of forces $\mathbf{P}_1 = \{ P_1, \overline{Q}_1, \overline{M_1} \}$ in the form $\mathbf{S}_x = \mathbf{C}'_x \mathbf{P}_1$.

2.2 Consider the free body of beam shown in Fig. 2.3. Derive the details of $\bar{P}_1^\circ$ and $\bar{P}_2^\circ$ for beam model taking p and q to be constants and occupying the whole span.

2.3 Repeat Ex. 2.2 by taking cantilever model.

2.4 Consider Eq. (2.4). By direct inversion obtain $S = \bar{C}_1^{-1} \bar{P}_1$. Verify the correctness by writing the equations of equilibrium directly.

2.5 We cannot invert relations (2.8). Apart from the fact that $\bar{C}$ is rectangular, what is the physical implication?

2.6 Interpret the elements of a given row of equilibrium matrix such as Eq. (a_9) of Sec 2.1, and show that the matrix multiplication implies superposition of effects. Also construct the elements of this matrix columnwise considering forces S, $\bar{M}_1$ and $\bar{M}_2$ one at a time.

2.7 Consider the following connectivity data between 7 members and five joints.

Member	End 1	End 2
A	D	E
B	D	B
C	B	E
D	B	C
E	C	E
F	A	C
G	A	B

Sketch the structural connectivity implied by the above data. Assume certain coordinates for the joints and fix up an actual shape (*see* Ex. 2.8).

2.8 The truss of the given geometrical proportions is according to the connectivity of Ex. 2.7. Assemble the equations of equilibrium for the pinconnected truss treating the reactive forces $\{R_1, R_2, R_3\}$ as shown, hinge at D and a roller at E. Take loads of 10 kN each vertically and horizontally, at all the nodes excluding D and E. If such forces do act at D and E would it make any difference in the analysis?

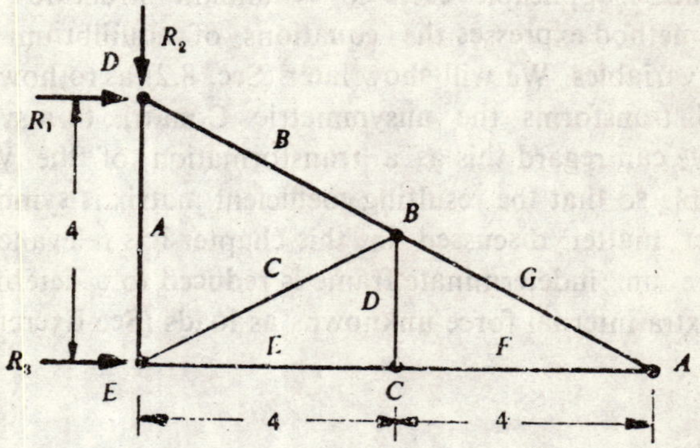

Exercise 2.8

2.9 Set up the equation of equilibrium at joint A of the two bar arrangement. Show that the static matrix becomes singular for $\alpha = 0$ (case of critical form).

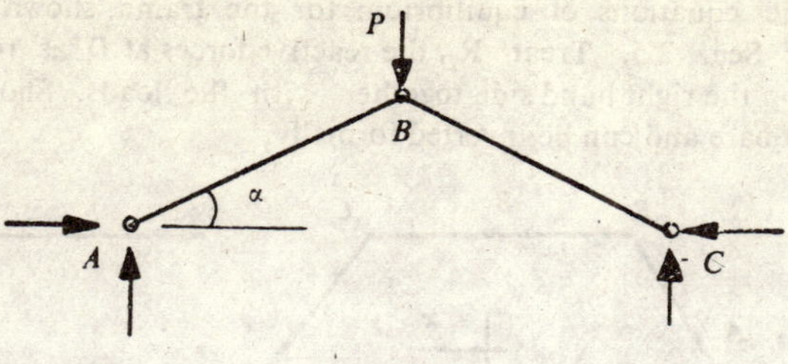

Exercise 2.9

2.10 Show that the regular hexagonal frame work has a critical form by proving that the static matrix is singular. See Ref [3] for scalar analysis.

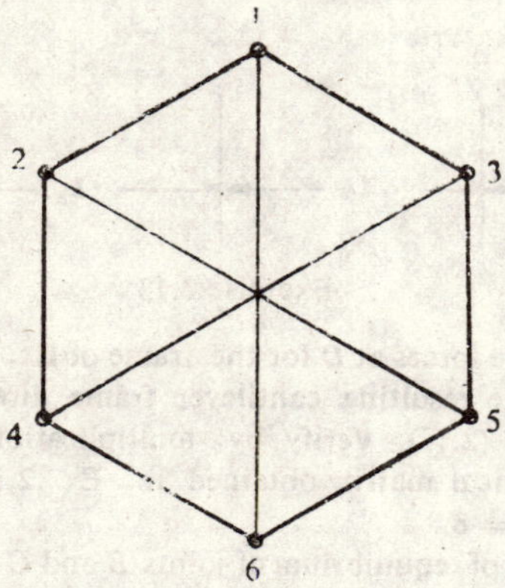

Exercise 2.10

2.11 Verify the equations of equilibrium denoted by the first two rows of Eq. (b) of Exam. 2.1.

2.12 Set up the equations of equilibrium between the bar forces $\mathbf{S} = \{S_A, S_B, S_C\}$ and the load $\mathbf{P} = \{ \mathbf{P}_A, \mathbf{Q}_A \}$ in the form $\mathbf{C\,S} = \mathbf{P}$. Supposing S_B is assumed to be an equivalent load as in force method, express $\mathbf{S}' = \{ S_A, S_C \}$ in terms of $P' = \{ P_A, Q_A, S_B \}$ in the form $C'\,S' = P'$. Invert the relations and obtain the internal forces in the form $S' = B'\,P'$.

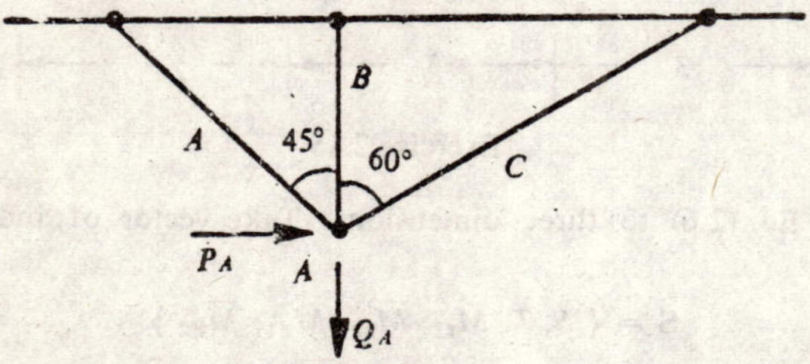

Exercise 2.12

2.13 Set up the equations of equilibrium for the frame shown in the form of Eq. (a_2) of Sec. 2.5. Treat $\mathbf{R}_D$ the reactive forces at D as redundants. Take this term on the right hand side together with the loads. Show that resulting matrix is square and can be inverted formally.

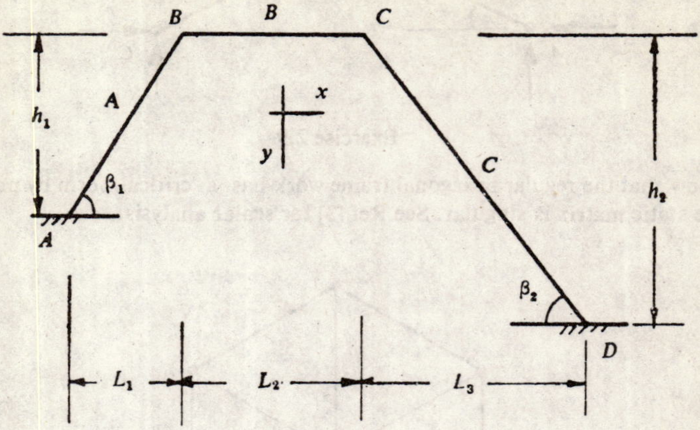

Exercise 2.13

2.14 Treating the reactive forces at D for the frame of Ex. 2.13 as known obtain the internal forces in the resulting cantilever frame directly from statics. Express them in the form (2.17). Verify by multiplication that this matrix is the inverse of the statical matrix obtained in Ex. 2.13. Take $L_1 = 3$, $L_2 = 5$. $L_3 = 6$, $h_1 = 4$, $h_2 = 8$.

2.15 Write the equations of equilibrium of joints B and C of the transversely loaded grid frame work. Treat A and D as fixed.

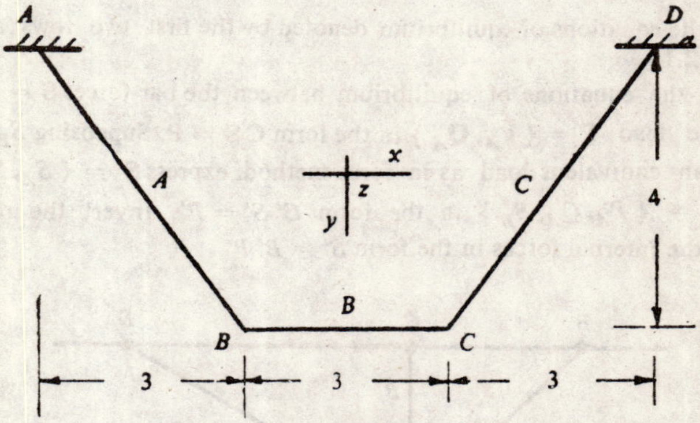

Exercise 2.15

2.16 Generalize Eq. (2.6) to three dimensions. Take vector of independent forces $\mathbf{S}$ as

$$\mathbf{S} = \{ S, T, \overline{M}_{z1}, \overline{M}_{z2}, \overline{M}_{y1}, \overline{M}_{y2} \}$$

where S is axial force, T the torque, $\overline{M}_{z1}$ and $\overline{M}_{z2}$ the bending moments about $\bar{z}$ axis, and $\overline{M}_{y1}$ and $\overline{M}_{y2}$ those about $\bar{y}$ axis. The member end forces are. for $i = 1, 2,$

$$\overline{P_i} = \{ \overline{P_{xi}}, \overline{P_{yi}}, \overline{P_{zi}}, \overline{M_{xi}}, \overline{M_{yi}}, \overline{M_{zi}} \}$$

Obtain $\overline{C_1}$ and $\overline{C_2}$ matrices. (See Fig. 8.5 and Eq. 8.40 of Chapter 8.)

2.17 Consider the closing remarks at the end of Sec. 2.1 regarding generalized member forces. The figure shows $S_0 = \{ P_0, Q_0, M_0 \}$ taken as internal forces at a point 0. Determine $\overline{C_1}$ and $\overline{C_2}$ matrices giving $\overline{P_1} = \overline{C_1}\ S$, and $\overline{P_2} = \overline{C_2}\ S$ as

$$\overline{C_1} = \begin{bmatrix} -1 & 0 & 0 \\ 0 & -1 & 0 \\ 0 & x_1 & -1 \end{bmatrix} ; \overline{C_2} = \begin{bmatrix} 1 & 0 & 0 \\ 0 & 1 & 0 \\ 0 & -x_2 & 1 \end{bmatrix}$$

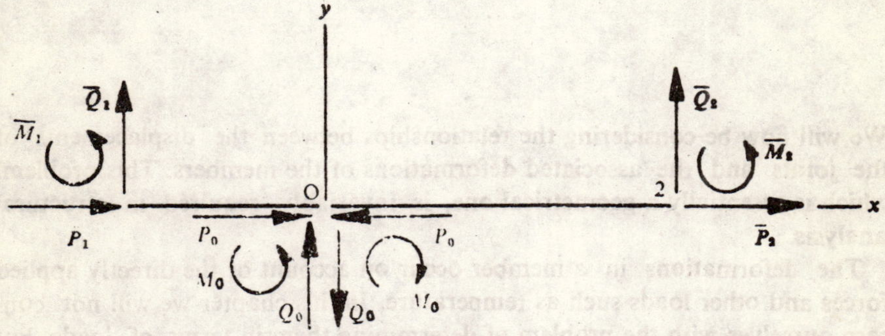

Exercise 2.17

Derive the cantilever model by taking the point 0 at end 2.

2.18 Consider the curved member imagined to be connected to point 0 through rigid arms. Consider internal forces at 0 as S and determine $\overline{C_1}$ and $\overline{C_2}$ similar to those in Ex. 2.17.

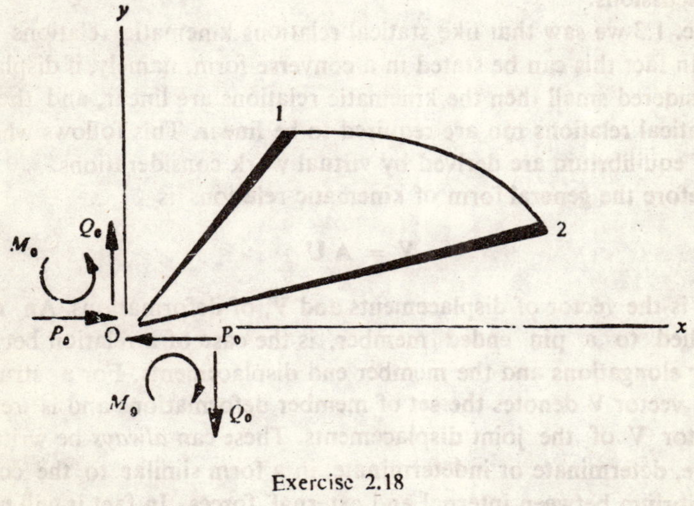

Exercise 2.18

Chapter 3

Kinematics

We will now be considering the relationships between the displacements of the joints and the associated deformations of the members. This problem, which is essentially a geometrical one, is invariably required in structural analysis.

The deformations in a member occur on account of the directly applied forces and other loads such as temperature. In this chapter we will not concern ourselves with the problem of determining them in terms of loads but take them as given and with the purely kinematical (geometrical) problem of the calculation of the joint displacements in terms of these given quantities. The determination of the member deformations in terms of the given loads is the subject matter of flexibility of a member and will be studied later. We shall consider here a matter purely of geometry which is very necessary in later discussions.

In Sec. 1.3 we saw that like statical relations kinematic relations too are linear. In fact this can be stated in a converse form, namely, if displacements are considered small then the kinematic relations are linear, and the consistent statical relations too are required to be linear. This follows when equations of equilibrium are derived by virtual work considerations.

Therefore the general form of kinematic relations is

$$\mathbf{V} = \mathbf{A}\,\mathbf{U}$$

Here **U** is the vector of displacements and **V**, of deformations. An example, as applied to a pin ended member, is the case of a relation between the member elongations and the member end displacements. For a structure as a whole vector **V** denotes the set of member deformations and is related to the vector **V** of the joint displacements. These can *always* be written for a structure, determinate or indeterminate, in a form similar to the conditions of equilibrium between internal and external forces. In fact it will be shown later that the kinematic matrix A is actually C^t. Alternatively if one has a kinematic relationship established then equilibrium equations automatically follow in terms of kinematic matrix A as $A^t S = P$.

The member deformations chosen always 'correspond' to the independent set of member forces such that they can be related later by flexibility or stiffness relations, $v = f S$ or $S = k v$. Recall the discussion of Sec. 1.4 leading to Eqs. (1.6) and (1.7).

3.1 DEFORMATIONS AND DISPLACEMENTS OF MEMBER

We define sets of deformations and displacements corresponding to the different set of forces introduced for a member or a structure.* To begin with let us define the set of deformations which correspond to the force system of Fig. 2.2(a). These are as shown in Fig. 3.1(a). We define the following sets:

$$\mathbf{v} = \{e, \varphi_1, \varphi_2\} \tag{3.1}$$

$$\bar{\mathbf{u}} = \{\bar{u}_1, \bar{v}_1, \bar{\theta}_1, \mid \bar{u}_2, \bar{v}_2, \bar{\theta}\} = \{\bar{\mathbf{u}}_1 \mid \bar{\mathbf{u}}_2\} \tag{3.2}$$

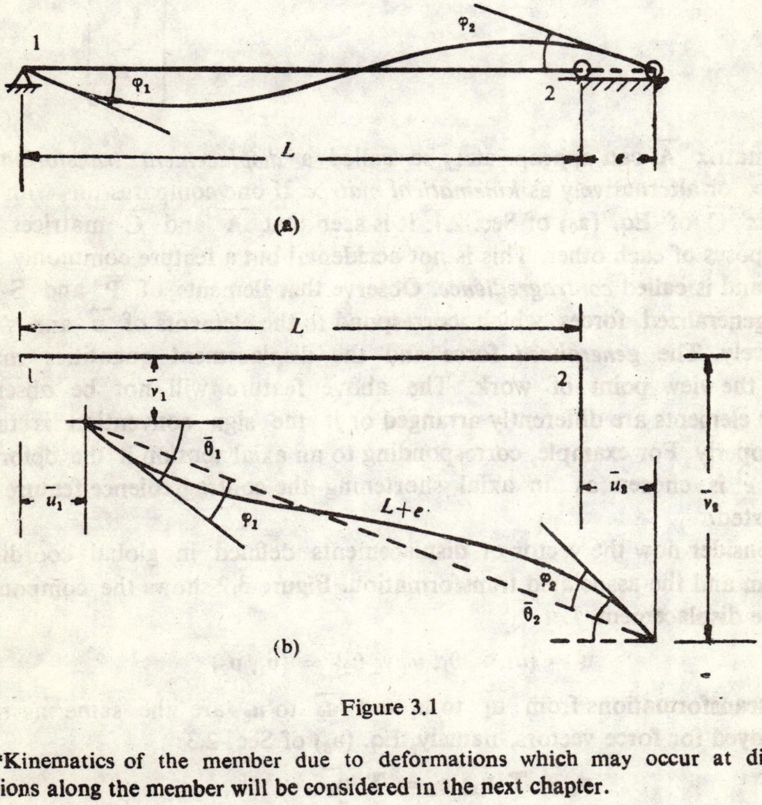

Figure 3.1

*Kinematics of the member due to deformations which may occur at different sections along the member will be considered in the next chapter.

The first column vector consists of an axial elongation e, and two end rotations φ_1 and φ_2. This vector will be called *deformations* in contrast to the second vector which we call, *displacements*, as they include rigid body movement of the bar. Observe that the quantities e, φ_1 and φ_2 are as a result of the accumulated strains in the member only, exclusive of any rigid body components.

To begin with let us concern ourselves with the problem of relating these two sets. From the geometry, we have, (Fig. 3.1b).

$$e = \bar{u}_2-\bar{u}_1; \ \varphi_1 = \bar{\theta}_1-(\bar{v}_2-\bar{v}_1)/L; \ \varphi_2 = \bar{\theta}_2-(\bar{v}_2-\bar{v}_1)/L \qquad (a_1)$$

Herein the displacements are assumed to be small. These can be stated as

$$\mathbf{v} = \begin{Bmatrix} e \\ \varphi_1 \\ \varphi_2 \end{Bmatrix} = \left[\begin{array}{ccc|ccc} -1 & 0 & 0 & 1 & 0 & 0 \\ 0 & 1/L & 1 & 0 & -1/L & 0 \\ 0 & 1/L & 0 & 0 & -1/L & 1 \end{array}\right] \begin{Bmatrix} \bar{u}_1 \\ \bar{v}_1 \\ \bar{\theta}_1 \\ \cdots \\ \bar{u}_2 \\ \bar{v}_2 \\ \bar{\theta}_2 \end{Bmatrix} = \bar{\mathbf{A}}\,\bar{\mathbf{u}} \qquad (3.3)$$

The matrix $\bar{\mathbf{A}}$, can appropriately be called a *displacement transformation* matrix, or alternatively as *kinematical matrix*. If one compares this with the matrix $\bar{\mathbf{C}}$ of Eq. (a_9) of Sec. 2.1, it is seen that $\bar{\mathbf{A}}$ and $\bar{\mathbf{C}}$ matrices are transposes of each other. This is not accidental but a feature commonly met with and is called *contragredience*. Observe that elements of $\bar{\mathbf{P}}$ and $\mathbf{S}$ are the generalized forces which correspond to the elements of $\bar{\mathbf{u}}$ and $\mathbf{v}$ respectively. The *generalized* force and the displacement quantities match from the view point of work. The above feature will not be observed if the elements are differently arranged or if the sign convention is taken improperly. For example, corresponding to an axial tension if the deformation e is chosen as an axial shortening the contragredience feature gets distorted.

Consider now the vector of displacements defined in global coordinate system and the associated transformation. Figure 3.2 shows the components of the displacement. Let

$$\mathbf{u} = \{u_1,\ v_1\ \theta_1,\ u_2\ v_2\ \theta_2\} = \{\mathbf{u}_1,\ \mathbf{u}_2\} \qquad (3.4)$$

The transformations from $\bar{\mathbf{u}}_1$ to $\mathbf{u}_1$ and $\bar{\mathbf{u}}_2$ to $\mathbf{u}_2$ are the same as those employed for force vectors, namely, Eq. (a_3) of Sec. 2.3:

$$\mathbf{u}_1 = \mathbf{T}\ \bar{\mathbf{u}}_1; \ \mathbf{u}_2 = \mathbf{T}\ \bar{\mathbf{u}}_2 \qquad (a_2)$$

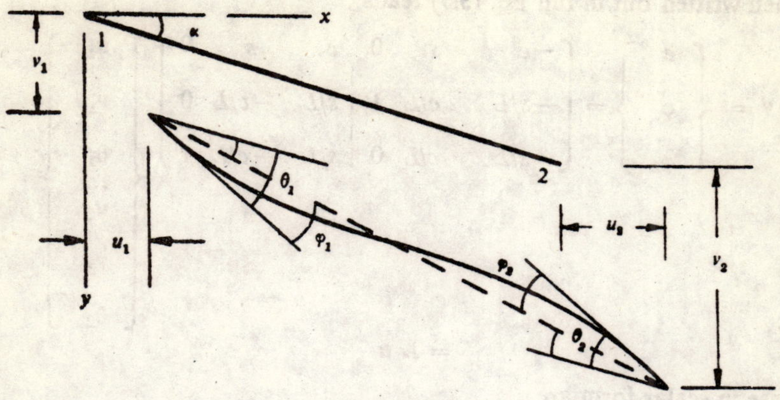

Figure 3.2

Let us rewrite Eq. (3.3) interms of submatrices :

$$\mathbf{v} = \begin{Bmatrix} e \\ \varphi_1 \\ \varphi_2 \end{Bmatrix} = \begin{bmatrix} \overline{\mathbf{A}_1} & \overline{\mathbf{A}_2} \end{bmatrix} \begin{Bmatrix} \overline{\mathbf{u}_1} \\ \overline{\mathbf{u}_2} \end{Bmatrix} \tag{a_3}$$

or

$$\mathbf{v} = \overline{\mathbf{A}_1}\ \overline{\mathbf{u}_1} + \overline{\mathbf{A}_2}\ \overline{\mathbf{u}_2} \tag{3.5}$$

with

$$\overline{\mathbf{A}_1} = \begin{bmatrix} -1 & 0 & 0 \\ 0 & 1/L & 1 \\ 0 & 1/L & 0 \end{bmatrix}, \quad \overline{\mathbf{A}_2} = \begin{bmatrix} 1 & 0 & 0 \\ 0 & -1/L & 0 \\ 0 & -1/L & 1 \end{bmatrix} \tag{3.6}$$

Thus Eq. (3.5) and Eq. (a_2) together give

$$\mathbf{v} = \mathbf{A}_1\,\mathbf{u}_1 + \mathbf{A}_2\,\mathbf{u}_2 = \mathbf{A}\,\mathbf{u} \tag{3.7}$$

where

$$\mathbf{A}_1 = \overline{\mathbf{A}_1}\,\mathbf{T}^t; \ \mathbf{A}_2 = \overline{\mathbf{A}_2}\,\mathbf{T}^t \tag{3.8}$$

Employing **T** matrices of Eq. (2.12) we get

$$\mathbf{A}_1 = \begin{bmatrix} -c & -s & 0 \\ -s/L & c/L & 1 \\ -s/L & c/L & 0 \end{bmatrix}; \quad \mathbf{A}_2 = \begin{bmatrix} c & s & 0 \\ s/L & -c/L & 0 \\ s/L & -c/L & 1 \end{bmatrix} \tag{3.9}$$

When written out in full Eq. (3.7) reads

$$\mathbf{v} = \begin{Bmatrix} e \\ \varphi_1 \\ \varphi_2 \end{Bmatrix} = \left[\begin{array}{ccc|ccc} -c & -s & 0 & c & s & 0 \\ -s/L & c/L & 1 & s/L & -c/L & 0 \\ -s/L & c/L & 0 & s/L & -c/L & 1 \end{array}\right] \begin{Bmatrix} u_1 \\ v_1 \\ \theta_1 \\ \cdots \\ u_2 \\ v_2 \\ \theta_2 \end{Bmatrix}$$

$$= \mathbf{A}\,\mathbf{u} \tag{a$_4$}$$

These in scalar form are

$$\begin{aligned} e &= (u_2-u_1)\cos\alpha+(v_2-v_1)\sin\alpha \\ \varphi_1 &= \theta_1+\frac{u_2-u_1}{L}\sin\alpha-\left(\frac{v_2-v_1}{L}\right)\cos\alpha \\ \varphi_2 &= \theta_2+\frac{u_2-u_1}{L}\sin\alpha-\left(\frac{v_2-v_1}{L}\right)\cos\alpha \end{aligned} \tag{a$_5$}$$

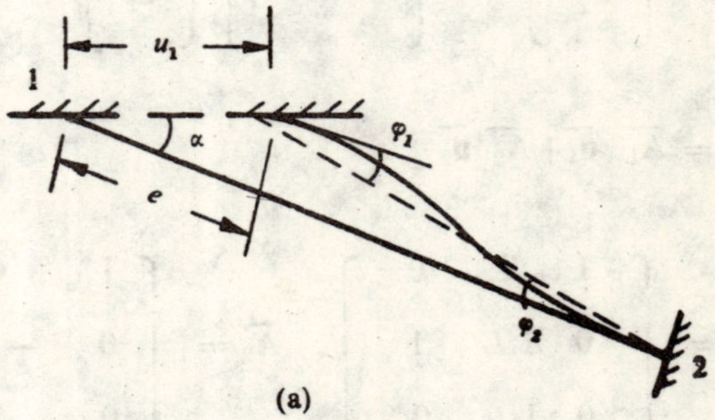

(a)

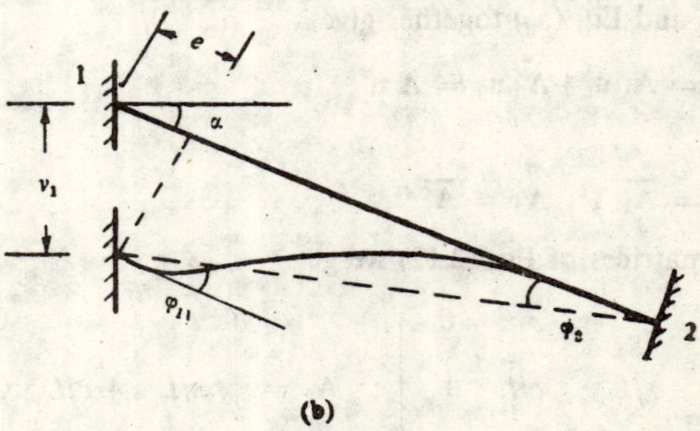

(b)

Figure 3.3

These relations can be verified by comparing the deformations shown in Fig. 3.1(a) with those shown in Fig. 3.2. It is seen that the verification is not quite obvious. Let us therefore *assemble* the elements of the kinematic matrix of Eq. (3.10) *directly*. In fact this would be *a more general procedure* adoptable to any new situation of having to derive a displacement transformation matrix.

Consider the interpretation of the elements of *each column* of the matrix of Eq. (a_4). The elements of first column denote the values of the elements of the left hand vector due to u_1 alone, other elements being zero, those of the second column denote the elements of left hand side vector due to v_1 alone others held at zero and so. Figures 3.3(a) and (b) show the deformed shape due to u_1 alone and due to v_1 alone respectively. Observe that e is the axial extension and φ_1 and φ_2 are rotations of ends measured (clockwise) from the chord. In Fig. 3.3(a) all these are negative, axial deformation being shortening and rotations anti-clockwise. Similarly one can arrive at the elements of the other columns of the kinematic matrix,

Cantilever Model

The deformations **v**, when end 1 is treated as fixed, are identical to $\bar{\mathbf{u}}_2$. In short $\mathbf{v} \equiv \bar{\mathbf{u}}_2$ when $\bar{\mathbf{u}}_1 \equiv 0$. We define for this model $\mathbf{v} = \{e, v, \varphi\}$ (see Fig. 3.4a). Here e is the elongation of bar, v the deflection of end 2 from the tangen at end 1, and φ the rotation of end 2 measured from tangent at end 1. Figure 3.4(b) shows the displaced shape of the beam. We can obtain from this figure,

$$e = \bar{u}_2 - \bar{u}_1;\ v = \bar{v}_2 - \bar{v}_1 - L\,\bar{\theta}_1;\ \varphi = \bar{\theta}_2 - \bar{\theta}_1 \tag{a_6}$$

These can be put in the form

$$\begin{Bmatrix} e \\ v \\ \varphi \end{Bmatrix} = \begin{Bmatrix} \bar{u}_2 \\ \bar{v}_2 \\ \bar{\theta}_2 \end{Bmatrix} + \begin{bmatrix} -1 & 0 & 0 \\ 0 & -1 & -L \\ 0 & 0 & -1 \end{bmatrix} \begin{Bmatrix} \bar{u}_1 \\ \bar{v}_1 \\ \bar{\theta}_1 \end{Bmatrix} \tag{a_7}$$

We will write this in the form (3.5):

$$\mathbf{v} = \begin{Bmatrix} e \\ v \\ \varphi \end{Bmatrix} = \begin{bmatrix} -1 & 0 & 0 \\ 0 & -1 & -L \\ 0 & 0 & -1 \end{bmatrix} \begin{Bmatrix} \bar{u}_1 \\ \bar{v}_1 \\ \bar{\theta}_1 \end{Bmatrix} + \begin{bmatrix} 1 & 0 & 0 \\ 0 & 1 & 0 \\ 0 & 0 & 1 \end{bmatrix} \begin{Bmatrix} \bar{u}_2 \\ \bar{v}_2 \\ \bar{\theta}_2 \end{Bmatrix}$$

$$= \bar{\mathbf{A}}_1\,\bar{\mathbf{u}}_1 + \bar{\mathbf{A}}_2\,\bar{\mathbf{u}}_2 \tag{3.10}$$

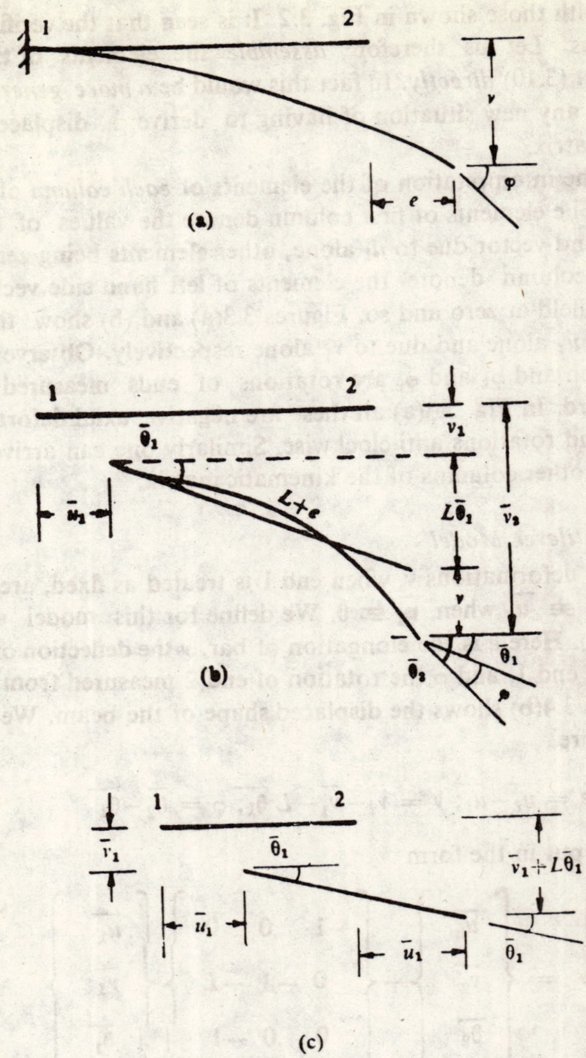

Figure 3.4

We thus see that form of Eq. (3.5) is a general one. In the cantilever model we have an advantage that $\overline{\mathbf{A}_2} \equiv \mathbf{I}$. Recall that $\overline{\mathbf{C}_2} \equiv \mathbf{I}$ (see Eq. 2.5).

We can derive (a_7) by superimposing on $\overline{\mathbf{u}_2}$ a rigid body movement such that total displacement $\overline{\mathbf{u}_1}$ becomes zero. Thus if at end 1 $\overline{u_1}$, $\overline{v_1}$ and $\overline{\theta_1}$ are rigid body movements, then at end 2 the rigid body movements are (Fig. 3.4 (c)) $\overline{u_2} = \overline{u_1}$, $\overline{v_2} = \overline{v_1} + L\,\overline{\theta_1}$ and $\overline{\theta_2} = \overline{\theta_1}$. If we subtract these from $\overline{u_2}$, $\overline{v_2}$ and $\overline{\theta_1}$ we have by definition e, v and φ respetively, as stated by Eq. (a_7).

Matrices $\overline{A}_1$ and $\overline{A}_2$ defined for transformations measured global axes, as stated by Eq. (3.8) also hold in this case. Observe that

$$A_1 = \overline{A}_1 T^t = \begin{Bmatrix} -c & -s & 0 \\ s & c & -L \\ 0 & 0 & -1 \end{Bmatrix} \qquad (a_8)$$

$$A_2 = I\, T^t = T^t = \begin{Bmatrix} c & s & 0 \\ -s & c & 0 \\ 0 & 0 & 1 \end{Bmatrix} \qquad (a_9)$$

3.2 KINEMATICS OF STRUCTURE

The next problem is one of relating the member deformations to joint displacements.

For a structure deforming within the elastic range the members are monolithic with one another and do not show discontinuities. When however plastic hinges develop at member ends the continuity would be destroyed. At this stage we are excluding such a behaviour. Thus for the frame shown in Fig. 3.5, the continuity relationship between the member end displacements and joint displacements are

$$\begin{aligned} u_{1A} &= u_A \;;\; u_{2A} = u_B \\ u_{1B} &= u_B \;;\; u_{2B} = u_C \\ u_{1C} &= u_C \;;\; u_{2C} = u_D. \end{aligned} \qquad (a_1)$$

The first relation states that displacements of end 1 of member A is the same as the displacement of joint A. The others convey similar information.

The more practical kinematic matrix is the one which relates the deformations $\mathbf{v}$ of the members to the joint displacements $\mathbf{u}$. Thus from the basic relations (3.7) as applied to each member, we have

$$\begin{aligned} v_A &= A_{1A}\, u_{1A} + A_{2A}\, u_{2A} = A_{1A}\, u_A + A_{2A}\, u_B \\ v_B &= A_{1B}\, u_{1B} + A_{2B}\, u_{2B} = A_{1B}\, u_B + A_{2B}\, u_C \\ v_C &= A_{1C}\, u_{1C} + A_{2C}\, u_{2C} = A_{1C}\, u_C + A_{2C}\, u_D \end{aligned} \qquad (a_2)$$

which in matrix form are

$$\underset{(9\times 1)}{\begin{Bmatrix} v_A \\ v_B \\ v_C \end{Bmatrix}} = \underset{(9\times 12)}{\begin{bmatrix} A_{1A} & A_{2A} & 0 & 0 \\ 0 & A_{1B} & A_{2B} & 0 \\ 0 & 0 & A_{1C} & A_{2C} \end{bmatrix}} \underset{(12\times 1)}{\begin{Bmatrix} u_A \\ u_B \\ u_C \\ u_D \end{Bmatrix}} \qquad (3.11)$$

This is the basic kinematic relationship for a structure as a whole. For further discussion, let us take $\mathbf{u}_A = \mathbf{u}_D = 0$, assuming that the supports at A and D are fully clamped. Thus deleting the first and the fourth columns, we have the reduced relationships,

$$\begin{Bmatrix} \mathbf{v}_A \\ \mathbf{v}_B \\ \mathbf{v}_C \end{Bmatrix} = \begin{bmatrix} \mathbf{A}_{2A} & 0 \\ \mathbf{A}_{1B} & \mathbf{A}_{2B} \\ 0 & \mathbf{A}_{1C} \end{bmatrix} \begin{Bmatrix} \mathbf{u}_B \\ \mathbf{u}_C \end{Bmatrix} \qquad (a_3)$$

which are of the form

$$\mathbf{V} = \mathbf{A}\,\mathbf{U} \qquad (3.12)$$

Before proceeding with the details of the above matrix let us construct the equilibrium relations for the same frame. Equilibrium equations are,

$$\mathbf{P}_{2A}+\mathbf{P}_{1B} = \mathbf{P}_B\ ; \quad \mathbf{P}_{2B}+\mathbf{P}_{1C} = \mathbf{P}_C$$

or, using Eq. (2.13),

$$\mathbf{C}_{2A}\,\mathbf{S}_A+\mathbf{C}_{1B}\,\mathbf{S}_B = \mathbf{P}_B;\ \mathbf{C}_{2B}\,\mathbf{S}_B+\mathbf{C}_{1C}\,\mathbf{S}_C = \mathbf{P}_C$$

Therefore,

$$\mathbf{C}\,\mathbf{S} = \begin{bmatrix} \mathbf{C}_{2A} & \mathbf{C}_{1B} & 0 \\ 0 & \mathbf{C}_{2B} & \mathbf{C}_{1C} \end{bmatrix} \begin{Bmatrix} \mathbf{S}_A \\ \mathbf{S}_B \\ \mathbf{S}_C \end{Bmatrix} = \begin{Bmatrix} \mathbf{P}_B \\ \mathbf{P}_C \end{Bmatrix} = \mathbf{P} \qquad (3.13)$$

We have seen that **C** and **A** matrices of a member are transposes of each other. Comparing now the matrices of Eq. (3.12) and (3.13), we see that **C** and **A** matrices as applied to a structure are also transposes of each other, thus displaying again a contragredient relationship.

Let us now write in detail the elements of the kinematic matrix as applied to the frame shown in Fig. 3.5.

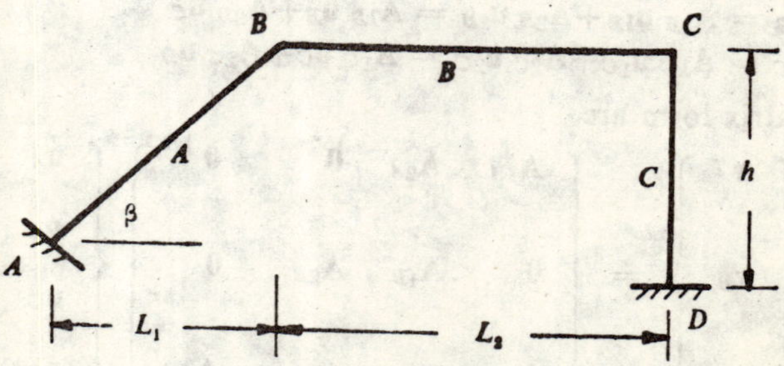

Figure 3.5

The kinematic matrix in terms of the submatrices A_1, A_2 is the same as in Eq. (a_3). For the three members the direction cosines with y axis downward are

Member	$\cos\alpha$	$\sin\alpha$
A	$\cos\beta$	$-\sin\beta$
B	1	0
C	0	1

(a_4)

The elements of A_1 and A_2' are as in Eq. (3.9). Equation (a_3) then reads as follows :

$$\mathbf{V} = \left\{\begin{array}{c} e_A \\ \varphi_{1A} \\ \varphi_{2A} \\ \hline e_B \\ \varphi_{1B} \\ \varphi_{2B} \\ \hline e_C \\ \varphi_{1C} \\ \varphi_{2C} \end{array}\right\} = \left[\begin{array}{ccc|ccc} \cos\beta & -\sin\beta & 0 & & & \\ -\dfrac{\sin\beta}{L_A} & -\dfrac{\cos\beta}{L_A} & 0 & & 0 & \\ -\dfrac{\sin\beta}{L_A} & -\dfrac{\cos\beta}{L_A} & 1 & & & \\ \hline -1 & 0 & 0 & 1 & 0 & 0 \\ 0 & \dfrac{1}{L_B} & 1 & 0 & -\dfrac{1}{L_B} & 0 \\ 0 & \dfrac{1}{L_B} & 0 & 0 & -\dfrac{1}{L_B} & 1 \\ \hline & & & 0 & -1 & 0 \\ & 0 & & -\dfrac{1}{L_C} & 0 & 1 \\ & & & -\dfrac{1}{L_C} & 0 & 0 \end{array}\right] \left\{\begin{array}{c} u_B \\ v_B \\ \theta_B \\ \hline u_C \\ v_C \\ \theta_C \end{array}\right\} \qquad (a_5)$$

$$= \mathbf{AU}$$

These equations give the deformations of members (three for each member), in terms of joint displacements. The elements of this matrix can be verified by constructing the matrix columnwise. The elements of each column represent the deformation of the members due to a unit value of the deformation component corresponding to the column. These can be derived by imposing unit joint displacement one at a time. Figure 3.6 shows the deformations due to v_B alone. The elements of the second column are the deformations due to unit value of this component. These may be verified from the figure. The axial shortening of A is seen to be $v_B \sin\beta$. and the chord rotations are $(v_B \cos\beta)/L_A$ in counter clockwise sense. All these are negative according to the convention adopted for e, φ_1 and φ_2. Thus the first three elements of the second column are verified. For the

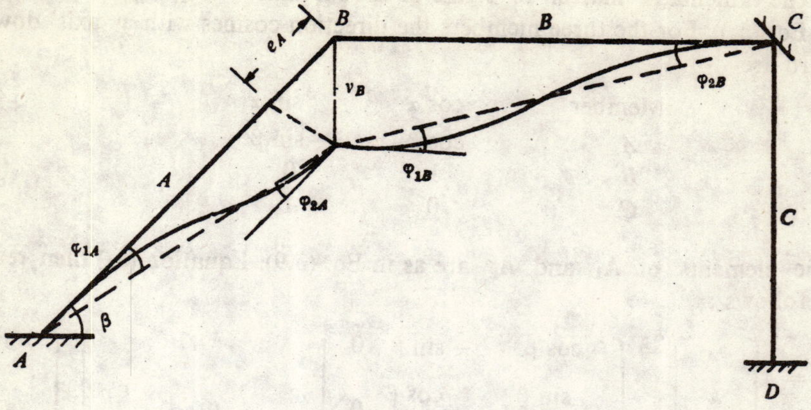

Figure 3.6

member B, it is seen that axial elongation is zero and the chord rotations φ_1 and φ_2 are v_B/L_B each in clockwise sense. Thus the next three elements of the column are verified. Similarly the last three elements are also verified: these are zero as there is no deformation of the member due to v_B alone.

3.3 INEXTENSIONAL BENDING

In the analysis of rigid jointed plane frames a common assumption is to neglect the axial deformations. The kinematics of such a structure is often required in the displacement methods of analysis. We will now take up this special problem which can be considered as a degenerate case of general kinematics.

The essential problem of the kinematics is to obtain the member deformations in terms of the joint displacements. When the axial deformations are ignored the member deformations are simply

$$\mathbf{v} = \{\varphi_1, \varphi_2\} \tag{a$_1$}$$

The general feature of inextensional bending is that the member deformations are not obtainable in terms of the displacements of the joints at the ends 1 and 2 of the member as stated by Eq. (3.5). They are in general controlled by the displacements of other remote joints also. The kinematic relationship can be established either by putting in matrix form the scalar equations determined physically from the displaced configuration or by a mere transformation of kinematic matrices. We will illustrate both these with respect to an example.

Figure 3.7 shows the inextensional bending mode. Since axial deformations are neglected, the independent displacements of joints are

$$\mathbf{U}' = \{\theta_B, \theta_C\ u_B\} \tag{a_2}$$

The six independent displacements are reduced to three by the three assumptions $e_A = e_B = e_C = 0$, with the reduced deformation vector for each member defined by Eq. (a_1). We have now to obtain the following kinematical relations:

$$\mathbf{V}' = \begin{Bmatrix} \varphi_{1A} \\ \varphi_{2A} \\ \varphi_{1B} \\ \varphi_{2B} \\ \varphi_{1C} \\ \varphi_{2C} \end{Bmatrix} = \begin{bmatrix} \\ \mathbf{A}' \\ (6\times 3) \\ \\ \end{bmatrix} \begin{Bmatrix} \theta_B \\ \theta_C \\ u_B \end{Bmatrix} = \mathbf{A}'\mathbf{U}' \tag{a_3}$$

The elements of the above matrix will now be obtained by imposing θ_B, θ_C, and $\mathbf{u}_B$, one at a time, and getting the elements of the first, second and third colums respectively. Figure 3.7 shows the deformed configuration due to u_B alone ($\theta_B = \theta_C = 0$). The deformed configuration has been obtained

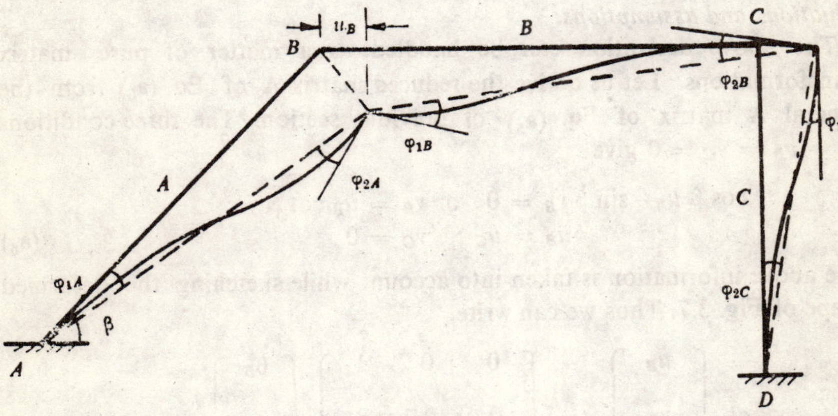

Figure 3.7

by taking into consideration that axial deformations in members are zero. For the deformed shape we observe:

$$\begin{aligned} \varphi_{1A} &= \varphi_{2A} = -(u_B/\sin\beta)/L_A = -u_B/L_A \sin\beta \\ \varphi_{1B} &= \varphi_{2B} = (u_B \cot{}_B)/L_B \\ \varphi_{1C} &= \varphi_{2C} = -\ (u_B/L_C) \end{aligned} \tag{a_4}$$

Similarly we derive the deformations due to θ_B and θ_C. When expressed in matrix form we have

$$\mathbf{V}' = \begin{Bmatrix} \phi_{1A} \\ \varphi_{2A} \\ \hline \varphi_{1B} \\ \varphi_{2B} \\ \hline \varphi_{1C} \\ \varphi_{2C} \end{Bmatrix} = \begin{bmatrix} 0 & 0 & -\dfrac{1}{L_A \sin\beta} \\ 1 & 0 & -\dfrac{1}{L_A \sin\beta} \\ \hline 1 & 0 & \dfrac{\cot\beta}{L_B} \\ 0 & 1 & \dfrac{\cot\beta}{L_B} \\ \hline 0 & 1 & -\dfrac{1}{L_C} \\ 0 & 0 & -\dfrac{1}{L_C} \end{bmatrix} \begin{Bmatrix} \theta_B \\ \theta_C \\ u_B \end{Bmatrix} = \mathbf{A}'\,\mathbf{U}' \qquad (a_5)$$

These are the reduced relations corresponding to the assumptions of inextensional bending. *The above procedure of assembling the kinematic matrix columnwise is quite general and can always be used to meet new situations and assumptions.*

The above derivation can be handled as a matter of pure matrix transformations. Let us derive the reduced matrix **A'** of Eq. (a_5) from the general **A** matrix of Eq. (a_5) of previous section. The three conditions $e_A = e_B = e_C = 0$ give

$$\cos\beta\, u_B - \sin\beta\, v_B = 0 \quad \text{or } v_B = u_B \cot\beta$$
$$u_B = u_C\,; \quad v_C = 0 \qquad (a_6)$$

The above information is taken into account while sketching the deformed shape of Fig. 3.7. Thus we can write,

$$\mathbf{U} = \begin{Bmatrix} u_B \\ v_B \\ \theta_B \\ \hline u_C \\ v_C \\ \theta_C \end{Bmatrix} = \begin{bmatrix} 0 & 0 & 1 \\ 0 & 0 & \cot\beta \\ 1 & 0 & 0 \\ \hline 0 & 0 & 1 \\ 0 & 0 & 0 \\ 0 & 1 & 0 \end{bmatrix} \begin{Bmatrix} \theta_B \\ \theta_C \\ u_B \end{Bmatrix} = \mathbf{R}\,\mathbf{U}' \qquad (a_7)$$

where the matrix **R** is a transformation matrix relating **U** to **U'**. Therefore we have

$$\mathbf{V} = \mathbf{A}\mathbf{U} = (\mathbf{A}\,\mathbf{R})\mathbf{U}' = \mathbf{A}'\,\mathbf{U}' \qquad (a_8)$$

By carrying out the multiplication **A R** and considering only the rows corresponding to φ's we obtain **A'** matrix. However the method recommended previously is advisable as it is adoptable to more general kinematical definitions of deformations and displacements.

3.4 KINEMATICS OF PIN JOINTED FRAMES

The kinematical matrices $\mathbf{A}_1$ and $\mathbf{A}_2$ for a pin ended member reduce to very simple forms, as only the axial elongations are required to be expressed in terms of joint displacements. Thus considering the first row of Eq. (3.9), and deleting joint rotation terms, we have

$$\mathbf{v} \equiv (e) = [-c \;\; -s \mid c \;\; s] \begin{Bmatrix} u_1 \\ v_1 \\ u_2 \\ v_2 \end{Bmatrix} = \mathbf{A\,u}$$

$$= [\mathbf{A}_1 \;\; \mathbf{A}_2] \begin{Bmatrix} \mathbf{u}_1 \\ \mathbf{u}_2 \end{Bmatrix} = \mathbf{A}_1\,\mathbf{u}_1 + \mathbf{A}_2\,\mathbf{u}_2 \qquad (3.14)$$

The displaced position is as shown in Fig. 3.8. For a space frame member

$$\mathbf{v} = (e) = [-l \;\; -m \;\; -n \mid l \;\; m \;\; n] \begin{Bmatrix} u_1 \\ v_1 \\ w_1 \\ u_2 \\ v_2 \\ w_2 \end{Bmatrix} \qquad (3.15)$$

$$= \mathbf{A}_1\,\mathbf{u}_1 + \mathbf{A}_2\,\mathbf{u}_2$$

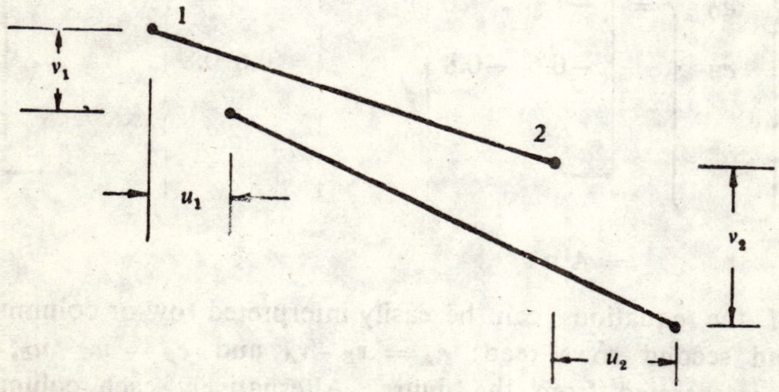

Figure 3.8

where u, v and w are the components of displacements along the global axes, and l, m and n are the direction cosines of member axis.

EXAMPLE 3.1

Consider the pinjointed plane frame of Exam. 2.2.

The **A** matrix for the structure is assembled row wise as discussed in Sec. 3.1, which in effect is applying Eq. (3.7) for each member in succession. This is as shown below:

$$\mathbf{V} = \begin{Bmatrix} \mathbf{v}_A \\ \mathbf{v}_B \\ \mathbf{v}_C \\ \mathbf{v}_D \\ \mathbf{v}_E \end{Bmatrix} = \begin{bmatrix} \mathbf{A}_{1A} & \mathbf{A}_{2A} & \cdot & \cdot \\ \cdot & \mathbf{A}_{1B} & \mathbf{A}_{2B} & \cdot \\ \cdot & \cdot & \mathbf{A}_{1C} & \mathbf{A}_{2C} \\ \mathbf{A}_{1D} & \cdot & \cdot & \mathbf{A}_{2D} \\ \mathbf{A}_{1E} & \cdot & \mathbf{A}_{2E} & \cdot \end{bmatrix} \begin{Bmatrix} \mathbf{u}_A \\ \mathbf{u}_B \\ \mathbf{u}_C \\ \mathbf{u}_D \end{Bmatrix} = \mathbf{A}\,\mathbf{U} \qquad (a_1)$$

Observe that the matrix in this equation is the transpose of the matrix **C** of Eq. (g) of Ex. 2.2, as the **A** and **C** matrices of each member are transposes of each other. (Consider only the part of the **C** matrix of Eq. (g).

We shall write the above equation in full as applied to the particular case given in Fig. (b), Ex. 2.2. We use the relations (3.14) for $\mathbf{A}_1$ and $\mathbf{A}_2$ matrices. Using the values of $\cos\alpha$ and $\sin\alpha$ listed in the table of Ex. 2.2 we have the details of Eq. (a_1) above as follows :

$$\mathbf{V} = \begin{Bmatrix} e_A \\ e_B \\ e_C \\ e_D \\ e_E \end{Bmatrix} = \left[\begin{array}{cc|cc|cc|cc} 0 & -1 & 0 & 1 & & & & \\ & & -1 & 0 & 1 & 0 & & \\ & & & & -0.6 & 0.8 & 0.6 & -0.8 \\ -1 & 0 & & & & & 1 & 0 \\ -0.6 & -0.8 & & & 0.6 & 0.8 & & \end{array}\right] \begin{Bmatrix} u_A \\ v_A \\ \hline u_B \\ v_B \\ \hline u_C \\ v_C \\ \hline u_D \\ v_D \end{Bmatrix}$$

$$= \mathbf{A}\,\mathbf{u} \qquad (a_2)$$

Any of the equations can be easily interpreted row or columnwise. The first and second rows read: $e_A = v_B - v_A$ and $e_B = u_C - u_B$, which are readily interpreted from the figure. Alternatively each column can be

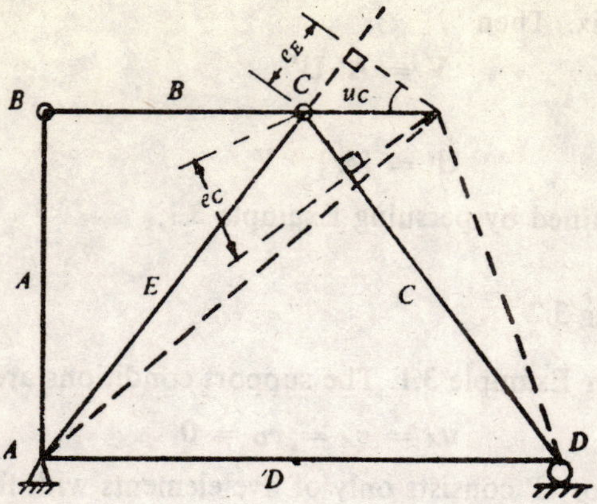

Example 3.1

interpreted to represent the bar elongations due to the unit value of the corresponding displacement, as explained earlier. Consider the fifth column : the elements of this column show that $e_B = u_C$, $e_C = -0.6\ u_C$ and $e_E = 0.6\ u_C$. These are also verified with the help of the sketch showing the displaced configuration due to u_C alone. Note that elongations are considered positive.

3.5 INVERSE FORM OF KINEMATICS—DISPLACEMENTS

The kinematics of structure considered so far may be described as *determinate* inasmuch as the member deformations are expressed in terms of the joint displacements. The problem is a determinate one since we can always obtain the deformations of a member when once its end displacements are prescribed. Writing of these kinematic relations is analogous to the writing of the *conditions* of equilibrium of forces which can always be written, statically determinate or otherwise.

The inverse problem, namely determining the displacements of the joints of a structure in terms of the member deformations, which may be due to loads or other causes such as temperature and support movements, is not direct but has to be obtained by *solving* the direct formulation of the kinematics discussed so far. The mechanics of occurence of the displacements depends on the connectivity of the members just as the force in a member. This problem is analogous to the *solution* of equilibrium equations of forces.

Before solving the kinematic matrix of a structure, it has to be reduced to a square one by incorporating the support conditions ensuring that rigid body movements are excluded. Let **U′** be the reduced displacement vector. The number of reductions will be the same as the number of reaction components, namely three for a plane frame and six for a space frame. Let A′ be the reduced matrix. Then

$$\mathbf{V} = \mathbf{A}'\,\mathbf{U}' \tag{3.16}$$

giving

$$\mathbf{U}' = [\mathbf{A}']^{-1}\,\mathbf{V} \tag{3.17}$$

This will be explained by persuing Example 3.1.

Example 3.2

Consider Example 3.1. The support conditions are

$$u_A = v_A = v_D = 0 \qquad (a_1)$$

The reduced vector U′ consists only of five elements with the above elements removed. The first, second and the last columns of the matrix of Eq. (a_2) of Example 3.1 are deleted to obtain the reduced square A′. The reduced kinematic relations are

$$\mathbf{V} = \begin{Bmatrix} e_A \\ e_B \\ e_C \\ e_C \\ e_E \end{Bmatrix} = \begin{bmatrix} 0 & 1 & \cdot & \cdot & \cdot \\ -1 & 0 & 1 & 0 & \cdot \\ \cdot & \cdot & -0.6 & 0.8 & 0.6 \\ \cdot & \cdot & \cdot & \cdot & 1 \\ \cdot & \cdot & 0.6 & 0.8 & \cdot \end{bmatrix} \begin{Bmatrix} u_B \\ v_B \\ u_C \\ v_C \\ u_D \end{Bmatrix} = \mathbf{A}'\,\mathbf{U}' \qquad (a_2)$$

This is of the form (3.16).

We can now solve for the joint displacements in terms of bar elongations in the form (3.17). Therefore we obtain, after inversion,

$$\begin{Bmatrix} u_B \\ v_B \\ u_C \\ v_C \\ u_D \end{Bmatrix} = \begin{bmatrix} \cdot & -1 & -0.833 & 0.5 & 0.833 \\ 1 & \cdot & \cdot & \cdot & \cdot \\ \cdot & \cdot & -0.833 & 0.5 & 0.833 \\ \cdot & \cdot & 0.625 & -0.375 & 0.625 \\ \cdot & \cdot & \cdot & 1 & \cdot \end{bmatrix} \begin{Bmatrix} e_A \\ e_B \\ e_C \\ e_D \\ e_E \end{Bmatrix} \qquad (a_3)$$

The first row shows that u_B is controlled by all but the bar elongation e_A. Except u_B and u_D which are controlled only by the elongation of the bar connected to the rigid foundation, others are generally influenced by the bar elongations of more than one member.

The solution (a_3) states the usual solution obtained by the Williot-Mohr construction used in the kinematics of pin-jointed frames. For the present we shall only note that the matrix in (a_3) above and the matrix in Eq. (m) of Ex. 2.2 are transposes of each other. Thus if we have solved the statical problem, the kinematic one need not be solved afresh.

3.6 CONCLUDING REMARKS

It is seen that the purely geometrical problems of relating displacements and deformations have been discussed. We started with the deformations of a member related to the member end displacements (Eq. 3.3) and using them we constructed kinematic relations of type (3.12) as applied to a structure.

In this chapter the deformations of members are taken to be known. The expressions for them in terms of forces causing them will involve knowledge of flexibility properties, to be yet dealt with. Chapter 6 continues the subject matter of this chapter after the next chapter which discusses the problem of flexibility and stiffness. This chapter will also involve considerations of the kinematics of a beam, namely beam deformations obtained as integrated effects of deformations of the beam differential elements.

EXERCISES

3.1 The kinematic relation relates the deformations **v** to the member end displacements **u**. The inverse definition, namely **u** in terms of **v** is not possible. Explain the physical implication, apart from the fact that the matrix **A** is rectangular. (Compare with Exercise 2.5).

3.2 Assume that the deformations e, φ_1 and φ_2 appearing in Eq. (3.3) are all zero. Then solve for $\{\overline{u}_2, \overline{v}_2, \overline{\theta}_2\}$ in terms of $\{\overline{u}_1, \overline{v}_1, \overline{\theta}_1\}$. Conclude that such a displacement pattern corresponds to a rigid body movement of the beam, since by definition, there are no 'deformations' ($\mathbf{v} \equiv 0$).

3.3 Construct the kinematic matrix for the frame of Ex. 2.8. Verify contragradience. (*See* Ex. 2.8)

3.4 Consider the kinematic relation for the frame of Ex. 2.9. Interpret the result when α tends tends to zero and identify the assumption made in small deflection theory that a small transverse movement of a member produces zero axial strains.

3.5 Construct the kinematic matrix (relating the axial elongations of members to the displacements of joint A) of the frame of Ex. 2.12. Verify contragradience (*See* Ex 2.12).

3.6 With the statical matrix derived in Ex 2.13, construct the kinematic matrix for the rigid jointed frame of Ex 2.13, by evoking contragradience; interpret each row of the matrix.

3.7 From the kinematic relations derived in Ex. 3.6, arrive at the relationship between the displacements of joints B and C when axial deformations are ignored. Verify them directly from the geometry of the structure.

3.8 Assume that the pin-jointed frame of Ex. 2.8 is pinned at D and supported on rollers at E permitting movement along DE. Reduce the kinematic matrix (Ex. 3.3) and solve for the displacements in terms of member elongations. Use the geometrical data and compute them in terms of e_A, e_B, etc.

3.9 Continue Ex. 3.5 to obtain the displacements of joint A in terms of member elongations.

3.10 Deduce using contragradiene the kinematic matrices for the problem set out in Ex. 2.17, and interpret physically. Repeat for Ex. 2.18.

Chapter 4

Virtual Work

Virtual work ideas form the most powerful and versatile tools of structural mechanics. They are used in a variety of ways and with advantage for linear as well as nonlinear problems. In this text too, it has been used in preference to the other alternative namely minimization of energy potentials (Castigliano's theorems).

The principle of virtual work is used chiefly for (a) kinematics and (b) equilibrium. The familiar unit (dummy) load method of finding the displacement in a structure has virtual work as the underlying principle. Geometrical (kinematic) considerations are avoided and in its place a virtual force system is selected and the equation of work applied. Also, writing the equations of compatibility as required in force method is usually accomplished using the principle of virtual work. On the other hand, in many situations considerations of the equations of equilibrium are replaced by kinematical considerations of a virtual displacement system and subsequent application of a work equation. An example is the wide use of this technique in collapse load analysis of structures. The Muller-Breslau principle of constructing influence lines is another example.

4.1 PRINCIPLE OF VIRTUAL WORK

Let a system of internal and external forces acting on a body be in equilibrium. Let a set of *small* virtual displacements be imagined in which the displacements and the deformations satisfy only the kinematical restraints and characteristics. This means for example, that for a beam the virtual displacement can be a smooth enough curve satisfying the support conditions. The principle of virtual work states that *the virtual work of the external forces through the displacements is equal to the virtual work of internal forces through the deformations*, or simply, *the external virtual work is equal to the internal virtual work*. The above principle is very general and applies to a small differential element, a member, or a structure. Hence its versatility and usefulness as a tool.

In this chapter it will be shown that a work equation can be written between a statically admissible set of forces and a kinematically consistent set of displacements. Thus using a statically admissible set of forces and the work equation we can derive the kinematic relations between the deformations and displacements. Alternatively by using a kinematically consistent set of displacements and the virtual work equation we can obtain the equations of equilibrium between the forces. This line of thinking has been well explained in reference [1] which appears to be somewhat less publicized literature. This has excellent academic as well as practical value.

Consider the case of the beam shown in Fig. 4.1. The force and displacement systems can be as follows :

Force System (Fig. 4.1):

Internal $\mathbf{S}_x = \{S_x, M_x, Q_x\}$ (a₁)

External* $\mathbf{P} = \{\mathbf{P}_1, \mathbf{P}_2\} = \{P_1, Q_1, M_1, P_2, Q_2, M_2\}$

$\mathbf{p}_x = \{p, q\}$

Deformation System (Fig 4.2):

Internal $\mathbf{d}\epsilon = \{du, d\theta\ dv'\}$ (a₂)

External $\mathbf{u} = \{u_1\ v_1\ \theta_1\ u_2\ v_2\ \theta_2\}$

$\mathbf{u}_x = \{u, v\}$

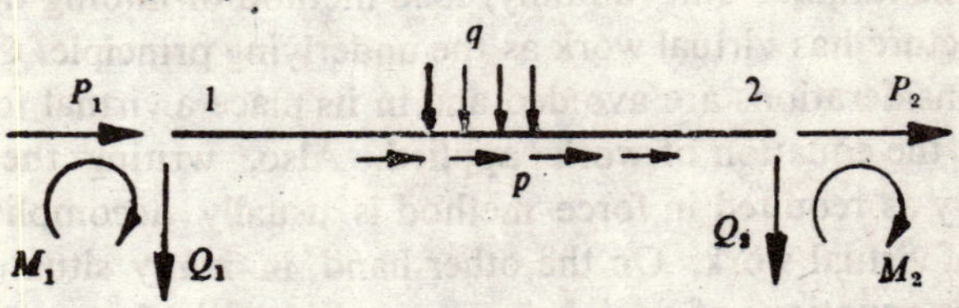

(a)

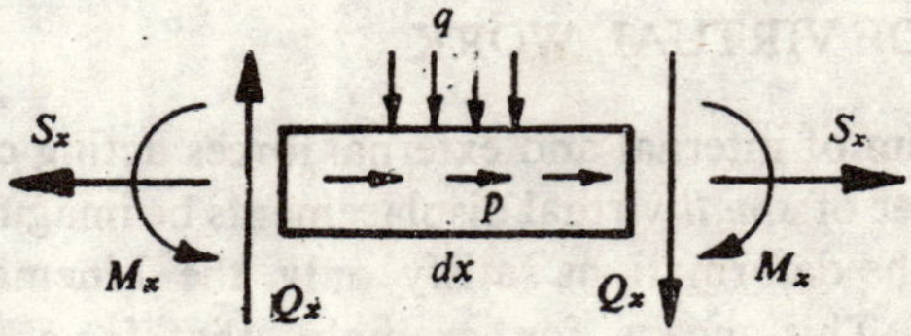

(b)

Figure 4.1.

*Here bars on top are dropped, for convenience.

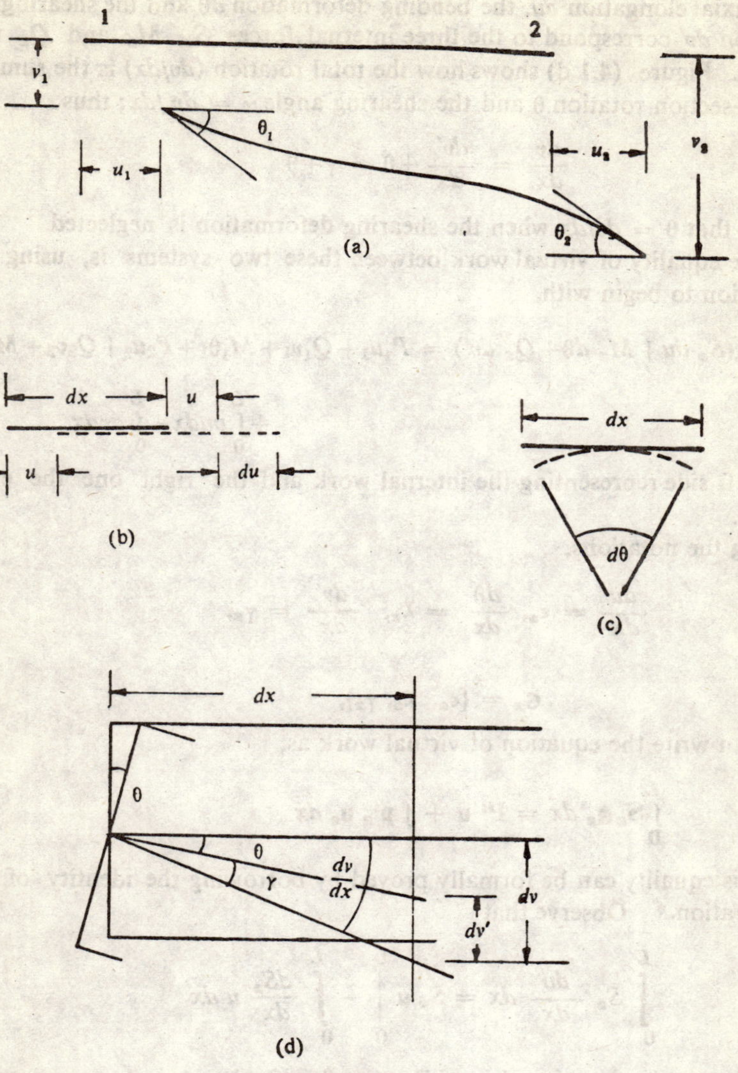

Figure 4.2.

The force system is required to satisfy only the conditions of statics, and the deformation system is required to satisfy only the kinematical characteristics, namely that support conditions are maintained and the deformed shape is smooth enough. Second, the two systems may not be related, namely the deformations *may not* be due to the force system. It is this feature that makes the method a versatile one.

The equilibrium relations are (Fig 4.1 b).

$$\frac{dS_x}{dx}+p=0;\ \frac{dM_x}{dx}+Q_x=0;\ \frac{dQ_x}{dx}+q=0 \qquad (a_3)$$

The axial elongation du, the bending deformation $d\theta$ and the shearing deformation dv' correspond to the three internal forces S_x, M_x and Q_x respectively. Figure (4.1 d) shows how the total rotation (dv/dx) is the sum of the cross-section rotation θ and the shearing angle $\gamma = dv'/dx$; thus

$$\frac{dv}{dx} = \frac{dv'}{dx} + \theta = \gamma + \theta \tag{a_4}$$

Note that $\theta = dv/dx$ when the shearing deformation is neglected.

The equality of virtual work between these two systems is, using scalar notation to begin with,

$$\int_0^L (S_x\, du + M_x\, d\theta + Q_x\, dv') = P_1u_1 + Q_1v_1 + M_1\theta_1 + P_2u_2 + Q_2v_2 + M_2\theta_2 + \int_0^L pudx + \int_0^L qvdx \tag{a_5}$$

the left side representing the internal work and the right one the external work.

Using the notations,

$$\frac{du}{dx} = \epsilon_x, \quad \frac{d\theta}{dx} = \chi_x, \quad \frac{dv'}{dx} = \gamma_x, \tag{a_6}$$

and

$$\boldsymbol{\epsilon}_x = \{\epsilon_x, \chi_x, \gamma_x\}, \tag{a_7}$$

we can write the equation of virtual work as,

$$\int_0^L \mathbf{S}_x \boldsymbol{\epsilon}_x\, dx = \mathbf{P}^t\, \mathbf{u} + \int \mathbf{p}^t_x\, \mathbf{u}_x\, dx \tag{4.1}$$

This equality can be formally proved by borrowing the identity of double integration.* Observe that

$$\int_0^L S_x \frac{du}{dx}\, dx = S_x u \Big|_0^L - \int_0^L \frac{dS_x}{dx} u\, dx \tag{a_8}$$

Herein use the first equation of Eq. (a_3) for dS_x/dx; also observe that

$$S_x(0) = -P_1, \qquad S_x(L) = P_2$$

$$u(0) = u_1 \quad \text{and} \quad u(L) = u_2. \quad \text{Therefore}$$

Eq. (a_8) gives

$$\int_0^L S_x \frac{du}{dx}\, dx = P_1u_1 + P_2u_2 + \int_0^L p\, u\, d\, x \tag{a_9}$$

* This is the degenerate case of the Green's theorem used in two and three dimensions.

Similarly we can deduce that

$$\int_0^L (M_x\chi_x+Q_x\gamma_x)\,dx = \int_0^L \left\{ M_x\frac{d\theta}{dx}+Q_x\left(\frac{dv}{dx}-\theta\right)\right\} dx$$

$$= (M_x\theta+Q_xv)\Big|_0^L +$$

$$\int_0^L \left(-\frac{dM_x}{dx}+Q_x\right)\theta\,dx - \int_0^L \frac{dQ_x}{dx}v\,dx. \qquad (a_{10})$$

This in view of the second and third equations of Eq. (a_3) and the notations,

$$M_x(0) = -M_1,\ M_x(L) = M_2,\ Q_x(0) = -Q_1,\ Q_x(L) = Q_2$$

because of the deformation and statical sign conventions, becomes,

$$\int (M_x\chi_x+Q_x\gamma_x)\,dx = M_1\theta_1+M_2\theta_2+Q_1v_1+Q_2v_2+\int q\,v\,d\,x. \qquad (a_{11})$$

Thus Eq. (a_9) and the above equation when added, lead to Eq. (4.1).

The main points to be noted are : (a) that only the equilibrium and kinematic conditions have been used, and (b) that there are no relations between the forces and the displacements. Only the double integration rule has been used to prove the principle. The significance of the virtual work equation lies in the fact that it only states a property which exists between a set of statically admissible forces and a kinematically admissible set of the displacements. It is this feature which renders the principle versatile, since the choices of the force and displacement systems are free. It is not unusual therefore to find many 'clever' solutions as applied to special problems in some books.

4.2. VIRTUAL WORK FOR STRUCTURES

The virtual work equation for a structure as a whole is written by treating members as units. It is supposed that the distributed lateral loads are replaced by equivalent loads. In this regard recall the discussion of Sec. 1.5 where these were first introduced. Thus for the purposes of structural (global) analysis we will consider loads directy applied to the joints. Thus Eq. (4.1) becomes

$$\int_0^L \mathbf{S}_x\mathbf{e}_x\,dx = \bar{\mathbf{p}}^t\bar{\mathbf{u}} \qquad (4.2)$$

We have now now introduced bar on top to distinguish that these are in member coordinates. Such a distinction was not necessary in the previous section.

We can transform the right hand side in terms of **S** and **v**. Recall Eqs. (2.6), (3.7) and the contragradient property that $\overline{\mathbf{C}}$ and $\overline{\mathbf{A}}$ are transposes of each other (Sec. 3.1). Thus

$$\overline{\mathbf{P}}^t\,\overline{\mathbf{u}} = (\overline{\mathbf{C}}\,\mathbf{S})^t\,\overline{\mathbf{u}} = \mathbf{S}^t\overline{\mathbf{C}}^t\overline{\mathbf{u}} = \mathbf{S}^t\overline{\mathbf{A}}\overline{\mathbf{u}} = \mathbf{S}^t\mathbf{v} \qquad (a_1)$$

Therefore we have the identity,

$$\int_0^L \mathbf{S}_x\,\boldsymbol{\epsilon}_x\,dx = \mathbf{S}^t\,\mathbf{v} \qquad (4.3)$$

Consider the frame of Fig. 3.5 as an example of a structure. We will treat this as an assembly of three members. Let a virtual displacement of the frame be considered. Let

$$\mathbf{V} = \begin{Bmatrix} \mathbf{v}_A \\ \mathbf{v}_B \\ \mathbf{v}_C \end{Bmatrix} = \begin{bmatrix} \mathbf{A}_{2A} & \mathbf{0} \\ \mathbf{A}_{1B} & \mathbf{A}_{2B} \\ \mathbf{0} & \mathbf{A}_{1C} \end{bmatrix} \begin{Bmatrix} \mathbf{u}_B \\ \mathbf{u}_C \end{Bmatrix} = \mathbf{AU} \qquad (a_2)$$

be a virtual system. (See Eq. a_3 of Sec. 3.2.) Let $\mathbf{P} = \{\mathbf{P}_B, \mathbf{P}_C\}$ be the vector of joint loads which corresponds to the joint displacement vector **U**. In summation form the virtual work equation is

$$\sum_{\text{members}} \int_0^L \mathbf{S}_x\boldsymbol{\epsilon}_x\,dx = \mathbf{P}^t\mathbf{U} \qquad (4.4)$$

We make use of the result of applying virtual work equations to each of the members, namely Eq. (4.3). Thus

$$\mathbf{s}_A{}^t\mathbf{v}_A + \mathbf{s}^t{}_B\mathbf{v}_B + \mathbf{s}_C{}^t\mathbf{v}_C = \mathbf{P}^t\mathbf{U} \qquad (a_3)$$

Denoting $\mathbf{S} = \{\mathbf{s}_A, \mathbf{s}_B, \mathbf{s}_C\}$
we see that

$$\mathbf{S}^t\mathbf{V} = \mathbf{P}^t\mathbf{U} \qquad (4.5)$$

This is a typical form. Here **S** is the vector of member forces and **V** that of the corresponding deformations. Their virtual work is equal to that of external loads (See Ex. 4.3).

4.3 EQUILIBRIUM BY VIRTUAL WORK

Let us suppose that a kinematic relationship

$$\mathbf{V} = \mathbf{A}\,\mathbf{U} \qquad (4.6)$$

such as shown by Eq. (a_2) of the previous section, is derived for a structure. Next let the virtual work equation, Eq. (4.5) be written. If we then substitute the relations (4.6) we obtain $\mathbf{S}^t\mathbf{AU} = \mathbf{P}^t\mathbf{U}$. Since the virtual displacement **U** is arbitrary we can cancel it on either side. This in physical term means that we can use unit magnitudes of displacements. Taking transpose of each side we find that

$$\mathbf{A}^t\,\mathbf{S} = \mathbf{P} \qquad (4.7)$$

This is the condition of equilibrium which is required to be satisfied between the internal and the external forces.

The number and type of equations will be dependent on the choice of the kinematic relations (4.6). We derive *as many* equilibrium equations as are elements in the vector **U**.

Therefore if *not all* the possible joint displacement are used and only a reduced set $\mathbf{U}_*$ is taken such that

$$\mathbf{V}^* = \mathbf{A}^* \mathbf{U}_* \tag{4.8}$$

are member deformations, then we obtain

$$\mathbf{A}^{*t} \mathbf{S} = \mathbf{P}_* \tag{4.9}$$

as the limited number of equilibrium equations where elements of $\mathbf{P}_*$ match with those of $\mathbf{V}_*$. If we chose only one displacement then we have a single equilibrium equation. This is the so-called *unit displacement* method [7].

Suppose we take, for the frame of Fig. 3.5,

$$\mathbf{V}^* = \begin{bmatrix} \mathbf{A}_{2A} \\ \mathbf{A}_{1B} \\ \mathbf{O} \end{bmatrix} \left\{ \mathbf{U}_B \right\} = \mathbf{A}^* \mathbf{U}^* \tag{a$_1$}$$

as a particular virtual pattern. This is a particular case of Eq. (a$_2$) of the previous section. Then the equation of work gives the equilibrium relations

$$\mathbf{A}^{*t} \mathbf{S} = [\mathbf{A}_{2A}\ \ \mathbf{A}^t_{1B}\ \ \mathbf{0}] \begin{Bmatrix} \mathbf{S}_A \\ \mathbf{S}_B \\ \mathbf{S}_C \end{Bmatrix} = \{\mathbf{P}_B\} = \mathbf{P}_* \tag{a$_2$}$$

If we write (4.7) for the frame using **A** of Eq. (a$_2$), of the previous section, we obtain,

$$\begin{bmatrix} \mathbf{A}^t_{2A} & \mathbf{A}^t_{1B} & 0 \\ 0 & \mathbf{A}^t_{2B} & \mathbf{A}^t_{1C} \end{bmatrix} \begin{Bmatrix} \mathbf{S}_A \\ \mathbf{S}_B \\ \mathbf{S}_C \end{Bmatrix} = \begin{Bmatrix} \mathbf{P}_B \\ \mathbf{P}_C \end{Bmatrix} \tag{a$_3$}$$

It is thus seen that Eq. (a$_2$) is only a particular case. We generally write the *full* kinematic matrix **A** and obtain *all* the equations of equilibrium.

The above procedure is a very general procedure. Recall the use of virtual displacement patterns used in collapse load analysis to derive equations of equlibrium between bending moments. Recall also the Muller-Breslau principle used in deriving influence lines.

EXAMPLE 4.1

In many structures certain assumptions are made with respect to deformations and the corresponding equilibrium equations are required to be derived. As an example consider the derivation of all the three equations of equilibrium for the frame shown in Fig. 3.5 in which axial deformations are neglected. The reduced kinematic relations are derived in Sec. 3.3 and are given in Eq. (a_5). To rewrite (see Figs. 3.5, 3.7), taking

$$L_A \sin\beta = h\ ; \qquad L_C = h\ ; \qquad L_B = L_2,$$

$$\mathbf{V}' = \begin{Bmatrix} \varphi_{1A} \\ \varphi_{2A} \\ \varphi_{1B} \\ \varphi_{2B} \\ \varphi_{1C} \\ \varphi_{2C} \end{Bmatrix} = \begin{bmatrix} 0 & 0 & -1/h \\ 1 & 0 & -1/h \\ \hline 1 & 0 & \cot\beta/L_2 \\ 0 & 1 & \cot\beta/L_2 \\ \hline 0 & 1 & -1/h \\ 0 & 0 & -1/h \end{bmatrix} \begin{Bmatrix} \varphi_B \\ \varphi_B \\ u_C \end{Bmatrix} = \mathbf{A}\,\mathbf{U}' \qquad (a_1)$$

The member forces are

$$\mathbf{S}' = \{M_{1A}\ M_{2A} \mid M_{1B}\ M_{2B} \mid M_{1C}\ M_{2C}\}$$

The relationship between the reduced displacement $\mathbf{U}'$ and the regular ones $\mathbf{U}$ was derived in Eq. (a_6) of Sec. 3.3 which followed as a result of neglecting the axial deformations. To rewrite,

$$\mathbf{U} = \begin{Bmatrix} u_B \\ v_B \\ \theta_B \\ u_C \\ v_C \\ \theta_C \end{Bmatrix} = \begin{bmatrix} 0 & 0 & 1 \\ 0 & 0 & \cot\beta \\ 1 & 0 & 0 \\ \hline 0 & 0 & 1 \\ 0 & 0 & 0 \\ 0 & 1 & 0 \end{bmatrix} \begin{Bmatrix} \theta_B \\ \theta_C \\ u_B \end{Bmatrix} = \mathbf{R}\,\mathbf{U}' \qquad (a_2)$$

The equation of virtual work is

$$\mathbf{S}'^t V' = \mathbf{P}^t \mathbf{U} \qquad (a_3)$$

Using Eqs. (a_1) and (a_2)

$$\mathbf{S}'^t \mathbf{A}\,\mathbf{U}' = \mathbf{P}^t \mathbf{R}\,\mathbf{U}'$$

giving, on cancellation of $\mathbf{U}'$ and transposition,

$$\mathbf{A}^t\,\mathbf{S}' = \mathbf{R}^t\,\mathbf{P} = \mathbf{P}' \qquad (a_4)$$

We have here also obtained the generalized load vector $\mathbf{P}'$ which corresponds to the displaeement vector $\mathbf{U}'$. Written out in full, this reads as follows:

$$\mathbf{P}' = \mathbf{R}^t\,\mathbf{P} = \begin{bmatrix} 0 & 0 & 1 & 0 & 0 & 0 \\ 0 & 0 & 0 & 0 & 0 & 1 \\ 1 & \cot\beta & 0 & 1 & 0 & 1 \end{bmatrix} \begin{Bmatrix} P_B \\ Q_B \\ M_B \\ P_C \\ Q_C \\ M_C \end{Bmatrix}$$

$$= \begin{Bmatrix} M_B \\ M_C \\ P_B+P_C+Q_B\cot\beta \end{Bmatrix} \qquad (a_5)$$

This is the generalized load vector corresponding to $\mathbf{U}'$. Observe that $\mathbf{P}'^t\,\mathbf{U}'$ is the external virtual work equal in magnitude to $\mathbf{P}^t\,\mathbf{U}$. The virtual work is

$$\mathbf{p}'^t\,\mathbf{U}' = M_B\theta_B+M_C\theta_C+(P_B+P_C+Q_B\cot\beta)\,u_B$$

In the last term, the quantity $Q_B \cot\beta\, u_B$ is equal to $Q_B\,(u_B\cot\beta) = Q_B v_B$ as can be seen from geometry. It represents the work of Q_B moving through v_B.

Equation (a_4) written in full reads as follows :

$$\begin{bmatrix} 0 & 1 & 1 & 0 & 0 & 0 \\ 0 & 0 & 0 & 1 & 1 & 0 \\ -\frac{1}{h} & -\frac{1}{h} & \frac{\cot\beta}{L_2} & \frac{\cot\beta}{L_2} & -\frac{1}{h} & -\frac{1}{h} \end{bmatrix} \begin{Bmatrix} M_{1A} \\ M_{2A} \\ \hline M_{1B} \\ M_{2B} \\ \hline M_{1C} \\ M_{2C} \end{Bmatrix}$$

$$= \begin{Bmatrix} M_B \\ M_C \\ P_B+P_C+Q_B\cot\beta \end{Bmatrix} \qquad (a_6)$$

The first two equations of equilibrium are obvious. The third equation is the so-called sway equation. This can be verified directly by statics which, as one can see is not quite straightforward. This is left as an exercise for the reader. (See Ex 4.5).

4.4 KINEMATICS BY VIRTUAL WORK

Derivation of geometrical relationships by consideration of a virtual force system is the familiar Unit Load or Dummy Load method used in structural mechanics. Setting up of the conditions of geometry, namely of compatibility of deformations as required in the force method of analysis of indeterminate frames using virtual work, is another example. Suppose a kinematic relation between a set of deformations **V** and a set of joint displacements $\mathbf{U}_*$ are to be derived. To achieve this we choose a virtual force system **S*** satisfying the equilibrium relations

$$\mathbf{S}^* = \mathbf{B}^* \mathbf{P}^* \tag{4.10}$$

Here the elements of the vector **P*** should match with those of the required vector **U***. The only requirement is of statics. The deformations of the virtual system is immaterial. Then, the virtual work equation, Eq. (4.5) becomes $(\mathbf{B}^*\mathbf{P}^*)^t\,\mathbf{V} = \mathbf{P}^{*t}\,\mathbf{U}^*$, which gives,

$$\mathbf{B}^{*t}\,\mathbf{V} = \mathbf{U}^* \tag{4.11}$$

for arbitrary virtual forces **P***. If only a unit force is chosen in **P*** then Eq. (4.11) denotes one single equation giving the corresponding single displacement. This is the Unit load (Dummy load) method. On the other hand if **P*** contains all the components matching with all possible displacements then

$$\mathbf{B}^* \equiv \mathbf{B}, \quad \mathbf{U}_* \equiv \mathbf{U}$$

and we have

$$\mathbf{B}^t\,\mathbf{V} = \mathbf{U} \tag{4.12}$$

Equation (4.11) (or 4.12) denotes the manner according to which the member deformations are to be 'added' to produce the displacements of the joints. This is the familiar problem of the Williot-Mohr construction. Recall the contents of Sec. 3.5 in which the inverse form of kinematics was discussed. $(\mathbf{A}')^{-1}$ of Eq. (3.17) is actually the present **B** matrix.

EXAMPLE 4.2

Consider the frame of Example 2.2. Eq. (m) gives the matrix relating the internal forces to external forces. Consider Example 3.2 of Chapter 3 in which the displacements of the joints were obtained. It is observed that $\mathbf{B}^t\,\mathbf{V} = \mathbf{U}$. Therefore all we need is to construct **B** matrix.

If only few displacements are needed, then correspondingly the reduced matrix $\mathbf{B}^*$ can be used. Thus if only u_B were required we construct $\mathbf{B}^*$ matrix relating S due to P_B alone.

$$\mathbf{S} = \{0, -1, -0.833, 0.5, 0.833\}\, P_B = \mathbf{B}^* \mathbf{P}^*$$

This is the first column of Eq. (m) of Example 2.2. Then

$$u_B = \mathbf{B}^{*t}\mathbf{V} = [0, -1, -0.833, 0.5, 0.833]\ \mathbf{V}$$

This is how the member defomations are to be added to obtain u_B.

EXAMPLE 4.3

Consider now an example illustrating the use of Eq. (4.3) as applied to the beam shown. Let us consider only the bending curvature χ as the deformations occurring due to some cause ; ($\epsilon_x \equiv \gamma_x = 0$). Let $\varphi = \{\varphi_1, \varphi_2\}$ the end rotations, be the quantities of interest. To determine this we choose a virtual system of forces

$$\mathbf{S}^* = \{M_1^*, M_2^*\} \tag{a$_1$}$$

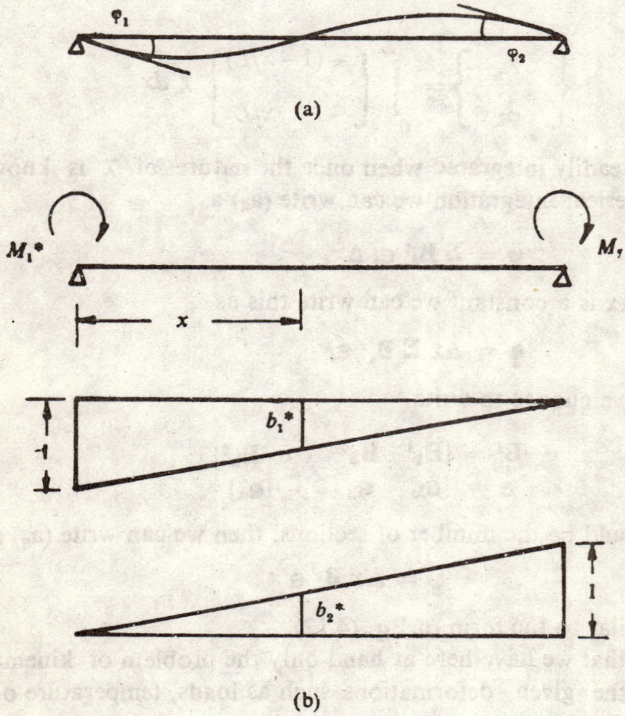

Example 4.3.

and the corresponding bending moments,

$$\mathbf{S}^*_x \equiv M_x \equiv \left[-\left(1-\frac{x}{L}\right) \quad \frac{x}{L} \right] \left\{ \begin{matrix} M_1^* \\ M_2^* \end{matrix} \right\}$$

$$= [b_1^* \; b_2^*] \, \{\mathbf{S}^*\} = \mathbf{B}^*\mathbf{S}^* \qquad (a_2)$$

Observe that b_1^* and b_2^* are the unit bending moments, namely moments due to $M_1^* = 1$ and $M_2^* = 1$ respectively. Equation of virtual work is,

$$\int M_x \chi_x \, dx = M_1^* \varphi_1 + M_2^* \varphi_2$$

or

$$\int \mathbf{S}_x^{*t} \boldsymbol{\epsilon}_x \, dx = \mathbf{S}^{*t} \boldsymbol{\varphi} \qquad (a_3)$$

Using (a_2) we obtain

$$\int (\mathbf{B}^* \mathbf{S}^*)^t \boldsymbol{\epsilon}_x \, dx = \mathbf{S}^{*t} \boldsymbol{\varphi} \qquad (a_4)$$

we can take $\mathbf{S}^*$ outside the integral and cancel the same with that on the right; on transposing we obtain

$$\boldsymbol{\varphi} = \int \mathbf{B}^{*t} \boldsymbol{\epsilon}_x \, dx \qquad (a_5)$$

These are the geometrical relations

$$\left\{ \begin{matrix} \varphi_1 \\ \\ \varphi_2 \end{matrix} \right\} = \int_0^L \left\{ \begin{matrix} -(1-x/L) \\ \\ x/L \end{matrix} \right\} \chi \, dx \qquad (a_6)$$

which are readily integrated when once the nature of χ is known, If we adopt numerical integration we can write (a_5) as

$$\boldsymbol{\varphi} = \Sigma \, \mathbf{B}_i^t \, \boldsymbol{\epsilon}_i \, \Delta x_i$$

If $\Delta x_i = \Delta x$ is a constant we can write this as

$$\boldsymbol{\varphi} = \Delta x \, \Sigma \, \mathbf{B}_i^t \, \boldsymbol{\epsilon}_i \qquad (a_8)$$

Further if we choose to write

$$\begin{aligned} \mathbf{B}^t &= [\mathbf{B}_1^t \quad \mathbf{B}_2^t \, . \, . \, . \, . \, \mathbf{B}_n^t] \\ \boldsymbol{\epsilon} &= \{\boldsymbol{\epsilon}_1, \quad \boldsymbol{\epsilon}_2 \, . \, . \, . \, . \, \boldsymbol{\epsilon}_n\} \end{aligned} \qquad (a_9)$$

where n would be the number of sections, then we can write (a_8) as

$$\boldsymbol{\varphi} = \Delta x \, \mathbf{B}^t \, \boldsymbol{\epsilon} \qquad (a_{10})$$

This is similar to the form of Eq. (4.12).

Observe that we have here at hand only the problem of kinematics. The causes of the given deformations such as loads, temperature or creep, are not involved.

4.5 CONCLUDING REMARKS

Use of virtual principle has been demonstrated for the two practical issues, namely establishing equations of equilibrium using an assumed (virtual) deformation model and deriving kinematic relations using an assumed (virtual) force model. The virtual displacements and deformations are required to be small. In the small displacement (linear) theory actual displacements can be used. However when we wish to include large displacements we have to consider small changes in the displacements rather than displacements themselves. We write $\mathbf{dv} = \mathbf{A}\,\mathbf{du}$, and since $\mathbf{A}$ is independent of displacements in small displacement theory, write $\mathbf{v} = \mathbf{Au}$.

Forms (4.9) and (4.11) are typical of equilibrium, and kinematic relations derived from virtual work considerations. These are in discrete form, as applied to a set of variables. Equation (a_5) of Example (4.3) is a variant of (4.11) as applied to an element in terms of strains as continuous functions of the coordinate. (See Eq. 5.16). We can define a similar analogue for Eq. (4.9) also (See Eq. 5.61 & 9.21).

EXERCISES

4.1 Consider the case of a small rigid movement as a virtual pattern. Show that the application of the virtual work principle establishes the overall equations of equilibrium between the external forces. In other words if $\bar{\mathbf{u}}_1$ and $\bar{\mathbf{u}}_2$ are member end displacements equivalent to a rigid body movement (see Ex. 3.2), then the relations obtained between $\bar{\mathbf{P}}_1$ and $\bar{\mathbf{P}}_2$ are the equations of equilibrium.

4.2 Take the case of simply supported beam of span L with a concentrated load P at centre. Consider a virtual displacement pattern $v^* = a \sin(\pi x/L)$. Verify the virtual work between the given force system and the assumed virtual displacement system.

4.3 Consider the pinjointed frame of Ex. 2.12. Assume a set of internal forces which are in equilibrium with the external forces at joint A. (Do not try to 'solve', but assume internal forces and arrive at the required forces at A.) Assume a virtual displacement pattern corresponding to a vertical displacement of A (See Ex. 3.5.). Verify the equality of virtual work (Eq. 4.5).

4.4 Assuming that the axial deformations of the members are negligible, derive the equations of equilibrium for the frame of Ex. 2.13, by using the principle of virtual work.

4.5 Derive Eq. (4.9) with respect to the frame of Ex. 2.12.

4.6 Solve Ex. 3.13 using virtual work.

Chapter 5

Member Analysis

In this chapter we will be dealing with the first stage of structural analysis, namely with the analysis of member. This analysis includes the study of a member (element) with a view to determine its flexibility, stiffness, and equivalent loads required in the subsequent problem of structural analysis. The latter is the analysis of the assembly of members (elements). In this connection recall the introductory remarks in Chapter 1 while describing the idea of structural analysis (see Sec. 1.1).

Essentially member analysis includes determination of (a) member flexibility and stiffness and (b) equivalent loads. The first is the essential problem of defining the response of a member to loads. The second is the determination of loads which are equivalent to the actual given loads and are the ones which enter into the formulation of the structural analysis problem. The loads such as the distributed loads acting along the length of a member are replaced by equivalent loads acting at the ends of the member and these form the equivalent loads in the formulation of the problem of global structural analysis. The loads not only include distributed loads but also loads such as initial strains due to temperature or lack of fit, initial stresses, or creep strains. We will now be considering these aspects in this chapter.

5.1 MATERIAL ELASTICITY

Most of the engineering materials show an elastic response in the intial stages of the stresses followed by a gradually decreasing stiffness. We will be considering only the initial stage for the present.

The elastic response of an isotropic body is characterized by two elastic constants. These are the modulii of elasticity for direct (normal) stress and for shear (tangential) stress. These are generally denoted by E and G. The choice of the Poisson ratio is generally made as an alternative to G since these three are related to one another. The elastic relationship can be expressed as

$$\boldsymbol{\sigma} = \begin{Bmatrix} \sigma \\ \tau \end{Bmatrix} = \begin{bmatrix} E & 0 \\ 0 & G \end{bmatrix} \begin{Bmatrix} \epsilon \\ \gamma \end{Bmatrix} = \mathbf{k}_0\, \boldsymbol{\epsilon} \tag{5.1}$$

These state the stiffness of the material. The matrix $\mathbf{k}_0$ will be called the *elastic modulus* matrix. This belongs to the class of stiffness matrices and at present refers to a small differential element of a material.

The flexibility relations stating the above in inverse form are

$$\boldsymbol{\epsilon} = \begin{Bmatrix} \epsilon \\ \gamma \end{Bmatrix} = \begin{bmatrix} 1/E & 0 \\ 0 & 1/G \end{bmatrix} \begin{Bmatrix} \sigma \\ \tau \end{Bmatrix} = \mathbf{k}_0^{-1}\, \boldsymbol{\sigma} = \mathbf{f}_0\, \boldsymbol{\sigma} \tag{5.2}$$

The matrix $\mathbf{f}_0$ is the *flexibility* matrix as applied to a material. (See Sec. 9.1 (*c*) for plane stress and plain strain cases.)

5.2 MEMBER ELASTICITY

We will now generalize the ideas of flexibility and stiffness to a differential element of a beam member. Here we deal in terms of stress resultants, namely forces and moments, and in terms of deformations, namely elongations and angular distortions.

Figures 4.1 and 4.2 show the forces and corresponding deformations of a beam element occurring in a plane. The internal forces will be represented by a vector

$$\mathbf{S}_x = \{S_x\ M_x\ Q_x\} \tag{5.3}$$

and the corresponding deformations

$$\mathbf{dv} = \{du,\ d\theta,\ dv'\} \tag{5.4}$$

We had an occasion to discuss these vectors earlier. Recall that dv' is the part of the total relative deflection due to shearing strain alone. Equation (a_4) of Sec 4.1, and Fig. 4.2(d) show the relationship between the (bending) rotation θ, the total rotation (dv/dx), and the shearing strain (dv'/dx)

We will write the deformations (5.4) as

$$\mathbf{dv} = \{\epsilon_x\ \chi_x\ \gamma_x\}\ dx = \boldsymbol{\epsilon}_x\ dx \tag{5.5}$$

by introducing the strains

$$\epsilon_x = \frac{du}{dx}, \quad \chi_x = \frac{d\theta}{dx}, \quad \gamma_x = \frac{dv'}{dx} = \frac{dv}{dx} - \theta \tag{5.6}$$

Here the first term denotes the axial strain, the second the bending curvature and the term (dv/dx) the total rotation of the beam axis. This is equal to the bending rotation θ plus the shearing rotation γ. Observe that in the usual bending theory of beams the shearing strain is taken to be zero, in which case $\theta = dv/dx$.

The elastic stiffness and flexibility relations consist of relations between the internal forces of Eq. (5.3) and the deformations (5.4). According to the simple theory of stretching and bending of bars we have, for the axial forces and deformations,

$$S_x = AE\left(\frac{du}{dx}\right) = (AE)\,\epsilon_x. \tag{5.7}$$

Here AE is the extensional rigidily, with A as the effective area of section resisting the axial force. For bending

$$M_x = EI\left(\frac{d\theta}{dx}\right) = (EI)\,\chi_x \tag{5.8}$$

with EI as the flexural rigidity and χ_x as the curvature. For shearing forces,

$$Q_x = A^*G\,\frac{dv'}{dx} = (A^*G)\,\gamma_x \tag{5.9}$$

where A^* is the effective area to take into account the nonuniform variation of shearing stresses over the section.

The above relations can be set in the form,

$$\mathbf{S}_x = \begin{Bmatrix} S_x \\ M_x \\ Q_x \end{Bmatrix} = \begin{bmatrix} AE & & \\ & EI & \\ & & A^*G \end{bmatrix} \begin{Bmatrix} \epsilon_x \\ \chi_x \\ \gamma_x \end{Bmatrix} = \mathbf{k}_x\,\mathbf{e}_x \tag{5.10}$$

where $\mathbf{k}_x$ is simply a diagonal matrix, and stands for the stiffness of the differential element of unit length. These are the analogue of Eq. (5.1) and will be used for beams.

To the above must be added the stiffness relation between the torsional moment and the corresponding twist. From the engineering theory of torsion, we have the relation

$$T_x = GJ\,\frac{d\psi}{dx} \tag{5.11}$$

where J is the torsional constant of the section and $(d\psi/dx)$ is the rate of twist.

The four stiffness constants AE, EI, A^*G and GJ introduced above are according to the simple theory generally applicable to normal cases where there is no interaction between these basic deformations. But, for instance, in a curved beam element, the axial force and the bending moment are both related to the curvature. In the thin walled sections too where warping effects are to be considered there is a complex interaction between the forces and the deformations. In such cases the stiffness matrix of Eq. (5.10) will not be a diagonal one. Observe that the diagonal matrix reflects the fact that the stiffnesses are *uncoupled*. Thus in general, there would be off-diagonal terms.

The inverse of relations (5.10) between strains and stress resultants :

$$\boldsymbol{\epsilon}_x = \begin{Bmatrix} \epsilon_x \\ \chi_x \\ \gamma_x \end{Bmatrix} = \begin{bmatrix} 1/AE & 0 & 0 \\ 0 & 1/EI & 0 \\ 0 & 0 & 1/A^*G \end{bmatrix} \begin{Bmatrix} S_x \\ M_x \\ Q_x \end{Bmatrix} = \mathbf{f}_x \, \mathbf{S}_x \qquad (5.12)$$

These would be used in building up the member flexibility.

5.3 MEMBER FLEXIBILITY AND STIFFNESS

Recall here the elementary ideas discussed in Sec. 1.3 and 1.4, in particular the kinematic considerations which gave Eqs. (1.4) and (1.5). Also, those who are facing the matrix notations for the first time may review Sec. 1.4 to note the notations and the interpretations of the flexibility coefficients f_{ij} and stiffness coefficients k_{ij} as given in Eq. (1.6) and (1.7) and explained in Fig. 1.6.

The *beam* and *cantilever* models showing the forces and the corresponding deformations are redrawn at one place as Figs. 5.1 and 5.2 for convenience. The general procedure is to evaluate the member flexibility matrix $\mathbf{f}$ relating the member forces to the member deformations,

$$\mathbf{v} = \mathbf{f}\,\mathbf{S} \qquad (5.13)$$

and by inversion the stiffness relations :

$$\mathbf{S} = \mathbf{f}^{-1}\,\mathbf{v} = \mathbf{k}\,\mathbf{v} \qquad (5.14)$$

These relations will be called the *basic stiffness* relations. The word basic has been chosen because these form the elementary relations from which more general relations can be obtained by mere transformations. These belong to the class of 'natural' stiffness matricies [17]. Basic stiffness relations can also be obtained directly provided we follow a displacement approach either by solving the differential equations of the beam or by assuming a deformed shape (see Secs. 5.8, 5.9).

Member flexibility relations are determined by 'summing' the deformations of the small (differential) elements (laminae) of the beam, since, the elongation and end rotations of beam are the integrated effect of the strains occuring in the beam. In simple cases we obtain them by direct considerations of the kinematics. (This for instance is what is involved in the moment area method.) However the more convenient approach is to adopt an algebraic one, based on virtual work principle. This is the basis of the well known Unit Load (virtual load) method. Having illustrated an example of the direct kinematic approach we will use this as a general principle for summing up the element strains.

Let ϵ be the strains occuring at an element in the beam. These may be due to any cause, the applied end forces, distributed lateral forces, temperature or shrinkage, etc. Let **v** be the desired member deformations, such as

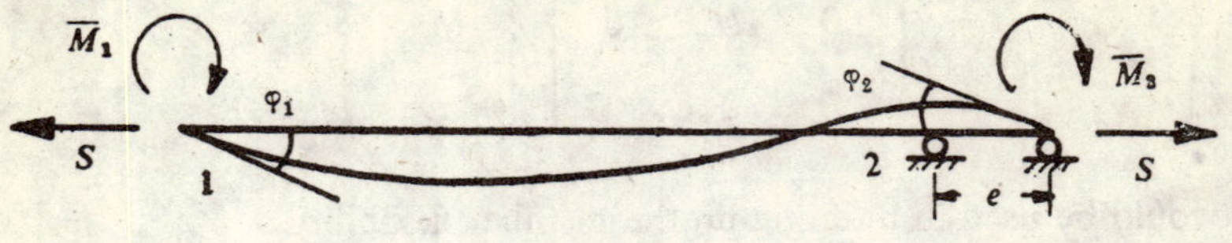

Fig. 5.1

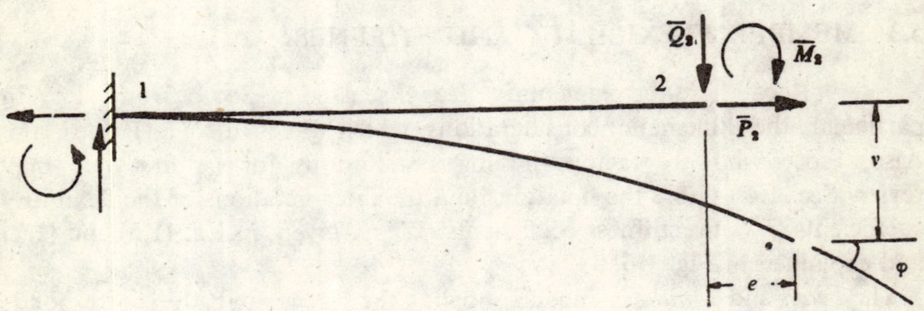

Fig. 5.2

$\{e, \varphi_1, \varphi_2\}$ shown in Fig. 5.1. We take a virtual force system $\mathbf{S}^*$ causing internal forces $\mathbf{S}^*_x$ at an element in the beam, such that $\mathbf{S}^*$ corresponds, in work sense, to the desired deformations **v**, namely the external virtual work would be $\mathbf{S}^{*t}\mathbf{v}$. Let the equilibrium relations between the internal forces $\mathbf{S}^*_x$ and $\mathbf{S}^*$ be

$$\mathbf{S}^*_x = \mathbf{B}^* \mathbf{S}^* \tag{5.15}$$

The equation of virtual work gives

$$\mathbf{S}^{*t}\, \mathbf{v} = \int \mathbf{S}^{*t}_x\, \epsilon \, dx$$

Substituting for $\mathbf{S}^*_x$, we obtain

$$\mathbf{S}^{*t}\, \mathbf{v} = \mathbf{S}^{*t} \int \mathbf{B}^{*t}\, \epsilon \, dx$$

We can cancel $\mathbf{S}^{*t}$ which in effect means that we could have taken unit values for the virtual forces. This is the matrix analogue of taking 'unit' load as in Unit Load method. Thus we obtain the desired deformations

$$\mathbf{v} = \int \mathbf{B}^{*t}\, \epsilon \, dx \tag{5.16}$$

This states the special way in which the element strains are summed up to give the deformations. (Recall Eq. a_5, Example 4.3.)

Equation (5.16) is the general form of kinematics obtained in terms of the state matrix **B** giving the internal forces (corresponding to given element strains) due to unit values of virtual forces (corresponding to the desired deformations).

In flexibility analysis we express the element strains in terms of given loads and Eq. (5.16) would read

$$\mathbf{v} = \mathbf{f}\,\mathbf{S} + \mathbf{v}^\circ \tag{5.17}$$

where S are the given forces and $\mathbf{v}^\circ$ 'free' deformations due to temperature creep or shrinkage. Figure 5.3 shows $\mathbf{v}^\circ$ due to lateral load effects alone.

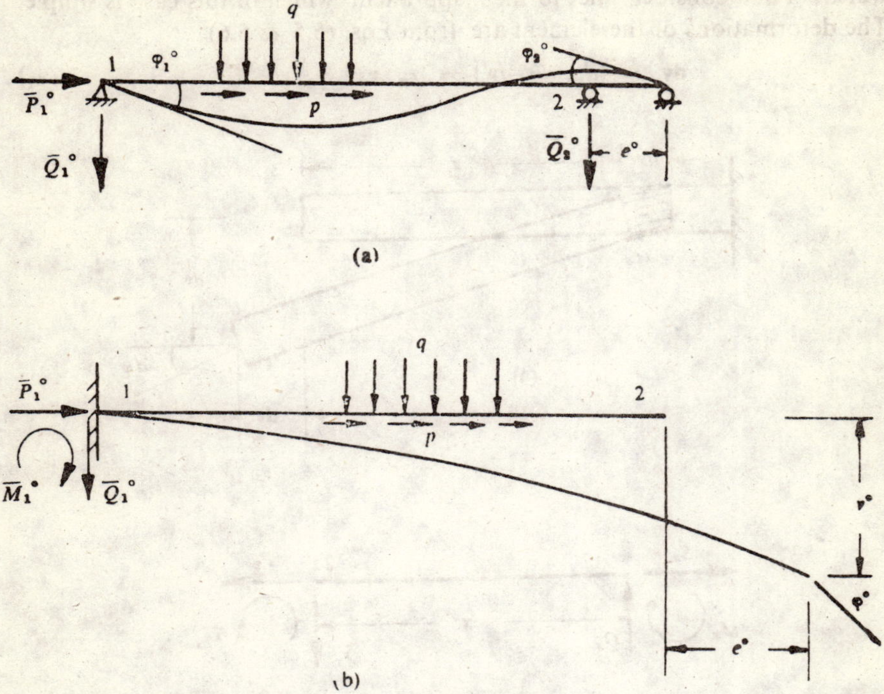

Fig. 5.3

5.4 EVALUATION OF MEMBER FLEXIBILITY

We will now consider the general procedure of evaluating the member flexibility. Consider the cantilever model to begin with in view of the relative simplicity of algebra involved.

a. Cantilever Model

The total displacements at the end 2 are equal to the integrated effect of the displacements caused there due to the deformation of small elements such as the typical one shown in Fig. 5.4 (a). The deformation of such an element is as a result of the internal forces $\{S_x, M_x, Q_x\}$. The deformations of the element are shown in Fig. 4.2 (b), (c) and (d). These are respectively the axial elongation, relative rotation of cross-section due to moments and the relative sliding of cross-sections due to the shearing forces. Figure 5.4 (a) shows the displacements at end 2 due to these deformations occuring at the typical element which alone is assumed elastic and other portions rigid. These can be obtained, either directly by the geometry of the movements taking place or by the application of the principle of virtual work. First consider the former approach, which in this case is simple. The deformations of the element are, from Eqs. (5.5 & 5.6)

$$\mathbf{dv}_x = \{du, d\theta, dv'\} = \{\epsilon_x \;\; \chi_x \;\; \gamma_x\}\, dx. \tag{a$_1$}$$

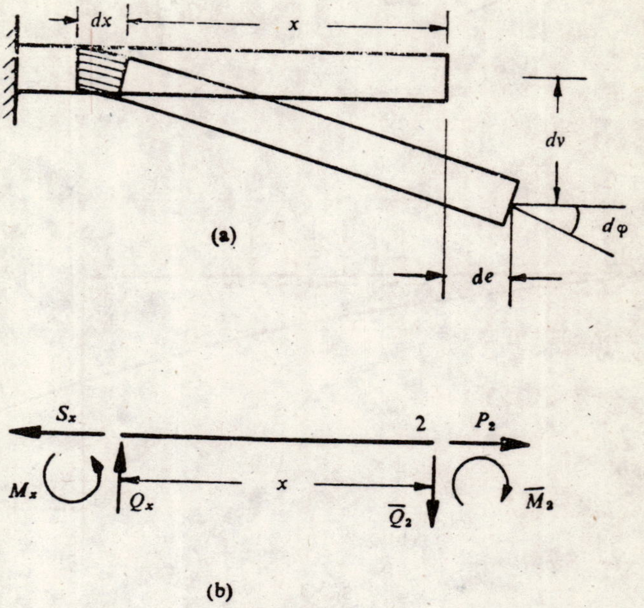

Fig. 5.4

The displacements $\{de, dv, d\varphi\}$ occuring at end 2 are

$$\mathbf{dv} = \begin{Bmatrix} de \\ dv \\ d\varphi \end{Bmatrix} = \begin{Bmatrix} du \\ x d\theta + dv' \\ d\theta \end{Bmatrix} = \begin{Bmatrix} \epsilon_x \\ x\chi_x + \gamma_x \\ \chi_x \end{Bmatrix} dx \tag{a$_2$}$$

Observe that the deflection at end 2, is $xd\theta$ as a result of bending rotation plus dv' due to relative sliding of sections. The equation can be written as

$$\mathbf{dv} = \begin{Bmatrix} de \\ dv \\ d\varphi \end{Bmatrix} = \begin{bmatrix} 1 & 0 & 0 \\ 0 & x & 1 \\ 0 & 1 & 0 \end{bmatrix} \begin{Bmatrix} \epsilon_x \\ \chi_x \\ \gamma_x \end{Bmatrix} dx \tag{a$_3$}$$

This will be written as

$$\mathbf{dv} = \mathbf{B}^t \boldsymbol{\epsilon}_x \, dx \tag{a$_4$}$$

where*

$$\mathbf{B} = \begin{bmatrix} 1 & 0 & 0 \\ 0 & x & 1 \\ 0 & 1 & 0 \end{bmatrix} \tag{a$_5$}$$

The second step is to write the statical relations giving the forces $\mathbf{S}_x$ in terms of the forcs $\mathbf{S}$. From Fig. 5.4 (b) we directly obtain,

$$\mathbf{S}_x = \begin{Bmatrix} S_x \\ M_x \\ Q_x \end{Bmatrix} = \begin{bmatrix} 1 & 0 & 0 \\ 0 & x & 1 \\ 0 & 1 & 0 \end{bmatrix} \begin{Bmatrix} \overline{P_2} \\ \overline{Q_2} \\ \overline{M_2} \end{Bmatrix} = \mathbf{B}\,\mathbf{S} \tag{a$_6$}$$

Take note again of the contragredient relationship between these statical relations and the kinematical relation (a$_4$).

The third step is to consider the elasticity relations (5.12) for the beam, to rewrite

$$\boldsymbol{\epsilon}_x = \begin{Bmatrix} \epsilon_x \\ \chi_x \\ \gamma_x \end{Bmatrix} = \begin{bmatrix} 1/AE & 0 & 0 \\ 0 & 1/EI & \\ 0 & 0 & 1/A^*G \end{bmatrix} \begin{Bmatrix} S_x \\ M_x \\ Q_x \end{Bmatrix} = \mathbf{f}_x\,\mathbf{S}_x \tag{a$_7$}$$

Equations (a$_4$), (a$_6$) and (a$_7$) determine the required flexibility. Therefore

$$d\mathbf{v} = \mathbf{B}^t\,\mathbf{f}_x\,\mathbf{S}_x\,dx = (\mathbf{B}^t\,\mathbf{f}_x\,\mathbf{B}\,dx)\,\mathbf{S} \tag{a$_8$}$$

Thus the total displacements obtained on integration over the length of the beam are,

$$\mathbf{v} = [\int_L (\mathbf{B}^t\,\mathbf{f}_x\,\mathbf{B})\,dx]\,\mathbf{S} = \mathbf{f}\,\mathbf{S}$$

Thus the flexibility matrix is

$$\mathbf{f} = \int \mathbf{B}^t\,\mathbf{f}_x\,\mathbf{B}\,dx \tag{5.18}$$

This is a general form expressing the physical feature that the flexibility of the member is the sum of the flexibilities of the differential elements of the beam. The integration extends over the total length of the beam. The matrices $\mathbf{B}$ and $\mathbf{f}_x$ are functions of x and the integral can thus be evaluated either in a closed form, or numerically in the form

$$\mathbf{f} = \sum_{i=1}^{n} (\mathbf{B}_i^t\,\mathbf{f}_{xi}\,\mathbf{B}_i)\,\Delta x_i \tag{5.19}$$

We will be considering the details later in Sec. 5.11.

* $\mathbf{B}$ is seen to be symmetrical; this is only casual. It can in general be unsymmetric and even rectangular. See Eq. (a$_{16}$) and Eq. 5.28.

Consider now the second method namely the unit load method or the use of virtual work principle. This avoids the geometrical considerations which gave Eq. (a_3). Imagine that the deformation shown in Fig. 5.4 (a) are a set of virtual deformation occuring on the force system shown in Fig. 5.2 (a). The external work is $\mathbf{S}^t\mathbf{dv}$ and the internal work is $\mathbf{S}_x\, \epsilon_x\, dx$. Note that for the member, the end forces are the extetnal forces and **dv** are the corresponding displacements. Therefore the equation of the virtual work is,*

$$\mathbf{S}^t\,\mathbf{dv} = \mathbf{S}_x{}^t\, \epsilon_x\, dx \tag{a_9}$$

Use now the conditions of statics, Eq. (a_6), and obtain,

$$\mathbf{S}^t\,\mathbf{dv} = (\mathbf{BS})^t\, \epsilon_x\, dx$$

Cancelling the forces $\mathbf{S}^t$ from both sides we directly obtain (a_4). Thus the geometrical consideration is avoided. Observe that what we essentially need is the matrix **B**.

Features of the Flexibility Matrix

Consider now the features of the flexibility matrix given by Eq. (5.18). Writing this in full

$$\mathbf{f} = \int_0^L \begin{bmatrix} 1 & 0 & 0 \\ 0 & x & 1 \\ 0 & 1 & 0 \end{bmatrix} \begin{bmatrix} 1/AE & 0 & 0 \\ 0 & 1/EI & 0 \\ 0 & 0 & 1/A^*G \end{bmatrix} \begin{bmatrix} 1 & 0 & 0 \\ 0 & x & 1 \\ 0 & 1 & 0 \end{bmatrix} dx$$

$$= \int_0^L \begin{bmatrix} \dfrac{1}{AE} & 0 & 0 \\ 0 & \dfrac{x^2}{EI} + \dfrac{1}{A^*G} & \dfrac{x}{EI} \\ 0 & \dfrac{x}{EI} & \dfrac{1}{EI} \end{bmatrix} dx \tag{a_{10}}$$

The coefficients can be evaluated if the variation of the rigidities *EI*, *AE* etc. are known. For a beam of ***uniform*** section we can readily integrate these. Thus the flexibility relation for a cantilever of uniform section are given by

* **Recall Eq. (4,3)**

$$\mathbf{v} = \begin{Bmatrix} e \\ v \\ \varphi \end{Bmatrix} = \begin{bmatrix} \frac{L}{AE} & 0 & 0 \\ 0 & \frac{L^3}{3EI}+\frac{L}{A^*G} & \frac{L^2}{2EI} \\ 0 & \frac{L^2}{2EI} & \frac{L}{EI} \end{bmatrix} \begin{Bmatrix} \overline{P}_2 \\ \overline{Q}_2 \\ \overline{M}_2 \end{Bmatrix} = \mathbf{f}\,\mathbf{S} \qquad (5.20)$$

The elements of the matrix are easily verified for this case using area moment theorem and noting that the term $(L/A^*G)\ Q_2$ represents the transverse displacement at end 2 due to the shearing deformation. The term Q_2/A^*G represents the shearing strain, which being constant throughout gives the above contribution to the transverse displacement.

We shall now, interpret the flexibility matrix given by Eq. (5.18) in general terms. The statical matrix **B** of Eq. (a_6) in general may be written as

$$\mathbf{S}_x = \begin{Bmatrix} S_x \\ M_x \\ Q_x \end{Bmatrix} = \begin{bmatrix} s_1 & s_2 & s_3 \\ m_1 & m_2 & m_3 \\ q_1 & q_2 & q_3 \end{bmatrix} \begin{Bmatrix} \overline{P}_2 \\ \overline{Q}_2 \\ \overline{M}_2 \end{Bmatrix} = \mathbf{B}\,\mathbf{S} \qquad (5.21)$$

Each column represents the internal forces due to corresponding unit loads applied at end 2. In terms of these the matrix product of Eq. (5.18) can be written as follows :

$$\mathbf{f} = \int_0^L \begin{bmatrix} m_1^2 & m_1m_2 & m_1m_3 \\ m_2m_1 & m_2^2 & m_2m_3 \\ m_3m_1 & m_3m_2 & m_3^2 \end{bmatrix} \frac{dx}{EI} + \int_0^L \begin{bmatrix} q_1^2 & q_1q_2 & q_1q_3 \\ q_2q_1 & q_2^2 & q_2q_3 \\ q_3q_1 & q_3q_2 & q_3^2 \end{bmatrix} \frac{dx}{A^*G}$$

$$+ \int_0^L \begin{bmatrix} s_1^2 & s_1s_2 & s_1s_3 \\ s_2s_1 & s_2^2 & s_2s_3 \\ s_3s_1 & s_3s_2 & s_3^2 \end{bmatrix} \frac{dx}{AE}. \qquad (5.22)$$

This is of the form

$$\mathbf{f} = \mathbf{f}_m + \mathbf{f}_q + \mathbf{f}_s \qquad (5.23)$$

This reflects the *superposition* of the flexibilities due to bending, shearing and axial deformations considered separately. If the bending action alone is considered, which is by far a significant proportion of the total, we could have proceeded as follows :

$$\epsilon_x = \{\chi_x\} = \frac{M_x}{EI} \ ; \ M_x = [m_1\ m_2\ m_3]\ \{\mathbf{S}\} = \mathbf{B}\,\mathbf{S} \qquad (a_{11})$$

$$\mathbf{f}_m = \int \begin{bmatrix} m_1 \\ m_2 \\ m_3 \end{bmatrix} \left(\frac{1}{EI}\right) [m_1\ m_2\ m_3]\ dx \qquad (a_{12})$$

The multiplications will then lead to the first term of Eq. (5.23).

The appearance of the familiar integrals such as m_i, m_j, dx/EI reflects the basic ideas of the unit load method for finding deflection. The symmetry of the matrix is also apparent as

$$\int m_i\, m_j \frac{dx}{EI} = \int m_j\, m_i \frac{dx}{EI}, \text{ etc.}$$

Use of Castigliano's Theorem

Many times the strain energy expression is written and the flexibility coefficients obtained by applying the Castigliano's theorem. This method however has no computational advantage. We shall consider this approach too for the purposes of seeing the relationships between the fundamental principles of structural analysis.

The elastic strain energy of the typical differential element is

$$(S_x \epsilon_x + M_x \chi_x + Q_x \gamma_x)\, dx/2$$

We write this as $\mathbf{S}_x^t \boldsymbol{\epsilon}_x dx/2$ therefore using Eq.(a_7) we write,

$$dW = \tfrac{1}{2}\, [\mathbf{S}_x^t\, \mathbf{f}_x\, \mathbf{S}_x]\, dx \qquad (a_{13})$$

The total strain energy is obtained as the integral of this over the length of the beam. Thus

$$W = \int_L \tfrac{1}{2}[\mathbf{S}_x^t\, \mathbf{f}_x\, \mathbf{S}_x]\, dx \qquad (5.24)$$

Written out in full, we have the familiar expression

$$W = \int_L \tfrac{1}{2} \left(S_x^2 \frac{dx}{AE} + M_x^2 \frac{dx}{EI} + Q_x^2 \frac{dx}{A^*G} \right) dx \qquad (a_{14})$$

Substituting $\mathbf{S}_x$ from Eq. (a_6) we have

$$W = \tfrac{1}{2}\, \mathbf{S}^t \left[\int_L \mathbf{B}^t\, \mathbf{f}_x\, \mathbf{B}\, dx \right] \mathbf{S}$$

Let us write this as

$$W = \tfrac{1}{2}\, \mathbf{S}^t\, \mathbf{f}'\, \mathbf{S} \qquad (a_{15})$$

At this stage we still do not know that the matrix $\mathbf{f}'$ is the flexibility matrix. It will be shown now that it is actually the required flexibility matrix. Let us write the above products in full as follows :

$$W = \tfrac{1}{2}\, [\bar{P}_2, \bar{Q}_2, \bar{M}_2] \begin{bmatrix} f'_{11} & f'_{12} & f'_{13} \\ f'_{21} & f'_{22} & f'_{23} \\ f'_{31} & f'_{32} & f'_{33} \end{bmatrix} \begin{Bmatrix} P_2 \\ \bar{Q}_2 \\ \bar{M}_2 \end{Bmatrix} \qquad (b_1)$$

After carrying out the multiplication we obtain

$$\mathbf{W} = \tfrac{1}{2}[\, \overline{P}_2\, (\overline{P}_2 f'_{11} + \overline{Q}_2 f'_{12} + \overline{M}_2 f'_{13}) + \overline{Q}_2\, (\overline{P}_2 f'_{21} + \overline{Q}_2 f'_{22} + \overline{M}_2 f'_{23}) + \text{etc}]$$

The Castigliano theorem requires

$$\frac{\partial \mathbf{W}}{\partial \overline{P}_2} = e \,; \quad \frac{\partial \mathbf{W}}{\partial \overline{Q}_2} = v \,; \quad \frac{\partial \mathbf{W}}{\partial \overline{M}_2} = \varphi \qquad (b_2)$$

Before performing the differentiation we can introduce the fact that the flexibility matrix **f'** is symmetric. Thus on differentiation and inclusion of the relation that $f'_{12} = f'_{21}$

$$\begin{aligned} e &= \overline{P}_2 f'_{11} + \overline{Q}_2 f'_{12} + \overline{M}_2 f'_{13}, \\ v &= \overline{P}_2 f'_{21} + \overline{Q}_2 f'_{22} + \overline{M}_2 f'_{23}, \\ \varphi &= \overline{P}_2 f'_{31} + \overline{Q}_2 f'_{32} + \overline{M}_2 f'_{33}, \end{aligned} \qquad (b_3)$$

which in matrix form lead to the relations $\mathbf{v} = \mathbf{f}'\,\mathbf{s}$. Thus the matrix **f'** is the required flexibility matrix.

Observe the fact that the flexibility matrix is determined in terms of two matrices **B** and $\mathbf{f}_x$ only. The Castigliano's theorem does not give any computational advantage. It affords another method of derivation.

b. BEAM MODEL

We shall evaluate the basic flexibility of beam member taking the forces and displacements as in Fig. 5.1 for the beam model.

The equations of equilibrium are,

$$\mathbf{S}_x = \begin{Bmatrix} S_x \\ M_x \\ Q_x \end{Bmatrix} = \begin{bmatrix} 1 & 0 & 0 \\ 0 & -\left(1-\dfrac{x}{L}\right) & \dfrac{x}{L} \\ 0 & -\dfrac{1}{L} & -\dfrac{1}{L} \end{bmatrix} \begin{Bmatrix} S \\ \overline{M}_1 \\ \overline{M}_2 \end{Bmatrix} = \mathbf{B}\,\mathbf{S} \qquad (a_{16})$$

The element flexibility matrix $\mathbf{f}_x$ is the same as before, (Eq. a_7). Therefore the flexibility matrix is $\int \mathbf{B}^t\, \mathbf{f}_x\, \mathbf{B}\, dx$ with **B** as given by the above equation. The general result of the matrix multiplication is general terms has been given in Eqs. (5.21, 5.22 , 5.23). In the present case

$$\begin{aligned} s_1 &= 1 \,; \; s_2 = s_3 = 0 \\ q_1 &= 0 \,; \; q_2 = q_3 = -1/L \\ m_1 &= 0 \,; \; m_2 = -\left(1-\frac{x}{L}\right) ; \; m_3 = \frac{x}{L} \end{aligned} \qquad (a_{17})$$

We have therefore the flexibility matrix

$$f = \int \begin{bmatrix} 0 & 0 & 0 \\ 0 & m_2^2 & m_2m_3 \\ 0 & m_3m_2 & m_3^2 \end{bmatrix} \frac{dx}{EI} + \int \begin{bmatrix} s_1^2 & 0 & 0 \\ 0 & 0 & 0 \\ 0 & 0 & 0 \end{bmatrix} \frac{dx}{AE}$$

$$+ \int \begin{bmatrix} 0 & 0 & 0 \\ 0 & q_2^2 & q_2q_3 \\ 0 & q_3q_2 & q_3^2 \end{bmatrix} \frac{dx}{A^*G} \qquad (a_{18})$$

For uniform beam,

$$\int m_2^2 \left(\frac{dx}{EI}\right) = \int \left(1 - \frac{x}{L}\right)^2 \frac{dx}{EI} = \frac{L}{3EI}$$

$$\int m_3^2 \frac{dx}{EI} = \frac{L}{3EI}\ ; \int m_2m_3 \frac{dx}{EI} = -\frac{L}{6EI} \qquad (a_{19})$$

$$\int \frac{s_1^2\, dx}{AE} = \frac{L}{AE}$$

Each of the integrals involving q is

$$\int \frac{1}{L^2 A^*G}\, dx = \frac{1}{LA^*G} \qquad (a_{20})$$

Thus the flexibility matrix is

$$\mathbf{f} = \begin{bmatrix} \dfrac{L}{AE} & 0 & 0 \\ 0 & \dfrac{L}{3EI}(1+\beta) & -\dfrac{L}{6EI}(1-2\beta) \\ 0 & -\dfrac{L}{6EI}(1-2\beta) & \dfrac{L}{3EI}(1+\beta) \end{bmatrix} \qquad (5.25)$$

with

$$\beta = \frac{3EI}{L^2A^*G} \qquad (5.26)$$

Observe that the factor β includes the effect of shearing deformations. When this effect is ignored we have $\beta = 0$.

c. Plane Grid Member

To derive the basic flexibility for the member of a transversely loaded grid, we have to include the torsional rigidity of the member as given in Eq. (5.11). We will now neglect the effect of shearing deformations. Thus the element flexibility $\mathbf{f}_x$ is taken as follows. (See Fig. 5.5 d.)

$$\boldsymbol{\epsilon}_x = \begin{Bmatrix} \dfrac{d\psi}{dx} \\ \dfrac{d\theta}{dx} \end{Bmatrix} = \begin{bmatrix} 1/GJ & 0 \\ 0 & 1/EI \end{bmatrix} \begin{Bmatrix} T \\ M \end{Bmatrix} = \mathbf{f}_x \mathbf{S}_x \qquad (5.27)$$

We will now derive the basic flexibility of the cantilever model. We will relate the displacements at end 2 to the forces thereat (see Fig 5.5 a,b).

$$\mathbf{v} = \begin{Bmatrix} \bar{\theta}_{x2} \\ \bar{\theta}_{y2} \\ \bar{W}_2 \end{Bmatrix} = \begin{bmatrix} f_{11} & f_{12} & f_{13} \\ f_{21} & f_{22} & f_{23} \\ f_{31} & f_{32} & f_{33} \end{bmatrix} \begin{Bmatrix} \bar{M}_{x2} \\ \bar{M}_{y2} \\ \bar{Q}_2 \end{Bmatrix} = \mathbf{f}\,\mathbf{S} \qquad (a_{21})$$

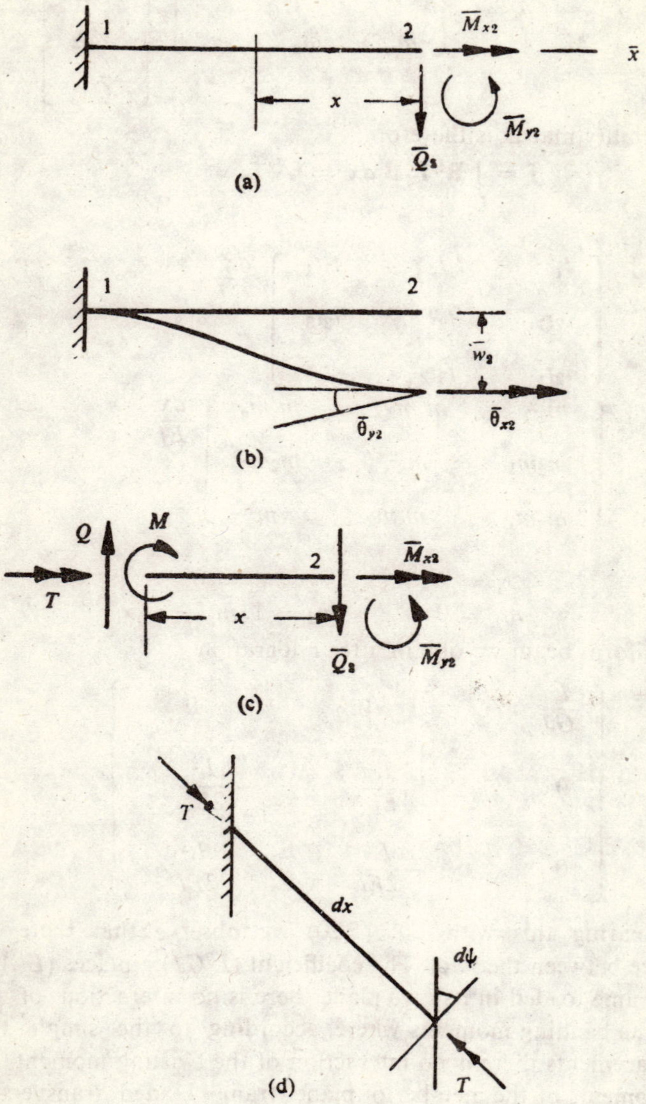

Fig. 5.5

We need to derive the statical matrix relating $\mathbf{S}_x$ to $\mathbf{S}$. From Fig. 5.5(c)

$$\mathbf{S}_x = \begin{Bmatrix} T \\ M \end{Bmatrix} = \begin{bmatrix} -1 & 0 & 0 \\ 0 & 1 & -x \end{bmatrix} \begin{Bmatrix} \overline{M}_{x2} \\ \overline{M}_{y2} \\ \overline{Q}_2 \end{Bmatrix} = \mathbf{B}\,\mathbf{S} \tag{5.28}$$

Let us rewrite this as

$$\mathbf{S}_x = \begin{Bmatrix} T \\ M \end{Bmatrix} = \begin{bmatrix} t_1 & t_2 & t_3 \\ m_1 & m_2 & m_3 \end{bmatrix} \begin{Bmatrix} \overline{M}_{x2} \\ \overline{M}_{y2} \\ \overline{Q}_2 \end{Bmatrix} = \mathbf{B}\,\mathbf{S} \tag{5.29}$$

The flexibility matrix is therefore

$$\mathbf{f} = \int_L \mathbf{B}^t\, \mathbf{f}_x\, \mathbf{B}\, dx = \mathbf{f}_t + \mathbf{f}_m \tag{a$_{23}$}$$

Where,

$$\mathbf{f}_t = \int \begin{bmatrix} t_1^2 & t_1t_2 & t_1t_3 \\ t_2t_1 & t_2^2 & t_2t_3 \\ t_3t_1 & t_3t_2 & t_3^2 \end{bmatrix} \frac{dx}{GJ} \tag{a$_{24}$}$$

$$\mathbf{f}_m = \int \begin{bmatrix} m_1^2 & m_1m_2 & m_1m_3 \\ m_2m_1 & m_2^2 & m_2m_3 \\ m_3m_1 & m_3m_2 & m_3^2 \end{bmatrix} \frac{dx}{EI} \tag{a$_{25}$}$$

Here

$$\begin{aligned} t_1 &= -1\,; \qquad t_2 = t_3 = 0 \\ m_1 &= 0\,; \qquad m_2 = 1,\ m_3 = -x \end{aligned} \tag{a$_{26}$}$$

For a uniform beam we obtain after integration

$$\mathbf{f} = \begin{bmatrix} \dfrac{L}{GJ} & 0 & 0 \\ 0 & \dfrac{L}{EI} & -\dfrac{L^2}{2EI} \\ 0 & -\dfrac{L^2}{2EI} & \dfrac{L^3}{3EI} \end{bmatrix} \tag{5.30}$$

By comparing this with Eq. (5.20) we observe that there is a close resemblance between the two. The coefficient (L/GJ) replaces (L/AE) term. As in the frame loaded in its own plane there is no interaction of the axial load with the bending moments where, according to the simple theory of small displacements there is no interaction of the twisting moment with the bending moments of the member of plane frame loaded transversely. The flexibility coefficients corresponding to the bending of the member are the same in each case. (Observe the appearance of the minus signs in f_{23} and f_{32} which is a result of conventions adopted for the forces and deformations.) This analogy is used while drawing computer codes for plane frames under in plane as well as transverse loads.

5.5 FLEXIBILITY TRANSFORMATIONS

We shall now consider some basic ideas about the transformation operations on flexibility matrices. These transformations are frequently needed in structural analysis.

Coordinate Transformation

To begin with consider the coordinate transformation. Let the forces and displacements be measured in frame coordinates which have the following relations with the corresponding quantities measured in member coordinates :

$$\mathbf{P}_1 = \mathbf{C}_1 \mathbf{S} = \mathbf{T} \overline{\mathbf{C}}_1 \mathbf{S}; \mathbf{P}_2 = \mathbf{C}_2 \mathbf{S} = \mathbf{T} \overline{\mathbf{C}}_2 \mathbf{S}$$

$$\mathbf{V} = \mathbf{A}_1 \mathbf{u}_1 + \mathbf{A}_2 \mathbf{u}_2 = \mathbf{C}^t{}_1 \mathbf{u}_1 + \mathbf{C}^t{}_2 \mathbf{u}_2 \qquad (a_1)$$

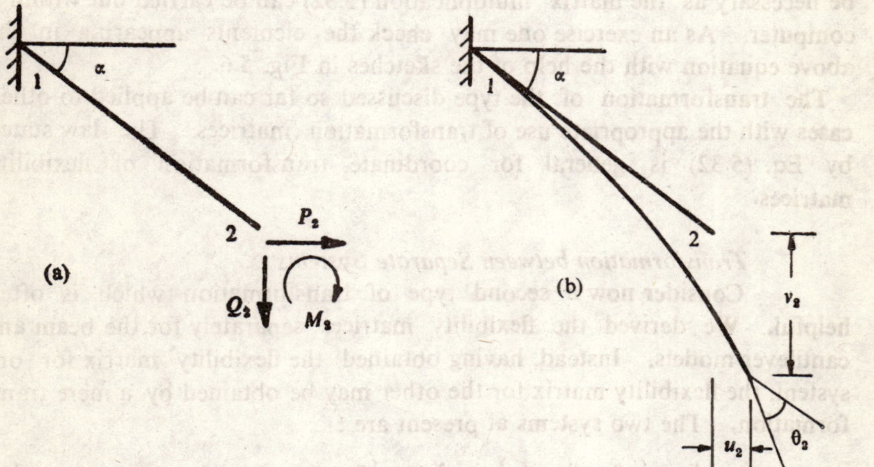

Fig. 5.6

Figure 5.6 shows the forces and displacements for the cantilever model. Let the flexibility relation between the displacements $\mathbf{u}_2$ and $\mathbf{P}_2$ be

$$\mathbf{u}_2 = \mathbf{f}' \mathbf{P}_2 \qquad (5.31)$$

This can be derived by transforming the relation $\mathbf{v} = \bar{\mathbf{f}}\, \mathbf{s}$. For the cantilever model* we have $\mathbf{u}_1 = 0$ and $\overline{\mathbf{C}}_2 = \mathbf{I}$ since by definition $\overline{\mathbf{P}}_2 \equiv \mathbf{S}$.

* See Eq. (a_9), Sec. 3.1.

Therefore from Eq. (a_1) we have

$$\mathbf{P}_2 = \mathbf{T}\,\mathbf{S}\ ;\quad \mathbf{v} = \mathbf{T}^t\,\mathbf{u}_2$$

Since T is orthogonal we have

$$\mathbf{S} = \mathbf{T}^t\,\mathbf{P}_2;\ \mathbf{u}_2 = \mathbf{T}\,\mathbf{v}$$

Therefore premultiplying the relation $\mathbf{v} = \mathbf{f}\,\mathbf{s}$ by $\mathbf{T}$ we obtain

$$\mathbf{u}_2 = \mathbf{T}\,\mathbf{v} = \mathbf{T}\,\bar{\mathbf{f}}\,\mathbf{s} = (\mathbf{T}\,\bar{\mathbf{f}}\,\mathbf{T}^t)\,\mathbf{P}_2 = \mathbf{f}'\,\mathbf{P}_2 \qquad (a_2)$$

Therefore we have the result

$$\mathbf{f}' = \mathbf{T}\,\mathbf{f}\,\mathbf{T}^t \qquad (5.32)$$

This matrix product written in full reads

$$\mathbf{f}' = \begin{bmatrix} c^2 f_{11} + s^2 f_{22} & cs\,(f_{22} - f_{11}) & 0 \\ cs\,(f_{22} - f_{11}) & s^2 f_{11} + c^2 f_{22} & cf_{23} \\ 0 & cf_{32} & f_{33} \end{bmatrix} \qquad (a_3)$$

Here c and s stand for sin α and cos α, and fij are the ones appearing Eq. (5.20). One may check that when $\alpha = 0$ the above matrix coincides with the f matrix. The derivation of the individual elements as above may not be necessary as the matrix multiplication (5.32) can be carried out within a computer. As an exercise one may check the elements appearing in the above equation with the help of the sketches in Fig. 5.6.

The transformation of the type discussed so far can be applied to other cases with the appropriate use of transformation matrices. The law stated by Eq. (5.32) is general for coordinate transformation of flexibility matrices.

Transformation between Separate Systems

Consider now a second type of transformation which is often helpful. We derived the flexibility matrices separately for the beam and cantilever models. Instead, having obtained the flexibility matrix for one system, the flexibility matrix for the other may be obtained by a mere transformation. The two systems at present are :

$$\mathbf{v}_c = \{\, e,\ \ v,\ \ \varphi \,\}\ ;\quad \mathbf{v}_b = \{e,\ \ \varphi_1,\ \ \varphi_2\,\} \qquad (a_4)$$

$$\mathbf{S}_c = \left\{ \overline{P_2},\ \overline{Q_2},\ \overline{M_2} \right\}\ ;\quad \mathbf{S}_b = \left\{\ S,\ \overline{M_1},\ \overline{M_2}\ \right\} \qquad (a_5)$$

The subscripts b and c denote the beam and cantilever systems respectively. The relation between the forces are

$$\mathbf{S}_b = \begin{Bmatrix} S \\ M_1 \\ M_2 \end{Bmatrix} = \begin{bmatrix} 1 & 0 & 0 \\ 0 & -L & -1 \\ 0 & 0 & 1 \end{bmatrix} \begin{Bmatrix} \overline{P_2} \\ \overline{Q_2} \\ \overline{M_2} \end{Bmatrix} = \mathbf{R}\,\mathbf{S}_c \qquad (a_6)$$

The relations between the corresponding displacements can be either directly written or obtained by virtual work principle. It pays to do it both ways: directly, we have (see Fig. 5.7)

$$e_c = e_b\,; \quad v = -L\varphi_1; \quad \varphi = \varphi_2 - \varphi_1 \tag{a_7}$$

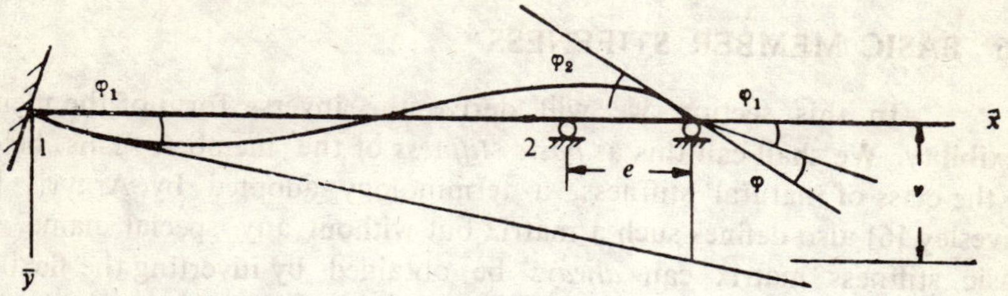

Fig. 5.7

These in matrix notation are,

$$\mathbf{v}_c = \begin{Bmatrix} e \\ v \\ \varphi \end{Bmatrix} = \begin{bmatrix} 1 & 0 & 0 \\ 0 & -L & 0 \\ 0 & -1 & 1 \end{bmatrix} \begin{Bmatrix} e \\ \varphi_1 \\ \varphi_2 \end{Bmatrix} = \mathbf{R}^t\,\mathbf{v}_b \tag{a_8}$$

It is thus seen that matrix relating the displacements is the transpose of the corresponding statical matrix. This is another instance of the contragradience.

We shall now derive Eq. (a_8) by the application of principle of virtual work. The systems of beam and cantilever models are two ways of defining the same set of end forces of a beam in equilibrium and a deformed shape of the beam (see Fig. 5.1). If this deformed shape is assumed as the virtual deformation to occur, then taking cantilever model, the equation of work is

$$W_i = \mathbf{S}_c^t\,\mathbf{v}_c \tag{a_9}$$

and with the beam model

$$W_i = \mathbf{S}_b^t\,\mathbf{v}_b \tag{a_{10}}$$

Since the deformed shape is the same, the internal work is the same in both the cases, giving

$$\mathbf{S}_c^t\,\mathbf{v}_c = \mathbf{S}_b^t\,\mathbf{v}_b \tag{a_{11}}$$

Using Eq. (a_6) for S_b,

$$\mathbf{S}_c^t\,\mathbf{v}_c = (\mathbf{R}\,\mathbf{S}_c)^t\,\mathbf{v}_b \tag{a_{12}}$$

which again leads to Eq. (a_8)

Consider now the flexibility relations

$$\mathbf{v}_b = \mathbf{f}_b\,\mathbf{S}_b \tag{a_{13}}$$

Eqs. (a_8 and a_6) then give

$$\mathbf{v}_c = \mathbf{R}^t\,\mathbf{f}_b\,\mathbf{S}_b = (\mathbf{R}^t\,\mathbf{f}_b\,\mathbf{R})\,\mathbf{S}_c$$

Thus

$$\mathbf{f}_c = \mathbf{R}^t\,\mathbf{f}_b\,\mathbf{R} \tag{5.33}$$

This is the required transformation to obtain the flexibility of the cantilever model from the flexbility $\mathbf{f}_b$ of the beam model and the transformation matrix $\mathbf{R}$ for the forces of the two systems. Alternatively we can obtain $\mathbf{f}_b$ from $\mathbf{f}_c$. This may be taken as an exercise.

5.6 BASIC MEMBER STIFFNESS

In this section we will derive the inverse form of the member flexibility. We shall call this as *basic stiffness* of the member. This belongs to the class of 'natural' stiffness, a terminology adopted by Argyris [17]. Livesley [6] also defines such a matrix but without any special name. The basic stiffness matrix can *always* be obtained by inverting the flexibility matrix. This, as we will see later is the fundamental way of obtaining stiffness ; for instance, when we encounter beams of variable section we cannot solve for the stiffness exactly since the governing differential equation does not possess any convenient solution. Further this is the only approach available in constructing stiffnesses of members while dealing with the situations of elastoplastic behaviour, beam-column interaction, creep effects, or cracking of concrete. Thus the inverses of matrices appearing in Eqs. (5.20) and (5.25) will lead to the corresponding stiffness matrices for the cantilever and the beam model respectively. These are as given below:

Cantilever Model

$$\mathbf{S} = \begin{Bmatrix} \bar{P}_2 \\ \bar{Q}_2 \\ \bar{M}_2 \end{Bmatrix} = \begin{bmatrix} \dfrac{AE}{L} & 0 & 0 \\ 0 & \dfrac{12EI}{L^3}\,\dfrac{1}{1+4\beta} & -\dfrac{6EI}{L^2}\,\dfrac{1}{1+4\beta} \\ 0 & -\dfrac{6EI}{L^2}\,\dfrac{1}{1+4\beta} & \dfrac{4EI}{2}\,\dfrac{1+\beta}{1+4\beta} \end{bmatrix} \begin{Bmatrix} e \\ v \\ \varphi \end{Bmatrix} = \mathbf{k}\,\mathbf{v} \tag{5.34}$$

Beam Model

$$\mathbf{S} = \begin{Bmatrix} S \\ M_1 \\ M_2 \end{Bmatrix} = \begin{bmatrix} \dfrac{AE}{L} & 0 & 0 \\ 0 & \dfrac{4EI}{L}\,\dfrac{1+\beta}{1+4\beta} & \dfrac{2EI}{L}\,\dfrac{1-2\beta}{1+4\beta} \\ & \dfrac{2EI}{L}\,\dfrac{1-2\beta}{1+4\beta} & \dfrac{4EI}{L}\,\dfrac{1+\beta}{1+4\beta} \end{bmatrix} \begin{Bmatrix} e \\ \varphi_1 \\ \varphi_2 \end{Bmatrix} = \mathbf{k}\,\mathbf{v} \tag{5.35}$$

The transformations indicated for the flexibility hold for stiffness also. Thus similar to Eq. (5.32) we have,

$$\mathbf{k}' = \mathbf{T}\,\mathbf{k}\,\mathbf{T}^t \tag{5.36}$$

which is the transformation with respect to coordinates. The general transformation similar to the one represented in Eq. (5.33) also holds.

5.7 GENERAL MEMBER STIFFNESS

The *general stiffness** relations for a member relate the member end forces $\bar{\mathbf{P}}_1$ and $\bar{\mathbf{P}}_2$ in terms of its end displacements $\bar{\mathbf{u}}_1$ and $\bar{\mathbf{u}}_2$. These are expressed in the following notations

$$\bar{\mathbf{P}} = \begin{Bmatrix} \bar{\mathbf{P}}_1 \\ \hline \bar{\mathbf{P}}_2 \end{Bmatrix} = \left[\begin{array}{c|c} \bar{\mathbf{K}}_{11} & \bar{\mathbf{K}}_{12} \\ \hline \bar{\mathbf{K}}_{21} & \bar{\mathbf{K}}_{22} \end{array}\right] \begin{Bmatrix} \bar{\mathbf{u}}_1 \\ \hline \bar{\mathbf{u}}_2 \end{Bmatrix} = \bar{\mathbf{K}}\,\bar{\mathbf{u}} \tag{5.37}$$

The sizes of the vectors and the submatrices will depend on the member in question. However we can, without entering into details, discuss the general nature of these relations.

First, observe that $\bar{\mathbf{K}}_{11}$ and $\bar{\mathbf{K}}_{21}$ are the forces at the ends of the member due to $\bar{\mathbf{u}}_1$ only, and similarly $\bar{\mathbf{K}}_{12}$ and $\bar{\mathbf{K}}_{22}$ are those due to $\bar{\mathbf{u}}_2$ only. Second, on account of symmetry we always have

$$\bar{\mathbf{K}}_{21} = \bar{\mathbf{K}}_{12}{}^{t} \tag{5.38}$$

Third, the stiffness matrix is singular. In other words it cannot be inverted. This also means physically that the displacements $\bar{\mathbf{u}}_1$ and $\bar{\mathbf{u}}_2$ cannot be solved in terms of the end loads $\bar{\mathbf{P}}_1$ and $\bar{\mathbf{P}}_2$. This is true because the displacements $\bar{\mathbf{u}}_1$ and $\bar{\mathbf{u}}_2$ include possible rigid body movements which can be arbitrary. However observe that the basic matrix $\mathbf{k}$ of Eq. (5.34) or (5.35) is nonsingular as the beam is prevented from rigid body motion. Observe that it is only in such structures that we can evaluate displacements.

General Stiffness by Transformation

We can always derive the general stiffness from the basic stiffness by mere transformations though direct methods are available. This procedure which is simpler seems to have been well explained in reference [6], while most texts seem to adopt a direct procedure. The forces $\bar{\mathbf{P}}_1$ and $\bar{\mathbf{P}}_2$ are related to $\mathbf{S}$ by equilibrium relations (2.6) :

$$\bar{\mathbf{P}} = \begin{Bmatrix} \mathbf{P}_1 \\ \mathbf{P}_2 \end{Bmatrix} = \begin{bmatrix} \bar{\mathbf{C}}_1 \\ \bar{\mathbf{C}}_2 \end{bmatrix} \mathbf{S} \tag{a$_1$}$$

and the displacements $\bar{\mathbf{u}}_1$ and $\bar{\mathbf{u}}_2$ by the kinematic relations (3.5) :

$$\mathbf{v} = \bar{\mathbf{A}}\,\bar{\mathbf{u}} = \bar{\mathbf{A}}_1\,\bar{\mathbf{u}}_1 + \bar{\mathbf{A}}_2\,\bar{\mathbf{u}}_2 \tag{a$_2$}$$

*These can be alternatively termed 'direct stiffness' relations. They are used in 'direct stiffness method'. See Sec. 8.2.

The contragredient property gives

$$\overline{C}_1 = \overline{A}_1^t \; ; \; \overline{C}_2 = \overline{A}_2^t \; ; \; \overline{C} = \overline{A}^t \tag{a$_3$}$$

Therefore using the basic stiffness $\mathbf{S} = \mathbf{k\,v}$ and the above relations,

$$\overline{\mathbf{P}} = \overline{\mathbf{A}}^t\,\mathbf{S} = \overline{\mathbf{A}}^t\,\mathbf{k\,v} = (\overline{\mathbf{A}}^t\,\mathbf{k}\,\overline{\mathbf{A}})\,u \tag{a$_4$}$$

Thus we obtain

$$\overline{\mathbf{K}} = \overline{\mathbf{A}}^t\,\mathbf{k}\,\overline{\mathbf{A}} \tag{5.39}$$

And in terms of submatrices,

$$\overline{\mathbf{K}} = \begin{Bmatrix} \overline{\mathbf{A}}^t_1 \\ \overline{\mathbf{A}}_2^t \end{Bmatrix} \mathbf{k}\,[\mathbf{A}_1\;\mathbf{A}_2]$$

$$= \begin{bmatrix} \overline{\mathbf{A}}^t_1\,\mathbf{k}\,\overline{\mathbf{A}}_1 & \overline{\mathbf{A}}_1^t\,\mathbf{k}\,\overline{\mathbf{A}}_2 \\ \overline{\mathbf{A}}_2^t\,\mathbf{k}\,\overline{\mathbf{A}}_1 & \overline{\mathbf{A}}_2^t\,\mathbf{k}\,\overline{\mathbf{A}}_2 \end{bmatrix} = \begin{bmatrix} \overline{\mathbf{K}}_{11} & \overline{\mathbf{K}}_{12} \\ \overline{\mathbf{K}}_{21} & \overline{\mathbf{K}}_{22} \end{bmatrix} \tag{5.40}$$

Thus we can wrtte the submatrices, with $i = j = 1, 2$

$$\overline{\mathbf{K}}_{ij} = \overline{\mathbf{A}}_i^t\,\mathbf{k}\,\overline{\mathbf{A}}_j \tag{5.41}$$

Observe further that the suiffness relations (5.37) can also be written in frame coordinates. In such a case

$$\mathbf{P} = \begin{Bmatrix} \mathbf{P}_1 \\ \mathbf{P}_2 \end{Bmatrix} = \begin{bmatrix} \mathbf{K}_{11} & \mathbf{K}_{12} \\ \mathbf{K}_{21} & \mathbf{K}_{22} \end{bmatrix} \begin{Bmatrix} \mathbf{u}_1 \\ \mathbf{u}_2 \end{Bmatrix} = \mathbf{K\,u} \tag{5.42}$$

Proceeding as before

$$\mathbf{K}_{ij} = \mathbf{A}^t_i\,\mathbf{k}\,\mathbf{A}_j \tag{5.43}$$

Noting that $\mathbf{A}_i = \overline{\mathbf{A}}_i\;\;\mathbf{T}^t$ (Eq. 3.8), we can write

$$\mathbf{K}_{ij} = (\overline{\mathbf{A}}_i\,\mathbf{T})^t\,\mathbf{k}\,(\mathbf{A}_j\,\mathbf{T}^t) = \mathbf{T}\overline{\mathbf{K}}_{ij}\,\mathbf{T}^t \tag{5.44}$$

It is thus seen that the stiffness matrix $\mathbf{K}_{ij}$ too transforms similar to the flexibility (see Eq. 5.32).

To make the above transformation it will be more convenient to adopt the cantilever model and use basic stiffness given in Eq. (5.34). We have from Eq. (3.10)

$$\overline{\mathbf{A}}_1 = \overline{\mathbf{C}}_1^t = \begin{bmatrix} -1 & 0 & 0 \\ 0 & -1 & -L \\ 0 & 0 & -1 \end{bmatrix} ; \overline{\mathbf{A}}_2 \equiv \mathbf{I} \tag{a$_5$}$$

By carrying out the multiplications we obtain the general stiffness of a plane frame member of uniform section as follows :

$$\begin{Bmatrix} \bar{P}_1 \\ \bar{Q}_1 \\ \bar{M}_1 \\ \hdashline \bar{P}_2 \\ \bar{Q}_2 \\ \bar{M}_2 \end{Bmatrix} = \left[\begin{array}{ccc:ccc} \frac{AE}{L} & 0 & 0 & -\frac{AE}{L} & 0 & 0 \\ 0 & \frac{12EI}{L^3} & \frac{6EI}{L^2} & 0 & -\frac{12EI}{L^3} & \frac{6EI}{L^2} \\ 0 & \frac{6EI}{L^2} & \frac{4EI}{L} & 0 & -\frac{6EI}{L^2} & \frac{2EI}{L} \\ \hdashline -\frac{AE}{L} & 0 & 0 & \frac{AE}{L} & 0 & 0 \\ 0 & -\frac{12EI}{L^3} & -\frac{6EI}{L^2} & 0 & \frac{12EI}{L^3} & -\frac{6EI}{L^2} \\ 0 & \frac{6EI}{L^2} & \frac{2EI}{L} & 0 & -\frac{6EI}{L^2} & \frac{4EI}{L} \end{array}\right] \begin{Bmatrix} \bar{u}_1 \\ \bar{v}_1 \\ \bar{\theta}_1 \\ \hdashline \bar{u}_2 \\ \bar{v}_2 \\ \bar{\theta}_2 \end{Bmatrix} \tag{5.45}$$

Here the shearing deformation factor β has been set equal to zero. One may take it as an exercise to obtain the above matrix using the beam model and using appropriate **k**, $\mathbf{A}_1$ and $\mathbf{A}_2$ matrices (Eqs. 5.35 and 3.6).

The general stiffness matrix too is symmetrical. The second, third, fifth and the sixth equations are the slope-deflection equations. The first and the fourth equations state

$$\bar{P}_1 = -\frac{AE}{L}e; \quad \bar{P}_2 = \frac{AE}{L}e = -\bar{P}_1 \tag{a$_6$}$$

or

$$\begin{Bmatrix} \bar{P}_1 \\ \bar{P}_2 \end{Bmatrix} = \begin{bmatrix} \frac{AE}{L} & -\frac{AE}{L} \\ -\frac{AE}{L} & \frac{AE}{L} \end{bmatrix} \begin{Bmatrix} \bar{u}_1 \\ \bar{u}_2 \end{Bmatrix} \tag{a$_7$}$$

since $e = \bar{u}_2 - \bar{u}_1$

These readily show that the matrix is singular, as the determinant can be verified to be zero. In fact both the equations are identical in view of the equilibrium relation $\bar{P}_1 + \bar{P}_2 = 0$. When once the end displacements are prescribed the deformation in the member are prescribed and hence also the corresponding forces.

We will postpone considering the details of the more general form (5.42) to Chapter 8 where structural effects are considered in global coordinates.

General Stiffness—Beam of Variable Section

The method of obtaining the general stiffness from the basic stiffness described so far proves very convenient in handling beams of variable sections. We first derive the flexibility and by inverting it, the basic stiffness. Let the basic stiffness be*

$$\mathbf{k} = \begin{bmatrix} d & 0 & 0 \\ 0 & a & b \\ 0 & b & c \end{bmatrix} \tag{a$_8$}$$

Then transforming according to (5.41) leads to the following general stiffness, which the reader can verify as an exercise :

$$\mathbf{k} = \left[\begin{array}{ccc|ccc} d & 0 & 0 & -d & 0 & 0 \\ 0 & a_1 & a_2 & 0 & -a_1 & a_3 \\ 0 & a_2 & a & 0 & -a_2 & b \\ \hline -d & 0 & 0 & d & 0 & 0 \\ 0 & -a_1 & -a_2 & 0 & a_1 & -a_3 \\ 0 & a_3 & b & 0 & -a_3 & c \end{array}\right] \tag{a$_9$}$$

where

$$a_1 = \frac{a+2b+c}{L}, \quad a_2 = \frac{a+b}{L}, \quad a_3 = \frac{b+c}{L} \tag{a$_{10}$}$$

Observe that this matrix reduces to the one given in Eq. (5.45) if the beam is of uniform section.

Evaluation of Stiffness

We obtained the basic and general stiffness by a transformation of a known flexibility relation. The direct determination of the stiffness relations (either basic or general) can be achieved by adopting any of the following methods :

a. Solution of the governing differential equations

b. Methods based on assumed displacements.

The first method is common and convenient for beams of uniform section where the differential equations are easily solved. The second method is inevitably necessary in cases of beams of variable section and in two and three dimensional elements such as plate or solid elements as used in finite element method. Such elements, in structural parlance are statically indeterminate and the static matrix **B** cannot be exactly written in terms of the element end forces. Also the exact solution of the differential equa-

* See Example 5.7

tion is not possible. We will be illustrating these two general methods in subsequent sections.

5.8 STIFFNESS FROM DIFFERENTIAL EQUATION SOLUTION

In cases where the governing differential equations can be solved exactly, the stiffness matrix for such a member can always be determined by imposing the required (geometrical) boundary conditions. We will now derive the stiffness of the plane frame member in bending using this approach.

The differential equation for the beam is derived using the following basic relations :

(a) Equation of equilibrium Fig.(4.1)b with $p = q = 0$:

$$\frac{dS_x}{dx} = 0; \quad \frac{dM_x}{dx} = -Q_x; \quad \frac{dQ_x}{dx} = 0 \tag{5.46}$$

(b) Kinematic relations (Eq. 5.6) :

$$\epsilon_x = \frac{du}{dx}; \quad \chi_x = \frac{d\theta}{dx}; \quad \theta = \frac{dv}{dx} - \gamma_x \tag{5.47}$$

(c) Elasticity relations (Eq. 5.10)

$$S_x = AE\,\epsilon_x; \quad M_x = EI\,\chi_x; \quad Q_x = A^*G\,\gamma_x \tag{5.48}$$

Here we are not including the effect of shearing deformation pending its inclusion in an example later. Relations (5.47) and (5.48) together give

$$S_x = AE\frac{du}{dx}; \quad M_x = EI\frac{d}{dx}\left(\frac{dv}{dx} - 0\right) = EI\frac{d^2v}{dx^2} \tag{5.49}$$

Therefore Eqs. (5.46) give,

$$AE\frac{d^2u}{dx^2} = 0; \quad EI\frac{d^4v}{dx^4} = 0 \tag{5.50}$$

These are the governing equations of the problem.

The solution of these are

$$u = c_1 + c_2x \tag{a$_1$}$$

$$v = c_3 + c_4x + c_5x^2 + c_6x^3 \tag{a$_2$}$$

We will now be writing these in an alternative form before deriving the stiffness.

We shall write these as

$$\mathbf{u}_x = \begin{Bmatrix} u \\ v \end{Bmatrix} = \begin{bmatrix} 1 & x & 0 & 0 & 0 & 0 \\ 0 & 0 & 1 & x & x^2 & x^3 \end{bmatrix} \begin{Bmatrix} c_1 \\ c_2 \\ c_3 \\ c_4 \\ c_5 \\ c_6 \end{Bmatrix} = \mathbf{X}\,\mathbf{C} \tag{5.51}$$

The boundary conditions three at each end, will from the necessary conditions to evaluate the arbitrary constants. Let us determine the *general* stiffness. The boundary conditions for the member are,

$$\begin{aligned} &\text{At} \quad x = 0, \quad u = \bar{u}_1; \quad v = \bar{v}_1; \quad \frac{dv}{dx} = \bar{\theta}_1 \\ &\text{At} \quad x = L, \quad u = \bar{u}_2; \quad v = \bar{v}_2; \quad \frac{dv}{dx} = \bar{\theta}_2 \end{aligned} \tag{5.52}$$

Using the displacement functions given in Eq. (5.51) we can write these as follows :

$$\bar{\mathbf{u}} = \begin{Bmatrix} \bar{u}_1 \\ \bar{v}_1 \\ \bar{\theta}_1 \\ \bar{u}_2 \\ \bar{v}_2 \\ \bar{\theta}_2 \end{Bmatrix} = \left[\begin{array}{cc|cccc} 1 & 0 & . & . & . & . \\ . & . & 1 & . & . & . \\ . & . & . & 1 & . & . \\ 1 & L & . & . & . & . \\ . & . & 1 & L & L^2 & L^3 \\ . & . & . & 1 & 2L & 3L^2 \end{array}\right] \begin{Bmatrix} c_1 \\ c_2 \\ c_3 \\ c_4 \\ c_5 \\ c_6 \end{Bmatrix} = \mathbf{L}\,\mathbf{C} \tag{5.53}$$

On solution, we obtain

$$\mathbf{C} = \mathbf{L}^{-1}\bar{\mathbf{u}} \tag{5.54}$$

and on substituting back into Eq. (5.51) we obtain the displacement function in terms of $\bar{\mathbf{u}}$ as follows :

$$\mathbf{u}_x = \begin{Bmatrix} u \\ v \end{Bmatrix} = \begin{bmatrix} N_{11} & 0 & 0 & N_{14} & 0 & 0 \\ 0 & N_{22} & N_{23} & 0 & N_{25} & N_{26} \end{bmatrix} \begin{Bmatrix} \bar{u}_1 \\ \bar{v}_1 \\ \bar{\theta}_1 \\ \bar{u}_2 \\ \bar{v}_2 \\ \bar{\theta}_2 \end{Bmatrix} = \mathbf{N}\bar{\mathbf{u}} \tag{5.55}$$

Here N_{ij} are functions of x and will be called *shape functions*. These are :

$$N_{11} = 1 - \frac{x}{L}\,; \quad N_{14} = \frac{x}{L}$$

$$N_{22} = 1 - 3\left(\frac{x}{L}\right)^2 + 2\left(\frac{x}{L}\right)^3$$

$$N_{23} = L\left(\left(\frac{x}{L}\right) - 2\left(\frac{x}{L}\right)^2 + \left(\frac{x}{L}\right)^3\right) = L\left(\frac{x}{L}\right)\left(1 - \frac{x}{L}\right)^2$$

$$N_{25} = 3\left(\frac{x}{L}\right)^2 - 2\left(\frac{x}{L}\right)^3 \tag{5.56}$$

$$N_{26} = L\left(-\left(\frac{x}{L}\right)^2 + \left(\frac{x}{L}\right)^3\right)$$

Observe that the arbitrary constants are linearly trasformed to the member end displacements by applying the boundary conditions. Equation (5.54) shows the transformation. The functions 1, x, x^2, etc. appearing in Eq. (5.51) can be called *arbitrary functions,* which in this case are polynomials. The transformed functions are called *shape functions* and have very convenient physical meaning. Observe that N_{ij} gives the functional variation of the ith displacement due to unit value of the jth displacement. Figure 5.8 shows these functions. Functions N_{11} and N_{14} denote the axial displacements due to $\bar{u}_1 = 1$ alone and $\bar{u}_2 = 1$ alone, respectively. It is a linear variation. The shape functions due to bending are seen to be *cubic* functions in general.

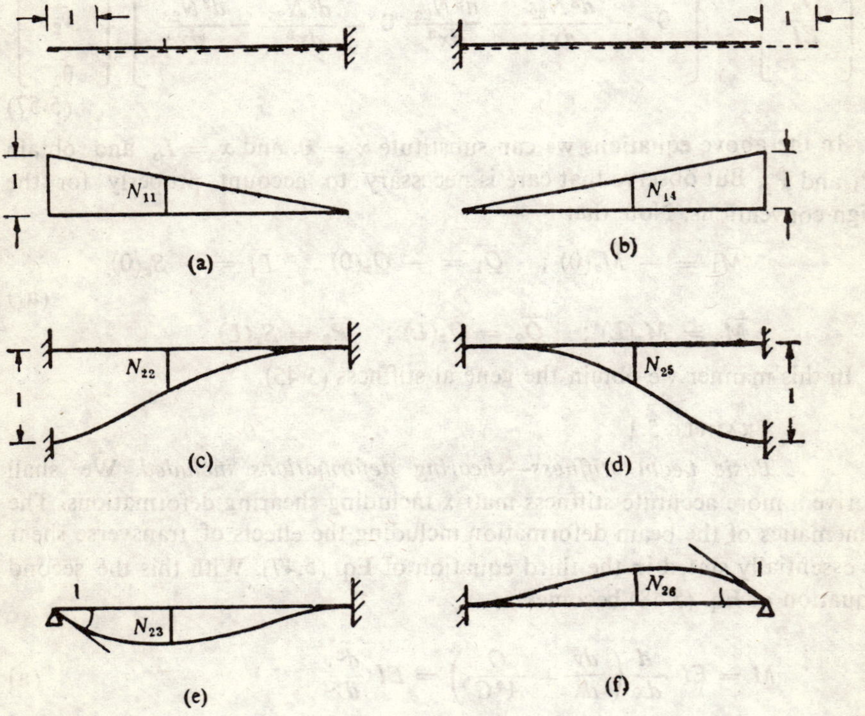

Fig. 5.8

Solving for each of these shape functions is relatively easier than the simultaneous solution as required by Eq. (5.54). Let us derive N_{22} ; the boundary conditions to determine this are

$$\text{At} \quad x = 0, \quad \frac{dv}{dx} = 0, \quad v = \bar{v}_1 = 1 \tag{a$_3$}$$

$$\text{At} \quad x = L, \quad v = 0, \frac{dv}{dx} = 0.$$

As all but one have zero on their right hand sides it is relatively easier to evaluate the constants. We will pursue this topic of shape functions in greater detail in Chapter 9.

Having determined the displacements in Eq. (5.55), the bending moments, shear, etc. are obtainable from Eqs. (5.48). Therefore

$$\begin{Bmatrix} \dfrac{S_x}{AE} \\ \dfrac{M_x}{EI} \\ \dfrac{Q_x}{EI} \end{Bmatrix} = \begin{bmatrix} \dfrac{dN_{11}}{dx} & 0 & 0 & \dfrac{dN_{14}}{dx} & 0 & 0 \\ 0 & \dfrac{d^2N_{22}}{dx^2} & \dfrac{d^2N_{23}}{dx^2} & 0 & \dfrac{d^2N_{25}}{dx^2} & \dfrac{d^2N_{26}}{dx^2} \\ 0 & -\dfrac{d^3N_{22}}{dx^3} & -\dfrac{d^3N_{23}}{dx^3} & 0 & -\dfrac{d^3N_{25}}{dx^3} & -\dfrac{d^3N_{26}}{dx^3} \end{bmatrix} \begin{Bmatrix} \bar{u}_1 \\ \bar{v}_1 \\ \bar{\theta}_1 \\ \bar{u}_2 \\ \bar{v}_2 \\ \bar{\theta}_2 \end{Bmatrix} \tag{5.57}$$

In the above equations we can substitute $x = 0$, and $x = L$, and obtain $\mathbf{P}_1$ and $\mathbf{P}_2$. But observe that care is necessary to account properly for the sign conventions. Note that

$$\bar{M}_1 = -M_x(0)\ ; \quad \bar{Q}_1 = -Q_x(0)\ ; \quad \bar{P}_1 = -S_x(0)$$
$$\bar{M}_2 = M_x(L)\ ; \quad \bar{Q}_2 = Q_x(L)\ ; \quad \bar{P}_2 = S_x(L) \tag{a$_4$}$$

In this manner we obtain the general stiffness (5.45).

EXAMPLE 5.1

Basic beam stiffness—shearing deformations included. We shall derive a more accurate stiffness matrix including shearing deformations. The kinematics of the beam deformation including the effects of transverse shear is essentially stated in the third equation of Eq. (5.47). With this the second equation of Eq. (5.48) becomes

$$M = EI\,\frac{d}{dx}\left(\frac{dv}{dx} + \frac{Q}{A^*G}\right) = EI\,\frac{d^2v}{dx^2} \tag{a}$$

as the second term vanishes on account of third equation of Eq. (5.46). Thus the expression for the bending moment remains the same, and hence the differential equation in v also remains unchanged. The effect of shearing deformation enters through the boundary conditions.

Let us derive the *basic* stiffness matrix. Taking the cantilever model,

$$\text{At} \quad x = 0, \qquad u = v = \theta = 0$$

$$\text{At} \quad x = L \qquad u = \bar{u}_2 \; ; \; v = \bar{v}_2 \; ; \; \theta = \bar{\theta}_2$$

(Note that we can use the notations e, v and φ respectively for $\bar{u}_2$ and $\bar{\theta}_2$)
The expression for θ in terms of v is,

$$\theta = \frac{dv}{dx} - \frac{Qx}{A^*G} = \frac{dv}{dx} + \frac{EI}{A^*G}\frac{d^3v}{dx^3} \tag{c}$$

These give

$$C_3 = 0 \; ; \; C_4 = -\frac{2\beta}{1+4\beta}\left(\bar{\theta}_2 - 2\frac{\bar{v}_2}{L}\right)$$

$$C_5 = \frac{1}{1+4\beta}\left(3\frac{\bar{v}_2}{L^2} - (1-2\beta)\frac{\bar{\theta}_2}{L}\right) \tag{d}$$

$$C_6 = \frac{1}{1+4\beta}\left(\frac{\bar{\theta}_2}{L^2} - 2\frac{\bar{v}_2}{L^3}\right)$$

with

$$\beta = \frac{3EI}{A^*GL^2}$$

The expression for forces will be

$$M_x = \frac{EI}{1+4\beta}\left\{\left(6\frac{x}{L^2} - \frac{2(1-2\beta)}{L}\right)\bar{\theta}_2 + \left(\frac{6}{L^2} - 12\frac{x}{L^3}\right)\bar{v}_2\right\}$$

$$Q_x = \frac{EI}{1+4\beta}\left(-\frac{6}{L^2}\bar{\theta}_2 + \frac{12}{L^3}\bar{v}_2\right) \tag{f}$$

In the above relations we can substitute $x = L$ and obtain $\bar{M}_2$ and $\bar{Q}_2$. The stiffness coefficients corresponding to the axial forces and the axial displacements remain the same. Thus with $x = L$

$$\bar{M}_2 = \frac{EI}{1+4\beta}\left\{\frac{4(1+\beta)}{L}\bar{\theta}_2 - \frac{6}{L_2}\bar{v}_2\right\}$$

$$\bar{Q}_2 = \frac{EI}{1+4\beta}\left\{-\frac{6}{L^2}\bar{\theta}_2 + \frac{12}{L^3}\bar{v}_2\right\} \tag{g}$$

$$\bar{P}_2 = \frac{AE}{L}\bar{u}_2$$

These agree with Eqs. (5.34).

Example 5.2

Stiffness with higher order (beam-column) effects. **Let us derive the stiffness of a plane frame member taking into account, to an approximate degree, higher order effects of axial loads. This is the so-called *beam-column***

effect which accounts for the effect of axial load in equilibrium considerations only.

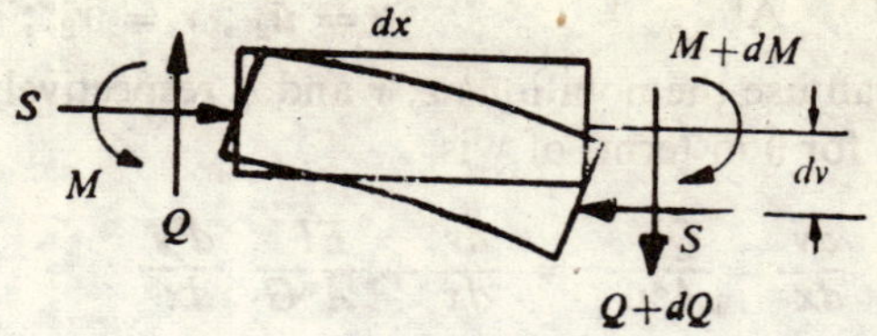

Example 5.2

The figure shows equilibrium of the differential element. While writing the moment equilibrium equation, we include the additional moment due to the axial load S due to the eccentricity dv. Assuming S to be a *compressive* force, we have

$$dM+Qdx+Sdv = 0$$

giving

$$\frac{dM}{dx}+Q+S\frac{dv}{dx} = 0 \tag{a$_1$}$$

The equilibrium relation $(dQ/dx) = 0$ is again assumed by neglecting the component of S due to the slope (dv/dx). Thus, these equations together with the elastic relation $M = EI\,(d^2v/dx^2)$ lead to the following differential equation :

$$EI\frac{d^4v}{dx^4}+S\frac{d^2v}{dx^2} = 0 \tag{a$_2$}$$

The solution of this equation is

$$v = C_1 \cos \beta x+C_2 \sin \beta x+C_3x+C_4 \tag{a$_3$}$$

with

$$\beta^2 = \left(\frac{S}{EI}\right) \tag{a$_4$}$$

Let us derive the stiffness for the beam model. The relationship between the axial load S and the axial elongation e is considered to remain the same, though higher order effects because of the curvature can be included. We will be excluding such considerations at this stage. Let us now derive the stiffness relations,

$$\begin{bmatrix} M_1 \\ M_2 \end{bmatrix} = \begin{bmatrix} k_{11} & k_{12} \\ k_{21} & k_{22} \end{bmatrix} \begin{Bmatrix} \varphi_1 \\ \varphi_2 \end{Bmatrix} \tag{a$_5$}$$

The boundary conditions are,

$$\begin{aligned} &\text{At} \quad x=0: \quad v=0\,; \quad \frac{dv}{dx}=\varphi_1 \\ &\text{At} \quad x=L: \quad v=0\,; \quad \frac{dv}{dx}=\varphi_2 \end{aligned} \tag{a$_6$}$$

These give

$$C_1 = -C_4 \; ; \; C_3 + C_2\,\beta = \varphi_1$$
$$C_1 \cos\beta L + C_2 \sin\beta L + C_3 L + C_4 = 0 \qquad (a_7)$$
$$-C_1\,\beta \sin\beta L + C_2\,\beta \cos\beta L + C_3 = \varphi_2$$

These are to be solved in terms of φ_1 and φ_2. Thus we obtain

$$\begin{Bmatrix} C_2 \\ C_4 \end{Bmatrix} = \frac{1}{\Delta}\begin{bmatrix} 1-\cos\beta L - \beta L \sin\beta L & -(1-\cos\beta L) \\ -\sin\beta L + \beta L \cos\beta L & \sin\beta L - \beta L \end{bmatrix}\begin{Bmatrix} \varphi_1 \\ \varphi_2 \end{Bmatrix} \qquad (a_8)$$

with $$\Delta = 2\beta \sin\beta L \left\{ \tan\frac{\beta L}{2} - \beta L \right\} \qquad (a_9)$$

The bending moments when expressed in terms of curvatures give,

$$\begin{Bmatrix} M_1 \\ M_2 \end{Bmatrix} = EI\beta^2 \begin{bmatrix} 0 & -1 \\ -\sin\beta L & \cos\beta L \end{bmatrix}\begin{Bmatrix} C_2 \\ C_4 \end{Bmatrix} \qquad (a_{10})$$

By substituting for C_2 and C_4, and simplifying we obtain the stiffness coefficients as required in (a_5). Thus

$$k_{11} = k_{22} = \left(\frac{EI}{L}\right)\psi_1 \; ; \; k_{12} = k_{21} = \left(\frac{EI}{L}\right)\psi_2 \qquad (a_{11})$$

where

$$\psi_1 = \frac{\alpha\,(1 - 2\alpha \cos 2\alpha)}{\tan\alpha - \alpha} \qquad (a_{12})$$

$$\psi_2 = \frac{\alpha\,(2\alpha \operatorname{co sec} 2\alpha - 1)}{\tan\alpha - \alpha} \qquad (a_{13})$$

$$\alpha = \frac{\beta L}{2} = \sqrt{\frac{SL^2}{4EI}} = \frac{\pi}{2}\sqrt{\frac{S}{S_E}} \qquad (a_{14})$$

$$S_E = \frac{\pi^2 EI}{L^2}$$

Functions ψ_1 and ψ_2 depend on the magnitude of the axial compressive force S, and the slenderness ratio of the beam. These are the stability functions used in critical load analysis of frame works [9]. Since the stiffness coefficients depend on the magnitude of the force S, this in an indeterminate frame will be an unknown quantity. The stiffness matrix is not strictly explicit, but involves the forces.

The stiffnesses tend to zero as the axial load approaches the Euler critical load $\pi^2 EI/L^2$, *i.e.* when $\alpha = \pi/2$. We will persue this later in Chapter 8 (see Sec. 8.8).

Axial Tensile Force

The case of axial tensile force can be similarly considered. The above formulae can be used by replacing S by $-S$. Thus from Eq. (a_4) we deduce that we can replace β by $i\beta$. Therefore also α can be replaced by $-i\alpha$. Thus

$$\psi_1 = \frac{\alpha\,(1-2\alpha\coth 2\alpha)}{\tanh\alpha-\alpha}$$

$$\psi_2 = \frac{\alpha\,(2\alpha\,\mathrm{cosech}\,2\alpha-1)}{\tanh\alpha-\alpha} \qquad (a_{15})$$

In this case the stiffness coefficients increase with the magnitude of the force S. See Ref [11].

Flexibility Coefficients

These can be directly derived from Eq. (a_2). More easily, relations (a_5) can be inverted to obtain the flexibility matrix

$$\begin{Bmatrix} \varphi_1 \\ \varphi_2 \end{Bmatrix} = \begin{bmatrix} f_{11} & f_{12} \\ f_{21} & f_{22} \end{bmatrix} \begin{Bmatrix} M_1 \\ M_2 \end{Bmatrix} \qquad (a_{16})$$

After some algebraic manipulations we obtain, for compressive axial force

$$f_{11} = f_{22} = \frac{L}{3EI}\left(\frac{3}{2\alpha}\left(\frac{1}{2\alpha}-\frac{1}{\tan 2\alpha}\right)\right) = \frac{L}{3EI}\psi'_1$$

$$f_{12} = f_{21} = -\frac{L}{6EI}\left(\frac{3}{\alpha}\left(\frac{1}{\sin 2\alpha}-\frac{1}{2\alpha}\right)\right) = -\frac{L}{6EI}\psi'_2 \qquad (a_{17})$$

Such flexibility connficients, can be found in the text of Timoshenko [10] with few applications.

EXAMPLE 5.3

Beam on elastic foundation. The stiffness for a beam on elastic foundation (Winkler foundation) can also be determined from the solution of its governing equation.

The elastic springs offer resistive forces proportional to the compression they they undergo (see Fig. a). If the stiffness of the spring is k_f then $k_f v$ is the force in the spring and $-k_f v$ is the upward force on the beam. Thus the differential equation for the beam (see second equation of Eq. 5.50) becomes

$$EI\frac{d^4v}{dx^4} = -k_f v \qquad (a_1)$$

or

$$\frac{d^4v}{dx^4} + \left(\frac{k_f}{EI}\right)v = 0 \qquad (a_2)$$

With

$$\xi = \frac{x}{L}, \quad 4\lambda^4 = \frac{k_f L^4}{EI}, \qquad (a_3)$$

we have
$$\frac{d^4v}{d\xi^4}+4\lambda^4 v=0 \tag{a$_4$}$$

This has a solution [11]
$$v=f_1c_1+f_2c_2+f_3c_3+f_4c_4 \tag{a$_5$}$$
where
$$\begin{aligned} f_1 &= \cosh\lambda\xi\,\cos\lambda\xi; \quad f_2=\cosh\lambda\xi\,\sin\lambda\xi\\ f_3 &= \sin\lambda\xi\,\cos\lambda\xi; \quad f_4=\sin\lambda\xi\,\sin\lambda\xi \end{aligned} \tag{a$_6$}$$

The arbitrary constants can be evaluated in terms of prescribed support displacement as required for stiffness. Thus the conditions

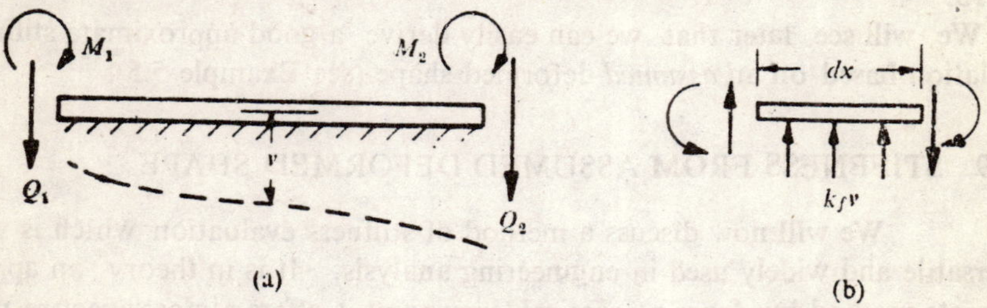

Example 5.3

$$v(0)=v_1\;;\quad v(1)=v_2;\quad \frac{dv}{d\xi}(0)=L\theta_1$$
$$\frac{dv}{d\xi}(1)=L\theta_2 \tag{a$_7$}$$

determine the arbitrary constants in terms of the end displacements v_1, v_2, θ_1 and θ_2. Proceeding as in the previous example we obtain after a long algebra, the following stiffness relations.

$$\begin{Bmatrix} Q_1\\ M_1\\ Q_2\\ M_2 \end{Bmatrix}=\begin{bmatrix} \frac{EI}{L^3}\psi_5 & \frac{EI}{L^2}\psi_3 & -\frac{EI}{L^3}\psi_6 & \frac{EI}{L^2}\psi_4\\ & \frac{EI}{L}\psi_1 & -\frac{EI}{L^2}\psi_4 & \frac{EI}{L}\psi_2\\ & \text{(Sym)} & \frac{EI}{L^3}\psi_5 & -\frac{EI}{L^2}\psi_3\\ & & & \frac{EI}{L}\psi_1 \end{bmatrix}\begin{Bmatrix} v_1\\ \theta_1\\ v_2\\ \theta_2 \end{Bmatrix} \tag{a$_8$}$$

where,

$$\psi_1=\frac{2\lambda}{a}(\cosh\lambda\,\sinh\lambda-\cos\lambda\,\sin\lambda)$$

$$\psi_2=\frac{2\lambda}{a}(\cosh\lambda\,\sin\lambda-\sinh\lambda\,\cos\lambda)$$

$$\psi_3=\frac{2\lambda^2}{a}(\sinh^2\lambda+\sin^3\lambda)$$

$$\psi_4 = \frac{4\lambda^2}{a}(\sinh\lambda \, \sin\lambda) \qquad (a_9)$$

$$\psi_5 = \frac{4\lambda^3}{a}(\cosh\lambda \, \sinh\lambda + \cos\lambda \, \sin\lambda)$$

$$\psi_6 = \frac{4\lambda^3}{a}(\cosh\lambda \, \sin\lambda + \sinh\lambda \, \cos\lambda)$$

$$a = \sinh^2\lambda - \sin^2\lambda.$$

The stiffness matrix reduces to that of an ordinary beam as λ tends to zero.

We will see later that we can easily derive a good approximate stiffness relation based on an *assumed* deformed shape (see Example 5.5).

5.9 STIFFNESS FROM ASSUMED DEFORMED SHAPE

We will now discuss a method of stiffness evaluation which is very versatile and widely used in engineering analysis. It is in theory an approximate method but from a practical viewpoint it offers almost exact results, more so when used in conjunction with modern digital computers. In the mathematical literature it is commonly referred to as the Ritz [18] or Galerkin [19] method. This method is not quite necessary for a beam member as it is a determinate element. However in the finite element ***idialization,*** it plays an inevitable role as exact solutions are not possible. We can grasp the idea by considering the beam member.

We will first consider a simple example to bring out the essential idea keeping the algebra to a minimum. Figure 5.9 (a) shows a symmetrical beam with moments M_1 applied at each support causing rotations θ_1. The exact stiffness relation is

$$M_1 = \frac{2EI}{L}\,\theta_1 = k\,\theta_1 \qquad (a_1)$$

The deflected shape $v(x)$ is a parabola

$$v(x) = a(Lx - x^2) \qquad (a_2)$$

This has a constant curvature which when multiplied by the flexural rigidity gives constant bending moment as required.

Let us assume instead a sine curve :

$$v^* = a \sin\frac{\pi x}{L} \qquad (a_3)$$

which appears to be sufficiently smooth curve. Let us choose this curve such that the rotation at ends is θ_1. This gives $a = (L/\pi)$, and

$$v^* = \left(\frac{L}{\pi}\sin\frac{\pi x}{L}\right)\theta_1 = N_1\theta_1 \qquad (a_4)$$

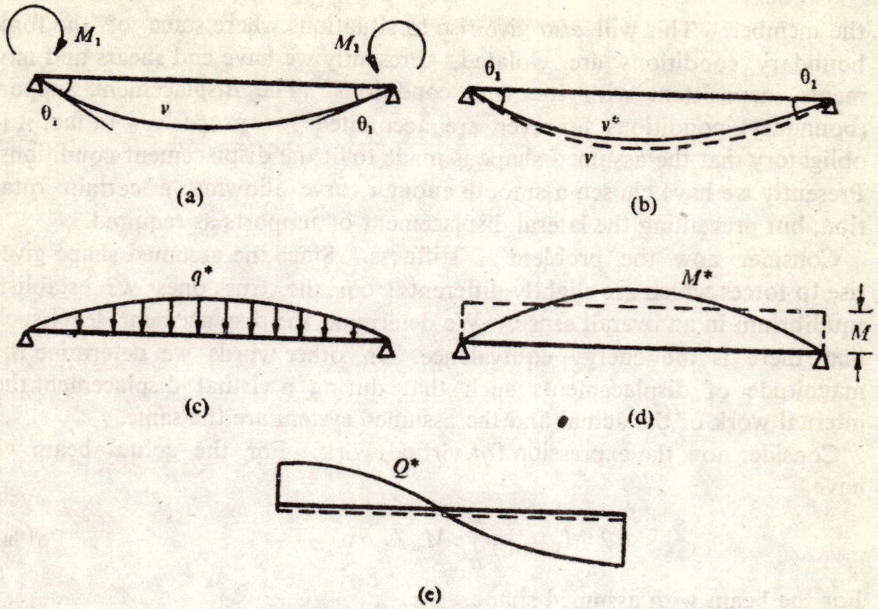

Fig. 5·9

Here N_1 is the shape function as introduced in the previous section.

The assumed shape determines a distribution of strains which together with the elastic law implies also a stress distribution. Further this also implies that we have a set of external loads equilibrant with this stress distribution. Thus the assumed shape implies curvatures

$$\chi^* = \frac{d^2v^*}{dx^2} = \frac{d^2N_1}{dx^2}\theta_1 = \left(-\frac{\pi}{L}\sin\frac{\pi x}{L}\right)\theta_1 \qquad (a_5)$$

bending moments,

$$M_x^* = EI\chi^* = EI\frac{d^2N_1}{dx^2}\theta_1 = -\frac{EI\pi}{L}\sin\frac{\pi x}{L}\theta_1 \qquad (a_6)$$

shearing forces

$$Q_x^* = -\frac{dM^*}{dx} = -EI\frac{d^3N_1}{dx^3}\theta_1 = \frac{EI\pi^2}{L^2}\cos\frac{\pi x}{L}\theta_1 \qquad (a_7)$$

and finally a lateral load,

$$q_x^* = -\frac{dQ_x^*}{dx} = EI\frac{d^4N_1}{dx^4}\theta_1 = \frac{EI\pi^3}{L^3}\sin\frac{\pi x}{L}\theta_1 \qquad (a_8)$$

The assumed shape is equivalent to a sinusoidally varying bending moments with zero values at ends. The beam is carrying some lateral loads q^* causing certain shears. Both these are required to be zero in the original structure. This is an inherent feature of an assumed shape that the differential equations of equilibrium are not satisfied from point to point within the member. This will *also* give rise to situations where some of the force boundary conditions are violated. Presently we have end shears and moments inconsistent with the true conditions. The displacement support (boundary) conditions however are accounted for properly. In fact it is obligatory that the assumed shape is made to fit the displacement conditions. Presently we have chosen a smooth enough curve allowing a certain rotation, but preventing the lateral displacement of supports as required.

Consider now the problem of stiffness. Since the assumed shape gives rise to forces which are slightly different from the true ones we establish equilibrium in an overall sense. We determine the displacement level such that there is the energy equivalence. In other words we determine the magnitude of displacements such that during a virtual displacement the internal work of the actual and the assumed system are the same.

Consider now the expression for virtual work. For the actual beam we have

$$2\,M_1\,\theta_1 = \int_0^L M_x\,\chi_x\,dx \qquad (a_9)$$

For the beam with assumed shape,

$$\int_0^L q_x^*\,v^*\,dx = \int_0^L M_x^*\,\chi_x^*\,dx \qquad (a_{10})$$

Note that in the beam with assumed shape, bending moments M_1 are zero at the two ends. We determine the stiffness by equating the right hand sides of the above two equations, which means that internal work in both the cases is the same. Thus

$$2M_1\,\theta_1 = \int_0^L M_x^*\,\chi_x^*\,dx \qquad (a_{11})$$

Also, equating the left hand sides we obtain

$$2\,M_1\,\theta_1 = \int q_x^*\,v^*\,dx \qquad (a_{12})$$

This relation defines 'equivalent loads'. The assumed shape gives rise to an external load q^* while as the given forces are the two end moments M_1. We determine the loads q^* such that it is equivalent to the loads M_1 in a work sense. In other words, they are energy equivalent loads.

Using Eqs. (a_5) and (a_6) in Eq. (a_{11}) we obtain

$$2\,M_1\,\theta_1 = \int_0^L \left(-\frac{EI\,\pi}{L}\sin\frac{\pi x}{L}\theta_1\right)\left(-\frac{\pi}{L}\sin\frac{\pi x}{L}\right)\theta_1\,dx$$

$$= \frac{EI\pi^2}{L^2}\theta_1^2\int_0^L \sin^2\frac{\pi x}{L}\,dx = \frac{EI\,\pi^2\,\theta_1^2}{L^2}\cdot\frac{L}{2}$$

$$= \frac{EI\,\pi^2}{2L}\theta_1^2$$

This gives

$$M_1 = \frac{\pi^2}{4}\left(\frac{EI}{L}\right)\theta_1 = 2.467\left(\frac{EI}{L}\right)\theta_1 \qquad (a_{13})$$

It is seen that this is in error by 23 per cent, which of course can be improved by a better choice of the assumed shape. Before proceeding to discuss this issue let us take note of the equivalent loads. Substituting for θ_1 in terms of M_1 in Eq. (a_8) we obtain

$$q_x^* = \frac{4\pi M_1}{L^2} \sin\frac{\pi x}{L} \qquad (a_{14})$$

This shows the equivalence between the given loads M_1 and the assumed loads q_x^* as implied in Eq. (a_{12}). Note as to how we establish equivalence between two sets of forces using equivalence in an energy sense, by requiring that during a virtual deformation the internal work generated is the same.

We observe by substituting θ_1 from Eq. (a_{13}) in Eq. (a_4)

$$v^* = \frac{M_1 L^2}{(\pi^3/4)} \sin\frac{\pi x}{L} = \frac{ML^2}{7.75} \sin\frac{\pi x}{L} \qquad (a_{15})$$

The exact value of the deflection at centre is $ML^2/8EI$, it is therefore seen that the error in estimating the deflection at centre is lesser than in bending moments. Since the deformed shape of a body is insensitive to the variation of loads the error in deflections is generally the least with increasing error in moments and shears.*

The question of improving accuracy is discussed in the example below.

EXAMPLE 5.4

We will derive now an improved estimate of the stiffness by assuming the deformed shape in the form,

$$v = a_1 \sin\frac{\pi x}{L} + a_2 \sin\frac{3\pi x}{L} \qquad (a_1)$$

which is a superposition of sinusoidal shapes as shown in **Figs. (a) and (b).** We will write this as

$$\mathbf{v} = (v) = [\, f_1 \quad f_2 \,] \begin{Bmatrix} a_1 \\ a_2 \end{Bmatrix} = \mathbf{f\,a} \qquad (a_2)$$

where f_1 and f_2 stand for $\sin \pi x/L$ and $\sin 3\pi x/L$ respectively. The rotation θ_1 is related to a_1 and a_2. The relationship is

In a simply supported beam loaded by a concentrated load the error, in assuming that the point of maximum deflection is at centre, is only about 3 percent

$$\theta_1 = (\theta_1) = \frac{dv}{dx} = \left[\frac{\pi}{L} \quad \frac{3\pi}{L} \right] \begin{Bmatrix} a_1 \\ a_2 \end{Bmatrix} = \mathbf{R\ a} \qquad (a_3)$$

where R can be considered as a transformation matrix. The curvatures and bending moments assumed are then,

$$\mathbf{M}_1 = (\mathbf{M}) = (EI)\chi = EI \frac{d^2v}{dx^2} \left[f_1'' \quad f_2'' \right] \{ a \} \qquad (a_4)$$

$$= EI \left[-\frac{\pi^2}{L^2} f_1 - \frac{9\pi^2}{L^2} f_2 \right] \{ a \}$$

$$= (EI)\ \mathbf{g\ a} \qquad (a_5)$$

Thus the assumed shape corresponds to zero moments at the ends thus violating the force boundary conditions at supports. Also all along the beam they give rise to shears and lateral loads both of which are required to be zero. But the approximate nature of analysis implies that these external forces are equivalent to the given forces so that during a virtual displacement the work of the external forces determined by the assumed shape is equal to the work of the actual (given) external forces. Each of

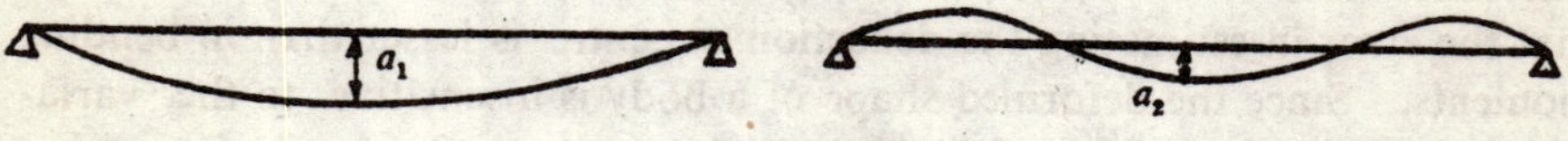

(a) (b)

Example 5.4

these is equal to the internal virtual work. Thus

$$\int \mathbf{q}^t\ \mathbf{v}\ dx = (2\mathbf{M}_1)^t\ \boldsymbol{\theta}_1 = \int (EI\ \chi)^t\ \chi\ dx \qquad (a_5)$$

The first pair of equality defines the equivalent loads and the second determines the stiffness. Therefore considering the latter,

$$(2\mathbf{M}_1)^t\ \mathbf{R\ a} = \int (EI\ \mathbf{g\ a})^t\ \mathbf{g\ a}\ dx \qquad (a_6)$$

giving after the usual transformations,

$$\left[\int \mathbf{g}^t\ (EI)\ \mathbf{g}\ dx \right] \mathbf{a} = \mathbf{R}^t\ (2\mathbf{M}_1) \qquad (a_7)$$

This can be written as

$$[\mathbf{K}_a]\ \{\mathbf{a}\} = \mathbf{P}_a \qquad (a_8)$$

and can be interpreted as a stiffness relation between the displacement variables a and the corresponding generalised forces $\mathbf{P}_a$. Presently using g from (a_4),

$$\mathbf{K}_a = EI \int_0^L \begin{bmatrix} f_1''^2 & f_1'' f_2'' \\ f_1'' f_2'' & f_2''^2 \end{bmatrix} dx \qquad (a_9)$$

In view of the orthogonal* nature of f_1 and f_2, we obtain a diagonal matrix as follows :

$$\mathbf{K}_a = EI \begin{bmatrix} \frac{\pi^4}{2L^3} & 0 \\ 0 & \frac{81\,\pi^4}{2L^3} \end{bmatrix} \qquad (a_{10})$$

Further, the generalized load $\mathbf{P}_a$ becomes

$$\mathbf{P}_a = \mathbf{R}^t\,(2\mathbf{M}_1) = \begin{Bmatrix} \pi/L \\ 3\pi/L \end{Bmatrix} 2\,M_1 = \begin{Bmatrix} 2\pi/L \\ 6\pi/L \end{Bmatrix} M_1 \qquad (a_{11})$$

We can now solve for a from (a_8). In view of the diagonnal nature of $\mathbf{K}_a$, we obtain readily

$$\mathbf{a} = \begin{Bmatrix} a_1 \\ a_2 \end{Bmatrix} = \frac{1}{EI} \begin{bmatrix} \frac{2L^4}{\pi^4} & 0 \\ 0 & \frac{2L^3}{81\pi^4} \end{bmatrix} \begin{Bmatrix} 2\pi/L \\ 6\pi/L \end{Bmatrix} M_1 = \frac{M_1}{EI} \begin{Bmatrix} \frac{4L^2}{\pi^3} \\ \frac{12L^2}{81\pi^3} \end{Bmatrix} \qquad (a_{12})$$

We can now substitute this in (a_3) and obtain,

$$\theta_1 = \mathbf{R}\,\mathbf{a} = \frac{40}{9\pi^2}\left(\frac{L}{EI}\right) M_1 \qquad (a_{13})$$

Thus the stiffness is now given by

$$M_1 = \frac{9\pi^2}{40}\left(\frac{EI}{L}\right)\theta_1 = 2.220\left(\frac{EI}{L}\right) \qquad (a_{14})$$

which is better than the previous value. If we assume a three term expression for the displacement,

$$v = a_1 \sin\frac{\pi x}{L} + a_2 \sin\frac{3\pi x}{L} + a_3 \sin\frac{5\pi x}{L} \qquad (a_{15})$$

we obtain the stiffness 2.1434 $(EI/L)\,\theta_1$

The reader may verify the first pair of equality of Eq. (a_5); one can work out the lateral load q by differentiating Eq. (a_4) twice.

General Expression for Stiffness

We will now develop a general expression for the stiffness of a member. Consider the derivation of the general stiffness of the beam element.

Let us suppose that the shape functions (5.56) appearing in Eq. (5.55) are assumed rather than derived. (Recall that these were derived as a solution

* Recall that functions f_1 and f_2 are said to be orthogonal if

$$\int_0^L f_1 f_2\, dx = 0$$

of the differential equation.) These determine strains along the member Considering only stretching and bending,

$$\begin{Bmatrix} \epsilon_x \\ \chi_x \end{Bmatrix} = \begin{bmatrix} \frac{dN_{11}}{dx} & 0 & 0 & \frac{dN_{14}}{dx} & 0 & 0 \\ 0 & \frac{d^2N_{22}}{dx^2} & \frac{d^2N_{23}}{dx^2} & 0 & \frac{d^2N_{25}}{dx^2} & \frac{d^2N_{26}}{dx^2} \end{bmatrix} \begin{Bmatrix} \bar{u}_1 \\ \bar{v}_1 \\ \bar{\theta}_1 \\ -- \\ \bar{u}_2 \\ \bar{v}_2 \\ \bar{\theta}_2 \end{Bmatrix} \quad (a_{16})$$

These are the kinematic relations

$$\epsilon_x = \bar{A}\,\bar{u} \quad (5.58)$$

relating the strains at any point in terms of the member end displacements. The elasticity relations then give the internal forces

$$S_x = \begin{Bmatrix} S_x \\ M_x \end{Bmatrix} = \begin{bmatrix} AE & 0 \\ 0 & EI \end{bmatrix} \begin{Bmatrix} \epsilon_x \\ \chi_x \end{Bmatrix} = k_x\,\epsilon_x \quad (5.59)$$

Here k_x is the diagonal matrix containing the elastic rigidities of the beam. Using the strains from Eq (5.58) we obtain the internal forces in terms of the member end forces :

$$S_x = k_x\,\bar{A}\,\bar{u} \quad (5.60)$$

The next step it to determine the stiffness relation using the principle of virtual work to establish overall equilibrium of the member. Thus

$$\bar{P}^t\,\bar{u} = \int_L S_x\,\epsilon_x\,dx$$

Using Eq. (5.58) we can write this as $\bar{P}^t\,\bar{u} = \int_L S_x\,\bar{A}\,\bar{u}\,dx$

On cancelling the arbitrary virtual displacements **u**, and taking transpose we obtain the equation of equilibrium as

$$\int_L A^t\,S_x\,dx = \bar{P} \quad (5.61)$$

Using next Eq. (5.60) we obtain

$$\bar{K}\,\bar{u} = [\int A^t\,k_x\,A\,dx]\,\bar{u} = \bar{P} \quad (5.62)$$

Thus we obtain stiffness matrix as

$$\bar{K} = \int_L A^t\,k_x\,A\,dx \quad (5.63)$$

This is the general form for the stiffness. Observe that this form is analogous to the form (5.18) for the flexibility Recall however the fact that the statical matrix **B** was directly obtained. It followed from statics alone. The form (5.18) is general and can be shown that it also affords an approximation when **B** matrix is approximate. For a statically determinate element such as a beam member there is no need for assuming the **B** matrix, where as, for a plate or solid element the matrix **B** can only be approximate. It can be verified that Eq. (5.63) leads to Eq. (5.45) for a uniform beam.

Let us now construct kinematic matrix A of (a_{16}) using N_{ij} from Eqs. (5.56). Thus

$$\epsilon_x : \begin{Bmatrix} \epsilon_x \\ \chi \end{Bmatrix} = \begin{bmatrix} -\frac{1}{L} & 0 & 0 & \frac{1}{L} & 0 & 0 \\ 0 & -\left(\frac{6}{L^2}+\frac{12x}{L^3}\right) & \left(-\frac{4}{L}+\frac{6x}{L^2}\right) & 0 & \frac{6}{L^2}-\frac{12x}{L^3} & \frac{6x}{L^2}-\frac{2}{L} \end{bmatrix} \begin{Bmatrix} \bar{u}_1 \\ \bar{v}_1 \\ \bar{\theta}_2 \\ \bar{u}_2 \\ \bar{v}_2 \\ \bar{\theta}_2 \end{Bmatrix} \tag{5.64}$$

Using $\mathbf{k}_x$ from Eq. (5.59) we can evaluate the product $\bar{\mathbf{A}}^t\ \bar{\mathbf{k}}_x\ \bar{\mathbf{A}}$ and integrate each term with respect to x between 0 and L. For beam of uniform section these lead to Eqs. (5.45).

For instance

$$k_{11}=\int_0^L \left(-\frac{1}{L}\right)\left(AE\right)\left(-\frac{1}{L}\right) dx=\frac{AE}{L}$$

$$k_{22}=\int_0^L \left(-\frac{6}{L^2}+\frac{12x}{L^3}\right)\left(EI\right)\left(-\frac{6}{L^2}+\frac{12x}{L^3}\right) dx=\frac{12EI}{L^3}, \text{ etc.}$$

We will now illustrate an example in which the assumed shape is only approximate.

EXAMPLE 5.5. *Beam on Elastic Foundation.*

We will now develop the stiffness matrix for the beam on elastic foundation by assuming an approximate deformed shape. The exact stiffness relationships for this case were derived from the solution of the governing differential equation. This was done in Example 5.3. The derivation entails considerable algebra, and what is more important the evaluation of stiffness requires evaluation of the hyperbolic functions. This will entail considerable computing time.

We can assume the same deflected shape as for a normal beam, namely a cubic instead of the exact form given by Eq. (a_5) of Example 5.3. In other words we may imagine that the hyperbolic functions are evaluated upto cubic terms alone. Thus considering only transverse effects,(axial terms excluded) we assume

$$v(x)=[N_1 \quad N_2 \quad N_3 \quad N_4]\begin{Bmatrix} v_1 \\ \theta_1 \\ v_2 \\ \theta_2 \end{Bmatrix}=\mathbf{NU} \qquad (a_1)$$

where ***Ni*** are the shape functions given in Eqs. (5.56). To rewrite, with the above notations,

$$\begin{aligned} N_1&=1-3\left(\frac{x}{L}\right)^2+2\left(\frac{x}{L}\right)^3 \\ N_2&=L\left(\frac{x}{L}\right)\left(1-\left(\frac{x}{L}\right)\right)^2 \\ N_3&=3\left(\frac{x}{L}\right)^2-2\left(\frac{x}{L}\right)^3 \\ N_4&=L\left(-\left(\frac{x}{L}\right)^2+\left(\frac{x}{L}\right)^3\right) \end{aligned} \qquad (a_2)$$

These determine strains in the beam and the elastic spring. Considering only bending deformations, we have for the beam,

$$\boldsymbol{\epsilon}_b\equiv\{\chi_x\}=\left[\frac{d^2N_1}{dx^2} \quad \frac{d^2N_2}{dx^2} \quad \frac{d^2N_3}{dx^2} \quad \frac{d^2N_4}{dx^2}\right]\{\mathbf{U}\}=\mathbf{A\,U} \qquad (a_3)$$

Here,

$$\frac{d^2N_1}{dx^2}=-\frac{d^2N_3}{dx^2}=-\frac{6}{L^2}+\frac{12x}{L^3}$$

$$\frac{d^2N_2}{dx^2}=-\frac{4}{L}+\frac{6x}{L^2}; \quad \frac{d^2N_4}{dx^2}=\frac{6x}{L^2}-\frac{2}{L} \qquad (a_4)$$

For the foundation spring, the shortening

$$\mathbf{e}_f\equiv\{v\}=\mathbf{NU} \qquad (a_5)$$

In writing the equation of virtual work we now include the work of the springs too. The internal force in the beam is

$$\mathbf{S}_b=\{M\}=\{EI\chi_x\}=\mathbf{k}_b\boldsymbol{\epsilon}_b=\mathbf{k}_b\,\mathbf{A\,U} \qquad (a_6)$$

And for the spring,

$$\mathbf{S}_f=\mathbf{k}_f\mathbf{e}_f=\mathbf{k}_f\,\mathbf{NU} \qquad (a_7)$$

Thus the equation of work, Eq. (5.61), extended to include the work of the springs, becomes,

$$\mathbf{P}^t\mathbf{U}=\int_0^L \mathbf{S}_b{}^t\boldsymbol{\epsilon}_b\,dx+\int_0^L \mathbf{S}_f{}^t\mathbf{e}_f dx \tag{a_8}$$

$$=\int_0^L (\mathbf{k}_b\mathbf{A}\mathbf{U})^t\,\mathbf{A}\mathbf{U}dx+\int (\mathbf{k}_f\mathbf{N}\mathbf{U})^t\mathbf{N}\mathbf{U}dx.$$

Cancelling the virtual displacement **u** and taking transpose

$$\mathbf{P}=[\mathbf{K}_b+\mathbf{K}_f]\mathbf{U}=\mathbf{K}\mathbf{U} \tag{a_9}$$

where

$$\mathbf{K}_b=\int \mathbf{A}^t\mathbf{k}_b\mathbf{A}dx\ ; \quad \mathbf{K}_f=\int_B^L \mathbf{N}^t\mathbf{K}_f\mathbf{N}dx \tag{a_{10}}$$

Thus it is seen that the stiffness is explicitly obtained as a sum of the stiffnesses of the beam and the foundation springs. The first one is the usual beam stiffness matrix given in Eq. (5.45), of course, the part which relates to only moments and shears, namely

$$\mathbf{K}_b=\left[\begin{array}{cc|cc} \frac{12EI}{L^3} & \frac{6EI}{L^2} & -\frac{12EI}{L^3} & \frac{6EI}{L^2} \\ & \frac{4EI}{L} & -\frac{6EI}{L^2} & \frac{2EI}{L} \\ \hline & & \frac{12EI}{L^3} & -\frac{6EI}{L^2} \\ \text{Sym.} & & & \frac{4EI}{L} \end{array}\right] \tag{a_{11}}$$

The stiffness $\mathbf{K}_f$ requires the evaluation of the integral

$\int_0^L \mathbf{N}^t k_f \mathbf{N}dx$, which reads

$$\mathbf{K}_f=k_f\int_O^L \begin{bmatrix} N_1^2 & N_1N_2 & N_1N_3 & N_1N_4 \\ & N_2^2 & N_2N_3 & N_2N_4 \\ & & N_3^2 & N_3N_4 \\ \text{Sym.} & & & N_4^2 \end{bmatrix} dx \tag{a_{12}}$$

Here $\mathbf{k}_f$ has been taken outside the integral assuming that it is uniform. Evaluation of the integrals is quite straightforward:

$$\int_0^L N_1^2 dx = \int_0^L N_3^2 dx = \frac{13}{35} L \ ; \ \int_0^L N_2^2 dx = \int_0^L N_4^2 dx = \frac{L^3}{105}$$

$$\int_0^L N_1 N_2 dx = -\int_0^L N_3 N_4 dx = \frac{11}{210} L^2 \ ; \ \int_0^L N_2 N_3 dx = -\int_0^L N_1 N_4 dx$$

$$= -\frac{13}{420} L^2$$

$$\int_0^L N_1 N_3 dx = \frac{9}{70} L \ ; \ \int_0^L N_2 N_4 dx = -\frac{L^3}{140} \qquad (a_{13})$$

The equality between certain integrals becomes obvious by looking at the shapes of the N_i diagrams. (See Figs. 5.8, Note, $N_1 \equiv N_{22}$; $N_2 \equiv N_{23}$, etc.)

Thus the total stiffness matrix obtained by adding (a_{11}) and (a_{12}) is as follows:

$$\begin{Bmatrix} Q_1 \\ M_1 \\ Q_2 \\ M_2 \end{Bmatrix} = \begin{bmatrix} \frac{EI}{L}\psi_5 & \frac{EI}{L^2}\psi_3 & -\frac{EI}{L^3}\psi_6 & \frac{EI}{L^2}\psi_4 \\ & \frac{EI}{L}\psi_1 & -\frac{EI}{L^2}\psi_4 & \frac{EI}{L}\psi_2 \\ & & \frac{EI}{L^3}\psi_5 & -\frac{EI}{L^2}\psi_3 \\ \text{Sym.} & & & \frac{EI}{L}\psi_1 \end{bmatrix} \begin{Bmatrix} v_1 \\ \theta_1 \\ v_2 \\ \theta_2 \end{Bmatrix} \qquad (a_{14})$$

Here ψ_1, ψ_2, etc., are functions similar to those introduced in Eq. (a_9) of Example 5.3.

$$\psi_1 = 4\left(1 + \frac{\lambda^4}{105}\right) ; \ \psi_2 = 2\left(1 - \frac{\lambda^4}{70}\right)$$

$$\psi_3 = 6\left(1 + \frac{11\lambda^4}{315}\right) ; \ \psi_4 = 6\left(1 - \frac{13\lambda^4}{630}\right) \qquad (a_{15})$$

$$\psi_5 = 12\left(1 + \frac{13}{105}\lambda^4\right) ; \ \psi_6 = 12\left(1 - \frac{3\lambda^4}{70}\right)$$

Here $\lambda^4 = k_f L^4/4EI$. These can be considered as approximation to ψ's of Eq. (a_9) of Example 5.3.

The above example shows the general nature of the possible approximations in deriving the stiffness relations. The above approximations are obviously valid for smaller values of λ, namely for lower foundation stiffness. For example, with $\lambda^4 = 1$, the exact and appropriate values are, as follows:

	ψ_1	ψ_2	ψ_3	ψ_4	ψ_5	ψ_6
Exact	4.040	1.968	6.202	5.700	11.452	11.512
Approx.	4.038	1.972	6.21	5.8764	13.490	11.489

These show that for the value of $\lambda^4=1$ the approximate expressions are quite satisfactory.

The concept of spring can be generalized. One can work with springs which offer moments proportional to the rotations. This is demonstrated in the next Example.

EXAMPLE 5.6. *Elastically Coupled Beam*

Figure (*a*) shows the case of a cantilever beam with elastic coupling of a different nature. The beam is elastically restrained by a series of cross beams which exert an axial load q and moment $m=qa$, per unit length. The free body of the cross beam (Fig. *b*) shows that these forces are related to its stiffnesses. Let

$$\mathbf{q}=\begin{Bmatrix} q \\ m \end{Bmatrix}=\begin{bmatrix} \lambda_1 & \lambda_2 \\ \lambda_2 & \lambda_3 \end{bmatrix}\begin{Bmatrix} u \\ \theta \end{Bmatrix}=\mathbf{k}_b\mathbf{U}_b \qquad (a_1)$$

where λ_1, λ_2 and λ_3 are the stiffness coefficients of the cross beam, and u and θ are the axial displacement and rotation of the main beam. Such a situation is encountered in *shear walls* of buildings and in *open web beams*. Presently it is a case of elastically supported beam similar to the beam on elastic foundation of the previous example. The shape functions (5.56) in terms of $\xi=x/L$, and different subscripts (as in previous example), are

$$N_5=1-\xi\ ;\ N_6=\xi,$$
$$N_1=1-3\xi^2+2\xi^3\ ;\ N_2=L\xi(1-\xi)^2$$
$$N_3=3\xi^2-2\xi^3\ ;\qquad N_4=L(\xi^3-\xi^2) \qquad (a_2)$$

The expression for $\mathbf{U}_b$ is

$$\mathbf{U}_b=\begin{Bmatrix} u \\ \theta \end{Bmatrix}=\begin{bmatrix} N_5 & 0 & 0 & N_6 & 0 & 0 \\ 0 & N_1' & N_2' & 0 & N_3' & N_4' \end{bmatrix}\begin{Bmatrix} u_1 \\ v_1 \\ \theta_1 \\ u_2 \\ v_2 \\ \theta_2 \end{Bmatrix}=\mathbf{N}_b\mathbf{U} \qquad (a_3)$$

with dash denoting differentiation with x.

Therefore, we have

$$\mathbf{q}=\mathbf{K}_b\mathbf{N}_b\mathbf{U} \qquad (a_4)$$

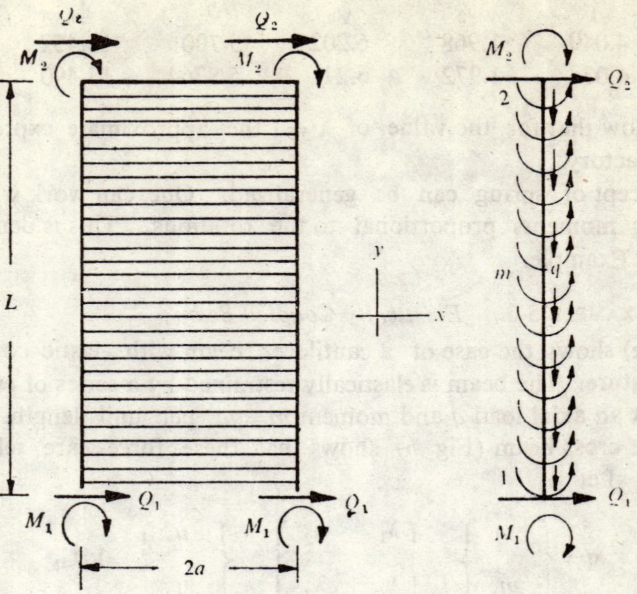

(a)

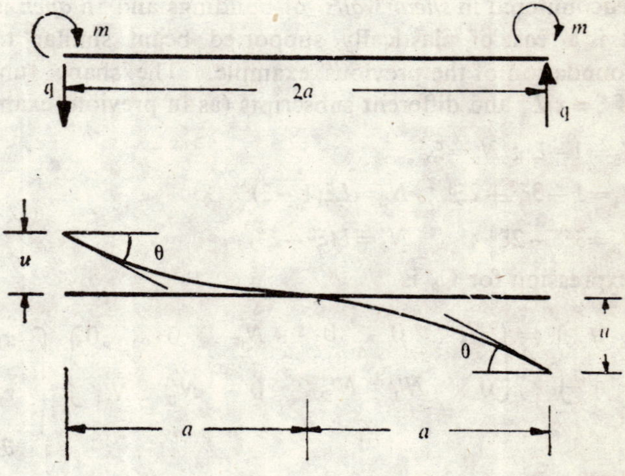

(b)

Example 5.6

The equation of virtual work becomes,*

$$\mathbf{P}^t\mathbf{U} - \int \mathbf{q}^t\mathbf{u}_b dx = \int \mathbf{S}^t \boldsymbol{\epsilon} dx \qquad (a_5)$$

* Observe that the forces q and m oppose the displacements u and O ; hence the negative work.

In this we use Eqs. (a_3) and (a_4); for $\boldsymbol{\epsilon}$ we have the general expression $\boldsymbol{\epsilon}=\mathbf{A\,U}$, for $\mathbf{S}$, the elastic relation $\mathbf{S}=\mathbf{k}\boldsymbol{\epsilon}=\mathbf{kAU}$. See Eqs. (5.58 and 5.59). In the present notation,

$$\boldsymbol{\epsilon}=\begin{Bmatrix}\dfrac{du}{dx}\\[2ex] \dfrac{d^2v}{dx}\end{Bmatrix}=\begin{bmatrix} N'_5 & 0 & 0 & N'_6 & 0 & 0\\ 0 & N_1'' & N_2'' & 0 & N_3'' & N_4''\end{bmatrix}\begin{Bmatrix}u_1\\ v_1\\ \vdots\\ \theta_2\end{Bmatrix}=\mathbf{AU} \qquad (a_6)$$

Thus Eq. (a_5) becomes

$$\mathbf{P}^t\mathbf{u}-\int(\mathbf{k}_b\mathbf{N}_b\mathbf{u})^t\mathbf{N}_b\mathbf{u}\,dx=\int(\mathbf{kAu})^t\mathbf{Au}\,dx \qquad (a_7)$$

On cancelling the virtual displacement $\mathbf{u}$ and transposing,

$$\mathbf{P}-[\int\mathbf{N}_b{}^t\mathbf{k}_b\mathbf{N}_b\,dx]\mathbf{u}=[\int\mathbf{A}^t\mathbf{kA}\,dx]\mathbf{u} \qquad (a_8)$$

This can be written as

$$[\mathbf{K}+\mathbf{k}_b]\mathbf{U}=\mathbf{P} \qquad (a_9)$$

where $\mathbf{K}=\int\mathbf{A}^t\mathbf{kA}\,dx$ is the ordinary stiffness matrix of the beam and $\mathbf{K}_b$ is the additional stiffness as a result of the elastic cross beams:

$$\mathbf{K}_b=\int\mathbf{N}_b{}^t\mathbf{k}_b\mathbf{N}_b\,dx \qquad (a_{10})$$

The triple product can be performed and each term integrated as in the previous example. Finally,

$$\mathbf{K}_b=\begin{bmatrix}
\dfrac{\lambda_1L}{3} & -\dfrac{\lambda_2}{2} & \dfrac{\lambda_2L}{12} & \dfrac{\lambda_1L}{6} & \dfrac{\lambda_2}{2} & -\dfrac{\lambda_2L}{12}\\[2ex]
 & \dfrac{6}{5}\,\dfrac{\lambda_3}{L} & \dfrac{\lambda_3}{10} & -\dfrac{\lambda_2}{2} & -\dfrac{6\lambda_3}{5L} & \dfrac{\lambda_3}{10}\\[2ex]
 & & \dfrac{2\,\lambda_3}{15} & -\dfrac{\lambda_2L}{12} & -\dfrac{\lambda_3}{10} & -\dfrac{\lambda_3L}{30}\\[2ex]
 & & & \dfrac{\lambda_1L}{3} & \dfrac{\lambda_2}{2} & -\dfrac{\lambda_2L}{12}\\[2ex]
 & & & & \dfrac{6}{5}\,\dfrac{\lambda_3}{L} & -\dfrac{\lambda_3}{10}\\[2ex]
 & \text{Sym.} & & & & \dfrac{2\lambda_3L}{15}
\end{bmatrix} \qquad (a_{11})$$

Adding this to the stiffness matrix given in Eq. (5.45) we obtain the complete stiffness as given below:

$$K=\begin{bmatrix} \frac{AE}{L}\phi_1 & -\frac{AE}{L}\phi_2 & \frac{AE}{L}\phi_3 & -\frac{AE}{L}\phi_4 & \frac{AE}{L}\phi_2 & -\frac{AE}{L}\phi_3 \\ & \frac{EI}{L^3}\phi_5 & \frac{EI}{L^2}\phi_6 & -\frac{AE}{L}\phi_2 & -\frac{EI}{L^3}\phi_5 & \frac{EI}{L^2}\phi_6 \\ & & \frac{EI}{L}\phi_7 & -\frac{AE}{L}\phi_3 & \frac{EI}{L^2}\phi_6 & \frac{EI}{L}\phi_8 \\ & & & \frac{AE}{L}\phi_1 & \frac{AE}{L}\phi_2 & -\frac{AE}{L}\phi_3 \\ & \text{Sym.} & & & \frac{EI}{L^3}\phi_5 & -\frac{EI}{L^2}\phi_6 \\ & & & & & \frac{EI}{L}\phi_7 \end{bmatrix} \qquad (a_{12})$$

with

$$\phi_1=\left(\frac{1+\lambda_1 L^2}{3AE}\right);\ \phi_2=\frac{\lambda_2 L}{2AE};\ \phi_3=\frac{\lambda_2 L^2}{12AE}$$

$$\phi_4=\left(1-\frac{\lambda_1 L^2}{6AE}\right);\ \phi_5=12\left(1+\frac{\lambda_3 L^2}{10EI}\right);\ \phi_6=6(1+\lambda_3 L^2/60EI)$$

$$\phi_7=4(1+\lambda_3 L^2/30EI);\ \phi_8=2(1-\lambda_3 L^2/60EI). \qquad (a_{13})$$

The constants λ_1, λ_2, λ_3, for a beam of uniform section can be easily determined. These are as in the following equation:

$$\begin{Bmatrix} q \\ m \end{Bmatrix}=\begin{bmatrix} 3EI/a^3 & -3EI/a^2 \\ -3EI/a^2 & 3EI/a \end{bmatrix}\begin{Bmatrix} u \\ \theta \end{Bmatrix} \qquad (a_{14})$$

5.10 EQUIVALENT LOADS AND COMPLETE GENERAL STIFFNESS

Recall the contents of Section 1.5 of Chapter 1, in which the definition of *equivalent loads* was introduced. The member analysis treats each member to be 'fixed' against all possible end displacements. The internal distributed loads, and also loads such as prescribed temperature shrinkage or creep strains require certain 'fixing end forces' at the ends of the member. Forces equal and opposite to these are the equivalent loads to be taken for the structural analysis which concerns itself with the determination of the joint displacements.

The complete basic stiffness for a member will now be derived as

$$\mathbf{S}=\mathbf{kv}+\mathbf{S}_F \qquad (5.65)$$

By the usual transformations we obtain the general stiffness relations.

To achieve this recall equilibrium relations (2.7) which we can rewrite as

$$\bar{\mathbf{P}} = \bar{\mathbf{C}}\,\mathbf{S} + \bar{\mathbf{P}}^o \equiv \bar{\mathbf{A}}^t\mathbf{S} + \bar{\mathbf{P}}^o \tag{a_1}$$

where the contragradient feature $\bar{\mathbf{A}} \equiv \bar{\mathbf{C}}^t$ has been utilized. The kinematic transformations (3.7) are $\mathbf{V} = \bar{\mathbf{A}}\,\bar{\mathbf{U}}$. Thus we obtain $\bar{\mathbf{P}}$ as

$$\bar{\mathbf{P}} = [\bar{\mathbf{A}}^t\mathbf{k}\bar{\mathbf{A}}]\bar{\mathbf{U}} + \bar{\mathbf{A}}^t\mathbf{S}_F + \bar{\mathbf{P}}^o \tag{a_2}$$

Thus the complete general stiffness for a member is

$$\bar{\mathbf{P}} = \bar{\mathbf{K}}\,\bar{\mathbf{U}} + \bar{\mathbf{P}}_F \tag{5.66}$$

Here $\bar{\mathbf{P}}_F = \bar{\mathbf{A}}^t\mathbf{S}_F + \bar{\mathbf{P}}^o$, is the vector of fixing end forces ($\bar{\mathbf{P}}_F \equiv \mathbf{P}$ when $\bar{\mathbf{U}} \equiv 0$). Procedures to determine $\mathbf{S}_F$ or $\bar{\mathbf{P}}_F$ are similar to those employed to determine $\mathbf{k}$ or $\bar{\mathbf{K}}$. These are as follows :

(*a*) Inversion of flexibility relations, or

(*b*) Direct evaluation based on assumed displacements.

Consider the first approach The complete basic flexibility relation will be first derived in the form

$$\mathbf{v} = \mathbf{f}\mathbf{S} + \mathbf{v}^o \tag{5.67}$$

and the stiffness relation (5.65) obtained as

$$\mathbf{S} = \mathbf{f}^{-1}(\mathbf{v} - \mathbf{v}^o) = \mathbf{k}\mathbf{v} - \mathbf{k}\mathbf{v}^o \tag{5.68}$$

Here $\mathbf{v}^o$ are the deformations in the member due to lateral loads alone ($\mathbf{S} \equiv 0$). Equation (5.67) simply states the superposition of effects due to the member forces and lateral loads each considered separately. We therefore see that

$$\mathbf{S}_F = -\mathbf{k}\mathbf{v}^o \tag{5.69}$$

Consider now the expression for the same in terms of the loads. Let $\mathbf{S}^o$ be the internal forces at any section within the beam due to the lateral loads alone ($\mathbf{S} \equiv 0$). Then the deformations occurring in a differential element are (see Eq. (a_7) of Sec. 5.4),

$$\boldsymbol{\epsilon}^o = \mathbf{f}_x\mathbf{S}^o \tag{a_3}$$

To compute $\mathbf{v}^o$ we adopt a virtual forces system and apply virtual work principle. The virtual force system is the system in which the member is acted on by forces $\mathbf{S}$ causing internal forces $\mathbf{B}\,\mathbf{S}$. Therefore we write

$$\mathbf{S}^t\mathbf{v}^o = \int(\mathbf{B}\mathbf{S})^t\boldsymbol{\epsilon}^o\,dx$$

Cancelling the arbitrary force $\mathbf{S}^t$ and using (a_3), we obtain

$$\mathbf{v}^o = \int\mathbf{B}^t\boldsymbol{\epsilon}^o\,dx = \int\mathbf{B}^t\mathbf{f}_x\mathbf{S}^o\,dx \tag{5.70}$$

As an application consider the case of uniform loads $p = p_o$ and $q = q_o$ (Fig. 5.10) and the beam model, The statical matrix $\mathbf{B}$ is as in Eq. (a_{16}) of See 5.4; to rewrite, neglecting the shearing force term

$$\mathbf{S}_x=\begin{Bmatrix} S_x \\ M_x \end{Bmatrix}=\begin{bmatrix} 1 & 0 & 0 \\ 0 & -\left(1-\frac{x}{L}\right) & \frac{x}{L} \end{bmatrix}\begin{Bmatrix} S \\ M_1 \\ M_2 \end{Bmatrix}=\mathbf{B}\,\mathbf{S} \quad (a_4)$$

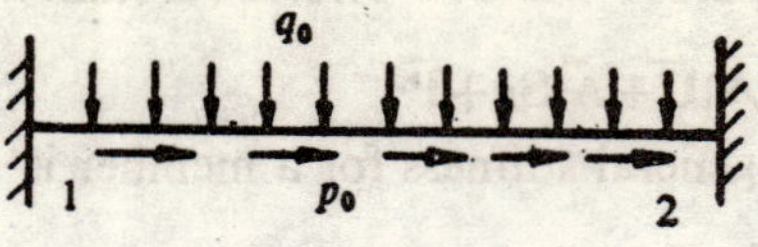

Fig. 5.10

The flexibility matrix $\mathbf{f}_x$ is given by

$$\boldsymbol{\epsilon}=\begin{Bmatrix} \epsilon_x \\ \chi_x \end{Bmatrix}=\begin{bmatrix} 1/AE & 0 \\ 0 & 1/EI \end{bmatrix}\begin{Bmatrix} S_x \\ M_x \end{Bmatrix}=\mathbf{f}_x\mathbf{S}_x$$

The expressions for $\mathbf{S}^o$ due to $\mathbf{p}_o$ and $\mathbf{q}_o$, are

$$\mathbf{S}^o=\begin{Bmatrix} S^o \\ M^o \end{Bmatrix}=\begin{bmatrix} x & 0 \\ 0 & -\frac{Lx}{2}\left(1-\frac{x}{L}\right) \end{bmatrix}\begin{Bmatrix} p_o \\ q_o \end{Bmatrix}$$

We therefore have

$$\boldsymbol{\epsilon}^o=\left\{\frac{x}{AE}\,p_o,\ -\frac{Lx}{2EI}\left(1-\frac{x}{L}\right)q_o\right\}$$

Equation (5.70) then becomes

$$\mathbf{v}^o=\begin{Bmatrix} e^o \\ \varphi_1{}^o \\ \varphi_2{}^o \end{Bmatrix}=\int_0^L\begin{bmatrix} 1 & 0 \\ 0 & -\left(1-\frac{x}{L}\right) \\ 0 & \frac{x}{L} \end{bmatrix}\begin{Bmatrix} \frac{x}{AE}\,p_o \\ -\frac{Lx}{2EI}\,q_o\left(1-\frac{x}{L}\right) \end{Bmatrix} \quad (a_5)$$

Quantities *AE* and *EI* are variable in a beam of variable section. In such a case we adopt numerical integration. We can verify that for a uniform beam we obtain, on integration,

$$\mathbf{v}^o=\left\{\frac{L^2p_o}{2AE},\ \frac{L^3}{24EI}q_o,\ -\frac{L^3}{24EI}q_o,\right\}$$

The fixing end forces are those which are required to remove these deformations, as stated by Eq. (5.69); taking $\mathbf{k}$ from Eq. (5.35), with $\beta=0$,

$$\mathbf{S_F} = -\begin{bmatrix} \frac{AE}{L} & 0 & 0 \\ 0 & \frac{4EI}{L} & \frac{2EI}{L} \\ 0 & \frac{2EI}{L} & \frac{4EI}{L} \end{bmatrix} \begin{Bmatrix} \frac{L^2}{2AE}\, p_0 \\ \frac{L^3}{24EI}\, q_0 \\ -\frac{L^3}{24EI}\, q_0 \end{Bmatrix} = \begin{Bmatrix} -p_0 L/2 \\ -\frac{q_0 L^2}{12} \\ \frac{q_0 L^2}{12} \end{Bmatrix} \tag{a$_6$}$$

Using $\bar{A}$ from Eq. (3.3) and noting that

$$\overline{\mathbf{P}^{\circ}} \equiv \left\{ -p_0 L,\ -\frac{q_0 L}{2}\, 0,\ 0,\ -\frac{q_0 L}{2},\ 0 \right\}$$

we obtain from Eq. (5.66)

$$\overline{\mathbf{P_F}} = \overline{\mathbf{A}}^t \mathbf{S_F} + \overline{\mathbf{P}^{\circ}}$$

$$= \left\{ \frac{-p_0 L}{2},\ \frac{-q_0 L}{2},\ \frac{-q_0 L^2}{12},\ -\frac{p_0 L}{2},\ -\frac{q_0 L}{2},\ \frac{q_0 L^2}{12} \right\} \tag{a$_7$}$$

The general case of a beam of nonuniform section involving numerical integration will be demonstrated after discussing the details of numerical integration.

Kinematic Approach

Consider now the second approach which is based on a purely kinematical approach which will be helpful in situations where the stiffness is derived from an assumed deformation pattern. This therefore finds application in finite element analysis.

Consider, before proceeding with the general set of forces, the simplest case of having to determine the fixing end moment M_{F1} for a beam of uniform section loaded by transverse load of $q(x)$ (Fig. 5.11 a). To determine this we adopt the principle of virtual work. We will convert this statical problem to one of geometry. This is essentially the Muller-Breslau principle as applied to the problem of constructing influence lines. To determine M_{F1} we choose a virtual displacement pattern v^*. This is the deformed shape due to a prescribed value (unity for convenience) of the displacement corresponding to the moment. Thus Fig. 5.11 (b) shows the deformed shape so that we have only at end 1, $\theta = \theta_1^*$. The true deformed shape can in this case be determined. In fact, this is the shape function N_{23} given in Eq. (5.56). Thus

$$v^* = L\left(\frac{x}{L}\right)\left(1 - \frac{x}{L}\right)^2 \theta_1^* = N_1\, \theta_1^* \tag{a$_8$}$$

The equation of virtual work is

$$M_{F1}\theta_1^* + \int q v^*\, dx = \int M_x (d^2 v^*/dx^2)\, dx \tag{a$_9$}$$

We will now show that tne right hand side is zero. To do this, we consider the work of the forces of the assumed virtual system and the deformation of the actual given system. Since the end displacements in the latter system are all zero, we have zerc external work

$$\int M_x^* \left(\frac{d^2v}{dx^2}\right) dx=0 \tag{a_{10}}$$

Since

$$M=EI\frac{d^2v}{dx^2} \quad \text{and} \quad M^*=EI\,\frac{d^2v^*}{dx^2}, \text{ we see that}$$

$$\int M_x^* \frac{d^2v}{dx^2}\, dx=\int M_x\frac{d^2v^*}{dx^2}\, dx=\int EI\frac{d^2v}{dx^2}\,\frac{d^2v^*}{dx^2}\, dx. \tag{a_{11}}$$

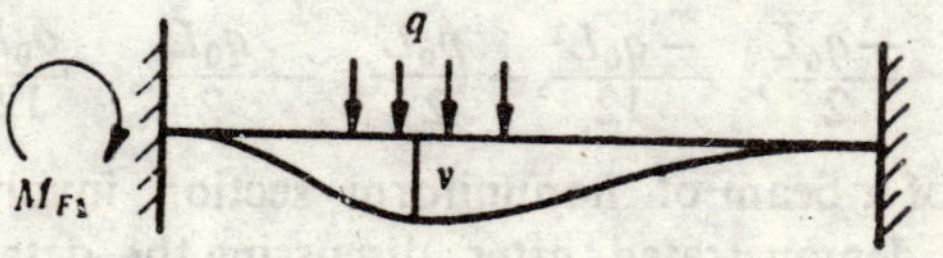

(a)

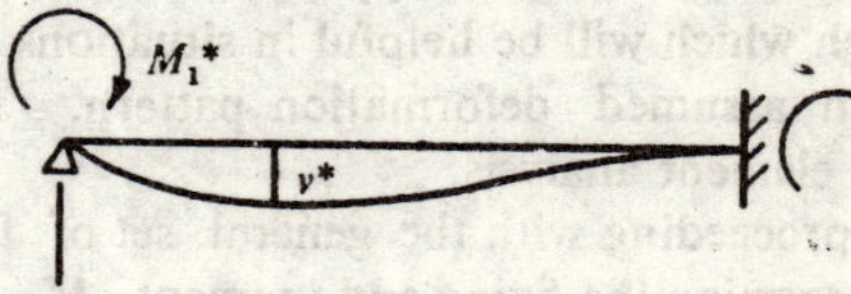

(b)

Fig. 5.11

Probably the reader may more readily appreciate the equality of the integrals if expressed as

$$\int MM^*\frac{dx}{EI}=\int M^*M\frac{dx}{EI} \tag{a_{12}}$$

Therefore setting the right hand side of Eq. (a_9) to zero, and using (a_8),

$$M_{F1}\times\theta_1^*=-\int qv^*\, dx=\left(-\int q\, N_1 dx\right)\theta_1^*$$

Cancelling θ_1^* (which is what is equivalent to considering a unit value of θ_1^*), we obtain

$$M_{F1}=-\int N_1 q\; dx \tag{a_{13}}$$

For a uniform load q_0

$$M_{F1} = -\int_0^L L\left(\frac{x}{L}\right)\left(1-\frac{x}{L}\right)^2 q_0 \; dx = -q_0 \frac{L^2}{12} \qquad (a_{14})$$

This agrees with the wellknown result. We can verify the case of a concentrated load at a distance x. In this case the virtual work will be the value of the load P multiplied by the value of v^* at x.

Thus

$$M_{F1} = -P \times v^* \qquad (a_{15})$$

This is the Muller-Breslau principle for the influence line for tne bending moment M_{F1}. Equation (a_{13}) is thus seen to be a generalization.

This form is useful where the deformed shape is assumed. Let us assume the following shape for the above problem:

$$v^* = \left(\frac{L}{\pi} \sin \frac{\pi x}{L} - \frac{L}{4\pi} \sin \frac{2\pi x}{L}\right) \theta_1^* = N_1\theta_1^* \qquad (a_{16})$$

Thus in place of Eq. (a_{14}) we have

$$M_{F1} = -\int \left(\frac{L}{2\pi} \sin \frac{\pi x}{L} - \frac{L}{4\pi} \sin \frac{2\pi x}{L}\right) q_0 \, dx = -\frac{q_0 L^2}{\pi} \qquad (a_{17})$$

This moment is the fixing end moment consistent with a stiffness of the beam derived by assuming deformed shape (a_{16})

General Member Stiffness

We will derive the general stiffness relations for a member carrying distributed load. Let

$$\mathbf{u}_x = \mathbf{N}\,\bar{\mathbf{U}} \qquad (a_{18})$$

be the deformed shape, either exact or an assumed one. Here $\mathbf{U}$ is the vector of member end displacements. These will determine strains

$$\boldsymbol{\epsilon} = \bar{\mathbf{A}}\,\bar{\mathbf{U}} \qquad (_{19})$$

Let $\boldsymbol{\epsilon}^\circ$ be the vector of *initial strains* due to temperature. The elastic strains because of the internal forces are $(\boldsymbol{\epsilon}-\boldsymbol{\epsilon}^\circ)$. Thus the elastic laws give

$$\mathbf{S}_x = \mathbf{k}_x(\boldsymbol{\epsilon}-\boldsymbol{\epsilon}^\circ) = \mathbf{k}_x(\bar{\mathbf{A}}\,\bar{\mathbf{U}}-\boldsymbol{\epsilon}^\circ) \qquad (a_{20})$$

Observe that when $\boldsymbol{\epsilon}=\mathbf{O}$, which means that total strains are completely prevented, internal forces $\mathbf{S}_x = -\mathbf{k}_x\,\boldsymbol{\epsilon}^\circ$ are required. These would be reduced somewhat as a result of elastic strains which occur due to the deformability of the body.

The loads on the beam, are the distributed loads

$$\mathbf{g} = \{\,p(x),\, q(x)\,\} \qquad (a_{21})$$

and the end forces $\overline{\mathbf{P}}$. The equation of virtual work becomes,

$$\overline{\mathbf{P}}^t\overline{\mathbf{U}} + \int \mathbf{g}^t\,\mathbf{u}_x\,dx = \int_L \mathbf{S}^t_x\,\boldsymbol{\epsilon}\,dx$$

Using (a_{18}) for $\mathbf{u}_x$ and (a_{19}) for $\boldsymbol{\epsilon}$, cancelling the arbitrary displacements $\overline{\mathbf{u}}$, and taking transpose, we obtain the equation of equilibrium

$$\overline{\mathbf{P}} + \int_L \mathbf{N}^t\mathbf{g}\,dx = \int \overline{\mathbf{A}}^t\,\mathbf{S}_x\,dx \qquad (a_{22})$$

We next use (a_{20}) for $\mathbf{S}_x$ and write this in general form as

$$\overline{\mathbf{P}} = \overline{\mathbf{K}}\,\overline{\mathbf{U}} + \overline{\mathbf{P}}_{Fg} + \overline{\mathbf{P}}_{F\epsilon^\circ} \qquad (5{\cdot}71)$$

with

$$\overline{\mathbf{K}} = \int \overline{\mathbf{A}}^t\,\mathbf{k}_x\,\overline{\mathbf{A}}\,dx \qquad (5{\cdot}72)$$

$$\overline{\mathbf{P}}_{Fg} = -\int \mathbf{N}^t\mathbf{g}\,dx \qquad (5{\cdot}73)$$

$$\overline{\mathbf{P}}_{F\epsilon^\circ} = -\int \overline{\mathbf{A}}^t\mathbf{k}_x\,\mathbf{E}^\circ\,dx \qquad (5.74)$$

Here $\overline{\mathbf{P}}_{Fg}$ and $\overline{\mathbf{P}}_{F\epsilon^\circ}$ are the fixing end forces due to the prescribed loads, and temperature respectively, since these are the values of $\overline{\mathbf{P}}$ when, $\overline{\mathbf{u}} = \mathbf{o}$, namely when end displacements are prevented completely. Equivalent loads are the loads equal and opposite to the above.

Form (5.74) admits a more general interpretation. $\boldsymbol{\epsilon}^\circ$ need not be necessarily due to temperature. They can be due to *shrinkage*, *creep*, or *plasticity*. The forces $\overline{\mathbf{P}}_F\boldsymbol{\epsilon}^\circ$ are the forces required to restrain the member against the free member end deformations produced by any of these causes. Let

$$\boldsymbol{\epsilon}^\circ = \{\,\boldsymbol{\epsilon}^\circ\ \chi^\circ\,\} \qquad (a_{23})$$

be a set of strains due to any of the above causes (See Fig. 5.12)

The $\overline{\mathbf{A}}$ matrix is as in Eq. (a_{16}) of the previous section; also same as Eq. 5.64. The general form is

$$\boldsymbol{\epsilon} = \begin{bmatrix} \overline{A}_{11} & \overline{A}_{12} & \cdots & \overline{A}_{16} \\ \overline{A}_{21} & \overline{A}_{22} & \cdots & \overline{A}_{26} \end{bmatrix} \begin{Bmatrix} \overline{u}_1 \\ \overline{v}_1 \\ \vdots \\ \overline{\theta}_2 \end{Bmatrix} = \overline{\mathbf{A}}\,\overline{\mathbf{U}} \qquad (a_{24})$$

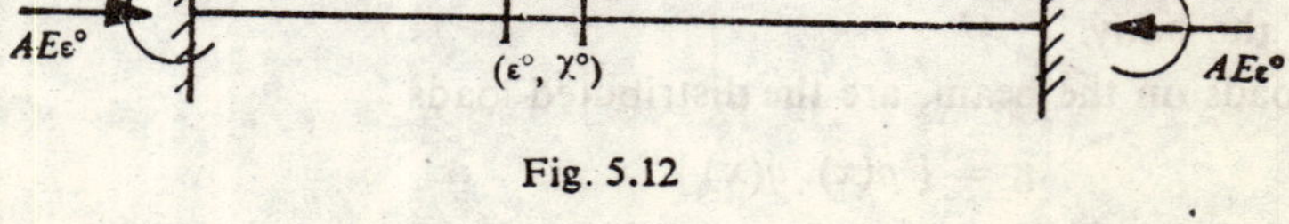

Fig. 5.12

Thus

$$\mathbf{P_F} = \int_0^L \begin{bmatrix} \bar{A}_{11} & \bar{A}_{21} \\ \bar{A}_{12} & \bar{A}_{22} \\ \cdots & \cdots \\ \cdots & \cdots \\ \bar{A}_{16} & \bar{A}_{26} \end{bmatrix} \begin{Bmatrix} A\,E\,\epsilon^\circ \\ E\,I\,\chi^c \end{Bmatrix} dx \tag{a$_{25}$}$$

Using $\bar{A}_{ij}$ from Eq.(5.64), and treating ϵ° and χ° as constants, we obtain, on integration,

$$\bar{P}_{F1} = -\int \left(-\frac{1}{2}\right) AE\epsilon^\circ\, dx = AE\epsilon^\circ$$

$$\bar{P}_{F2} = -\int \left(-\frac{6}{L^2} + \frac{12x}{L^3}\right) EI\,\chi^\circ\, dx = 0 \tag{a$_{26}$}$$

$$\bar{P}_{F3} = EI\,\chi^\circ;\quad \bar{P}_{F4} = -AE\,\epsilon^\circ$$

$$\bar{P}_{F5} = 0;\quad \bar{P}_{F6} = -EI\,\chi^\circ$$

These show that $\bar{P}_{F1}$ and $\bar{P}_{F4}$ are compressive forces equal to $AE\epsilon^\circ$. (Watch the sign conventions). These are axial compressive forces to prevent the longitudinal strains ϵ°. Due to (hogging) curvature χ° we have a constant moment throughout giving moments at end 1 and 2 equal to $EI\chi^\circ$; $\bar{P}_{F2}$ and $\bar{P}_{F5}$, which represent fixing end shears, are both seen to be zero as expected since the bending moments are uniform throughout.

The expressions indicated by Eqs. (5.72, 5.73 and 5.74) are general and used for any assumed displacement pattern. We will have occasion in Chapter 9 to return to these equations.

EQUIVALENT JOINT LOADS

The member equivalent loads summed over the members meeting at a joint gives the equivalent joint (nodal) load for the structural analysis of frame. This will be explained in Chapter 8.

5.11 NUMERICAL INTEGRATION

For members of non-uniform section the integrations required in determining member flexibility, stiffness, or equivalent loads are to be evaluated numerically. The integrals $\int \mathbf{A}^t\,\mathbf{k}\,\mathbf{A}\,dx$ or $\int \mathbf{B}^t\,\mathbf{f}\,\mathbf{B}\,dx$ will involve integrals of type

$$I=\int_0^L f(x)\,dx \qquad (a_1)$$

Of the many possible schemes, the Gauss schemes are commonly adopted in structural mechanics with good accuracy. In adopting such schemes it is usual, though not necessary to transform the integration limits from -1 to 1 by changing x to ξ :

$$x=\frac{1+\xi}{2}L \qquad (a_2)$$

Thus we consider the integration rule (Fig. 5.13 *a*).

$$I=\int_{-1}^{1} f(\xi)\,d\xi \qquad (a_3)$$

The simplest trapezoidal scheme assuming linear variation of $f(\xi)$ is

$$I=f(-1)+f(1) \qquad (a_4)$$

The Simpson rule is:

$$I=\tfrac{1}{3}\,(f(-1)+4\,f(0)+f(1)) \qquad (a_5)$$

A higher order Simpson rule is

$$I=\tfrac{1}{4}\,(f(-1)+3\,f(-\tfrac{1}{3})+3\,f(\tfrac{1}{3})+f(1)) \qquad (a_6)$$

All these rules involve the boundary values $f(-1)$ and $f(1)$.

The Gauss rules involve only values *within* the interval and are found to give better estimates. The general form is

$$I=\int_{-1}^{1} f(\xi)\,d\xi=\sum_{i=1}^{n} H_i\,f(\xi_i) \qquad (5.75)$$

Here H_i are the weighting factors to multiply the ordinates $f(\xi_i)$ evaluated at suitably chosen integrating points with abscissae ξ_i. The simplest two point rule, $(n=2)$, is

$$I=(1.0)\,f(-0.577)+1.0\,f(0.577) \qquad (a_7)$$

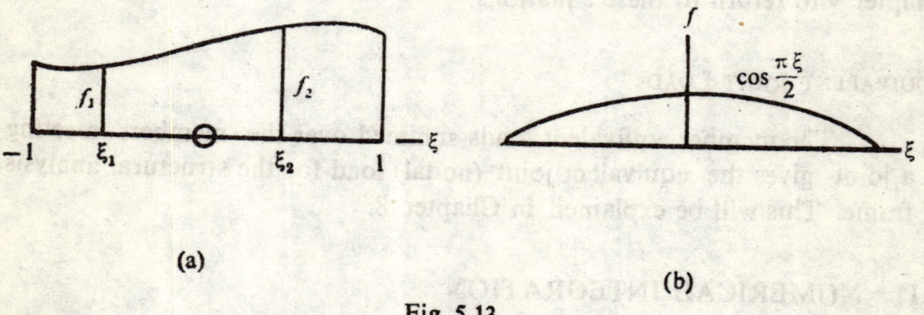

Fig. 5.13

The three point rule $(x=3)$, is

$$I=0.55555\,f\,(-0.77459)+0.88888\,f(0)+0.55555\,f(0.77459) \qquad (a_8)$$

Consider for illustration (Fig. 5.13*b*):

$$f(\xi)=\cos\frac{\pi\xi}{2} \qquad (a_9)$$

The exact value of the integral between -1 and 1 is $4/\pi = 1.2732395$. Eq. (5.75) becomes, for $n=2$,

$$I = 1.0 \times \cos\frac{\pi}{2}(-0.577350) + 1 \times \cos\frac{\pi}{2}(0.577350) \qquad (a_{10})$$

$$= 2\cos(51.96°) = 1.232382.$$

The higher order rule ($n=3$) gives

$$I = 0.555555\left\{\cos\frac{\pi}{2}(-0.774597) + \cos\frac{\pi}{2}(0.774597)\right\}$$

$$+0.888888\cos\frac{\pi}{2}(0) \qquad (a_{11})$$

$$= 0.555555 \times 2 \times 0.34671 + 0.88888 \times 1.0 = 1.27413$$

The abscissae and weight factors upto $n=5$ are as shown in Table 5. For more accurate and higher order rules see Ref. (12)

TABLE 5.1

Gauss Integration Rules

$$I = \int_{-1}^{1} f(\xi)\, d\xi = \sum_{1}^{n} H_i f(\xi_i).$$

	ξ_i	H_i
$n=2$	±0.577350	1.0
$n=3$	±0.774597	0.55555—
	0	0.88888
$n=4$	±0.861136	0.347855
	±0.339981	0.652145
$n=5$	±0.906180	0.236927
	±0.538469	0.478629
	0	0.568889

We may use interval as 0 to L and express the integral as follows :

$$\int_0^L f(x)\,dx = \frac{L}{2}\sum_{i=1}^{n} H_i f(x_i) \qquad (5.76)$$

with

$$x_i = \frac{L}{2}(1+\xi_i) \qquad (5.77)$$

Taking $f(x)=x$ as an example

$$I=\int x dx=\frac{L}{2}\sum_{i=1}^{n} H_i \frac{L}{2}(1+\xi_i)$$

which for $n=2$ becomes

$$I=\frac{L}{2}(1.0)\left(\frac{L}{2}\right)(1+0.57735)$$
$$+\frac{L}{2}(1.0)\frac{L}{2}(1-0.57735)=\frac{L^2}{2}$$

which agrees with the exact result. Thus we can calculate x_i from (5.17) using ξ_i from the Table. In other words, say for $n=2$, we have integrating points at $x_i=\frac{L}{2}(1\pm0.57735)$ namely of 0.211 L and 0.788 L.

The application of the numerical integration has been illustrated in the example below. The numerical integration is invariably used in finite element theory (Sec 9.5).

EXAMPLE 5.7 : *Beam of Variable Section*

Let us derive the complete flexibility and stiffness relations for the beam shown at (a), considering for simplicity, only the transverse effects and further neglecting the effect of shearing deformations. We will first derive the flexibility relations

$$\mathbf{v}=\begin{Bmatrix}\varphi_1\\ \varphi_2\end{Bmatrix}=\begin{bmatrix}f_{11} & f_{12}\\ f_{21} & f_{22}\end{bmatrix}\begin{Bmatrix}M_1\\ M_2\end{Bmatrix}+\begin{Bmatrix}\varphi_1^\circ\\ \varphi_2^\circ\end{Bmatrix}=\mathbf{f}\,\mathbf{S}+\mathbf{v}^\circ \qquad (a_1)$$

and later, by inversion, the stiffness relations

$$\mathbf{S}=\begin{Bmatrix}M_1\\ M_2\end{Bmatrix}=\begin{bmatrix}\mathbf{f}^{-1}\end{bmatrix}\begin{Bmatrix}\varphi_1-\varphi_1^\circ\\ \varphi_2-\varphi_2^\circ\end{Bmatrix}=\mathbf{k}\,\mathbf{v}+\mathbf{S}_F \qquad (a_2)$$

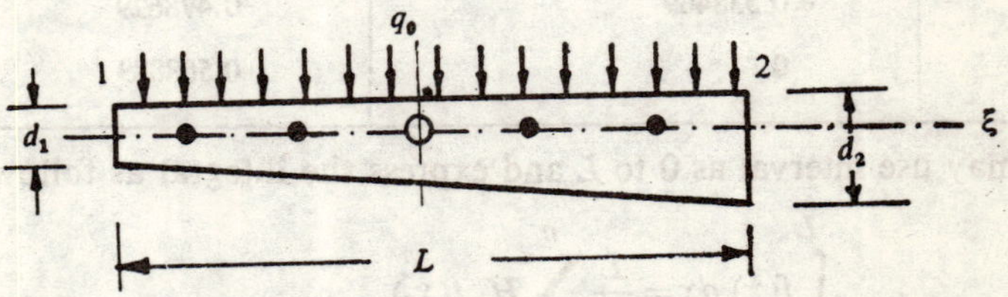

Example 5.7

The statical relation giving the internal force $\mathbf{S}_x=\{M_x\}$ to the member forces $\{M_1, M_2\}$ and due to lateral loads is

$$\mathbf{S}_x = \{M_x\} = \left[-\left(1-\frac{x}{L}\right) \;\; \frac{x}{L} \right] \left\{ \begin{matrix} M_1 \\ M_2 \end{matrix} \right\} + \left\{ -\frac{Lx}{2}\left(1-\frac{x}{L}\right) q^\circ \right\}$$
$$= \mathbf{B}\,\mathbf{S}+\mathbf{S}^\circ \qquad (a_3)$$

The element strain, consisting at present only of the curvature, is given by

$$\boldsymbol{\varepsilon} = \{\chi\} = \left(\frac{1}{EI}\right) M_x = \mathbf{f}_x\,\mathbf{S}_x \qquad (a_4)$$

The usual virtual work considerations as shown in Sec. 5.10 gives

$$\mathbf{f} = \int \mathbf{B}^t\,\mathbf{f}_x\,\mathbf{B}\,dx; \qquad \mathbf{v}^\circ = \int \mathbf{B}^t\,\boldsymbol{\varepsilon}^\circ\,dx \qquad (a_5)$$

These will now be integrated numerically by taking the depth of the beam to vary linearly as shown in Fig. (a). This is given by

$$d = \left(1-\frac{x}{L}\right) d_1 + \frac{x}{L}\, d_2 \qquad (a_6)$$

which then gives the moment of interia $I(x)$, as

$$I(x) = I_2 \left\{ \frac{1}{\alpha}\left(1-\frac{x}{L}\right) + \frac{x}{L} \right\}^3 = I_2\, C(x) \qquad (a_7)$$

where $C(x)$ is a function of x ; here $\alpha = d_2/d_1$, the ratio of depth at end 2 to that at end 1. (The ratio I_2/I_1 is α^3.) Presently $\mathbf{f}_x$ is $1/EI$ which can be written as

$$\mathbf{f}_x = \frac{1}{EI} = \frac{1}{EI_2}\left(\frac{1}{C(x)}\right) \qquad (a_8)$$

Consider the flexibility matrix :

$$\mathbf{f} = \int \mathbf{B}^t\,\mathbf{f}_x\,\mathbf{B}\,dx = \int_0^L \left\{ \begin{matrix} -\left(1-\frac{x}{L}\right) \\ \frac{x}{L} \end{matrix} \right\} \frac{1}{EI_2\,C(x)} \left[-\left(1-\frac{x}{L}\right),\; \frac{x}{L} \right] dx \qquad (a_9)$$

$$= \int_0^L \begin{bmatrix} \left(1-\frac{x}{L}\right)^2 & -\left(1-\frac{x}{L}\right)\frac{x}{L} \\ -\frac{x}{L}\left(1-\frac{x}{L}\right) & \left(\frac{x}{L}\right)^2 \end{bmatrix} \frac{dx}{EI_2\,C(x)} \qquad (a_{10})$$

To use the Gauss quadrature values, we first set the above functions in terms of $\xi = (2x/L)-1$ or $x = (L/2)(1+\xi)$ (See Eq. 5.77); thus the above relations become

$$\mathbf{f} = \int_{-1}^{1} \begin{bmatrix} \left(\frac{1-\xi}{2}\right)^2 & -\frac{1-\xi^2}{4} \\ -\frac{1-\xi^2}{4} & \left(\frac{1+\xi}{2}\right)^2 \end{bmatrix} \frac{1}{C(\xi)}\,\frac{1}{EI_2}\left(\frac{L}{2}\,d\xi\right) \qquad (a_{11})$$

The function $C(x)$ of Eq. (a_3), becomes

$$C(\xi)=\left[\frac{1}{\alpha}\left(\frac{1-\xi}{2}\right)+\frac{1+\xi}{2}\right]^3 \tag{a_{12}}$$

and function $S°$ of Eq. (a_3), becomes

$$S°(\xi)=-\frac{1-\xi^2}{4}\frac{qL^2}{2} \tag{a_{13}}$$

Therefore

$$\mathbf{v}°=\begin{Bmatrix}\varphi_1°\\ \varphi_2°\end{Bmatrix}=\int\begin{bmatrix}-\left(\frac{1-\xi}{2}\right)\\ \frac{1-\xi}{2}\end{bmatrix}\frac{1}{EI_2}\frac{1}{C(\xi)}\left(-\frac{1-\xi^2}{4}\frac{qL^2}{2}\right)\frac{L}{2}\,d\xi \tag{a_{14}}$$

The terms appearing in this and in Eq. (a_{11}) are to be evaluated using the numerical integration set out in Eq. 5.75.

Consider the integration rule with $n=4$. Here $\alpha^3=0.1$ ($\alpha=0.464$) has been assumed. The evaluation of the terms, for the different ξ_i, is set out in the Table 5.2 (a). The summation of the terms in columns 9 to 13 will give the required flexibility terms; including the appropriate signs and multiplying factors left out in the table

$$\begin{Bmatrix}\varphi_1\\ \varphi_2\end{Bmatrix}=\frac{L}{2\,EI_2}\begin{bmatrix}0.114436 & -0.10088\\ -0.10088 & 0.36297\end{bmatrix}\begin{Bmatrix}M_1\\ M_2\end{Bmatrix}\begin{Bmatrix}+0.03892\\ -0.06193\end{Bmatrix}\frac{qL^3}{4\,EI_2} \tag{a_{15}}$$

On simplifying,

$$\begin{Bmatrix}\varphi_1\\ \varphi_2\end{Bmatrix}=\frac{L}{EI_2}\begin{bmatrix}0.0572 & -0.0504\\ -0.0504 & 0.1815\end{bmatrix}\begin{Bmatrix}M_1\\ M_2\end{Bmatrix}+\begin{Bmatrix}0.009730\\ -0.01548\end{Bmatrix}\frac{qL^3}{EI_2} \tag{a_{16}}$$

On inversion we obtain

$$\begin{Bmatrix}M_1\\ M_2\end{Bmatrix}=\frac{EI_2}{L}\begin{bmatrix}23.14 & 6.43\\ 6.43 & 7.29\end{bmatrix}\begin{Bmatrix}\varphi_1\\ \varphi_2\end{Bmatrix}+\begin{Bmatrix}-0.1256\\ 0.0502\end{Bmatrix}qL^2 \tag{a_{17}}$$

Before these can be accepted with any reasonable sense of confidence, one has to repeat with a different order rule for integration. Table 5.2 (b) gives the details of integration with $n=3$. The values obtained are seen to be quite close. Proceeding as before, we obtain

$$\begin{Bmatrix}M_1\\ M_2\end{Bmatrix}=\frac{EI_2}{L}\begin{bmatrix}23.416 & 6.611\\ 6.611 & 7.415\end{bmatrix}\begin{Bmatrix}\varphi_1\\ \varphi_2\end{Bmatrix}+\begin{Bmatrix}-0.1215\\ 0.0534\end{Bmatrix}qL^2 \tag{a_{18}}$$

Table 5.2 (a) : Example 5.7

1	2	3	4	5	6	7	8	9	10	11	12	13
ξ_i	H_i	$\left(\frac{1-\xi}{2}\right)^2$	$\frac{1-\xi^2}{4}$	$\left(\frac{1+\xi}{2}\right)^2$	$\frac{1-\xi}{2}$	$\frac{1+\xi}{2}$	$C(\xi)$	Δf_{11}	Δf_{12}	Δf_{22}	$\Delta\varphi_1°$	$\Delta\varphi_2°$
								$2\times3\times8$	$2\times4\times8$	$2\times5\times8$	$2\times6\times8$	$2\times7\times8$
+0.861136	0.347855	0.0048	0.0646	0.8659	0.0694	0.9306	1.2604	0.00133	0.01783	0.23897	0.00123	0.01659
+0.339981	0.652145	0.1089	0.2211	0.4489	0.3300	0.6700	2.6351	0.02695	0.05472	0.11109	0.01805	0.03665
−0.339981	0.652145	0.4489	0.2211	0.1089	0.6700	0.3300	5.5826	0.05244	0.02582	0.01272	0.01730	0.00852
−0.861136	0.347855	0.8659	0.0646	0.00483	0.9306	0.0694	8.9340	0.03371	0.00251	0.00019	0.00233	0.00017
							$\sum$	0.114436	0.10088	0.36297	0.03892	0.06193

Table 5.2 (b) : Example 5.7

1	2	3	4	5	6	7	8	9	10	11	12	13
ξ_i	H_i	$\left(\frac{1-\xi}{2}\right)^2$	$\frac{1-\xi^2}{4}$	$\left(\frac{1+\xi}{2}\right)^2$	$\frac{1-\xi}{2}$	$\frac{1+\xi}{2}$	$C(\xi)$	Δf_{11} $(2\times3\times8)$	Δf_{12} $(2\times4\times8)$	Δf_{22} $(2\times5\times8)$	$\Delta \varphi_1^\circ$ $(2\times6\times8)$	$\Delta \varphi_2^\circ$ $(2\times7\times8)$
+0.774597	0.55555	0.01270	0.1	0.7873	0.1127	0.8873	1.4436	0.00488	0.03848	0.30298	0.00433	0.03415
0	0.88888	0.25	0.25	0.25	0.5	0.5	3.9264	0.05659	0.05659	0.05659	0.02829	0.02829
−0.774597	0.55555	0.7873	0.1	0.01270	0.8873	0.1127	8.3038	0.05267	0.00669	0.00085	0.005936	0.00754
							$\sum$	0.11415	0.10176	0.360426	0.038556	0.0632

A computer subroutine can be written with higher order rule and incorporated in the program of analysing plane frames. Observe further that we can adopt this procedure when E may vary, as is the case of elastoplastic behaviour or due to creep of beams in which certain parts of the beam might have decreased values as a result of plasticity or creep. In such a case we use an equivalent (EI) for the section as a whole.

5.12 FURTHER TOPICS

a. TRANSFER MATRIX

There is a mixed way in which the member end forces can be related to the member end displacements. This consists of expressing the forces and displacements of end 2 in terms of those at end 1:

$$\mathbf{S}_2 = \begin{Bmatrix} \overline{\mathbf{P}}_2 \\ \hline \overline{\mathbf{u}}_2 \end{Bmatrix} = \begin{bmatrix} \overline{\mathbf{g}}_{11} & \overline{\mathbf{g}}_{12} \\ \hline \overline{\mathbf{g}}_{21} & \overline{\mathbf{g}}_{22} \end{bmatrix} \begin{Bmatrix} \overline{\mathbf{P}}_1 \\ \hline \overline{\mathbf{u}}_1 \end{Bmatrix} = \overline{\mathbf{G}}\,\mathbf{S}_1 \tag{5.78}$$

The mixed vectors will be called *state vectors* and the $\overline{\mathbf{G}}$ matrix as the *transfer** matrix. This is found useful in certain type of problems.

The equilibrium relation for a member giving $\overline{\mathbf{P}}_2$ in terms of $\overline{\mathbf{P}}_1$ leads to the first row of Eq. 5.78. This can be directly written from Fig. 2.1 of Chapter 2.

$$\overline{\mathbf{P}}_2 = \begin{Bmatrix} \overline{P}_2 \\ \overline{Q}_2 \\ \overline{M}_2 \end{Bmatrix} = \begin{bmatrix} -1 & 0 & 0 \\ 0 & -1 & 0 \\ 0 & L & -1 \end{bmatrix} \begin{Bmatrix} \overline{P}_1 \\ \overline{Q}_1 \\ \overline{M}_1 \end{Bmatrix} = \overline{\mathbf{C}'}\,\overline{\mathbf{P}}_1 \tag{a$_1$}$$

Therefore we find that $\overline{\mathbf{g}}_{11} \equiv \overline{\mathbf{C}'}$ and $\overline{\mathbf{g}}_{12} \equiv 0$. The kinematic relation and the basic flexibility relations are now to be combined. The displacement $\overline{\mathbf{u}}_2$ at end 2 can be considered as the sum of the displacements there due to the deformations as a cantilever action $(\overline{\mathbf{u}}_1 \equiv 0)$, and the displacements due to the rigid body motion due to displacements $\overline{\mathbf{u}}_1$ alone. Due to the cantilever action the displacements at end 2 are given by $\mathbf{f}\ \overline{\mathbf{P}}_2$; due to the rigid body motion $\overline{\mathbf{u}}_1$ we obtain, (see Fig. 3.4).

$$\overline{\mathbf{u}}^{\circ}_2 = \begin{bmatrix} 1 & 0 & 0 \\ 0 & 1 & L \\ 0 & 0 & 1 \end{bmatrix} \begin{Bmatrix} \bar{u}_1 \\ \overline{v}_1 \\ \overline{\theta}_1 \end{Bmatrix} \tag{a$_2$}$$

Let this be written as

$$\overline{\mathbf{u}}^{\circ}_2 = \overline{\mathbf{A}'}\,\overline{\mathbf{u}}_1 \tag{a$_3$}$$

* Alternatively the word *distribution matrix* is also used [13, 14].

Thus we have,

$$\overline{\mathbf{u}_2} = \mathbf{f}\,\overline{P_2} + \overline{\mathbf{A}'}\,\overline{\mathbf{u}_1} = \mathbf{f}\,\overline{\mathbf{C}'}\,\overline{\mathbf{P}_1} + \overline{\mathbf{A}'}\,\overline{\mathbf{u}_1} \qquad (a_4)$$

This gives $\overline{\mathbf{g}_{21}} = \mathbf{f}\mathbf{C}'$ and $\overline{\mathbf{g}_{22}} \equiv \overline{\mathbf{A}'}$ Finally therefore

$$\mathbf{S}_2 = \begin{Bmatrix} \overline{P}_2 \\ \overline{\mathbf{u}}_2 \end{Bmatrix} = \begin{bmatrix} \overline{\mathbf{C}'} & 0 \\ \mathbf{f}\,\overline{\mathbf{C}'} & \overline{\mathbf{A}} \end{bmatrix} \begin{Bmatrix} \overline{P}_1 \\ \overline{\mathbf{u}}_1 \end{Bmatrix} = \mathbf{G}\,\mathbf{S}_1 \qquad (5.79)$$

This, when written out in full, reads,

$$\begin{Bmatrix} \overline{P}_2 \\ \overline{Q}_2 \\ \overline{M}_2 \\ \bar{u}_2 \\ \overline{v}_2 \\ \overline{\theta}_2 \end{Bmatrix} = \left[\begin{array}{ccc|ccc} -1 & 0 & 0 & 0 & 0 & 0 \\ 0 & -1 & 0 & 0 & 0 & 0 \\ 0 & L & -1 & 0 & 0 & 0 \\ \hline -\dfrac{L}{AE} & 0 & 0 & 1 & 0 & 0 \\ 0 & \dfrac{L^3}{6EI} & -\dfrac{L^2}{2EI} & 0 & 1 & L \\ 0 & \dfrac{L^2}{2EI} & -\dfrac{L}{EI} & 0 & 0 & 1 \end{array}\right] \begin{Bmatrix} \overline{P}_1 \\ \overline{Q}_1 \\ \overline{M}_1 \\ \bar{u}_1 \\ \overline{v}_1 \\ \overline{\theta}_1 \end{Bmatrix} \qquad (5.80)$$

We could reverse the order and obtain the state vector at 1 in terms of the one at end 2.

Structural analysis using transfer matrices proves efficient only in certain types of problems. Extensive applications have been demonstrated in References [6, 15] for frames and [13,14, 16] for continuum problems respectively. At this stage Eq. (5.80) can be considered as an alternative statement of relations between member end forces and displacements.

b. CURVED MEMBER

Properties of curved members like arches can be determined by adopting numerical integration since closed form integrations are possible only in simple cases. We will show in this section some details of this problem and special transformations generally adopted.

Figure 5.14(a) shows a curved member. The approach will be to first construct the complete basic flexibility in the form (5.67), and to obtain next the complete basic stiffness relations in the form (5.68). Observe that we do not have here the 'member axes' and that we have to directly derive these in a global set of axes as shown in Fig. 5.14(a).

We will obtain the complete flexibility relation between the deformations $\mathbf{v} = \{u_1, v_1, \theta_1\}$ at end 1, the forces $\mathbf{S} = \{P_1, Q_1, M_1\}$, and the distributed loads, in the form (5.67), treating end 2 as fixed (cantilever model). The first

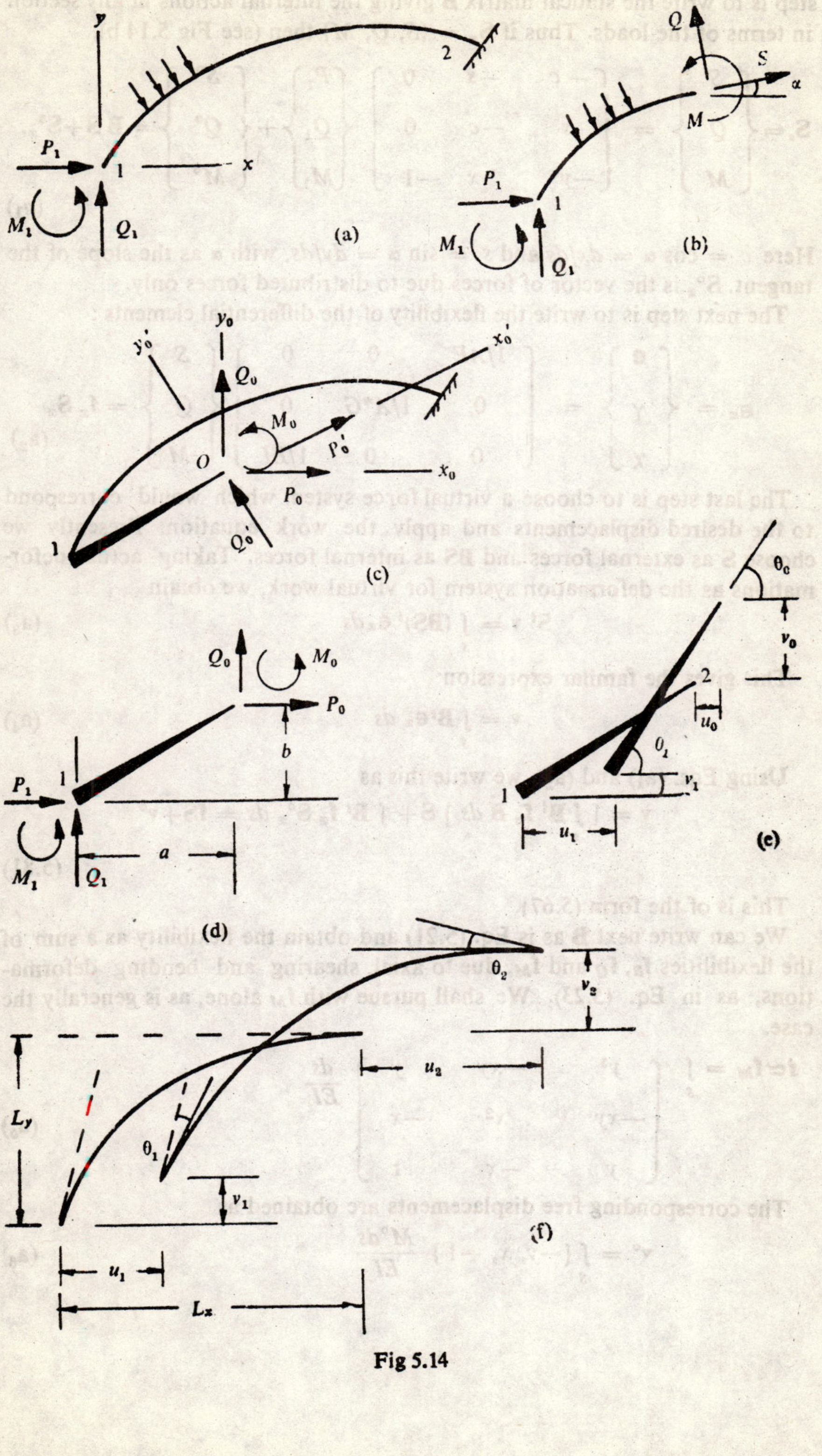
y
2
P_1
1
x
M_1
Q_1
(a)
Q
S
α
M
P_1
1
M_1
Q_1
(b)
y_0
y_0'
x_0'
Q_0
M_0
P_0'
O
x_0
P_0
Q_0'
1
(c)
Q_0
M_0
P_0
b
P_1
1
M_1
Q_1
a
(d)
θ_0
v_0
2
u_0
θ_1
v_1
1
u_1
(e)
θ_2
v_2
u_2
L_y
θ_1
v_1
u_1
L_x
(f)

Fig 5.14

step is to write the statical matrix **B** giving the internal actions at any section in terms of the loads. Thus if $\mathbf{S}_x = \{S, Q, M\}$ then (see Fig 5.14 b),

$$\mathbf{S}_x = \begin{Bmatrix} S \\ Q \\ M \end{Bmatrix} = \begin{bmatrix} -c & -s & 0 \\ s & -c & 0 \\ -y & x & -1 \end{bmatrix} \begin{Bmatrix} P_1 \\ Q_1 \\ M_1 \end{Bmatrix} + \begin{Bmatrix} S^\circ \\ Q^\circ \\ M^\circ \end{Bmatrix} = \mathbf{B}\,\mathbf{S} + \mathbf{S}^\circ_x. \qquad (a_1)$$

Here $c = \cos\alpha = dx/ds$ and $s = \sin\alpha = dy/ds$, with α as the slope of the tangent. $\mathbf{S}^\circ_x$ is the vector of forces due to distributed forces only.

The next step is to write the flexibility of the differential elements :

$$\boldsymbol{\epsilon}_x = \begin{Bmatrix} \epsilon \\ \gamma \\ \chi \end{Bmatrix} = \begin{bmatrix} 1/AE & 0 & 0 \\ 0 & 1/A^*G & 0 \\ 0 & 0 & 1/EI \end{bmatrix} \begin{Bmatrix} S \\ Q \\ M \end{Bmatrix} = \mathbf{f}_x\, \mathbf{S}_x \qquad (a_2)$$

The last step is to choose a virtual force system which would correspond to the desired displacements and apply the work equation. Presently we choose **S** as external forces and **BS** as internal forces. Taking actual deformations as the deformation system for virtual work, we obtain

$$\mathbf{S}^t\, \mathbf{v} = \int_s (\mathbf{BS})^t\, \boldsymbol{\epsilon}_x\, ds \qquad (a_3)$$

This gives the familar expression

$$\mathbf{v} = \int_s \mathbf{B}^t \boldsymbol{\epsilon}_x\, ds \qquad (a_4)$$

Using Eqs. (a_1) and (a_2), we write this as

$$\mathbf{v} = \left[\int_s \mathbf{B}^t\, \mathbf{f}_x\, \mathbf{B}\, ds\right] \mathbf{S} + \int \mathbf{B}^t\, \mathbf{f}_x\, \mathbf{S}^\circ_x\, ds = \mathbf{fS} + \mathbf{v}^\circ \qquad (5.81)$$

This is of the form (5.67).

We can write next **B** as is Eq. (5.21) and obtain the flexibility as a sum of the flexibilities $\mathbf{f}_S$, $\mathbf{f}_Q$ and $\mathbf{f}_M$, due to axial, shearing and bending deformations, as in Eq. (5.23). We shall pursue with $\mathbf{f}_M$ alone, as is generally the case.

$$\mathbf{f} \simeq \mathbf{f}_M = \int_s \begin{bmatrix} y^3 & -xy & y \\ -xy & x^2 & -x \\ y & -x & 1 \end{bmatrix} \frac{ds}{EI} \qquad (a_5)$$

The corresponding free displacements are obtained as

$$\mathbf{v}^\circ = \int_s \{-y, x, -1\} \frac{M^\circ ds}{EI} \qquad (a_6)$$

We evaluate these numerically by subdividing the arch into a number of sections. A computer code can be written to integrate terms such as $\Sigma\, y_i^2\, \Delta S_i/(EI)_i$, $\Sigma\, M^\circ_i\, .\, \Delta S_i/(EI)_i$, etc.

Diagonalization-Elastic Centre

We can consider here a transformation, which leads to idea of elastic centre. Such a transformation is of limited interest since, in the context of computers, it hardly provides computational advantage. Moreover this is possible only under the restrictive assumption of considering only the bending deformations.

We can imagine* a rigid arm (Fig 5.14(c) connecting end 1 to a point 0, we can treat the forces $\mathbf{S}_0 = \{P_0, Q_0, M_0\}$ to be applied there, and determine the resulting displacements $\mathbf{v}_0 = \{u_0, v_0, \theta_0\}$ in the form

$$\mathbf{v}_0 = \mathbf{f}_0\, \mathbf{S}_0 + \mathbf{v}_0^{\,0} \qquad (a_7)$$

We can then choose the location of the point 0 and a suitable axes (x_0', y_0') such that $\mathbf{f}_0$ becomes a diagonal matrix. To determine this we need a transformation matrix **R** relating forces **S** to $\mathbf{S}_0$ and **v** to $\mathbf{v}_0$. Sketches (d) and (e) give,

$$\mathbf{S} = \begin{Bmatrix} P_1 \\ Q_1 \\ M_1 \end{Bmatrix} = \begin{bmatrix} 1 & 0 & 0 \\ 0 & 1 & 0 \\ -b & a & 1 \end{bmatrix} \begin{Bmatrix} P_0 \\ Q_0 \\ M_0 \end{Bmatrix} = \mathbf{R}\,\mathbf{S}$$

and

$$(a_8)$$

$$\mathbf{v}_0 = \begin{Bmatrix} u_0 \\ v_0 \\ \theta_0 \end{Bmatrix} = \begin{bmatrix} 1 & 0 & -b \\ 0 & 1 & a \\ 0 & 0 & 1 \end{bmatrix} \begin{Bmatrix} u_1 \\ u_1 \\ \theta_1 \end{Bmatrix} = \mathbf{R}^t\mathbf{v} \qquad (a_9)$$

The contragredience observed should not surprise the reader. As an exercise Eq. (a_9) can be derived from Eq. (a_8) (or vice-versa) using virtual work.

Thus Eq. (5.81) becomes

$$\mathbf{v}_0 = [\mathbf{R}^t\, \mathbf{f}\, \mathbf{R}]\, \mathbf{S} + \mathbf{R}^t\, \mathbf{v}^0 = \mathbf{f}_0\, \mathbf{S}_0 + \mathbf{v}_0^{\,0} \qquad (a_{10})$$

The product $\mathbf{R}^t\, \mathbf{f}\, \mathbf{R}$ reads as

$$\mathbf{f}_0 = \int_0^2 \begin{bmatrix} (y-b)^2 & -(x-a)(y-b) & (y-b) \\ -(x-a)(y-b) & (x-a)^2 & -(x-a) \\ (y-b) & -(x-a) & 1 \end{bmatrix} \frac{ds}{EI} \qquad (a_{11})$$

Note that integrations have to proceed from point 0. Since the arm is rigid $1/EI$ is zero and integration reduces to the original one over the

*See last para of Sec 2.1, and also Ex. 2.18

arch length. We note that $(x-a)$ and $(y-b)$ denote x_0 and y_0, measured from point 0. Thus

$$\mathbf{f}_0 = \int_1^2 \begin{bmatrix} y_0^2 & -x_0 y_0 & y_0 \\ -x_0 y_0 & x_0^2 & -x_0 \\ y_0 & -x_0 & 1 \end{bmatrix} \frac{ds}{EI} \tag{a$_{12}$}$$

If a and b are so chosen that $\int x_0 ds/EI$ and $\int y_0 ds/EI$ are both zero, namely,

$$a = \int \frac{x ds}{EI} \Big/ \int \frac{ds}{EI} \ ; \quad b = \int y \, ds/EI \Big/ \int \frac{ds}{EI} \tag{a$_{13}$}$$

We obtain some simplification :

$$\mathbf{f}_0 = \int_1^2 \begin{bmatrix} y_0^2 & -x_0 y_0 & 0 \\ -x_0 y_0 & x^2 & 0 \\ 0 & 0 & 1 \end{bmatrix} \frac{ds}{EI} \tag{a$_{14}$}$$

Further if the section happens to be *symmetrical* about x_0 or y_0 then, $\int x_0 \, y_0 \, ds/EI$ vanishes. For such a section

$$\mathbf{f}_0 = \int_1^2 \begin{bmatrix} y_0^2 & 0 & 0 \\ 0 & x_0^2 & 0 \\ 0 & 0 & 1 \end{bmatrix} \frac{ds}{EI} \tag{a$_{15}$}$$

This represents a simplification since we can directly obtain

$$u_0 = \left(\int \frac{y_0^2 \, ds}{EI}\right) P_0 \ ; \quad v_0 = \left(\int \frac{x_0^2 ds}{EI}\right) Q_0$$

$$\theta_0 = (\int ds/EI) \, M_0 \tag{a$_{16}$}$$

The point 0 is called the *elastic centre*. In hand calculations the above diagonalization affords some convenience.

In the general case, we can further choose a set of axes $(x_0' \ y_0')$ at an angle β such that $\int x_0 \, y_0 \, ds/EI$ also vanishes. (Determination of β in terms of $\int x_0^2 \, ds/EI$ etc. is left as an exercise to the reader.) Recall that the saving in labour in determining this may not be worthwhile.

Complete General Stiffness Relations

By inverting relations (5.81) we obtain first the complete basic stiffness relations as

$$\mathbf{S} = \mathbf{f}^{-1} (\mathbf{v} - \mathbf{v}^0) = \mathbf{k}(\mathbf{v} - \mathbf{v}^0) = \mathbf{k}\mathbf{v} + \mathbf{S}_F \tag{5·82}$$

We then obtain the general stiffness relations by transformations. Observe that $\mathbf{v} = \mathbf{u}_1$ only when $\mathbf{u}_2 \equiv 0$; when $\mathbf{u}_1$ and $\mathbf{u}_2$ both occur we need to obtain $\mathbf{v}$ as $\mathbf{u}_1$ relative to end 2, namely, end 2 treated as fixed. Thus, superimpose the effect of a rigid body movement such that at end 2, $\mathbf{u}_2$ becomes zero. If the rigid body movement is such that at end 2 we have $\mathbf{u}_2$, then at end 1,

$$\mathbf{u}_1 = \begin{Bmatrix} u_1 \\ v_1 \\ \theta_1 \end{Bmatrix} = \begin{bmatrix} 1 & 0 & L_y \\ 0 & 1 & -L_x \\ 0 & 0 & 1 \end{bmatrix} \begin{Bmatrix} u_2 \\ v_2 \\ \theta_2 \end{Bmatrix} \quad (a_{17})$$

where L_y and L_x are projections of the arch on x and y axes. (See Fig. 5.15 f). Thus if the total displacements at end 2 should be zero, we subtract the above from $\mathbf{u}_1$ to obtain $\mathbf{v}$:

$$\mathbf{v} = \begin{Bmatrix} u_1 \\ v_1 \\ \theta_1 \end{Bmatrix} - \begin{bmatrix} 1 & 0 & L_y \\ 0 & 1 & -L_x \\ 0 & 0 & 1 \end{bmatrix} \begin{Bmatrix} u_2 \\ v_2 \\ \theta_2 \end{Bmatrix} \quad (a_{18})$$

We write this in the familiar from :

$$\mathbf{v} = \begin{bmatrix} 1 & 0 & 0 \\ 0 & 1 & 0 \\ 0 & 0 & 1 \end{bmatrix} \begin{Bmatrix} u_1 \\ v_1 \\ \theta_1 \end{Bmatrix} + \begin{bmatrix} -1 & 0 & -L_y \\ 0 & -1 & L_x \\ 0 & 0 & 1 \end{bmatrix} \begin{Bmatrix} u_2 \\ v_2 \\ \theta_2 \end{Bmatrix}$$

$$= \mathbf{A}_1\mathbf{u}_1 + \mathbf{A}_2\mathbf{u}_2 \quad (a_{19})$$

with $\mathbf{A}_1 \equiv \mathbf{I}$. The statical relations are

$$\mathbf{P}_1 \equiv \mathbf{S} = \mathbf{A}_1{}^t\mathbf{S} \; ; \mathbf{P}_2 = \mathbf{A}^t{}_2\mathbf{S} + \mathbf{P}^0{}_2$$

where $\mathbf{P}^0{}_2$ are the forces at end 2 due to distributed loads alone.

We can now combine Eqs. 5.82 , (a_{19}) and the above equations :

$$\mathbf{P}_1 = \mathbf{A}_1{}^t\mathbf{k}\,(\mathbf{A}_1\mathbf{u}_1 + \mathbf{A}_2\mathbf{u}_2) + \mathbf{A}_1{}^t\mathbf{S}_F$$
$$\mathbf{P}_2 = A_2{}^t\mathbf{k}\,(\mathbf{A}_1\mathbf{u}_1 + \mathbf{A}_2\mathbf{u}_2) + \mathbf{A}_2{}^t\mathbf{S}_F + \mathbf{P}^0{}_2$$

These are of the usual form

$$\begin{Bmatrix} \mathbf{P}_1 \\ \mathbf{P}_2 \end{Bmatrix} = \begin{bmatrix} \mathbf{K}_{11} & \mathbf{K}_{12} \\ \mathbf{K}_{21} & \mathbf{K}_{22} \end{bmatrix} \begin{Bmatrix} \mathbf{u}_1 \\ \mathbf{u}_2 \end{Bmatrix} + \begin{Bmatrix} \mathbf{P}_{F1} \\ \mathbf{P}_{F2} \end{Bmatrix} \quad (5.83)$$

where

$$\mathbf{K}_{ij} = \mathbf{A}_i{}^t\,\mathbf{k}\mathbf{A}_j \qquad i, j = 1.\ 2$$
$$\mathbf{P}_{F1} = \mathbf{A}_1{}^t\,\mathbf{S}_F \; ; \; \mathbf{P}_{F2} = \mathbf{A}_2{}^t\,\mathbf{S}_F + \mathbf{P}_2{}^0 \quad (5.84)$$

Equation (5.83) is of the form (5.66) discussed earlier. Note that we have here handled all quantities in global coordinates. Note further that form (5.83) is the same regardless of the shape of the arch. It can also be applied to segmental members like a portal or gable frame.

EXERCISES

5.1 Express the material elasticity given by Eqs. (5.1) and (5.2) in terms of E and the Poisson ration ν.

5.2 Show that symmetry in the flexibility (or stiffness) matrix follows from the validity of Maxwell reciprocal theorem.

5.3 Repeat the derivation of basic member flexibility of Sec 5.4 considering only bending without the axial load. Further neglect the work of shearing deformations.

5.4 Verify the other terms of flexibility matrix given in Eq. (a_3) of Sec. 5.5 by drawing sketches similar to Fig. 5.6.

5.5 Drive the flexibility $\mathbf{f}_b$ from $\mathbf{f}_e$ similar to the one by Eq. (5.33).

5.6 Derive the general stiffness matrix of Eq. (5.42) directly without depending on the transformations of the basic stiffness matrix and see if it saves any labour. You may use simple means as area moment theorem, conjugate beam or Castigliano's theorem. Assume uniform beam section. What other procedure can you follow for a beam of variable section ?

5.7 Derive the basic member stiffness from the solution of the differential equation as in Sec 5.8. Do the shape functions agree with those in Eq. (5.55) ?

5.8 Derive only shape function N_{25} of Eq. (5.55) directly from the solution of the differential equation.

5.9 Relate the stability functions ψ_1 and ψ_2 of Eqs. (a_{12}) and (a_{13}) of Example 5.2, with the carry over factor used in moment distribution method (see Ref [9]).

5.10 Derive Eqs. (a_{15}) of Example 5.2 proceeding directly with an axial tensile force, (See Ref [11]).

5.11 Repeat the contents of Sec 5.9 by assuming v in the form

$$v = a\frac{16}{5L^4}\ (L^3x-2\ Lx^3+x^4)$$

which is the form of deflection due to a uniform load.

5.12 Determine the stiffness of a cantilever of span L by assuming a suitable cosine curve for the transverse deflected shape.

5.13 Evaluate the fixing end moments using the shape of Ex. 5.11 by taking: (a) a uniform load w ; and (b) a concentrated load P at centre.

5.14 Determine the basic member stiffness of a uniform beam in bending by using numerical integration and compare the accuracies obtained by adopting $n = 1$, 2 and 3.

5.15 Treat the curved member of Fig 5.14 (a) to be a parabolic arch with secant variation of I, namely $I = I_0$ Sec α. Derive the basic flexibility, stiffness and general stiffness relations.

Chapter 6

Displacements

In this chapter we shall be combining the considerations of the flexibilities of members with those of the kinematics of a structure and thus obtain the displacement of the joints. The contents of the chapter are the generalization of the unit load method of finding displacements. They prepare the necessary background for the force method of analysing redundant structures.

6.1 DISPLACEMENT BY VIRTUAL WORK

The general steps involved in determining the displacements of the joints of a *determinate* frame are as follows :

(*a*) Determine the member deformations in terms of loads using the basic flexibility relations for each member ;

(*b*) choose a system of virtual forces corresponding to the desired displacements; and

(*c*) apply the principle of virtual work and obtain the desired displacements.

Let,

$$\mathbf{V}=\{\mathbf{v}_A \ \mathbf{v}_B \ \mathbf{v}_C \ldots\} \tag{a$_1$}$$

be the deformation vector listing the deformations of all members. Here $\mathbf{v}_A$, $\mathbf{v}_B$ etc. are submatrices of appropriate sizes, viz, (3×1) for a plane frame, (6×1) for space frame, etc. These deformations may be the ones occuring due to the applied loads or free deformations such as due to temperature or prescribed displacements such as lack of fit and support movements. Assume that displacements are required at few joints, say *K*, *M* and *R*. Let the corresponding displacement vector be

$$\mathbf{U}_1=\{\mathbf{u}_K \quad \mathbf{u}_M \quad \mathbf{u}_R\}. \tag{a$_2$}$$

Assume virtual loads corresponding to these displacements to act on the structure and denote by $\mathbf{P}^*$:

$$\mathbf{P}^*=\{\mathbf{P}^*_K, \quad \mathbf{P}^*_M, \quad \mathbf{P}^*_R\} \tag{a_3}$$

Let the corresponding internal forces be

$$\mathbf{S}^*=\mathbf{B}^*\mathbf{P}^* \tag{6.1}$$

Write now the equation of virtual work between the forces of the virtual load system (a_3) and the displacement of the real system (a_2) :

$$\mathbf{P}^{*t}\,\mathbf{U}_1=\mathbf{S}^{*t}\,\mathbf{V} \tag{a_4}$$

Using (6.1), we obtain

$$\begin{aligned}\mathbf{P}^{*t}\,\mathbf{U}_1&=(\mathbf{B}^*\,\mathbf{P}^*)^t\,\mathbf{V}\\&=\mathbf{P}^{*t}\,\mathbf{B}^{*t}\,\mathbf{V}\end{aligned}$$

Cancelling the arbitrary virtual loads, we obtain

$$\mathbf{U}_1=\mathbf{B}^{*t}\,\mathbf{V} \tag{6.2}$$

We note that we require $\mathbf{B}^*$ matrix alone. The cancelling of the virtual load is equivalent to the fact that it is only the $\mathbf{B}^*$ matrix (which gives internal forces in terms of unit values of the virtual loads $\mathbf{P}^*$) that matters. (Recall Eq. 4.11.)

Equation (6.2) is the generalization of the unit load method. In this equation we have to substitute the expression for the member deformations. Due to the member end loads we have, for each member the elastic relations :

$$\mathbf{v}_A=\mathbf{f}_A\,\mathbf{S}_A\ ;\quad \mathbf{v}_B=\mathbf{f}_B\ \mathbf{S}_B,\ \text{etc.} \tag{a_5}$$

These collectively can be represented as

$$\mathbf{V}=\begin{Bmatrix}\mathbf{v}_A\\ \mathbf{v}_B\\ \cdot\\ \cdot\\ \cdot\end{Bmatrix}=\begin{bmatrix}\mathbf{f}_A & & & & \\ & \mathbf{f}_B & & & \\ & & \cdot & & \\ & & & \cdot & \\ & & & & \cdot\end{bmatrix}\begin{Bmatrix}\mathbf{S}_A\\ \mathbf{S}_B\\ \cdot\\ \cdot\\ \cdot\end{Bmatrix}=\mathbf{f}\,\mathbf{S} \tag{a_6}$$

To conserve space we shall write now and henceforth the equations of this type as,

$$\mathbf{V}=\lceil\, \mathbf{f}_A\ \ \mathbf{f}_B\ \ \mathbf{f}_C\ \ \cdot\ \ \cdot\ \ \cdot\ \ \cdot \rfloor\mathbf{S}=\lceil\, \mathbf{f}\, \rfloor\,\mathbf{S} \tag{6.3}$$

using the special brackets for the diagonal matrix. The diagonal matrix $\mathbf{f}$ is also called the *unassembled flexibility matrix* of the structure. It is only a listing of flexibilities of the members and does not take into account the connectivity of the members. Let

$$V^0 = \{V^0{}_A, V^0{}_B \cdot \cdot \cdot\} \tag{6.4}$$

be the vector of free member deformations due to lateral loads, temperature or prescribed member strains. The total deformations of members are

$$V = f\ S + V^0 \tag{6.5}$$

Let the internal forces due to the applied load **P** be

$$S = B\ P \tag{6.6}$$

Therefore, we have the member deformations as

$$V = f\ B\ P + V^\circ \tag{6.7}$$

Equation (6.2) then gives the required displacements as :

$$U_1 = B^{*t}(f\ B\ P + V^\circ) = (B^{*t}\ f\ B)P + B^{*t}\ V^\circ \tag{6.8}$$

Let us write Eqs. (6.1) and (6.2) in full and interpret the results :

$$S^* = \begin{Bmatrix} S^*{}_A \\ S^*{}_B \\ S^*{}_C \\ \cdot \\ \cdot \\ \cdot \end{Bmatrix} = \begin{bmatrix} B^*{}_{AK} & B^*{}_{AM} & B^*{}_{AR} \\ B^*{}_{BK} & B^*{}_{BM} & B^*{}_{BR} \\ B^*{}_{CK} & B^*{}_{CM} & B^*{}_{CR} \\ \cdot & \cdot & \cdot \\ \cdot & \cdot & \cdot \\ \cdot & \cdot & \cdot \end{bmatrix} \begin{Bmatrix} P^*{}_K \\ P^*{}_M \\ P^*{}_R \end{Bmatrix} = B^*P^* \tag{6.1a}$$

$$U_1 = \begin{Bmatrix} u_K \\ u_M \\ u_R \end{Bmatrix} = \begin{bmatrix} B^{*t}{}_{AK} & B^{*t}{}_{BK} & \cdot & \cdot & \cdot \\ B^{*t}{}_{AM} & B^{*t}{}_{BM} & \cdot & \cdot & \cdot \\ B^{*t}{}_{AR} & B^{*t}{}_{BR} & \cdot & \cdot & \cdot \end{bmatrix} \begin{Bmatrix} v_A \\ v_B \\ \cdot \\ \cdot \\ \cdot \end{Bmatrix} = B^{*t}\ V \tag{6.2 (a)}$$

Thus the relations (6.2) are, in the present case, equivalent to three sets of equations. For any typical displacement say U_K, we can write

$$U_K = \sum_{A}^{\text{Members}} B_A{}^{*t}\ v_A \tag{6.9}$$

The meaning of the elements of **B*** should be identified and kept in mind : the elements of the first column are the forces due to the unit P^*_K, that of second column due to unit P^*_M and so on. Oberving the equivalance of the matrix product to the summation over the members, we identify that the expression on the right hand side is analogous to the one used in pin-jointed frames, if one takes **v** to be the bar elongation, and **B*** the forces in the members due to unit (virtual) load. Since **v**, in general, stands for a vector of deformations the matrix form is more general than the one used for pinjointed frames.

Consider the above equations as applied in the pinjointed plane frame shown in Fig. 6.1. Assume that the conditions of equilibrium have been solved and the internal forces are obtained in terms of the joint loads. These would be in the following form :

$$\mathbf{S}=\begin{Bmatrix} S_A \\ S_B \\ S_C \\ S_D \\ S_E \end{Bmatrix}=\begin{bmatrix} B_{11} & B_{12} & . & . & B_{15} \\ B_{21} & B_{22} & & & . \\ . & & & & . \\ . & & & & . \\ B_{51} & .. & & . & B_{55} \end{bmatrix}\begin{Bmatrix} P_B \\ Q_B \\ P_C \\ Q_C \\ P_D \end{Bmatrix}=\mathbf{B}\,\mathbf{P} \qquad (a_7)$$

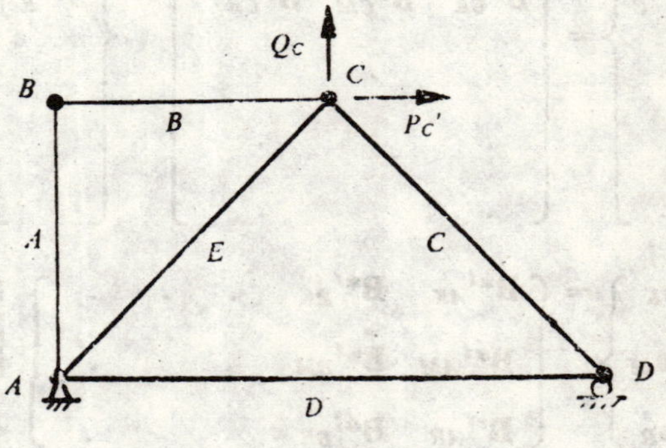

Fig. 6.1

Let the displacements

$$\mathbf{U}_1=\{\, \mathbf{v}_B \quad \mathbf{u}_D \,\} \qquad (a_8)$$

only be required, namely the vertical displacements of joint *B* and the horizontal movement of *D*. To obtain these the elements of **B*** matrix have to be obtained. **B*** matrix corresponds to the internal force due to unit values of Q_B and P_D. The second and the fifth columns of the above **B** matrix stand for the internal forces due to $Q_B=1$ and $P_D=1$ respectively. Thus

$$\mathbf{B}^* = \begin{bmatrix} B_{12} & B_{15} \\ B_{22} & B_{25} \\ . & . \\ . & . \\ B_{52} & B_{55} \end{bmatrix} \equiv \begin{bmatrix} S^*_A & S^{**}_A \\ S^*_B & S^{**}_B \\ . & . \\ . & . \\ S^*_E & S^{**}_E \end{bmatrix} \qquad (a_9)$$

The elastic relations are

$$\mathbf{V} = \lceil f_A \quad f_B \quad f_C \quad f_D \quad f_E \rfloor \, \mathbf{S} \qquad (a_{10})$$

The displacements due to the applied loads alone, are given by Eq. (6.8) in which V° terms are dropped :

$$\mathbf{U} = \mathbf{B}^{*t}\,\mathbf{f}\,\mathbf{S} = [\,\mathbf{B}^{*t}]\,\lceil \mathbf{f} \rfloor\,\{\mathbf{S}\}$$

$$= \begin{bmatrix} B_{12} \cdot B_{22} & . & . & . & B_{52} \\ B_{15} & B_{25} & . & . & B_{55} \end{bmatrix} \begin{Bmatrix} f_A S_A \\ f_B S_B \\ . \\ . \\ . \\ f_E S_E \end{Bmatrix}$$

$$= \begin{bmatrix} \Sigma S^*_A \, f_A S_A \\ \Sigma S^{**}_A \, f_A S_A \end{bmatrix} \qquad (6.10)$$

Here the summation proceeds over the members.

The reader can now identify the working formulae for the deflections of pinjointed frame S. S_A^* and S^{**}_A are the forces in a typical member A due to unit loads along v_B and u_D respectively. These are the elements of the second and fifth column of the **B** matrix.

The fact that the deflections are as a result of the flexibility of all the members may be noted. Thus the flexibilities of members are seen to be in a way added. $\mathbf{B}^{*t}\,\mathbf{f}\,\mathbf{B}$ is the modified flexibility of each member contributing towards the total deflection. (Compare with Eq. 5.18.)

EXAMPLE 6.1

Consider the pinjointed plane frame of Example 2.2. Assume that the members are all of same area of cross section. Let the flexibilities of the members be as follows :

$$f_A = 4f_0\ ;\ f_B = 3f_0\ ;\ f_C = f_E = 5f_0\ ;\ f_D = 6f_0 \text{ with } f_0 = 1/AE$$

Let us determine the horizontal and the vertical displacements of joint C. (Fig.6.1)Eq.(*m*, *j*, *n*) of Example 2.2, give the **B**, **P** and **S** = **BP**. To rewrite:

$$\mathbf{S}=\begin{Bmatrix} S_A \\ S_B \\ S_C \\ S_D \\ S_E \end{Bmatrix}=\begin{bmatrix} 0 & 1 & 0 & 0 & 0 \\ -1 & 0 & 0 & 0 & 0 \\ -0.833 & 0 & -0.833 & 0.625 & 0 \\ 0.5 & 0 & 0.5 & -0.375 & 1 \\ 0.833 & 0 & 0.833 & 0.625 & 0 \end{bmatrix}\begin{Bmatrix} 10 \\ -5 \\ 0 \\ -5 \\ 5 \end{Bmatrix}$$

$$=\begin{Bmatrix} -5 \\ -10 \\ -11.458 \\ 11.875 \\ 5.205 \end{Bmatrix} kN \qquad \text{(a)}$$

To obtain the horizontal and vertical displacements of joint *C* we consider the virtual force shown in Fig. 6.1. The third and the fourth column constitute the columns of the required **B*** matrix relating to the virtual forces in the form **S*** = **B* P***. Thus,

$$\mathbf{S}^*=\begin{Bmatrix} S^*_A \\ S^*_B \\ S^*_C \\ S^*_D \\ S^*_E \end{Bmatrix}=\begin{bmatrix} 0 & 0 \\ 0 & 0 \\ -0.833 & 0.625 \\ 0.6 & -0.375 \\ 0.833 & 0.625 \end{bmatrix}\begin{Bmatrix} P^*_C \\ Q^*_C \end{Bmatrix}=\mathbf{B}^*\,\mathbf{P}^* \qquad \text{(b)}$$

The flexibilities of the members are taken in the form

$$\mathbf{f} = [\,4 \quad 3 \quad 5 \quad 6 \quad 5\,](f_0) \qquad \text{(c)}$$

The member elongation are given by **f S** (Eq. 6.5). Note that premultiplying by a diagonal matrix simply reduces to multiplying the row of the matrix by the corresponding diagonal element.

Thus

$$\mathbf{V}=[4,\ 3,\ 5,\ 6,\ 5\,](f_0)\{-5, -10, -11.458, 11.875, 5.205\}$$
$$=\{-20.0, -30.0, -57.29, 71.25, 26.025\}\,(f_0). \qquad \text{(d)}$$

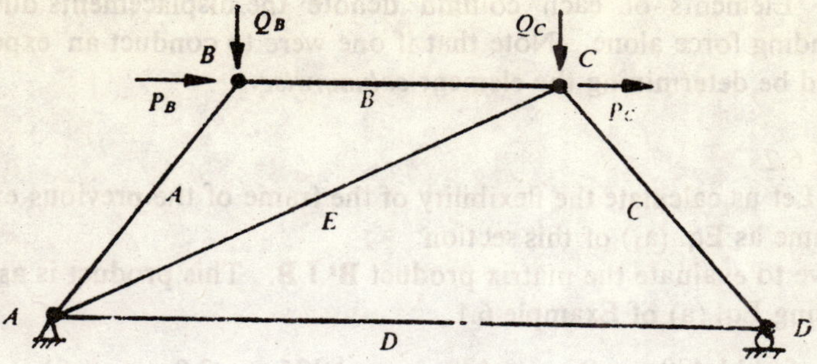

Fig. Example 6.1

The displacements are given by $\mathbf{B}^{*t}\mathbf{V}$ (Eq. 6.2) :

$$\mathbf{U}_1 = \mathbf{B}^{*t}\mathbf{V} = \begin{bmatrix} 0 & 0 & -0.833 & 0.6 & 0.833 \\ 0 & 0 & 0.625 & -0.375 & 0.625 \end{bmatrix} \begin{Bmatrix} 20.0 \\ 30.0 \\ -57.29 \\ 71.25 \\ 26.025 \end{Bmatrix} (f_0) \tag{e}$$

$$= \begin{Bmatrix} 112.17 \\ -29.07 \end{Bmatrix} \frac{1}{\mathbf{AE}} \tag{f}$$

6.2 FLEXIBILITY OF THE STRUCTURE

Set of relations between *all* the displacements of the joints of a structure and the corresponding forces is the (complete) flexibility relation for the structure. The $\mathbf{B}^*$ matrix will now be the $\mathbf{B}$ matrix itself, since displacements are required at all the joints. Therefore,

$$\mathbf{U} = (\mathbf{B}^t\, \mathbf{f}\, \mathbf{B})\mathbf{P} = \mathbf{F}\,\mathbf{P} \tag{6.11}$$

For the frame of Fig. 6.1 we have

$$\mathbf{U} = \begin{Bmatrix} u_B \\ v_B \\ u_C \\ v_C \\ u_D \end{Bmatrix} = \begin{bmatrix} F_{11} & F_{12} & \cdot & \cdot & F_{15} \\ F_{21} & F_{22} & \cdot & \cdot & \cdot \\ \cdot & & & & \cdot \\ \cdot & & & & \cdot \\ F_{51} & \cdot & \cdot & \cdot & F_{55} \end{bmatrix} \begin{Bmatrix} P_B \\ Q_B \\ P_C \\ Q_C \\ P_D \end{Bmatrix} = \mathbf{F}\,\mathbf{P} \tag{a}$$

The flexibility matrix will be a symmetrical matrix. The physical meaning of $\mathbf{F}_{ij}$ may again be noted. Each row states the superposition of the effects of loads. Elements of each column denote the displacements due to the corresponding force alone. Note that if one were to conduct an experiment one would be determining the element *columnwise*.

EXAMPLE 6.2

Let us calculate the flexibility of the frame of the previous example. This is same as Eq. (a_1) of this section.

We have to evaluate the matrix product $\mathbf{B}^t\,\mathbf{f}\,\mathbf{B}$. This product is as shown below, using Eq. (a) of Example 6.1.

$$\mathbf{F} = \begin{bmatrix} 11.439 & 0 & 8.439 & -1.125 & 3.0 \\ 0 & 4.0 & 0 & 0 & 0 \\ 8.439 & 0 & 8.439 & -1.125 & 3.0 \\ -1.125 & 0 & -1.125 & 4.75 & -2.25 \\ 3.0 & 0 & 3.0 & -2.25 & 6.0 \end{bmatrix} (f_0) \qquad (a_2)$$

Some of the elements of the flexibility matrix are easily verified. The several zero's in the second row and columns are easily interpreted. The elements of the second column show that there would be a deflection of $4f_0$ vertically at B due to unit load acting vertically at B, and no deflections anywhere else. The elements of the second row indicate that there will be no vertical deflection at B due to loads other than the vertical load at B. These are easily checked.

6.3 CONDITIONS OF COMPATIBILITY

Equation (6.2) is the general expression giving the displacements. If these displacements are known to be zero then,

$$\mathbf{B}^{*t}\,\mathbf{V} = \mathbf{0} \qquad (6.12)$$

These state the conditions of compatibility that the displacements $\mathbf{U} \equiv \mathbf{0}$. Such an equation will result if the displacements are relative virtual load systems. In such a case the chosen system is a self-equilibrating one, namely $\mathbf{P}^* \equiv \mathbf{0}$. This kind of virtual force systems are chosen in the force method of analysis. We shall be considering this topic in the next chapter after the ideas of redundant forces are discussed.

6.4 CASTIGLIANO'S THEOREMS

The application of Castigliano's theorem to determine the displacements does not in any way provide special advantage from the computational point of view. This was shown in Sec. 5.4 while deriving the flexibility of a member. A similar argument holds for a structure as a whole. The expression (6.8) remains unchanged. This shows that, what all we need is the static matrix **B*** to determine the displacements from the given member deformations. See Eq. (5.16) also.

EXERCISES

6.1 Determine the vertical and horizontal displacements of joint A of the pinconnected truss of Ex. 2.8.

6.2 Determine the vertical component of the displacement of joint A of Ex. 2.12. Take $P_A = 10\ kN$; $Q_A = 5\ kN$.

6.3 Determine the conditions of compatibility of member elongations of the frame of Ex. 2.12.

6 4 Determine the vertical component of the displacement of joint A of Ex. 2.12 due to temperature strain in (a) member B alone and (b) in all members.

6.5 Verify the remark of Sec 6.4 by taking a suitable example either Ex. 6.1 or 6.2.

6.6 Consider that support of member B of the frame of Ex. 2.12 undergoes a movement. Determine the internal forces in the bar.

Chapter 7

Force Method

In the force method, forces are treated as unknowns to be solved for. All the structural quantities and requirements are expressed in terms of these unknowns and the known loads, so that when once these force unknowns are determined all the required quantities are also determined. The conditions of equilibrium and the elasticity relations for members are first satisfied with the help of suitably chosen forces as unknown parameters. The condition of the compatibility between the resulting deformations of the members is then used to solve the unknown force parameters. For this reason the method is also called the *compatibility method.* The coefficients involved in this condition represent flexibility parameters. The name *flexibility method* is also used for the method on account of this feature.

The force method is suitable for certain type of problems. The method was originally developed for aircraft analysis [7]. The recent developments in the applications of this method have demonstrated its efficiency in developing general programs for the solution of complex redundant structures [18]. Considerable thinking and effort have been directed towards developing automatic selection of redundant systems [19] which has rendered the method considerably attractive for drawing general programs. However the *displacement* method still possesses several advantages from the practical viewpoints of developing more straight forward computer programs.

The force and the displacement methods are the dual methods of structural analysis which differ in incorporating the equilibrium and kinematical requirements. In the force method, the kinematical requirements are satisfied as the last stage in the analysis process, while in the displacement method the conditions of equilibrium are satisfied as the last stage. For a student of structural analysis study of both these methods, at least from the points of view of basic principles and understanding structural analysis is necessary and paying.

7.1 BASIC IDEAS OF THE FORCE METHOD

Before proceeding to the general equations applicable to any structure, the basic steps involved in the force method will be identified with respect to a simple example. Readers familiar with the use of matrix notations can directly proceed to Sec. 7.2.

Consider the three bar system shown in Fig. 7.1 (a), which is the simplest model of a redundant (statically indeterminate) structure. Let f be the flexibility (L/AE) of a bar. For simplicity we will assume $f_A = f_B$, leading to symmetric behaviour. The first step is to consider the equilibrium problem. In this case, the equations of equilibrium available are, with $\beta = 30°$,

$$\frac{\sqrt{3}}{2}\, S_A + \frac{\sqrt{3}}{2}\, S_B + S_C = P \ ; \ S_A = S_B \qquad (a_1)$$

The number of available equations of equilibrium are seen to be lesser than the number of unknowns by one. This is the basic feature and in general a structure is said to be statically redundant or indeterminate to the degree equal to the difference between the unknown forces and the available equations of equilibrium between them. In the present case, this is seen to be one. Suitable and convenient forces equal to the degree of redundancy of a structure are chosen as special unknowns out of the total unknown forces and called *redundants*. These redundants are taken as temporary unknowns during the process of solving the equations of compatibility. If the redundants are grouped together with the given loads, then the number of unknown forces becomes equal to that of the available equations of equilibrium between these unknowns and the effective combined loads consisting of the given loads and the unknown redundants. Equations of equilibrium can then be formally solved in terms of the given loads and the redundants. The latter are equivalent loads, unknown in magnitude, but known in directions of action. Presently, any one of S_A, S_B or S_C can be taken as a redundant quantity. Let S_C, the force in the central bar be taken as this quantity :

$$S_C = R \equiv \mathbf{R} \qquad (a_2)$$

We can now solve for S_A and S_B to obtain

$$S_A = S_B = \frac{1}{\sqrt{3}}(P - R) \qquad (a_3)$$

This process of solution is equivalent to a fact that the given loads and the redundant forces are taken to act on a determinate frame. Such a frame associated with the given set of redundants is called the *basic determinate structure* or the *released structure*. Equation (a_3) shows that only S_A and S_B are being solved ; the structure being solved is the one in which $S_C = 0$, which physically means that there is no action of S_C on the group of bars.

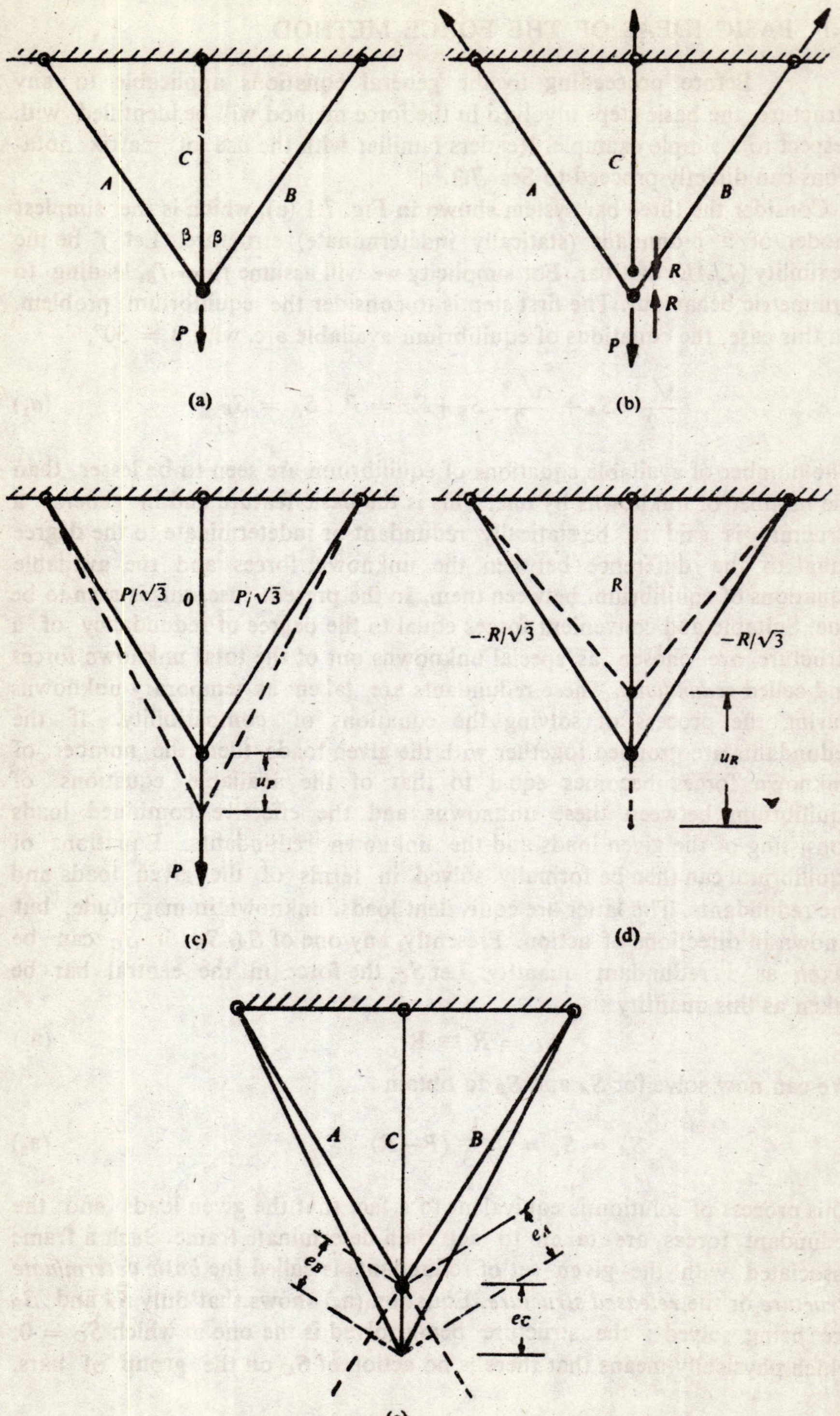
A
C
B
β
β
P
(a)
A
C
B
R
R
P
(b)
P/√3
0
P/√3
u_P
P
(c)
R
−R/√3
−R/√3
u_R
(d)
A
C
B
e_A
e_B
e_C
(e)

This is imagined to be achieved by associating that there is a 'cut' or 'release' introduced, between bar C and the joint A so that the bars A and B alone act as a two bar arrangement. Thus, a hypothetical determinate structure is imagined with the help of this release between the bar and the joint. Figure 7.1(b) shows that frame acted on by loads P and R developing forces S_A and S_B as in Eq. (a_3). For convenience, the loads and the redundants may be taken to act separately and the forces S_A and S_B obtained in each case.

Equations (a_2) and (a_3) together can be written as

$$\begin{Bmatrix} S_A \\ S_B \\ S_C \end{Bmatrix} = \begin{Bmatrix} 1/\sqrt{3} \\ 1/\sqrt{3} \\ 0 \end{Bmatrix} P + \begin{Bmatrix} -1/\sqrt{3} \\ -1/\sqrt{3} \\ 1 \end{Bmatrix} (R) \tag{a_4}$$

or

$$\mathbf{S} = \mathbf{BP} + \mathbf{bR} \tag{a_5}$$

These show that the internal forces in all members can be considered as a superimposition of two statically admissible internal force systems. The first system is the set of forces statically equivalent to the given loads with the redundant force equal to zero ; the second is the system of internal forces statically equivalent to zero external loads but to the redundant force alone. Equations written this way permit the interpretation of the superimposition of two internal force distribution in all the members (Fig 7.1c and d).

Consider now the elasticity effect. Corresponding to the internal forces, we have the elongations

$$e_A = f_A S_A \; ; \quad e_B = f_B S_B \; ; \quad e_C = f_C S_C$$

or collectively

$$\mathbf{V} = \{e_A, e_B, e_C\} = \lceil f_A \; f_B \; f_C \rfloor \{S_A, S_B, S_C\} = \mathbf{f}\,\mathbf{S}. \tag{a_6}$$

In this we can use Eq. (a_5), giving

$$\mathbf{V} = \mathbf{f}\,(\mathbf{BP} + \mathbf{bR}) \tag{a_7}$$

Written in full,

$$\mathbf{V} = \begin{Bmatrix} e_A \\ e_B \\ e_C \end{Bmatrix} = \begin{bmatrix} \frac{1}{\sqrt{3}} f_A \\ \frac{1}{\sqrt{3}} f_B \\ 0 \end{bmatrix} (P) + \begin{bmatrix} -\frac{1}{\sqrt{3}} f_A \\ -\frac{1}{\sqrt{3}} f_B \\ f_C \end{bmatrix} (R) \tag{a_8}$$

Thus, the member elongations are obtained in terms of the loads and the unknown redundant force.

The condition to determine the as yet unknown redundant is the condition of the kinematics. This is also termed as the *compatibility* condition. The behaviour of the structure puts certain restrictions on the member elongations. This condition, in the present case, is the consistent elongations of bars so that they remain jointed at A. The condition on the member elongations is, from Fig. 7.1(e),

$$e_A = \frac{\sqrt{3}}{2} e_C, \quad \text{or} \quad \frac{2}{\sqrt{3}} e_A - e_C = 0 \tag{a_9}$$

Using $e_A = e_B$, we can write this as

$$\left[-\frac{1}{\sqrt{3}} \quad -\frac{1}{\sqrt{3}} \quad 1\right]\{e_A, e_B, e_C\} = 0 \tag{a_{10}}$$

It is seen that this is same as

$$\mathbf{b}^t\ \mathbf{V} = 0 \tag{a_{11}}$$

where **b** is the statical matrix appearing in Eq. (a_4). This is not accidental but is always true as we will see later. In this equation we can now substitute for **V** from (a_7) and write,

$$\mathbf{b}^t\ \mathbf{f}\ (\mathbf{BP}+\mathbf{bR}) = 0 \tag{a_{12}}$$

or

$$\mathbf{F}_{RP}\ \mathbf{P}+\mathbf{F}_{RR}\ \mathbf{R} = 0 \tag{a_{13}}$$

where,

$$\mathbf{F}_{RP} = \mathbf{b}^t \mathbf{f}\ \mathbf{B}\ ;\ \mathbf{F}_{RR} = \mathbf{b}^t \mathbf{f}\ \mathbf{b} \tag{a_{14}}$$

The redundant **R** is thus found as :

$$\mathbf{R} = \mathbf{F}_{RR}^{-1}\ \mathbf{F}_{RP}\ \mathbf{P} = -(\mathbf{b}^t\ \mathbf{f}\ \mathbf{b})^{-1}(\mathbf{b}^t\ \mathbf{f}\ \mathbf{B}\ \mathbf{P}) \tag{a_{15}}$$

Presently,

$$\mathbf{F}_{RR}\ \mathbf{R} = \left[-\frac{1}{\sqrt{3}} \quad -\frac{1}{\sqrt{3}} \quad 1\right]\left[f_A\ f_B\ f_C\right]\left\{-\frac{1}{\sqrt{3}} \quad -\frac{1}{\sqrt{3}} \quad 1\right\} R$$

$$= \left(\frac{1}{3} f_A + \frac{1}{3} f_B + f_C\right) R = \left(\frac{2}{3} f_A + f_C\right) R \tag{a_{16}}$$

$$\mathbf{F}_{RP}\ \mathbf{P} = \left(-\frac{2}{3} f_A\right) P \tag{a_{17}}$$

Therefore Eq. (a_{13}) gives

$$R = \left(\frac{2}{3} f_A P\right) \Big/ \left(\frac{2}{3} f_A + f_C\right) \tag{a_{18}}$$

For $f_A = f_B = f_C = f_0$ we obtain $R = (2/5)\ P$. Then Eq. (a_4) gives

$$\mathbf{S} = \left\{\frac{\sqrt{3}}{5} \quad \frac{\sqrt{3}}{5} \quad \frac{2}{5}\right\} P. \tag{a_{19}}$$

It is thus seen that the compatibility condition (Eq. a_{11}), is the one from which the unknown redundant is solved for. The coefficients $\mathbf{F}_{RP}$ and $\mathbf{F}_{RR}$

are flexibility coefficients ((1 × 1) matrices presently) having definite meaning. This arises on account of the fact that the two internal force systems shown by Eq. (a_4) do not separately satisfy the conditions of the compatibility of the problem. For this reason they were called *statically admissible*.

Equation (a_{13}) can be written as

$$u_P + u_R = 0 \qquad (a_{20})$$

where the first term is the displacement occuring in the released frame due to external load **P**, with internal forces **BP**. (See Fig. 7.1 *c*.) The second term is the displacement in the basic determinate frame due to the load **R** and internal forces **bR**. (See Fig. 7.1 *d*.) In the first case the force in the bar *C* is zero and the tensile forces ($P/\sqrt{3}$) in bars *A* and *B* cause a gap of $\mathbf{u}_P$. In the second case there is a tensile force of *R* in bar *C* and compressure forces of $R/\sqrt{3}$ in bars *A* and *B*. U_R is the *relative* displacement between the bar *C* and the joint *A*. Thus the left hand side of Eq. (a_{20}) (which is the same as of Eqs. a_{11}, a_{12} and a_{13}) denotes the total relative gap between end of bar *C* and the joint *C* in the released frame. Thus these equations state that imaginary gap be zero. This is equivalent to the compatibility condition for the given frame where no such gaps are present. This can also be called as the continuity condition.

Use of Virtual Work

Consider now the application of the virtual work principle. Its chief use is in writing the equations of compatibility and in obtaining displacements. Equation (a_9) was obtained directly from Fig. 7.1(c). The property of the contragredience between the static matrix **b** appearing in Eq. (a_5) and the kinematic matrix appearing in Eq. (a_{11}) was noticed. This can be deduced from virtual work considerations. This procedure avoids the direct consideration of the geometry of the compatibility of deformations.

Consider the force system consisting of the internal forces **bR** and the external load **R**. The latter is a pair of forces *R* and $-R$ acting in the released frame. Observe that the resulting incompatible deformations are of no relevance in applying the virtual work rule, since the virtual force system need only be an equilibrant set, which presently is so. The actual deformations of the given frame are taken as the deformation system in writing the equation of virtual work. These form a compatible set. The equation of work is therefore

$$(\mathbf{bR})^t \mathbf{V} = 0 \qquad (a_{21})$$

It is thus seen that the external work is zero since the load *R* is an equal and opposite force-pair. In physical terms, we can consider that these consititute a zero external load and the corresponding internal forces **b R** to be thus self equilibrating. The above equation leads again to the required compatibility condition (a_{11}).

We will see as we proceed that the use of virtual work is quite convenient and avoids geometrical considerations.

EXAMPLE 7.1

Consider the pinjointed plane frome shown in Fig. (a). This is indeterminate to a single degree. We shall pursue this example from the point of view of showing how the matrix notations handle the equations which are normally handled using scalar algebra. In simple cases like this the **B** and **b** matrices of Eq. (a_5) can be directly constructed. Let us treat the force in the bar F as the redundant. The basic determinate structure is the one in which the force in this bar is zero. It is a determinate frame as shown in Fig. (b). The actual internal forces in the structure at Fig. (a) are the forces in the basic determinate frame acted on by the same loads and the redundant force R. In terms of the load P and the as yet unknown force R the forces in the members of the basic determinate structure are

$$S_A = S_B = S_C = S_D = (P - R)/\sqrt{2}$$
$$S_E = -(P - R) \tag{a}$$

By definition the force in the bar F is R. Thus for the original structure consisting of all members the forces in members are those given above and $S_F = R$. We write these as

$$\mathbf{S} = \begin{Bmatrix} S_A \\ S_B \\ S_C \\ S_D \\ S_E \\ S_F \end{Bmatrix} = \begin{Bmatrix} 1/\sqrt{2} \\ 1/\sqrt{2} \\ 1/\sqrt{2} \\ 1/\sqrt{2} \\ -1 \\ 0 \end{Bmatrix} (P) + \begin{Bmatrix} -1/\sqrt{2} \\ -1/\sqrt{2} \\ -1/\sqrt{2} \\ -1/\sqrt{2} \\ 1 \\ 1 \end{Bmatrix} (R) = \mathbf{B}\,\mathbf{P} + \mathbf{b}\,\mathbf{R} \tag{b}$$

The **B** and **b** are the (6×1) column matrices. The last row only states the choice of the redundant in the form of an equation. The internal forces in all the members can now be interpreted as a superimposition of two forces systems, one equivalent to the given loads and the internal force chosen as redundant set equal to zero, and the other system equivalent to zero external load but only the redundant internal force acting. Figures (c) and (d) show these force distributions. Obviously each of these systems do not yield member elogations which would be compatible with one another. The system at Fig. (c) will lead to zero elongation in member F, while as the joints A and C move outward as a result of the elasticity in other members. Similarly the force system in Fig. (d) also leads to incompatible member elogations.

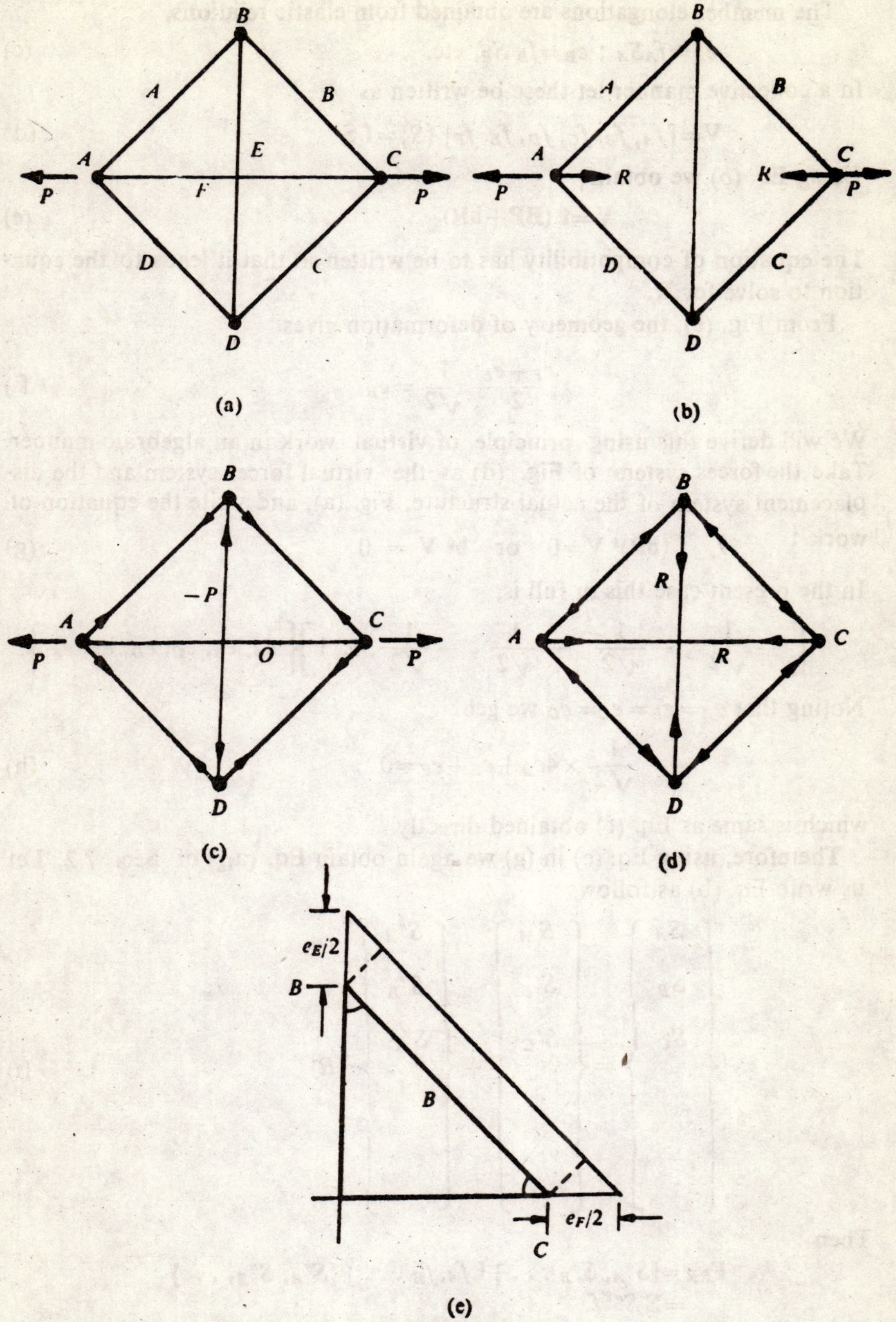

Example 7.1

The member elongations are obtained from elastic relations,

$$e_A = f_A S_A \ ; \ e_B = f_B S_B, \text{ etc.} \tag{c}$$

In a collective manner let these be written as

$$\mathbf{V} = [f_A, f_B, f_C, f_D, f_E, f_F] \{S\} = \mathbf{f\,S} \tag{d}$$

Using Eq. (*b*) we obtain

$$\mathbf{V} = \mathbf{f}\,(\mathbf{BP} + \mathbf{bR}) \tag{e}$$

The equation of compatibility has to be written so that it leads to the equation to solve for **R**.

From Fig. (e), the geometry of deformation gives,

$$\frac{e_F + e_E}{2} \frac{1}{\sqrt{2}} = e_B \tag{f}$$

We will derive this using principle of virtual work in an algebraic manner. Take the forces system of Fig. (d) as the virtual forces system and the displacement system of the actual structure, Fig. (a), and write the equation of work :

$$(\mathbf{bR})^t\,\mathbf{V} = 0 \quad \text{or} \quad \mathbf{b}^t\,\mathbf{V} = 0 \tag{g}$$

In the present case this in full is,

$$\left[-\frac{1}{\sqrt{2}}, -\frac{1}{\sqrt{2}}, -\frac{1}{\sqrt{2}}, -\frac{1}{\sqrt{2}}, 1, 1\right]\left\{e_A, e_B, e_C, e_D, e_E, e_F\right\}.$$

Noting that $e_A = e_B = e_C = e_D$ we get

$$= \frac{1}{\sqrt{2}} \times 4e_B + e_E + e_F = 0 \tag{h}$$

which is same as Eq. (f) obtained directly.

Therefore, using Eq. (e) in (g) we again obtain Eq. (a_{15}) of Sec. 7.2. Let us write Eq. (b) as follows

$$\begin{Bmatrix} S_A \\ S_B \\ S_C \\ \cdot \\ \cdot \\ \cdot \end{Bmatrix} = \begin{Bmatrix} S'_A \\ S'_B \\ S'_C \\ \cdot \\ \cdot \\ \cdot \end{Bmatrix} + \begin{Bmatrix} S''_A \\ S''_B \\ S''_C \\ \cdot \\ \cdot \\ \cdot \end{Bmatrix} (R) \tag{i}$$

Then

$$\mathbf{F}_{RR} = [S''_A, S''_B \ldots] \lceil f_A, f_B \ldots \rfloor \{S''_A, S''_B, \ldots\}$$
$$= \Sigma S''^2 f$$

Similarly

$$\mathbf{F}_{RP} = \Sigma S'\, S''\, f$$

Therefore Eq. (a_{13}) of Sec. 7.1 gives

$$\mathbf{R} \equiv R = \frac{\Sigma S' S'' f}{\Sigma S''^2 f} \tag{j}$$

Here the summation extends over *all* the members. This is in the familar form used in scalar formulation. Thus the evaluation of the product $\mathbf{b}^t \mathbf{f} \mathbf{b}$ and $\mathbf{b}^t \mathbf{f} \mathbf{B}$ are analogous to the evaluation of the denominator and numerator of Eq. (j) respectively.

Presently if all the members have the same area of cross-section we have the flexibilities of the diagonal bars $\sqrt{2}$ that of the side bars. Equation (j) then gives $R = P/\sqrt{2}$.

EXAMPLE 7.2

We shall consider another simple example but involving moments, Fig. (a). The continuous beam shown has redundancy of two degrees. For a member we will define the internal forces as the two end moments. Thus

$$\mathbf{S}_A = \{M_{1A}, M_{2A}\} \; ; \mathbf{S}_B = \{M_{1B}, M_{2B}\} \tag{a}$$

The corresponding member deformations are

$$\mathbf{v}_A = \{\varphi_{1A}, \varphi_{2A}\} \; ; \mathbf{v}_B = \{\varphi_{1B}, \varphi_{2B}\} \tag{b}$$

These are the same as the ones defined for beam model except that the axial force and the correspondence elongation terms are deleted. The basic flexibility relations between these, neglecting effect of shearing deformation, are,

$$\mathbf{v} = \begin{Bmatrix} \varphi_1 \\ \varphi_2 \end{Bmatrix} = \begin{bmatrix} L/3EI & -L/6EI \\ -L/6EI & L/3EI \end{bmatrix} \begin{Bmatrix} M_1 \\ M_2 \end{Bmatrix} = \mathbf{f}\,\mathbf{S} \tag{c}$$

The equations of equilibrium at the two joints B and C are :

$$\begin{aligned} &\text{At B :} \quad M_{2A} + M_{1B} = M_B \\ &\text{At C :} \quad M_{2B} = M_C \end{aligned} \tag{d}$$

The number of unknowns are the four member end moments. The degree of redundancy is therefore two. Note that even if we include the equation of equilibrium at A we will still have two redundancies, for, we would then obtain

$$M_{1A} = M_A \tag{e}$$

where M_A is the unknown reactive moment. Then we have three equations between the five unknown bending moments again leading to two redundancies.

We will chose M_{1A} and M_{2A} as redundants

$$M_{1A} = R_1 \; ; \qquad M_{2A} = R_2 \tag{f}$$

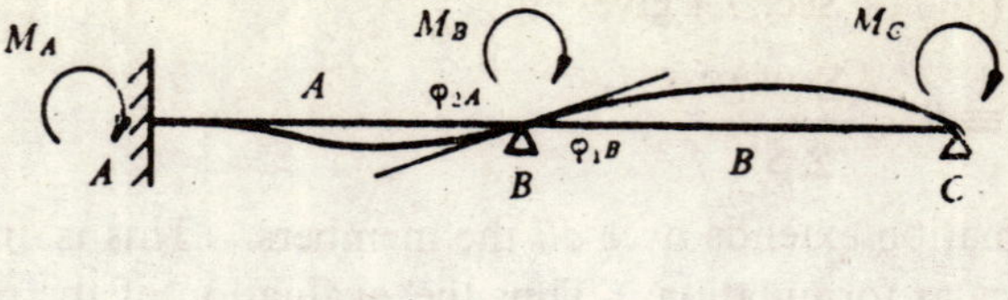

(a)

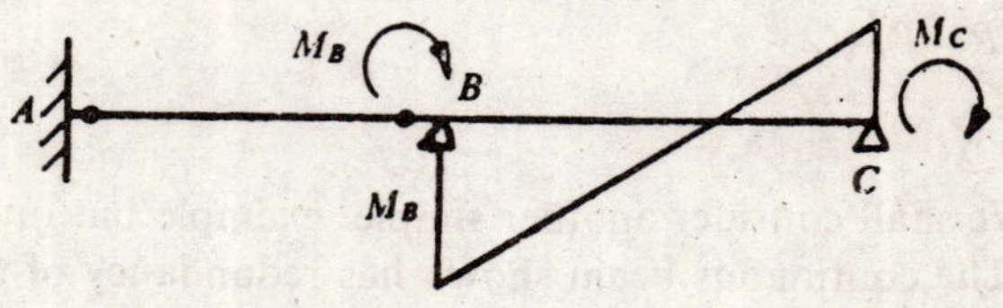

(b)

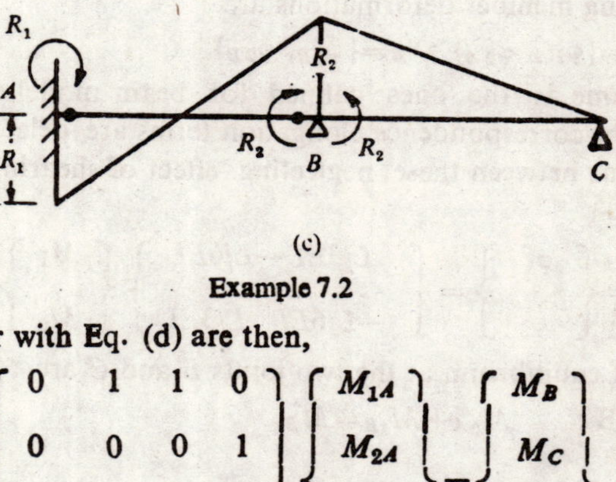

(c)

Example 7.2

These together with Eq. (d) are then,

$$\begin{bmatrix} 0 & 1 & 1 & 0 \\ 0 & 0 & 0 & 1 \\ 1 & 0 & 0 & 0 \\ 0 & 1 & 0 & 0 \end{bmatrix} \begin{Bmatrix} M_{1A} \\ M_{2A} \\ M_{1B} \\ M_{2B} \end{Bmatrix} = \begin{Bmatrix} M_B \\ M_C \\ R_1 \\ R_2 \end{Bmatrix} \tag{g}$$

These are of the form

$$\mathbf{C}'\,\mathbf{S} = \mathbf{P}' \tag{h}$$

We can *solve* for S and obtain,

$$\begin{Bmatrix} M_{1A} \\ M_{2A} \\ M_{1B} \\ M_{2B} \end{Bmatrix} = \begin{bmatrix} 0 & 0 & 1 & 0 \\ 0 & 0 & 0 & 1 \\ 1 & 0 & 0 & -1 \\ 0 & 1 & 0 & 0 \end{bmatrix} \begin{Bmatrix} M_B \\ M_C \\ R_1 \\ R_2 \end{Bmatrix} \tag{i}$$

These may be written as follows :

$$\mathbf{S}=\begin{Bmatrix} M_{1A} \\ M_{2A} \\ M_{1B} \\ M_{2B} \end{Bmatrix}=\begin{bmatrix} 0 & 0 \\ 0 & 0 \\ 1 & 0 \\ 0 & 1 \end{bmatrix}\begin{Bmatrix} M_B \\ M_C \end{Bmatrix}+\begin{bmatrix} 1 & 0 \\ 0 & 1 \\ 0 & -1 \\ 0 & 0 \end{bmatrix}\begin{Bmatrix} R_1 \\ R_2 \end{Bmatrix}$$

or

$$\mathbf{S}=\mathbf{B}\,\mathbf{P}+\mathbf{b}\,\mathbf{R} \tag{j}$$

The moments due to P and R are sketched in Figs. (b) and (c). The basic determinate structure is the beam with hinges at ends 1 and 2 of member A. The end moments are only statically admissible as will be seen further from the considerations of displacements.

Let us write the conditions of continuity (compatibility). From the deflected shape shown in Fig. (a) we have

$$\varphi_{1A}=0\,; \qquad \varphi_{2A}=\varphi_{1B}$$

These in matrix form are

$$\begin{bmatrix} 1 & 0 & 0 & 0 \\ 0 & 1 & -1 & 0 \end{bmatrix}\begin{Bmatrix} \varphi_{1A} \\ \varphi_{2A} \\ \varphi_{1B} \\ \varphi_{2B} \end{Bmatrix}=\begin{Bmatrix} 0 \\ 0 \end{Bmatrix} \tag{k}$$

It is seen by comparing with the second term of Eq. (j) that this relation is

$$\mathbf{b}^t\,\mathbf{v}=0 \tag{l}$$

as expected.

Therefore,

$$\mathbf{b}^t\,\mathbf{f}\,(\mathbf{BP}+\mathbf{bR})=0 \tag{m}$$

giving again,

$$\mathbf{R}=\mathbf{F}_{RR}^{-1}\,\mathbf{F}_{RP}\,\mathbf{P} \tag{n}$$

We shall consider beams A and B to be identical. We can write the above matrices in full as follows :

$$\mathbf{F}_{RR}=\mathbf{b}^t\,\mathbf{f}\,\mathbf{b}=\begin{bmatrix} 1 & 0 & 0 & 0 \\ 0 & 1 & -1 & 0 \end{bmatrix}\begin{bmatrix} \alpha & -\beta & \cdot & \cdot \\ -\beta & \alpha & \cdot & \cdot \\ \cdot & \cdot & \alpha & -\beta \\ \cdot & \cdot & -\beta & \alpha \end{bmatrix}\begin{bmatrix} 1 & 0 \\ 0 & 1 \\ 0 & -1 \\ 0 & 0 \end{bmatrix}$$

$$=\begin{bmatrix} \alpha & -\beta \\ -\beta & 2\alpha \end{bmatrix} \tag{o}$$

$$F_{RP}=\mathbf{b}^t\ \mathbf{f}\ \mathbf{B}=\begin{bmatrix} 0 & 0 \\ -\alpha & \beta \end{bmatrix} \tag{p}$$

with

$$\alpha=L/3EI\ ;\ \beta=L/6EI \tag{q}$$

Thus

$$F_{RR}^{-1}=\frac{6EI}{L}\begin{bmatrix} \frac{4}{7} & \frac{1}{7} \\ \frac{1}{7} & \frac{2}{7} \end{bmatrix} \tag{r}$$

Therefore the product $-\mathbf{F}_{RR}^{-1}\ \mathbf{F}_{RP}\ \mathbf{P}$ gives

$$\mathbf{R}=\begin{Bmatrix} R_1 \\ R_2 \end{Bmatrix}=\begin{bmatrix} \frac{2}{7} & -\frac{1}{7} \\ \frac{4}{7} & -\frac{2}{7} \end{bmatrix}\begin{Bmatrix} M_B \\ M_C \end{Bmatrix} \tag{s}$$

Substituting these in Eq. (j) we obtain the end moments as

$$\begin{Bmatrix} M_{1A} \\ M_{2A} \\ M_{1B} \\ M_{2B} \end{Bmatrix}=\frac{1}{7}\begin{bmatrix} 2 & -1 \\ 4 & -2 \\ 3 & 2 \\ 0 & 1 \end{bmatrix}\begin{Bmatrix} M_B \\ M_C \end{Bmatrix} \tag{t}$$

These can be checked, for instance, by pereforming the moment distribution. At B the distribution factors are 4/7 and 3/7 for members A and B respectively. Taking M_B and M_C one at a time, we can readily verify each column.

7.2 GENERAL EQUATIONS OF THE FORCE METHOD

We will derive the basic equations of the method in more general terms. We will again keep an example in view while explaining the meaning of the steps and the contents of the different matrices appearing in the equations. Figure 7.2 (a) shows a plane frame statically indeterminate to the third degree.

The starting point is essentially obtaining the internal forces in terms of

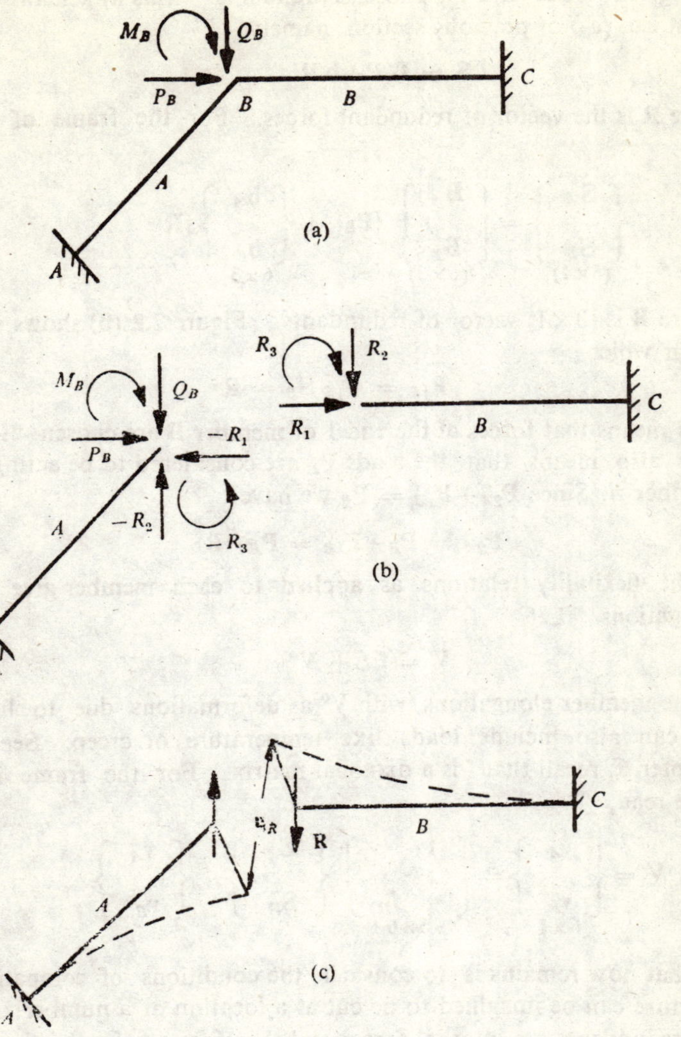

M_B
Q_B
P_B
B
B
C
A
A
(a)
R_3
R_2
M_B
Q_B
R_1
B
C
P_B
$-R_1$
A
$-R_2$
$-R_3$
A
(b)
e_R
R
C
B
A
A
(c)
R_2
B
C
R_3
A
A
R_1
(d)

Figure 7.2

the given loads and the chosen redundants. This in general will be of the form Eq. (a_5) of previous section, namely

$$\mathbf{S} = \mathbf{B\,P} + \mathbf{b\,R} \tag{7.1}$$

Here **R** is the vector of redundant forces. For the frame of Fig. 7.2, we have

$$\underset{(6\times 1)}{\begin{Bmatrix} \mathbf{S}_A \\ \mathbf{S}_B \end{Bmatrix}} = \underset{(6\times 3)}{\begin{bmatrix} \mathbf{B}_A \\ \mathbf{B}_B \end{bmatrix}} \{\mathbf{P}_B\} + \underset{(6\times 3}{\begin{Bmatrix} \mathbf{b}_A \\ \mathbf{b}_B \end{Bmatrix}} \{\mathbf{R}\} \tag{a_1}$$

where **R** is (3×1) vector of redundants. Figure 7.2 (b) shows one possible set in which

$$\mathbf{P}_{1B} = \mathbf{C}_{1B}\,\mathbf{S}_B = \mathbf{R} \tag{a_2}$$

This means that forces at the end 1 of member **B** are chosen as redundants. This also means that the loads $\mathbf{P}_B$ are considered to be acting at end 2 of member A. Since $\mathbf{P}_{2A}+\mathbf{P}_{1B} = \mathbf{P}_B$ we have

$$\mathbf{P}_{2A} = \mathbf{P}_B - \mathbf{P}_{1B} = \mathbf{P}_B - \mathbf{R} \tag{a_3}$$

The flexibility relations as applied to each member give the member elongations. Let

$$\mathbf{V} = \mathbf{f\,S} + \mathbf{V}^{\circ} \tag{7.2}$$

be the member elongations, with $\mathbf{V}^{\circ}$ as deformations due to lateral loads. We can also include loads like temperature or creep. See Eq. (6.5) of Chapter 6; recall that **f** is a diagonal matrix. For the frame in question, these read,

$$\mathbf{V} = \underset{6\times 1}{\begin{Bmatrix} \mathbf{v}_A \\ \mathbf{v}_B \end{Bmatrix}} = \underset{6\times 6}{\begin{bmatrix} \mathbf{f}_A & \\ & \mathbf{f}_B \end{bmatrix}} \begin{Bmatrix} \mathbf{S}_A \\ \mathbf{S}_B \end{Bmatrix} + \begin{Bmatrix} \mathbf{v}_A^{\circ} \\ \mathbf{v}_B^{\circ} \end{Bmatrix} \tag{a_4}$$

What now remains is to consider the conditions of compatibility. The structure can be imagined to be cut at a location or a number of locations. The redundants are internal forces and therefore are force pairs. These are the generalized forces corresponding to the imaginary 'releases' or 'cuts'. If the member is *fully* released at a section, the number of releases will be three in the case of a plane frame member, or six in the case of a space frame member. In Fig. 7.2 (b), corresponding to the redundants **R**, we have three release displacements, axial, transverse and rotational, which physically are equivalent to a complete cut of the member. Figure 7.2 (c) shows the release displacements schematically, which include a relative horizontal gap, a relative transverse gap and a relative angular gap. Figure 7.2 (d) shows the *angular releases* at ends 1 and 2 of member A and at end 2 of member B. These correspond to the redundants M_{1A}, M_{2A} and M_{2B} respectively. In the given structure such gaps are not there and therefore they are to be set equal to zero. Therefore these form the compatibility conditions. The generalized displacements corresponding to the redundants can be determined by the application of virtual work.

Consider the forces

$$\mathbf{S^*} = \mathbf{b\ R} \tag{7.3}$$

which correspond to zero external loads as can be seen from Eq. (7.1). In other words these are self-equilibrating. Therefore considering this force system as a virtual system and the actual deformation of the structure as the deformation system we have the equation of work in which the external work is zero. Thus $\mathbf{S^{*t}\ V} = \mathbf{0}$ giving $(\mathbf{b\ R})^t\ \mathbf{V} = \mathbf{0}$, or

$$\mathbf{b}^t\ \mathbf{V} = 0 \tag{7.4}$$

If this is not convincing enough we may take another virtual system and except to arrive at the same result. Let us consider the actual forces and actual deformation of the structure and write the work equation :

$$\mathbf{S}^t\ \mathbf{V} = \mathbf{P}^t\ \mathbf{U} \tag{a_5}$$

Substitute in this equation for **S** from Eq. (7.1) :

$$(\mathbf{B\ P} + \mathbf{b\ R})^t\ \mathbf{V} = \mathbf{P}^t\ \mathbf{U}$$

This becomes

$$\mathbf{P}^t\ \mathbf{B}^t\ \mathbf{V} + \mathbf{R}^t\ \mathbf{b}^t\ \mathbf{V} = \mathbf{P}^t\ \mathbf{U}$$

By comparing the coefficients of **P** we obtain

$$\mathbf{B}^t\ \mathbf{V} = \mathbf{U} \tag{7.5}$$

and similarly of **R**, leads again to Eq. (7.4).

Equation (7.5) above is not new but was the general form obtained in Chapter 6 for displacements. See Eq. (6.2). Recall that $\mathbf{B^*}$ can be **B**, if the displacements are desired at all points corresponding to **P**. We can now argue that if Eq. (7.5) is the general term to obtain displacements, then Eq. (7.4) can be interpreted to mean that the generalized displacements corresponding to the internal forces (**b R**) are zero. These denote the *imaginary* release displacements or relative 'gaps', which are evidently required to be zero in the given structure.

Equation (7.4) for the frame in question reads as follows :

$$\underset{(3\times 6)}{[\mathbf{b}^t_A\ \mathbf{b}^t_B]} \underset{(6\times 1)}{\begin{Bmatrix} \mathbf{v}_A \\ \mathbf{v}_B \end{Bmatrix}} = \underset{(3\times 1)}{\{\mathbf{0}\}} \tag{a_6}$$

Equations (7.1, 7.2, 7.4) are the basic equations ensuring equlibrium, elasticity and kinematical requirements. It now remains to combine them. Using (7.1) in (7.2),

$$\mathbf{V} = \mathbf{f\ B\ P} + \mathbf{f\ b\ R} + \mathbf{V}^\circ \tag{7.6}$$

Therefore Eq. (7.4) becomes

$$\mathbf{b}^t\,(\mathbf{f\,B\,P} + \mathbf{f\,b\,R} + \mathbf{V}^\circ) = 0$$

From this system we can solve for the redundants **R** :

$$\mathbf{R} = (\mathbf{b}^t\,\mathbf{f}\,\mathbf{b})^{-1}\,\{\mathbf{b}^t\,\mathbf{f\,B\,P} + \mathbf{b}^t\,\mathbf{V}^\circ\} \qquad (7.7)$$

Here $\mathbf{b}^t\,\mathbf{f}\,\mathbf{b}$ denotes the flexibility corresponding to the redundants **R**. In many structural problems the size of this matrix is small so that equation solving is not a major issue as in the displacement method. We will be discussing this and related issues in later sections.

Observe that matrices **B**, **b** and **f** are the basic matrices. The starting point of having to construct the **B** and **b** matrices requires proper choice of redundants and of course involves the main task of solving for internal forces in terms of loads and redundants. Though these are determinate systems they pose serious problems when it comes to solving a large system. We will discuss later this issue and indicate the methods available for constructing **B** and **b** matrices.

For the frame in question observe that

$$\begin{aligned}\mathbf{b}^t\,\mathbf{f}\,\mathbf{b} &= [\mathbf{b}^t_A\;\; \mathbf{b}^t_B]\;[\mathbf{f}_A\;\; \mathbf{f}_B]\;\{\mathbf{b}_A\;\; \mathbf{b}_B\}\\ &= (\mathbf{b}^t_A\,\mathbf{f}_A\,\mathbf{b}_A + \mathbf{b}^t_B\,\mathbf{f}_B\,\mathbf{b}_B) \qquad (a_7)\end{aligned}$$

This is an other example illustrating the addition of member flexibilities. (See Eq. 6.10, and subsequent comments.) Similarly the other matrices can be detailed out. This is pursued in the example below.

EXAMPLE 7.3

Consider the plane frame shown in Fig. (a). The equivalent loads shown correspond to a certain distributed loads (see Example 8.2 of Chapter 8).

We will consider the following cases : Case (a) : Redundants as in Fig. 7.2(b); Case (b) : Redundants as in Fig. 7.2(d).

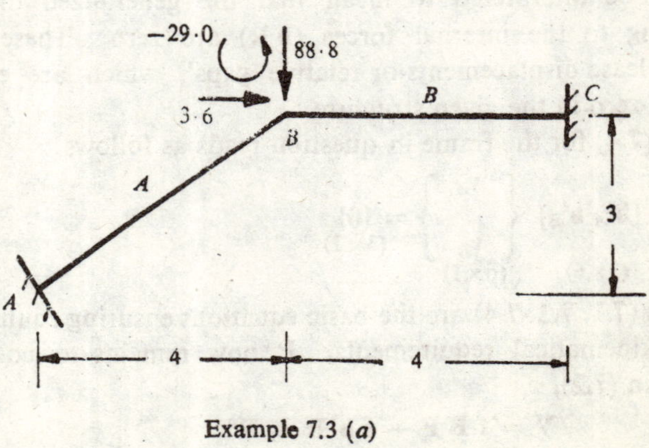

Example 7.3 (*a*)

Case (a)
Since a complete release is introduced at B, we choose the set of independent member forces $\mathbf{S}$, as

$$\mathbf{S} \equiv \overline{\mathbf{P}}_2 = \{\overline{P}_2, \overline{Q}_2, \overline{M}_2\} \tag{a$_1$}$$

We have, in other words the cantilever model in view. We will be employing the beam model for Case (b), to be taken up later.

The first step is to arrive at the internal forces $\mathbf{S}$, in terms of the loads and the redundants. This itself entails calculations though straight forward. Presently we have two cantilevers as shown in Fig. 7.2 (b). The forces $\overline{\mathbf{P}}_2$ at ends 2 of the members can be calculated. These are, for member A,

$$\mathbf{S}_A = \begin{Bmatrix} \overline{P}_{2A} \\ \overline{Q}_{2A} \\ \overline{M}_{2A} \end{Bmatrix} = \begin{bmatrix} 0.8 & -0.6 & 0 \\ 0.6 & 0.8 & 0 \\ 0 & 0 & 1 \end{bmatrix} \begin{Bmatrix} P_B - R_1 \\ Q_B - R_2 \\ M_B - R_3 \end{Bmatrix} \tag{a$_2$}$$

Here $\cos\alpha$ and $\sin\alpha$ for the member are 0.8 and -0.6 respectively. Similarly for member B

$$\mathbf{S}_B = \begin{Bmatrix} \overline{P}_{2B} \\ \overline{Q}_{2B} \\ \overline{M}_{2B} \end{Bmatrix} = \begin{bmatrix} -1 & 0 & 0 \\ 0 & -1 & 0 \\ 0 & 4 & -1 \end{bmatrix} \begin{Bmatrix} R_1 \\ R_2 \\ R_3 \end{Bmatrix} \tag{a$_3$}$$

We therefore have the internal forces expressed in the desired manner :

$$\mathbf{S} = \begin{Bmatrix} \mathbf{S}_A \\ \mathbf{S}_B \end{Bmatrix} = \begin{bmatrix} \mathbf{B}_A \\ \mathbf{B}_B \end{bmatrix} \mathbf{P} + \begin{bmatrix} \mathbf{b}_A \\ \mathbf{b}_B \end{bmatrix} \mathbf{R} = \mathbf{B}\,\mathbf{P} + \mathbf{b}\,\mathbf{R} \tag{a$_4$}$$

where,

$$\mathbf{B} = \left[\begin{array}{ccc} 0.8 & -0.6 & 0 \\ 0.6 & 0.8 & 0 \\ 0 & 0 & 1 \\ \hline & & \\ & \mathbf{0} & \\ & & \end{array}\right]; \quad \mathbf{b} = \left[\begin{array}{ccc} -0.8 & 0.6 & 0 \\ -0.6 & -0.8 & 0 \\ 0 & 0 & -1 \\ \hline -1 & 0 & 0 \\ 0 & -1 & 0 \\ 0 & 4 & -1 \end{array}\right] \tag{a$_5$}$$

The flexibility matrix for a member is the one relating the displacements at end 2 of a cantilever clamped at end 1, namely the basic flexibility derived in Eq. (5.20). Considering only bending deformations,

$$\mathbf{v}=\begin{Bmatrix}\bar{u}_2\\ \bar{v}_2\\ \bar{\theta}_2\end{Bmatrix}=\begin{bmatrix}L/AE & 0 & 0\\ 0 & L^3/3EI & L^2/2EI\\ 0 & L^2/2EI & L/EI\end{bmatrix}\begin{Bmatrix}\bar{P}_2\\ \bar{Q}_2\\ \bar{M}_2\end{Bmatrix}=\mathbf{f}\,\mathbf{S} \qquad (a_6)$$

Assume the following data for the members :

Member	EI	$\frac{L}{r}$	$\frac{L}{AE}$	$\frac{L^3}{3EI}\times EI_o$	$\frac{L^2}{2EI}\times EI_o$	$\frac{L}{EI}\times EI_o$
A	$12.5\,EI_o$	20	0.025	3.3333	2.6666	0.4
B	$8.0\,EI_o$	20	0.02	1.0	1.0	0.5

Here r is the radius of gyration. Observe that, in terms of this we have,

$$\frac{L}{AE}=\frac{L^3}{EI}\cdot\frac{I}{AL^2}=\frac{L^3}{EI}\Big/\left(\frac{L}{r}\right)^2 \qquad (a_7)$$

It is thus clear that axial flexibility is smaller compared to the flexural one. The flexibility matrices are,

$$\mathbf{f}_A=\frac{1}{EI_0}\begin{bmatrix}0.025 & 0 & 0\\ 0 & 3.333 & 1.0\\ 0 & 1.0 & 0.4\end{bmatrix};\ \mathbf{f}_B=\frac{1}{EI_0}\begin{bmatrix}0.02 & 0 & 0\\ 0 & 2.6666 & 1.0\\ 0 & 1.0 & 0.5\end{bmatrix} \qquad (a_8)$$

We have to modify these according to $\mathbf{b}^t\,\mathbf{f}\,\mathbf{b}$ and arrive at the flexibility matrix $F_{RR}=\mathbf{b}^t\,\mathbf{f}\,\mathbf{b}$ and load term $-\mathbf{F}_{RP}\mathbf{P}=-(\mathbf{b}^t\,\mathbf{f}\,\mathbf{B})\,\mathbf{P}$; (Eq. 7.7).

In view of the diagonal nature of f, $\mathbf{F}_{RP}=\mathbf{b}^t\,\mathbf{f}\,\mathbf{B}=\mathbf{b}^t{}_A\,\mathbf{f}_A\,\mathbf{B}_A+\mathbf{b}^t{}_B\,\mathbf{f}_B\,\mathbf{B}_B$, which in this case reduces to $\mathbf{b}^t{}_A\,\mathbf{f}_A\,\mathbf{b}_A$.

After carrying out the product,

$$\mathbf{b}^t\,\mathbf{f}\,\mathbf{B}\,\mathbf{P}=1/EI_0\begin{bmatrix}-1.2160 & -1.5880 & -0.6\\ -1.5880 & -2.1423 & -0.8\\ -0.6 & -0.8 & -0.4\end{bmatrix}\begin{Bmatrix}3.6\\ 88.8\\ -29.0\end{Bmatrix}$$

$$=\begin{Bmatrix}-127.9920\\ -172.7530\\ -61.6\end{Bmatrix}\left(\frac{1}{EI_0}\right) \qquad (a_9)$$

The vector on the right hand side denotes the gap at the release due to the loads above. Similarly we compute

$$\mathbf{b}^t\,\mathbf{f}\,\mathbf{b}=\mathbf{b}^t{}_A\,\mathbf{f}_A\,\mathbf{b}_A+\mathbf{b}^t{}_B\,\mathbf{f}_B\,\mathbf{b}_B=\left(\frac{1}{EI_0}\right)\begin{bmatrix}1.2360 & 1.5880 & 0.6\\ 1.5884 & 4.8089 & -0.2\\ 0.6 & -0.2 & 0.9\end{bmatrix} \qquad (a_{10})$$

This is the flexibility matrix with respect to the redundants **R**. They determine the 'gaps' or the release displacements due to the redundants. The total gap displacements due to loads and the redundants are set equal to zero. Thus the equation to be solved for **R** are :

$$\frac{1}{EI_o}\begin{bmatrix} 1.2360 & 1.5880 & 0.6 \\ 1.5880 & 4.8089 & -0.2 \\ 0.6 & -0.2 & 0.9 \end{bmatrix}\begin{Bmatrix} R_1 \\ R_2 \\ R_3 \end{Bmatrix} = -\begin{Bmatrix} -127.9920 \\ -172.7530 \\ -61.6 \end{Bmatrix} \quad (a_{11})$$

Thus on solution we obtain

$$\mathbf{R} = \{91.3870, \quad 6.1150, \quad 8.8787\} \quad (a_{12})$$

Equations (a_4) then determine the internal forces. Thus on carrying out the multiplications involved in (a_4), we obtain

$$\mathbf{S}_A = \begin{Bmatrix} \overline{P_{2A}} \\ \overline{Q_{2A}} \\ \overline{M_{2A}} \end{Bmatrix} = \begin{Bmatrix} -119.8406 \\ 13.4758 \\ -37.8787 \end{Bmatrix}; \; \mathbf{S}_B = \begin{Bmatrix} \overline{P_{2B}} \\ \overline{Q_{2B}} \\ \overline{M_{2B}} \end{Bmatrix} = \begin{Bmatrix} -91.3870 \\ -6.1150 \\ 15.5813 \end{Bmatrix} \quad (a_{13})$$

To obtain the forces at the end 1 we can use the equation of equilibrium $\overline{\mathbf{P}}_1 = \overline{\mathbf{C}}_1 \, \mathbf{S}$, where (see Eq. 2.4).

$$\overline{\mathbf{C}_1} = \begin{bmatrix} -1 & 0 & 0 \\ 0 & -1 & 0 \\ 0 & -L & -1 \end{bmatrix} \quad (a_{14})$$

Thus

$$\overline{\mathbf{P}}_{1A} = \begin{Bmatrix} \overline{P}_{1A} \\ \overline{Q_{1A}} \\ \overline{M}_{1A} \end{Bmatrix} = \begin{bmatrix} -1 & 0 & 0 \\ 0 & -1 & 0 \\ 0 & -5 & -1 \end{bmatrix}\begin{Bmatrix} -119.8406 \\ 13.4758 \\ -37.8787 \end{Bmatrix}$$

$$= \begin{bmatrix} 119.8406 \\ -13.4758 \\ -29.500 \end{bmatrix} \quad (a_{15})$$

Similarly

$$\overline{\mathbf{P}}_{1B} = \begin{bmatrix} -1 & 0 & 0 \\ 0 & -1 & 0 \\ 0 & -4 & 0 \end{bmatrix}\begin{Bmatrix} -91.3870 \\ -6.1150 \\ 15.5813 \end{Bmatrix} = \begin{Bmatrix} 91.3870 \\ 6.1150 \\ 8.8787 \end{Bmatrix} \quad (a_{16})$$

Axial Deformations Neglected

If we wish to neglect axial deformations we have only to set the axial flexibility term, (L/AE) to zero. Thus setting these terms (.025 and .02) in Eq. (a_8) to zero and repeating the steps, we obtain, instead of (a_9),

$$\mathbf{b}^t\,\mathbf{f}\,\mathbf{B}\ \mathbf{P} = 1/EI_0 \begin{bmatrix} -1.2 & -1.6 & -0.6 \\ -1.6 & -2.1333 & -0.8 \\ -0.6 & -0.8 & -0.4 \end{bmatrix} \begin{Bmatrix} 3.6 \\ 88.8 \\ -29.0 \end{Bmatrix}$$

$$= \begin{Bmatrix} -129.0 \\ -172.0 \\ -61.6 \end{Bmatrix} (1/EI_0) \qquad (a_{17})$$

And, in place of Eq. (a_{10}),

$$\mathbf{b}^t\,\mathbf{f}\,\mathbf{b} = \begin{bmatrix} 1.2 & 1.6 & 0.6 \\ 1.6 & 4.8 & -0.2 \\ 0.5 & -0.2 & 0.9 \end{bmatrix} \left(\frac{1}{EI_0}\right) \qquad (a_{18})$$

Thus on solution we obtain the redundants

$$\mathbf{R} = \{120.3888, \quad -4.8333, \quad -12.8889\} \qquad (a_{19})$$

The member forces obtained are as follows :

$$\begin{Bmatrix} \bar{P}_{2A} \\ \bar{Q}_{2A} \\ \bar{M}_{2A} \end{Bmatrix} = \begin{Bmatrix} -149.6103 \\ 4.8333 \\ -16.1112 \end{Bmatrix};\ \begin{Bmatrix} \bar{P}_{2B} \\ \bar{Q}_{2B} \\ \bar{M}_{2B} \end{Bmatrix} = \begin{Bmatrix} -120.3888 \\ 4.8333 \\ -6.444 \end{Bmatrix} \qquad (a_{20})$$

$$\begin{Bmatrix} \bar{P}_{1A} \\ \bar{Q}_{1A} \\ \bar{M}_{1A} \end{Bmatrix} = \begin{Bmatrix} 149.6103 \\ -4.8338 \\ -8.0578 \end{Bmatrix};\ \begin{Bmatrix} \bar{P}_{1B} \\ \bar{Q}_{1B} \\ \bar{M}_{1B} \end{Bmatrix} = \begin{Bmatrix} +120.3888 \\ -4.8333 \\ -12.8889 \end{Bmatrix} \qquad (a_{21})$$

It is seen that the effect of axial deformations, is relatively large as a result of our choice of (L/r), rather too small compared to common values.

Case (b)

We will now repeat the problem by introducing three moment releases, as in Fig. 7.2(d). Thus the redundants chosen are,

$$M_{1A} = R_1;\quad M_{1B} = R_2;\quad M_{2B} = R_3 \qquad (b_1)$$

In this case determination of the internal forces involves more calculations. We will write the equations of equilibrium and solve. $\mathbf{C}_1$ and $\mathbf{C}_2$ matrices (as per Eq. (2.14)) applicable in this case are,

$$\mathbf{C}_{2A} = \begin{bmatrix} 0.8 & -\frac{0.6}{5} & -\frac{0.6}{5} \\ -0.6 & -\frac{0.8}{5} & -\frac{0.8}{5} \\ 0 & 0 & 1 \end{bmatrix} = \begin{bmatrix} 0.8 & -0.12 & -0.12 \\ -0.6 & -0.16 & -0.16 \\ 0 & 0 & 1 \end{bmatrix} \qquad (b_2)$$

$$\mathbf{C}_{1B} = \begin{bmatrix} -1 & 0 & 0 \\ 0 & \frac{1}{4} & \frac{1}{4} \\ 0 & 1 & 0 \end{bmatrix} = \begin{bmatrix} -1 & 0 & 0 \\ 0 & 0.25 & 0.25 \\ 0 & 1.0 & 0 \end{bmatrix} \qquad (b_3)$$

The equations of equilibrium of joint B are

$$\mathbf{C}_{2A}\,\mathbf{S}_A + \mathbf{C}_{1B}\,\mathbf{S}_B = \mathbf{P}_B \qquad (b_4)$$

or

$$\left[\begin{array}{ccc|ccc} 0.8 & -0.12 & -0.12 & -1 & 0 & .0 \\ -0.6 & -0.16 & -0.16 & 0 & 0.25 & 0.25 \\ 0 & 0 & 1 & 0 & 1 & 0 \end{array}\right] \left\{\begin{array}{c} S_A \\ \hline |M_{1A}| \\ \hline M_{2A} \\ \hline S_B \\ \hline |M_{1B}| \\ \hline |M_{2B}| \end{array}\right\} = \left\{\begin{array}{c} P_B \\ Q_B \\ M_B \end{array}\right\} \qquad (b_5)$$

In this we introduce the definions of redundants as already chosen, namely as in Eq. (b_1). Thus *shifting* these to the right hand side, we write

$$\begin{bmatrix} 0.8 & -0.12 & -1 \\ -0.6 & -0.16 & 0 \\ 0 & 1 & 0 \end{bmatrix} \begin{Bmatrix} S_A \\ M_{2A} \\ S_B \end{Bmatrix} = \begin{Bmatrix} P_B \\ Q_B \\ M_B \end{Bmatrix} + \begin{bmatrix} 0.12 & 0 & 0 \\ 0.16 & -0.25 & -0.25 \\ 0 & -1 & 0 \end{bmatrix} \begin{Bmatrix} R_1 \\ R_2 \\ R_3 \end{Bmatrix} \qquad (b_6)$$

In fact this form permits an obvious interpretation. The structure being analysed is a three hinged determinate frame; the loads consist of the given loads and redundants. Thus we can determine the stresses due to given loads alone, and then due to the redundants alone. The procedure discussed above is general and can be programmed. The obvious disadvantage is the necessity of having to work with unsymmetric solvers. (See Sec. 7.3.)

On solving, these, together with definitions (b_1), can be written as follows:

$$\mathbf{S} = \begin{Bmatrix} S_A \\ M_{1A} \\ M_{2A} \\ \hline S_B \\ M_{1B} \\ M_{2B} \end{Bmatrix} = \begin{bmatrix} 0 & -1.6666 & -0.2666 \\ 0 & 0 & 0 \\ 0 & 0 & 1 \\ \hline -1 & -1.3333 & -0.3333 \\ 0 & 0 & 0 \\ 0 & 0 & 0 \end{bmatrix} \begin{Bmatrix} P_B \\ Q_B \\ M_B \end{Bmatrix}$$

$$+ \begin{bmatrix} -0.2666 & 0.6833 & 0.4166 \\ 1 & 0 & 0 \\ 0 & -1 & 0 \\ \hline -0.3333 & 0.6666 & 0.3333 \\ 0 & 1 & 0 \\ 0 & 0 & 1 \end{bmatrix} \begin{Bmatrix} R_1 \\ R_2 \\ R_3 \end{Bmatrix} \qquad (b_7)$$

These are of the form

$$\mathbf{S} = \begin{bmatrix} \mathbf{B}_A \\ \mathbf{B}_B \end{bmatrix} \mathbf{P} + \begin{bmatrix} \mathbf{b}_A \\ \mathbf{b}_B \end{bmatrix} \mathbf{R} \qquad (b_8)$$

The flexibility matrix, corresponding to the beam model (Eq. 5.25) is

$$\mathbf{v} = \begin{Bmatrix} e \\ \phi_1 \\ \phi_2 \end{Bmatrix} = \begin{bmatrix} L/AE & 0 & 0 \\ 0 & L/3EI & -L/6EI \\ 0 & -L/6EI & L/3EI \end{bmatrix} \begin{Bmatrix} S \\ M_1 \\ M_2 \end{Bmatrix} = \mathbf{f}\,\mathbf{S} \qquad (b_9)$$

For the present problem, the flexibility matrices are

$$\mathbf{f}_A = \frac{1}{EI_0} \begin{bmatrix} 0.025 & 0 & 0 \\ 0 & 0.13333 & -0.0666 \\ 0 & -0.06666 & 0.1333 \end{bmatrix}$$

$$\mathbf{f}_B = \frac{1}{EI_0} \begin{bmatrix} 0.02 & 0 & 0 \\ 0 & 0.1666 & -0.0833 \\ 0 & -0.0833 & 0.1666 \end{bmatrix} \qquad (b_{10})$$

Proceeding as before with the above matrices, we obtain

$$\mathbf{b}^t\,\mathbf{f}\,\mathbf{B}\,\mathbf{P} = 1/EI_0 \begin{bmatrix} 0.0066 & 0.02 & -0.0626 \\ -0.01333 & -0.0462 & -0.1423 \\ -0.0066 & -0.0262 & -0.005 \end{bmatrix} \begin{Bmatrix} 3.6 \\ 88.8 \\ -29.0 \end{Bmatrix}$$

$$= \begin{Bmatrix} 3.6151 \\ -0.0238 \\ -2.2053 \end{Bmatrix} (1/EI_0) \qquad (b_{11})$$

$$\mathbf{b}^t\,\mathbf{f}\,\mathbf{b} = \begin{bmatrix} 0.1373 & 0.0577 & -0.005 \\ 0.0577 & 0.3205 & -0.0717 \\ -0.005 & -0.0717 & 0.1731 \end{bmatrix} (1/EI_0) \qquad (b_{12})$$

Thus on solving $\mathbf{b}^t\,\mathbf{f}\,\mathbf{b}\,\mathbf{R} = -\,\mathbf{b}^t\,\mathbf{f}\,\mathbf{B}\,\mathbf{P}$, we obtain

$$\mathbf{R} = \{-29.48, \quad 8.866, \quad 15.56\} \qquad (b_{13})$$

And on back substitution into Eq. (b_7) we obtain

$$\begin{Bmatrix} S_A \\ M_{1A} \\ M_{1B} \end{Bmatrix} = \begin{Bmatrix} -119.8096 \\ -29.48 \\ -37.866 \end{Bmatrix}; \begin{Bmatrix} S_B \\ M_{1B} \\ M_{2B} \end{Bmatrix} = \begin{Bmatrix} -91.409 \\ 8.866 \\ 15.56 \end{Bmatrix} \qquad (b_{14})$$

Some of these can be compared with the ones obtained previously, as given in Eq. (a_{13}). The member end forces $\overline{\mathbf{P}}_1$ and $\overline{\mathbf{P}}_2$ can be obtained by premultiplying the vector S by $\overline{\mathbf{C}}_1$ or $\overline{\mathbf{C}}_2$.

Axial Deformations Neglected

Using the flexibility matrices given in Eq. (b_{10}), in which L/AE terms are set equal to zero, and repeating, we obtain

$$\mathbf{b}^t\,\mathbf{f}\,\mathbf{B}\,\mathbf{P} = \frac{1}{EI_0} \begin{bmatrix} 0 & 0 & -0.0666 \\ 0 & 0 & -0.1333 \\ 0 & 0 & 0 \end{bmatrix} \begin{Bmatrix} 3.6 \\ 88.8 \\ -29.0 \end{Bmatrix}$$

$$= \begin{Bmatrix} 1.9333 \\ 3.8666 \\ 0 \end{Bmatrix} (1/EI_0) \qquad (b_{15})$$

$$\mathbf{b}^t\,\mathbf{f}\,\mathbf{b} = \frac{1}{EI_0} \begin{bmatrix} 0.1333 & 0.0666 & 0 \\ 0.0666 & 0.3 & -0.0833 \\ 0 & -0.0833 & 0.1666 \end{bmatrix} \qquad (b_{16})$$

$$\mathbf{R} = \{8.0560, \quad -12.8889, \quad -6.444\} \tag{b$_{17}$}$$

The member forces are

$$\begin{Bmatrix} S_A \\ M_{1A} \\ M_{2A} \end{Bmatrix} = \begin{Bmatrix} -149.6066 \\ -8.0560 \\ -16.111 \end{Bmatrix}; \begin{Bmatrix} S_B \\ M_{1B} \\ M_{2B} \end{Bmatrix} = \begin{Bmatrix} -120.3883 \\ -12.8889 \\ -6.4444 \end{Bmatrix} \tag{b$_{18}$}$$

which stand in comparison to those in Eq, (a_{20}, a_{21}).

It should be evident from this example that there is some arbitrariness in selecting the redundants. For larger problems the initial step of determining the internal forces in terms of loads and redundants itself presents a problem. We will be discussing this issue together with others in the subsequent section.

Distributed Loads

Recall the complete flexibility relation (7.2). This includes the term $\mathbf{V}^\circ$ to account for lateral loads and loads such as temperature. The method of analysis remains essentially the same. We have then, in Eq. (7.7), only the term $\mathbf{V}^\circ$. Note that $\mathbf{b}^t\,\mathbf{f}\,\mathbf{B}\,\mathbf{P}$ denotes the displacements at the release due to the applied loads.

7.3 COMPUTER APPLICATIONS

As can be seen from the previous sections, the essential problem in the force method is the one of determining the basic matrices **B** and **b**. This depends on the choice of the redundants. While the total number, (the degree of redundancy) is fixed, the actual selection admits certain freedom. Let us assume that the user of a program would be specifying these through data. Even then the main issue of constructing **B** and **b** matrices, though straightforward, involves inversion of an unsymmetric set of equilibrium equations. This has been demonstrated in the examples we have discussed. The methods employed in the simple cases of Examples 7.2 and 7.3 can however be programmed. The method employed in Example 7.2, which is the one given in Reference [6], consists of adding additional rows corresponding to the redundants thus rendering the equilibrium matrix square. Equation (d) and (f) together were written as Eq. (g) and inverted to give **S** as in Eq. (j). For small size problems this can be well adopted. This procedure should be compared with the one adopted in Chapter 2 where the rectangular equilibrium matrix was made square by including reactive forces in the vector of internal forces. In a redundant structure even after adding the reactive forces the equilibrium matrix will be rectangular, with lesser rows than columns. We can add rows which are equivalent to specifying redundants and thus make a square matrix. Thus taking the frame of Fig. 7.2 as a typical case, the equations of equilibrium

$$\mathbf{C}_{2A}\,\mathbf{S}_A + \mathbf{C}_{1B}\,\mathbf{S}_B = \mathbf{P}_B \tag{a$_1$}$$

are solved together with the statements of the choice of redundants. Thus the choice $\mathbf{P}_{1B} = \mathbf{R}$ is written (see Eq. (a_2) of previous sęction), as,

$$\mathbf{C}_{1B}\,\mathbf{S}_B = \mathbf{R} \tag{a$_2$}$$

Thus the above two sets are combined :

$$\mathbf{C}'\,\mathbf{S} = \begin{bmatrix} \mathbf{C}_{2A} & \mathbf{C}_{1B} \\ \mathbf{0} & \mathbf{C}_{1B} \end{bmatrix} \begin{Bmatrix} \mathbf{S}_A \\ \mathbf{S}_B \end{Bmatrix} = \begin{Bmatrix} \mathbf{P}_B \\ \mathbf{R} \end{Bmatrix} = \mathbf{P}' \tag{7.8}$$

Here the augumented matrix $\mathbf{C}'$ is square and can be inverted to give $\mathbf{S}$ in terms of $\mathbf{P}_B$ and $\mathbf{R}$:

$$\mathbf{S} = \mathbf{B}'\,\mathbf{P}' = [\mathbf{B}\;\;\mathbf{b}] \begin{Bmatrix} \mathbf{P}_B \\ \mathbf{R} \end{Bmatrix} \tag{7.9}$$

The process of obtaining $\mathbf{B}'$ from $\mathbf{C}'$ is accomplished by procedures based on Gauss-Jordon elimination. Denke [18] and Robinson [18, 19] present methods which are based on this type of elimination. The chief feature is the one of automatic selection of redundants. The interested reader can find discussions of these in reference [4] also. The method adopted in Example 7.3, Case (b), also deserves attention and can be employed in developing small programs in which the user specifies the choice of redundants by inspecting his problem. Thus the user specifies through data, the numbers of the forces which are to be used as redundants. Thus a program can be written to modify the corresponding rows and columns.

Consider Eq. (b_5) of Example 7.3. Here M_{1A}, M_{1B} and M_{2B} are selected as redundants. The columns corresponding to these, namely second, fifth and the sixth, are transferred to the right, as shown in (b_6), and the determinate forces solved by inverting the reduced matrix. By rearranging the columns according to the selection of redundancies we can write a set of equilibrium equations as

$$\begin{bmatrix} \mathbf{C}_0 & \mathbf{C}_1 \\ \mathbf{0} & \mathbf{I} \end{bmatrix} \begin{Bmatrix} \mathbf{S}_0 \\ \mathbf{S}_1 \end{Bmatrix} = \begin{Bmatrix} \mathbf{P} \\ \mathbf{R} \end{Bmatrix} \tag{7.10}$$

Here $\mathbf{S}_0$ and $\mathbf{S}_1$ are respectively the determinate forces and those forces which are taken as redundants. We can solve for $\mathbf{S}_0$ and obtain it as

$$\mathbf{S}_0 = \mathbf{C}_0^{-1}\,\{\mathbf{P} - \mathbf{C}_1\,\mathbf{R}\} \tag{7.11}$$

Therefore the internal forces are obtained as

$$S' = \begin{Bmatrix} S_0 \\ S_1 \end{Bmatrix} = \begin{bmatrix} C_0^{-1} & -C_0^{-1}C_1 \\ 0 & I \end{bmatrix} \begin{Bmatrix} P \\ R \end{Bmatrix}$$

$$= [B' \; b'] \begin{Bmatrix} P \\ R \end{Bmatrix} \tag{7.12}$$

Here the dashes have been used to remind the reader that these are rearranged vectors so as to separate the forces into S_0 and S_1. In fact Eq. (7.10) is the rearranged version of Eq. (7.8). This rearrangement ensures that the invease C_0 exists. This in physical terms means that we have identified a set of forces S_0 which correspond to a determinate structure and therefore determinable. The procedures of references [18, 19] adopted a similar procedure but improvised to automatically select a determinate set S_0 and the corresponding C_0 and C_1 matrices. All these are handled employing the Gauss-Jordon elimination procedure.

For a given type of structure we can evolve methods of directly writing **B** and **b** matrices. The expressions for released structures, which essentially are cantilevers ('tree' structures), can be developed. This has been given by Livesley [6]. To add to this, a special scheme of redundants are chosen so that the flexibility matrix ($\mathbf{b}^t\,\mathbf{f}\,\mathbf{b}$) becomes 'banded' similar to stiffness matrix. These can be very conveniently adopted in developing programs for a given type of structure, such as building frames.

EXERCISES

7.1 Repeat the contents of Sec. 7.1 as applied to the pinconnected plane frame shown having two redundancies. Take $AE/L = k_0$ same for each bar and if felt necessary some numericals for P_A and Q_A.

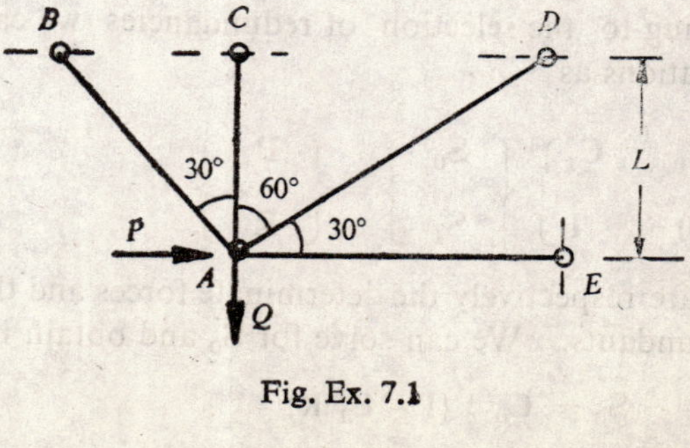

Fig. Ex. 7.1

7.2 Repeat the procedure of Example 7.2 as applied to the frame shown. Treat M_{1A}, M_{2A} and M_{2B} as redundants.

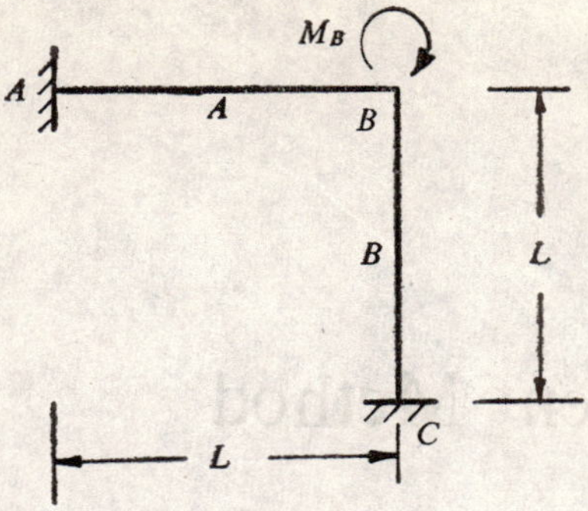

Fig. Ex. 7.2

7.3 Analyse the frame of Exercise 2.13 on the lines of Example 7.3. Treat (a) forces at end 2 of member C as redundants. Take geometrical data as in Exercise 2.13 and $P_B = Q_B = Q_C = 10$ kN. Assume (EI/L) to be the same for all members and the slenderness ratio (l/r) to be 20. Repeat the above cases by neglecting axial deformations.

7.4 Analyse the grid frame of Ex. 2.15. Take a transverse load Q_B alone to be acting. Assume (EI/L) to be the same for all members and the ratio (EI/GJ) to be 10.

7.5 Construct Eqs. (7.8) for the frame of Ex. 7.1 by treating forces in AC and AE as redundants. Solve for the internal forces in the form (7.9) and check your answers. You can use solution in the form given by Eq. (7.10).

7.6 Derive the internal forces of the frame of Ex. 7.3 using the formulation of Eq. (7.8) and treating moments M_{1A}, M_{1D} and M_{2D} as redundants. Solve in the form suggested by Eq. (7.10).

7.7 Draw the outline of a computer program which would follow a procedure suggested by Eq. (7.10). The user specifies the names of the forces to be selected as redundants. Think of analysing pinconnected frame with a maximum of 100 members and a maximum of 20 redundances. The equations of equilibrium would be assembled using the procedure suggested in Sec. 2.5.

Chapter 8

Displacement Method

In the displacement method the unknowns involved are the displacements of the joints. The choice of displacements for the unknowns renders the method more versatile than the force method for adoption to machine computations.

If the joint displacements are taken as the unknown parameters in terms of which all the structural requirements are formulated, the method then becomes straightforward. The member deformations are readily expressed in terms of joint displacements since it is purely a matter of kinematics. The stiffness laws then determine the internal forces in the members. These internal forces are required to be in equilibrium with the given external loads. These conditions then form the necessary equations to solve for the displacements. For this reason the method is also called the *equilibrium method*. The coefficients appearing in the equations for the displacements happen to be stiffness parameters. On this account the term *stiffness method* is also very popularly used.

The method being straightforward lends itself to development of general pragrams. In the force method there is some arbitrariness in the choice of the redundants which depends on the analyst. However in the displacement method the governing equations are unique for a given structure. They are completely determined by the displacement unknowns, namely the *degrees of freedom* of the structure. This offers relatively straightforward algorithms. The finite element methods dealing with continuum problems almost exclusively adopt the displacement method.

8.1 DISPLACEMENT METHOD

There are essentially two principal ways in which the equations of equilibrium, kinematics and elasticity can be combined so as to lead to a set of equations in displacement unknowns. Both these approaches obviously have to lead to the same set of final equations. The difference is only in the manner of constructing these equations. The two approaches will be termed as:

(a) Basic Stiffness Method
(b) Direct Stiffness Method.

In the first method [7, 17] we work with the *basic* stiffness matrices of members and in the second [20] with the *general* stiffness of members. The latter is more efficient from the programming point of view. The former is however very general and can always be used in situations where certain assumptions are made in the geometrical behaviour deviating from the general pattern. We will discuss both but with more bias given to the latter in view of its simplicity and practical advantage.

The common objective is to finally obtain the equations of equilibrium for the structure expressed in terms of the displacements of the joints:

$$\mathbf{KU} = \mathbf{P} \qquad (a_1)$$

Here $\mathbf{U}$ is the vector of displacements and $\mathbf{P}$ the vector of the corresponding joint forces. The matrix $\mathbf{K}$ is the *stiffness matrix* of the structure. This will be a square *symmetrical* matrix and can be inverted, provided the displacements are properly chosen so as to avoid rigid body motion of the structure. Thus the displacements can be formally solved for in terms of the forces:

$$\mathbf{U} = \mathbf{K}^{-1}\,\mathbf{P} \qquad (a_2)$$

With these calculated displacements the member analysis can be performed for each member, namely determining the end forces and, if required, the forces at any intermediate point along the length of the member.

Basic Stiffness Method

The stiffness of structure is expressed in the form

$$\mathbf{KU} = \begin{bmatrix} \mathbf{K}_{AA} & \mathbf{K}_{AB} & \mathbf{K}_{AC} & \cdots \\ \mathbf{K}_{BA} & \mathbf{K}_{BB} & \mathbf{K}_{BC} & \cdots \\ \mathbf{K}_{CA} & \mathbf{K}_{CB} & \mathbf{K}_{CC} & \cdots \\ \vdots & \vdots & \vdots & \vdots \end{bmatrix} \begin{Bmatrix} \mathbf{U}_A \\ \mathbf{U}_B \\ \mathbf{U}_C \\ \vdots \end{Bmatrix} = \begin{Bmatrix} \mathbf{P}_A \\ \mathbf{P}_B \\ \mathbf{P}_C \\ \vdots \end{Bmatrix} = \mathbf{P} \qquad (a_3)$$

where $\mathbf{U}_A$, $\mathbf{U}_B$ etc. are the possible joint displacements and $\mathbf{P}_A$, $\mathbf{P}_B$ etc. are the corresponding forces. The matrices $\mathbf{K}_{AA}$, $\mathbf{K}_{BB}$, etc. are obtained in terms of the basic stiffnesses $\mathbf{k}$ of members. The above relations state the equations of equilibrium in terms of joint displacements. We will adopt kinematic considerations and virtual work to establish the above equations of equilibrium.

The deformations of the members are related to the displacements of the joints through a kinematic matrix which is dependent on the structural connectivity. Consider the assembly of five members shown in Fig. 8.1 as an example. For each member we write the kinematic relations (Eq. 3.7); to rewrite, say taking member A,

$$\begin{aligned} \mathbf{v}_A &= \mathbf{A}_{1A}\mathbf{U}_{1A}+\mathbf{A}_{2A}\mathbf{U}_{2A} \\ &= \mathbf{A}_{1A}\mathbf{U}_A+\mathbf{A}_{2A}\mathbf{U}_B \end{aligned} \tag{8.1}$$

All such relations collectively are expressed in the form

$$\mathbf{V} = \begin{Bmatrix} \mathbf{v}_A \\ \mathbf{v}_B \\ \mathbf{v}_C \\ \mathbf{v}_D \\ \mathbf{v}_E \end{Bmatrix} = \begin{bmatrix} \mathbf{A}_{1A} & \mathbf{A}_{2A} & \cdot & \cdot \\ \cdot & \mathbf{A}_{1B} & \mathbf{A}_{2B} & \cdot \\ \cdot & \cdot & \mathbf{A}_{1C} & \mathbf{A}_{2C} \\ \mathbf{A}_{1D} & \cdot & \cdot & \mathbf{A}_{2D} \\ \mathbf{A}_{1E} & \cdot & \mathbf{A}_{2E} & \cdot \end{bmatrix} \begin{Bmatrix} \mathbf{U}_A \\ \mathbf{U}_B \\ \mathbf{U}_C \\ \mathbf{U}_D \end{Bmatrix} = \mathbf{A}\,\mathbf{U} \quad (a_4)$$

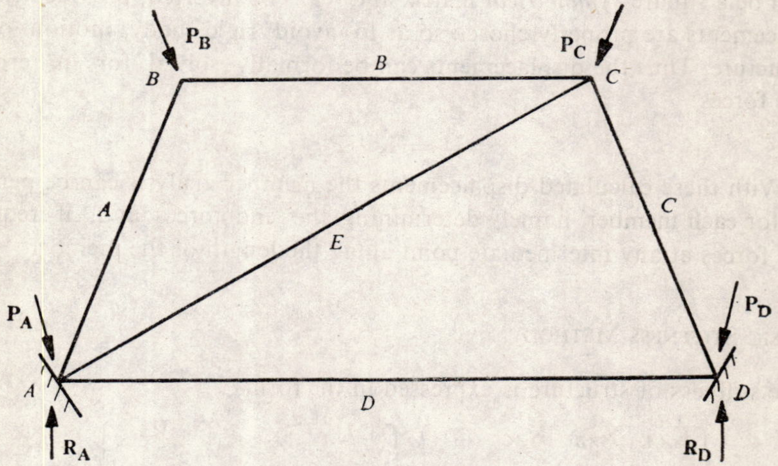

Fig. 8.1

Thus for a structure, in general, we have

$$\mathbf{V} = \mathbf{AU} \tag{8.2}$$

as the kinematical relations between member deformations and joint displacements.

These determine the internal forces in the members. For each member we have, (see Eqs. 5.14, 5.34, 5.35), $\mathbf{S}_A = \mathbf{k}_A\mathbf{v}_A$, $\mathbf{S}_B = \mathbf{k}_B\mathbf{v}_B$ and so on. We can express these collectively as

$$\mathbf{S} = \begin{Bmatrix} S_A \\ S_B \\ S_C \\ S_D \\ S_E \end{Bmatrix} = \begin{bmatrix} \mathbf{k}_A & & & & \\ & \mathbf{k}_B & & & \\ & & \mathbf{k}_C & & \\ & & & \mathbf{k}_D & \\ & & & & \mathbf{k}_E \end{bmatrix} \begin{Bmatrix} \mathbf{v}_A \\ \mathbf{v}_B \\ \mathbf{v}_C \\ \mathbf{v}_D \\ \mathbf{v}_E \end{Bmatrix} = \mathbf{kV} \quad (a_5)$$

In general therefore

$$\mathbf{S} = \mathbf{kV} \tag{8.3}$$

are the elastic relations available. Here $\mathbf{k}_A$, $\mathbf{k}_B$ etc. are the basic stiffnesses of the members (Eq. 5.14).

The next step is to write the equations of equilibrium. We write the virtual work equation taking the actual possible deformations as virtual system, namely those of Eq. (8.3). Thus the virtual work equation is

$$\mathbf{S}^t\,\mathbf{V} = \mathbf{P}^t\,\mathbf{U} \tag{8.4}$$

where **P** is the load vector corresponding to the joint displacement vector. Using Eq. (8.3) we obtain the equilibrium relation for the structure,

$$\mathbf{A}^t\,\mathbf{S} = \mathbf{P} \tag{8.5}$$

It therefore remains to combine Eqs. (8.2), (8.3) and (8.5). The first two combine to give

$$\mathbf{S} = \mathbf{kAU} \tag{8.6}$$

which together with the third leads to

$$\mathbf{KU} = \mathbf{P} \tag{8.7}$$

where the stiffness matrix **K** of the structure is obtained as

$$\mathbf{K} = \mathbf{A}^t\,\mathbf{k}\,\mathbf{A} \tag{8.8}$$

This form is similar to the form (6.11) for the flexibility of the structure. We will interpret the above expression as an addition of the stiffnesses of the various members just as the form $\mathbf{B}^t\,\mathbf{f}\,\mathbf{B}$ was interpreted as the addition of the flexibilities of the member.

The displacements are solved simultaneously from the set of Eqs. (8.7). Formally

$$\mathbf{U} = \mathbf{K}^{-1}\,\mathbf{P} = \mathbf{F}\,\mathbf{P} \quad (a_6)$$

Here then, F is the flexibility of the structure. The above form is only formal. It only means that we solve for the displacements in terms of the loads rather than actually inverting the **K** matrix and premultiplying the load vector. It may be recalled that inverting a matrix entails almost three times the computational labour involved in solving with a given right hand side.

Having solved for the displacements we compute the internal forces using Eq. (8.6). Instead of actually carrying out the matrix multiplication, we evaluate the member deformations $\mathbf{v}_A$, $\mathbf{v}_B$ etc. from Eq. (8.2), memberwise one after the other. Then premultiplying by $\mathbf{k}_A$, $\mathbf{k}_B$ etc., as in Eq. (8.3), we obtain the internal forces.

This in brief is the basic stiffness method. We will now demonstrate the application through an example. The method is preferable to elementary or unconventional problems in view of its general form.

EXAMPLE 8.1

Reconsider the frame considered in Example 4.1. See Fig. 3.5. Here the axial deformations of the members are ignored. In Sec. 4.3 the kinematic relations between the member deformations and the joint displacements were obtained in Eq. (a_1); to rewrite

$$\mathbf{V}' = \begin{Bmatrix} \varphi_{1A} \\ \varphi_{2A} \\ \varphi_{1B} \\ \varphi_{2B} \\ \varphi_{1C} \\ \varphi_{2C} \end{Bmatrix} = \begin{bmatrix} 0 & 0 & -1/h \\ 1 & 0 & -1/h \\ \hline 1 & 0 & \cot\beta/L_2 \\ 0 & 1 & \cot\beta/L_2 \\ \hline 0 & 1 & -1/h \\ 0 & 0 & -1/h \end{bmatrix} \begin{Bmatrix} \theta_B \\ \theta_C \\ u_B \end{Bmatrix} = \mathbf{AU}' \qquad (a_1)$$

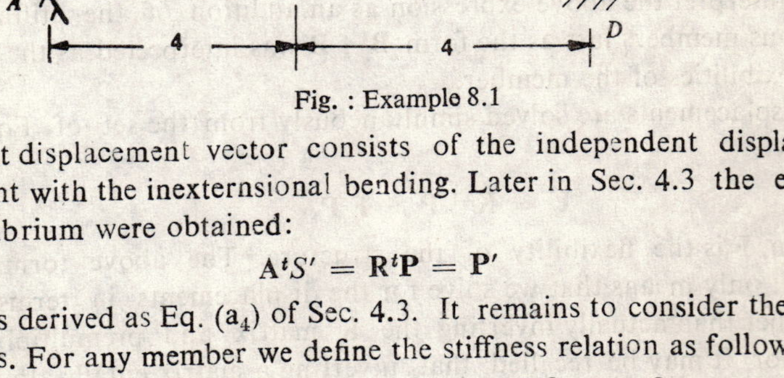

Fig. : Example 8.1

The joint displacement vector consists of the independent displacements consistent with the inexternsional bending. Later in Sec. 4.3 the equations of equilibrium were obtained:

$$\mathbf{A}^t S' = \mathbf{R}^t \mathbf{P} = \mathbf{P}' \qquad (a_2)$$

This was derived as Eq. (a_4) of Sec. 4.3. It remains to consider the stiffness relations. For any member we define the stiffness relation as follows:

$$\mathbf{S}' = \begin{Bmatrix} M_1 \\ M_2 \end{Bmatrix} = \begin{bmatrix} k_{11} & k_{12} \\ k_{21} & k_{22} \end{bmatrix} \begin{Bmatrix} \varphi_1 \\ \varphi_2 \end{Bmatrix} = \mathbf{kv}' \qquad (a_3)$$

Here $k_{11} = (4EI/L) = k_{22}$; $k_{12} = 2EI/L = k_{21}$. Here an uniform section is assumed.

For the frame as a whole we define

$$\mathbf{S}' = \begin{Bmatrix} \mathbf{S}'_A \\ \mathbf{S}'_B \\ \mathbf{S}'_C \end{Bmatrix} = \begin{bmatrix} \mathbf{k}_A & & \\ & \mathbf{k}_B & \\ & & \mathbf{k}_C \end{bmatrix} \begin{Bmatrix} \mathbf{V}'_A \\ \mathbf{V}'_B \\ \mathbf{V}'_C \end{Bmatrix} = \mathbf{k}\,\mathbf{V}' \qquad (a_4)$$

We can now combine Eqs. (a_1), (a_2) and this equation to give

$$[\mathbf{A}^t\,\mathbf{k}\,\mathbf{A}]\,\mathbf{U}' = \mathbf{R}^t\,\mathbf{P} = \mathbf{P}' \qquad (a_5)$$

Let the k_1 and k_2 values for the different members be as follows:

Member	EI/EI_0	k_1/EI_0	k_2/EI_0
A	12.5	10	5
B	8	8	4
C	6.75	9	4.5

(a_6)

Thus the unassembled stiffness matrix is

$$\mathbf{k} = \left[\begin{array}{cc|cc|cc} 10 & 5 & & & & \\ 5 & 10 & & & & \\ \hline & & 8 & 4 & & \\ & & 4 & 8 & & \\ \hline & & & & 9 & 4.5 \\ & & & & 4.5 & 9 \end{array}\right] (EI_0) \qquad (a_7)$$

The stiffness matrix $\mathbf{A}^t\mathbf{k}\mathbf{A}$ can now be worked out using, $L_1 = L_2 = 4$; $h = 3$; $\cot\beta = 4/3$. After multiplication,

$$\mathbf{A}^t\mathbf{k}\,\mathbf{A} = \begin{bmatrix} 18 & 4 & -1 \\ 4 & 17 & -0.5 \\ -1 & -0.5 & 9 \end{bmatrix} (EI_0) \qquad (a_8)$$

Thus we have the stiffness relations

$$(EI_0) \begin{bmatrix} 18 & 4 & -1 \\ 4 & 17 & -0.5 \\ -1 & -0.5 & 9 \end{bmatrix} \begin{Bmatrix} \theta_B \\ \theta_C \\ u_B \end{Bmatrix} = \mathbf{R}^t\,\mathbf{P} = \mathbf{P}' \qquad (a_9)$$

The right hand side is

$$\mathbf{R}^t\,\mathbf{P} = \begin{bmatrix} 0 & 1 & 1 & | & 0 & 0 & 0 \\ 0 & 0 & 0 & | & 0 & 0 & 1 \\ 1 & 4/3 & 0 & | & 1 & 0 & 0 \end{bmatrix} \begin{Bmatrix} P_B \\ Q_B \\ M_B \\ P_C \\ Q_C \\ M_C \end{Bmatrix} = \begin{Bmatrix} M_B \\ M_C \\ P_B + P_C + \frac{4}{3} Q_B \end{Bmatrix} \tag{a$_{10}$}$$

Let us take

$$P_B = P_C = 0;\ \tfrac{4}{3}\, Q_B = 100 \tag{a$_{11}$}$$

This means we are considering only the load $Q_B = 75$. The stiffness relations (a$_9$) are

$$(EI_0) \begin{bmatrix} 18 & 4 & -1 \\ 4 & 17 & -0.5 \\ -1 & -0.5 & 9 \end{bmatrix} \begin{Bmatrix} \theta_B \\ \theta_C \\ u_B \end{Bmatrix} = \begin{Bmatrix} 0 \\ 0 \\ 100 \end{Bmatrix} \tag{a$_{12}$}$$

On inversion we obtain

$$\begin{Bmatrix} \theta_B \\ \theta_C \\ u_B \end{Bmatrix} = \begin{bmatrix} 0.0592 & -0.0142 & 0.00578 \\ -0.0142 & 0.0622 & 0.00189 \\ 0.00578 & 0.00189 & 0.112 \end{bmatrix} \begin{Bmatrix} 0 \\ 0 \\ 100 \end{Bmatrix}$$

$$= \begin{Bmatrix} 0.578 \\ 0.189 \\ 11.2 \end{Bmatrix} (1/EI_0) \tag{a$_{13}$}$$

The internal forces can now be obtained using Eqs. (a$_4$) and (a$_1$).

$$\begin{Bmatrix} M_{1A} \\ M_{2A} \end{Bmatrix} = \begin{bmatrix} 10 & 5 \\ 5 & 10 \end{bmatrix} \begin{Bmatrix} \varphi_{1A} \\ \varphi_{2A} \end{Bmatrix} = \begin{bmatrix} 10 & 5 \\ 5 & 10 \end{bmatrix} \begin{bmatrix} 0 & 0 & -1/3 \\ 1 & 0 & -1/3 \end{bmatrix}$$

$$\begin{Bmatrix} 0.578 \\ 0.189 \\ 11.2 \end{Bmatrix} = \begin{Bmatrix} -53.11 \\ -50.2 \end{Bmatrix} \tag{a$_{14}$}$$

Similarly other internal forces are obtained:

$$\begin{Bmatrix} M_{1B} \\ \\ M_{2B} \end{Bmatrix} = \begin{bmatrix} 8 & 4 \\ & \\ 4 & 8 \end{bmatrix} \begin{bmatrix} 1 & 0 & 1/3 \\ & & \\ 0 & 1 & 1/3 \end{bmatrix} \begin{Bmatrix} 0.578 \\ 0.189 \\ 11.2 \end{Bmatrix} = \begin{Bmatrix} 50.2 \\ \\ 48.6 \end{Bmatrix} \quad (a_{15})$$

$$\begin{Bmatrix} M_{1C} \\ \\ M_{2C} \end{Bmatrix} = \begin{bmatrix} 9 & 4.5 \\ & \\ 4.5 & 9 \end{bmatrix} \begin{bmatrix} 0 & 1 & -1/3 \\ & & \\ 0 & 0 & -1/3 \end{bmatrix} \begin{Bmatrix} 0.578 \\ 0.189 \\ 11.2 \end{Bmatrix} = \begin{Bmatrix} -48.6 \\ \\ -49.55 \end{Bmatrix}$$

The member end shears and axial forces are obtained from the equilibrium of members or a part of a frame.

8.2 DIRECT STIFFNESS METHOD

We will now consider the direct stiffness method [20] of arriving at the load displacement relations (8.7). Handling of frames using the so called *slope-deflection* equations is in fact a direct stiffness method.

The direct stiffness method first constructs general stiffness matrices of members expressed in frame (global) coordinates. The equations of equilibrium are written between internal member forces expressed in frame coordinates and joint loads which, of course are always in frame coordinates. Since both these are in common coordinates the coefficients which control the equilibrium relations involve only zeroes and ones. The construction of the product $\mathbf{A}^t\mathbf{k}\mathbf{A}$ becomes a simpler process which appears as more 'direct'.

General Member Stiffness

The general stiffness* matrices for a member were derived in the form (5.42), to recall,

$$\begin{Bmatrix} \mathbf{P}_1 \\ \mathbf{P}_2 \end{Bmatrix} = \begin{bmatrix} \mathbf{K}_{11} & \mathbf{K}_{12} \\ \mathbf{K}_{21} & \mathbf{K}_{22} \end{bmatrix} \begin{Bmatrix} \mathbf{u}_1 \\ \mathbf{u}_2 \end{Bmatrix} \quad (8.9)$$

The $\mathbf{K}_{ij}$ $(i, j = 1, 2)$ matrices were obtained in terms of the kinematic matrices of the member and the coordinate transformation matrix; to rewrite,

$$\text{Eq. (5.44)}: \quad \mathbf{K}_{ij} = \mathbf{T}\,\overline{\mathbf{K}}_{ij}\,\mathbf{T}^t = \mathbf{A}_i^t\mathbf{k}\mathbf{A}_j \quad (8.10)$$

$$\text{Eq. (5.41)}: \quad \overline{\mathbf{K}}_{ij} = \overline{\mathbf{A}}_i\,\mathbf{k}\,\overline{\mathbf{A}}_j \quad (8.11)$$

*Complete stiffnesses to include lateral loads will be considered later in this section. See Eq. (8.16).

The $\overline{\mathbf{K}}_{ij}$ matrices were worked out for a plane frame of uniform section and are as given in Eq. (5.45). We shall rewrite them here using for brevity the following notations:

$$k_1 = 4EI/L;\ k_2 = 2EI/L;\ k_3 = 12EI/L^3;\ k_4 = 6EI/L^2;\ k_5 = AE/L \tag{8.12}$$

$$\overline{\mathbf{P}} = \begin{Bmatrix} \overline{P}_1 \\ \overline{Q}_1 \\ \overline{M}_1 \\ \overline{P}_2 \\ \overline{Q}_2 \\ \overline{M}_2 \end{Bmatrix} = \left[\begin{array}{ccc|ccc} k_5 & 0 & 0 & -k_5 & 0 & 0 \\ 0 & k_3 & k_4 & 0 & -k_3 & k_4 \\ 0 & k_4 & k_1 & 0 & -k_4 & k_2 \\ \hline -k_5 & 0 & 0 & k_5 & 0 & 0 \\ 0 & -k_3 & -k_4 & 0 & k_3 & -k_4 \\ 0 & k_4 & k_2 & 0 & -k_4 & k_1 \end{array}\right] \begin{Bmatrix} \bar{u}_1 \\ \overline{v}_1 \\ \overline{\theta}_1 \\ \bar{u}_2 \\ \overline{v}_2 \\ \overline{\theta}_2 \end{Bmatrix}$$

$$= \overline{\mathbf{K}}\,\overline{\mathbf{U}} \tag{8.13}$$

The third and the sixth equations are the familiar slope-deflection equations. The second and the fifth can be derived from the equilibrium conditions in terms of the end moments. The first and the fourth equations are the additional stiffness equations relating the axial forces to the axial displacements.

The general stiffness matrix in frame coordinates, given by Eq. (8.10), written out in full is as shown in Table 8.1, where c and s stand for cos α and sin α. Similar details can be obtained for other type of members. Of particular interest is the case of pin-ended member. The stiffness relations for this case are as follows:

$$\begin{Bmatrix} P_1 \\ Q_1 \\ P_2 \\ Q_2 \end{Bmatrix} = k_5 \begin{bmatrix} c^2 & cs & -c^2 & -cs \\ cs & s^2 & -cs & -s^2 \\ -c^2 & -cs & c^2 & cs \\ -cs & -s^2 & cs & s^2 \end{bmatrix} \begin{Bmatrix} \bar{u}_1 \\ \overline{v}_1 \\ \bar{u}_2 \\ \overline{v}_2 \end{Bmatrix} \tag{8.14}$$

These can be considered as the particular case of the general stiffness relations shown in Table 8.1 in which the bending stiffness are considered to be zero and the corresponding rows and columns excluded.

The physical meaning of the stiffness matrices should be kept in mind namely the meaning when one considers the elements of the matrix columnwise. The elements of the first column are the end forces due to $u_1 = 1$ alone, those of the second due to $v_1 = 1$ alone and so on. Figure 8.2 shows the displacement due to $u_1 = 1$ imposed and the corresponding forces required to achieve this deformation. These may be verified. Observe that axial and bending stiffness combine, as a result of the fact that a particular

TABLE 8.1. K_{ij} MATRICES—PLANE FRAME MEMBER

$k_5c^2+k_3s^2$	(k_5-k_3cs)	$-k_4s$	$-(k_5c^2+k_3s^2)$	$-(k_5-k_3)\,cs$	$-ks_4$
$(k_5-k_3)\,cs$	$k_3c^2+k_5s^2$	k_4c	$-(k_5-k_3)\,cs$	$-(k_3c^2+k_5s^2)$	k_4c
$-k_4s$	k_4c	k_1	k_4s	$-k_4c$	k_2
$-(k_5c^2+k_3s^2)$	$-(k_5-k_3)\,cs$	k_4s	$k_5c^2+k_3s^2$	$(k_5-k_3)\,cs$	k_4s
$-(k_5-k_3)\,cs$	$-(k_3c^2+k_5s^2)$	$-k_4c$	$(k_5-k_3)\,cs$	$k_3c^2+k_5s^2$	$-k_4c$
$-k_4s$	k_4c	k_2	k_4s	$-k_4c$	k_1

$k_1 = 4EI/L;\; k_2 = 2EI/L;\; k_3 = 12EI/L^3;\; k_4 = 6EI/L^2;\; k_5 = AE/L$

$c = \cos\alpha;\quad s = \sin\alpha$

displacement such as $u_1 = 1$ shown in the figure causes axial as well as bending deformations.

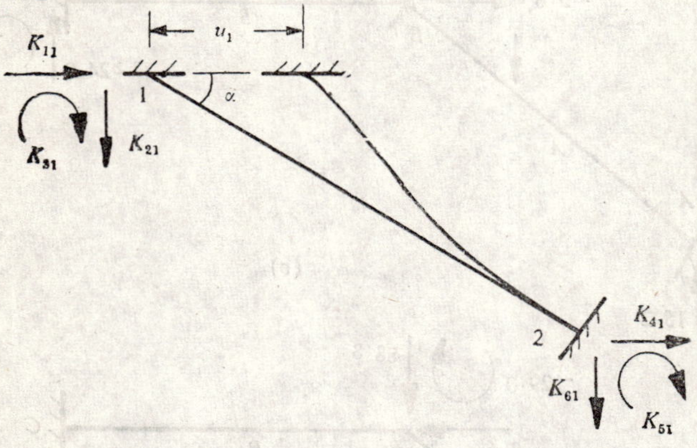

Fig. 8.2

Before proceeding with the general equations we give below an example to indicate the scope of the approach.

EXAMPLE 8.2

The basic principles of the direct stiffness approach can be illustrated with respect to the simple plane frame shown in Fig. Example 8.2(a). The preliminary step is to formulate the problem in terms of the equivalent joint loads supposed to be applied at the joints alone.* The frame shown in (a) is analysed in parts as shown in (b) and (c), and the results superimposed. In (b) the restraining forces are shown at the joint B and at (c) the equivalent joint loads are shown which are equal and opposite to

*Recall Sec. 1.5.

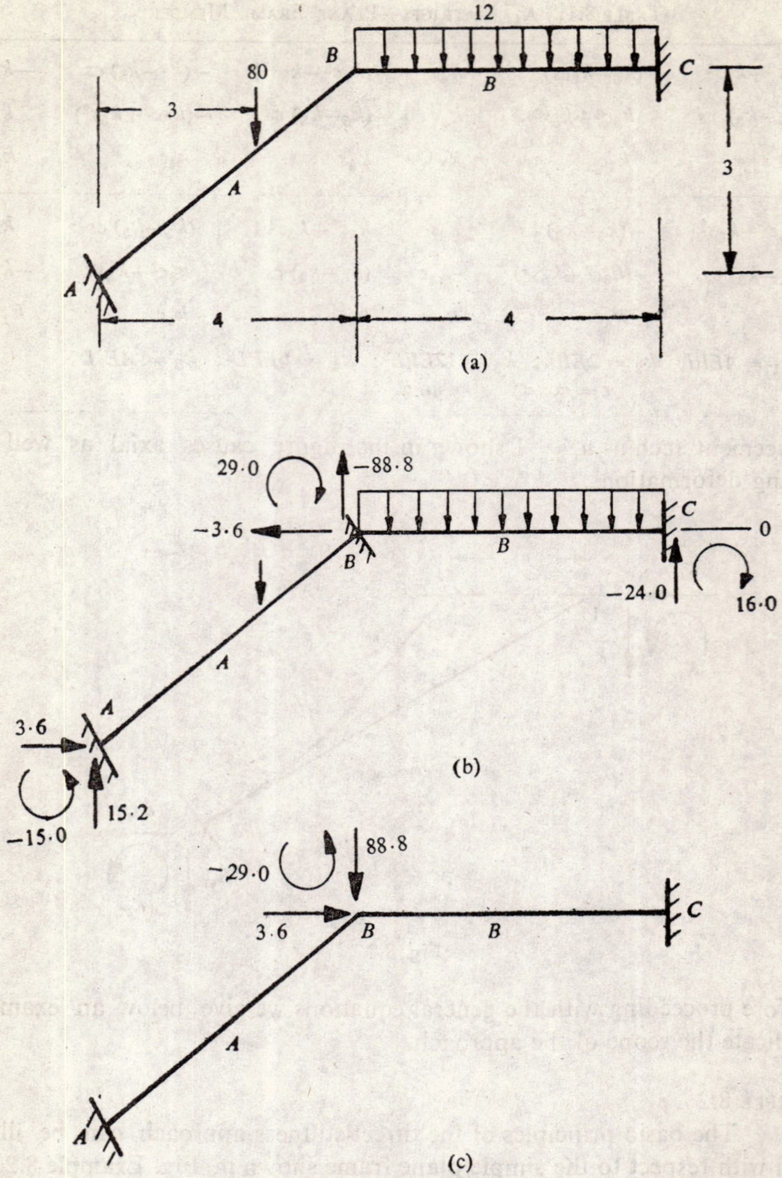

Fig. Example 8.2 (a to c)

the restraining forces. It is the latter part which is the main concern of structural analysis. The restraining joint forces are readily calculated by first calculating the restraining end forces required for each member and then summing such forces over the number of members meeting at the joint in question. Sketch (d) shows the analysis of member A. The vertical load is resolved along directions transverse and axial to the member and the effect of each component analysed separately. The restraining end forces are obtained by resolving back along the frame axes. Similarly sketch (e) shows the analysis of the member B. The superimposition of these will lead to the set of restraining forces as shown in (b). In the present case the joint B alone is required to be released to undergo possible displacements. Thus the

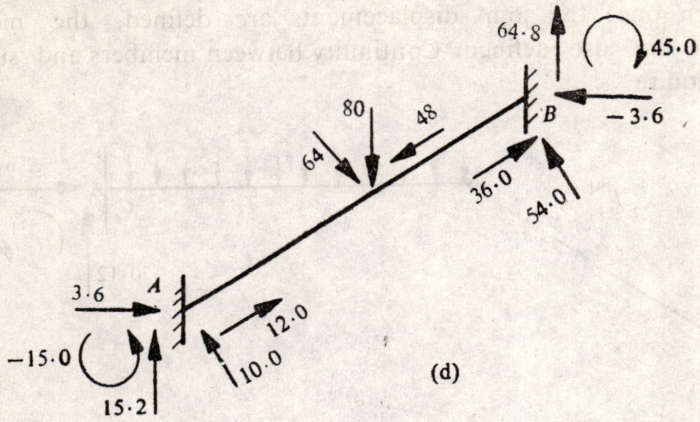

(d)

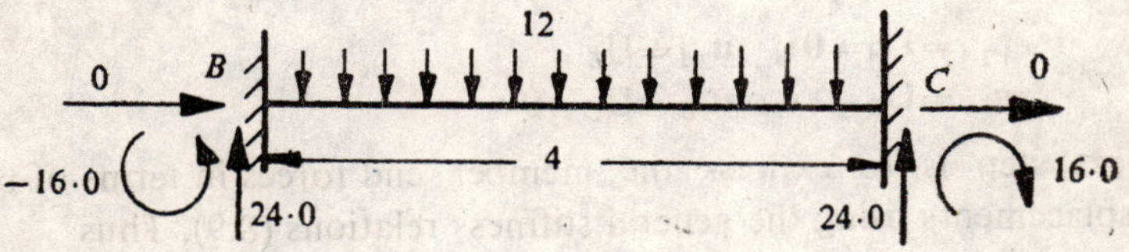

(e)

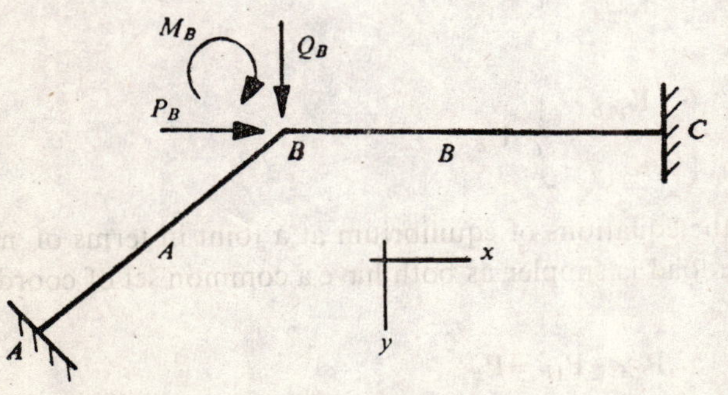

(f)

Fig. Example 8.2 (d to f)

equivalent joint loads are as shown in (c), in which the forces at joints A and C are the reactive forces. Thus the structural problem in general terms is as shown in sketch (f).

The first step in the analysis procedure is to identify the possible *displacement degrees of freedom*. Presently in view of the support conditions the possible joint displacements are only the ones at joint B, namely u_B, v_B, θ_B. When once the joint displacements are defined, the member displacements are also defined. Continuity between members and support conditions require

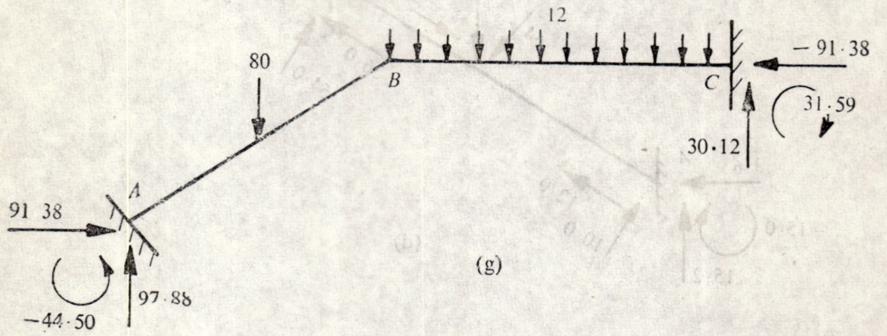

(g)

Fig. Example 8.2 (g)

$$\mathbf{u}_{1A}=\mathbf{U}_A=\mathbf{0}\ ;\quad \mathbf{u}_{2A}=\mathbf{U}_B$$
$$\mathbf{u}_{1B}=\mathbf{U}_B\ ;\quad \mathbf{u}_{2B}=\mathbf{U}_C=0 \qquad (a_1)$$

The next step is to express the member end forces in terms of member end displacements using the general stiffness relations (8.9). Thus

$$\begin{Bmatrix} \mathbf{P}_{1A} \\ \mathbf{P}_{2A} \end{Bmatrix}=\begin{bmatrix} \mathbf{K}_{11A} & \mathbf{K}_{12A} \\ \mathbf{K}_{21A} & \mathbf{K}_{12A} \end{bmatrix}\begin{Bmatrix} \mathbf{u}_{1A} \\ \mathbf{u}_{2A} \end{Bmatrix}=\begin{bmatrix} \mathbf{K}_{11A} & \mathbf{K}_{12A} \\ \mathbf{K}_{21A} & \mathbf{K}_{22A} \end{bmatrix}\begin{Bmatrix} \mathbf{0} \\ \mathbf{U}_B \end{Bmatrix}$$
$$=\begin{Bmatrix} \mathbf{K}_{12A} \\ \mathbf{K}_{22A} \end{Bmatrix}\mathbf{U}_B \qquad (a_2)$$

Similarly

$$\begin{Bmatrix} \mathbf{P}_{1B} \\ \mathbf{P}_{2B} \end{Bmatrix}=\begin{Bmatrix} \mathbf{K}_{11B} \\ \mathbf{K}_{21B} \end{Bmatrix}\mathbf{U}_B \qquad (a_3)$$

The writing of the equations of equilibrium at a joint in terms of member end forces and the load is simpler as both have a common set of coordinates. Thus at joint B,

$$\mathbf{P}_{2A}+\mathbf{P}_{1B}=\mathbf{P}_B$$

These represent a set of three equations of equilibrium. The next step is to express these in terms of displacements $\mathbf{U}_B$ using relations (a_2) and (a_3).

Therefore,

$$\mathbf{K}_{22A}\ \mathbf{U}_B+\mathbf{K}_{11B}\ \mathbf{U}_B=\mathbf{P}_B$$

or

$$[\mathbf{K}_{22A}+\mathbf{K}_{11B}]\{\mathbf{U}_B\}=[\mathbf{K}]\mathbf{U}_B=\mathbf{P}_B \tag{a$_4$}$$

Thus the stiffness matrix is seen to be a sum of the stiffness matrices of the members meeting at B. The matrix **K** denotes the stiffness of the joint B. It is seen to be the sum of the stiffness matrices of the members meeting at the joint B. This is a direct translation into numbers, of a fact qualitatively appealing.

On solving these equations we obtain the displacement $\mathbf{U}_B$. Substitution in Eqs. (a_2) and (a_3) then gives the member end forces in terms of the loads **P**. The member end forces measured in the member axes are then obtainable as

$$\bar{\mathbf{P}}_{1A}=\mathbf{T}^t{}_A\ \mathbf{P}_{1A}\ ; \qquad \bar{\mathbf{P}}_{2A}=\mathbf{T}^t{}_B\ \mathbf{P}_{1B} \tag{a$_5$}$$

etc, where **T** is the coordinate transformation matrix appropriate to the member.

For the frame in question let the member properties be as shown in Table a of Example 8.2. Here γ is the slenderness ratio of the member. In terms of this quantity, we have

$$k_5=AE/L=EI/L^3/(L/r)^2=(EI/L^3)/\gamma^2$$

Observe that stiffness k_5 is generally large compared to the flexural stiffness. Here a value of 20 has been assumed for γ, only to obtain some appreciable axial deformations. The values of EI are in terms of (EI_o), a suitable number (kN m² units) which can be chosen for the frame as a whole. Thus

$$EI_A=12.5\ EI_o,\ EI_B=8.0\ EI_o \text{ etc.}$$

TABLE a. MEMBER PROPERTIES

Member	*L*	cos α	sin α	EI/EI_o	*r*	K_1/EI_o	K_2/EI_o	K_3/EI_o	K_4/EI_o	K_5/EI,
A	5	0.8	—0.6	12.5	20	10	5	1.2	3	40
B	4	1	0	8.0	20	8	4	1.5	3	50

The elements of $\mathbf{K}_{ij}$ matrices can be obtained with the help of Table 8.1. For member A we have

$$k_5c^2+k_3s^2=40\times0.64+1.2\times0.36=26.032$$

$$k_5s^2+k_3c^2=40\times0.36+1.2\times0.64=15.168$$

$$(k_5-k_3)cs=(40-1.2)(0.8)(-0.6)=-18.624$$

$$k_4c=3.0\times0.8=2.4\ ;\ k_4s=3.0\times-0.6=-1.8$$

Thus the $\mathbf{K}_{ij}$ matrices are as follows :

$$\mathbf{K}_A = \left[\begin{array}{ccc|ccc} 26.032 & -18.624 & 1.8 & -26.032 & 18.624 & 1.8 \\ -18.624 & 15.168 & 2.4 & 18.624 & -15.168 & 2.4 \\ 1.8 & 2.4 & 10.0 & -1.8 & -2.4 & 5.0 \\ \hline -26.032 & 18.624 & -1.8 & 26.032 & -18.624 & -1.8 \\ 18.624 & -15.168 & -2.4 & -18.624 & 15.168 & -2.4 \\ 1.8 & 2.4 & 5.0 & -1.8 & -2.4 & 10.0 \end{array}\right] \quad (a_6)$$

For member B we have the member axes same as the frame axes. The coordinate transformation matrix reduces to a unit matrix leading to the matrices given in Eqs. (8.13) as the particular case of the matrices given in Table 8.1. Thus for member B we have the $\mathbf{K}_{ij}$ matrices as follows :

$$\mathbf{K}_B = \left[\begin{array}{ccc|ccc} 50.0 & 0 & 0 & -50.0 & 0 & 0 \\ 0 & 1.5 & 3.0 & 0 & -1.5 & 3.0 \\ 0 & 3.0 & 8.0 & 0 & -3.0 & 4.0 \\ \hline -50.0 & 0 & 0 & 50.0 & 0 & 0 \\ 0 & -1.5 & -3.0 & 0 & 1.5 & -3.0 \\ 0 & 3.0 & 4.0 & 0 & -3.0 & 8.0 \end{array}\right] \quad (a_7)$$

Equation (a_4) can now be constructed by adding $\mathbf{K}_{22A}$ and $\mathbf{K}_{11B}$. Thus for the given set of loads we obtain, on adding,

$$(EI_o)\begin{bmatrix} 76.032 & -18.624 & -1.8 \\ -18.624 & 16.668 & 0.6 \\ -1.8 & 0.6 & 18 \end{bmatrix}\begin{Bmatrix} u_B \\ v_B \\ \theta_B \end{Bmatrix} = \begin{Bmatrix} 3.6 \\ 88.8 \\ -29.0 \end{Bmatrix} \quad (a_8)$$

On solving,

$$\begin{Bmatrix} u_B \\ v_B \\ \theta_B \end{Bmatrix} = \begin{Bmatrix} 1.8276 \\ 7.4300 \\ -1.6760 \end{Bmatrix}\left(\frac{1}{EI_o}\right) \quad (a_9)$$

The member end forces are next obtained using Eqs. (a_2) and (a_3). Thus using the matrices from Eq. (a_6) and (a_7).

$$\begin{Bmatrix} P_{1A} \\ Q_{1A} \\ M_{1A} \end{Bmatrix} = \begin{bmatrix} -26.032 & 18.624 & 1.8 \\ 18.624 & -15.168 & 2.4 \\ -1.8 & -2.4 & 5.0 \end{bmatrix} \begin{Bmatrix} 1.8276 \\ 7.4300 \\ -1.6760 \end{Bmatrix} = \begin{Bmatrix} 87.78 \\ -82.68 \\ -29.50 \end{Bmatrix} \tag{a_{10}}$$

In a similar manner the other member end forces are obtained. These are

$$\begin{aligned} \{P_{2A}, Q_{2A}, M_{2A}\} &= \{-87.78,\ 82.68,\ -37.84\} \\ \{P_{1B}, Q_{1B}, M_{1B}\} &= \{\ 91.38,\ 6.117,\ 8.882\} \\ \{P_{2B}, Q_{2B}, M_{2B}\} &= \{-91.38,\ -6.117,\ 15.586\} \end{aligned} \tag{a_{11}}$$

To obtain the final end forces we have to superpose the end forces due to distributed loads when the displacements were prevented (sketches d, e.) Thus superposing these,

$$\begin{Bmatrix} P_{1A} \\ Q_{1A} \\ M_{1A} \end{Bmatrix} = \begin{Bmatrix} 87.78+3.6 \\ -82.68-15.2 \\ -29.50-15.0 \end{Bmatrix} = \begin{Bmatrix} 91.38 \\ -97.88 \\ -44.5 \end{Bmatrix}$$

$$\begin{aligned} \{\mathbf{P}_{2A}\} &= \{-91.38,\ 17.88,\ 7.12\} \\ \{\mathbf{P}_{1B}\} &= \{\ 91.38,\ -17.88,\ -7.12\} \\ \{\mathbf{P}_{2B}\} &= \{-91.38,\ -30.117,\ 31.586\} \end{aligned} \tag{a_{12}}$$

The member end forces obtained are as shown in sketch (g).

Complete General Stiffness Relations

Recall Eq. (5.66) which gives the total member end forces $\overline{\mathbf{P}}$ as a sum of fixing end forces $\overline{\mathbf{P}}_F$ due to lateral loads and $\overline{\mathbf{K}}\,\overline{\mathbf{U}}$ due to end displacements. Let us rewrite it as follows :

$$\overline{\mathbf{P}} = \begin{Bmatrix} \overline{\mathbf{P}}_1 \\ \overline{\mathbf{P}}_2 \end{Bmatrix} = \begin{bmatrix} \overline{\mathbf{K}}_{11} & \overline{\mathbf{K}}_{12} \\ \overline{\mathbf{K}}_{21} & \overline{\mathbf{K}}_{22} \end{bmatrix} \begin{Bmatrix} \overline{\mathbf{u}}_1 \\ \overline{\mathbf{u}}_2 \end{Bmatrix} + \begin{Bmatrix} \overline{\mathbf{P}}_{F1} \\ \overline{\mathbf{P}}_{F2} \end{Bmatrix} \tag{8.15}$$

We transform these to global system by noting that

$\mathbf{P}_1 = \mathbf{T}\,\overline{\mathbf{P}}_1$, $\overline{\mathbf{U}}_1 = \mathbf{T}^t\,\mathbf{U}_1$ etc., where T is the one given by Eq. (2.12). Thus

$$\begin{Bmatrix} \mathbf{P}_1 \\ \mathbf{P}_2 \end{Bmatrix} = \begin{bmatrix} \mathbf{K}_{11} & \mathbf{K}_{12} \\ \mathbf{K}_{21} & \mathbf{K}_{22} \end{bmatrix} \begin{Bmatrix} \mathbf{u}_1 \\ \mathbf{u}_2 \end{Bmatrix} + \begin{Bmatrix} \mathbf{P}_{F1} \\ \mathbf{P}_{F2} \end{Bmatrix} \tag{8.16}$$

where

$$\mathbf{K}_{ij} = \mathbf{T}\,\overline{\mathbf{K}}_{ij}\,\mathbf{T}^t\ ;\ \mathbf{P}_{Fi} = \mathbf{T}\,\overline{\mathbf{P}}_{Fi} \qquad i, j = 1, 2 \tag{8.16a}$$

Equations (8.16) are the extensions of Eq. (8.9) to include lateral load effects.

In the previous paragraphs it was assumed that equivalent joint loads would be computed separately. In a general procedure which can be programmed, this step is achieved memberwise, alongwith memberwise assembly of stiffness.

Equations of Equilibrium

The equations of equilibrium of a joint can be stated as

$$\Sigma \mathbf{P}_r = \mathbf{P}' \tag{a$_1$}$$

where $\mathbf{P}_r$ is the member end force of a typical member and the summation is taken over the number of members meeting at the joint. The suffix r can be either 1 or 2 depending on whether the end of the member at the joint is 1 or 2. $\mathbf{P}'$ are the joint forces *directly applied* to the joints. Using Eq. (8.16) we can write

$$\mathbf{P}_r = [\mathbf{K}_{r1}\mathbf{u}_1 + \mathbf{K}_{r2}\mathbf{u}_2] + \mathbf{P}_{Fr} \tag{8.17}$$

with $r=1$ or 2. Thus Eq. (a_1) can be written as

$$\sum(\mathbf{K}_{r1}\,\mathbf{u}_1 + \mathbf{K}_{r2}\,\mathbf{u}_2) = \mathbf{P}' - \Sigma\mathbf{P}_{Fr} = \mathbf{P}' + \mathbf{P}'' = \mathbf{P} \tag{8.18}$$

The right hand vector is the net equivalent joint load. The left hand side will be a linear equation involving the displacements of the joint in question and of those joints which are the 'other ends' of the members meeting at the joint. Thus for the joint C of Fig. 8.1,

$$\mathbf{P}_{2B} + \mathbf{P}_{2E} + \mathbf{P}_{1C} = \mathbf{P}_C \tag{a$_2$}$$

or $$(\mathbf{K}_{21B}\,\mathbf{u}_{1B} + \mathbf{K}_{22B}\,\mathbf{u}_{2B}) + (\mathbf{K}_{21E}\,\mathbf{u}_{1E} + \mathbf{K}_{22E}\,\mathbf{u}_{2E}) + (\mathbf{K}_{11C}\,\mathbf{u}_{1C} + \mathbf{K}_{12C}\,\mathbf{u}_{2C})$$
$$= \mathbf{P}_C' - (\mathbf{P}_{F2B} + \mathbf{P}_{F2E} + \mathbf{P}_{F1C}) \tag{a$_3$}$$

Replacing the end displacements by the appropriate joint displacements, and denoting the right hand side by $\mathbf{P}_C$,

$$(\mathbf{K}_{21B}\,\mathbf{U}_B + \mathbf{K}_{22B}\,\mathbf{U}_C) + (\mathbf{K}_{21E}\,\mathbf{U}_A + \mathbf{K}_{22E}\,\mathbf{U}_C)$$
$$+ (\mathbf{K}_{11C}\,\mathbf{U}_C + \mathbf{K}_{12C}\,\mathbf{U}_D) = \mathbf{P}_C \tag{a$_4$}$$

Collecting the terms,

$$(\mathbf{K}_{22B} + \mathbf{K}_{22E} + \mathbf{K}_{11C})\,\mathbf{U}_C + \mathbf{K}_{21B}\,\mathbf{U}_B + \mathbf{K}_{21E}\,\mathbf{U}_A + \mathbf{K}_{12C}\mathbf{U}_D = \mathbf{P}_C \tag{a$_4$}$$

All such equations together constitute the stiffness relation for the structure as a whole. Thus for the frame of Fig. 8.1 the complete set of equations are as follows :

$$\begin{bmatrix} \begin{pmatrix} K_{11A} \\ + \\ K_{11D} \\ + \\ K_{11E} \end{pmatrix} & K_{12A} & K_{12E} & K_{12D} \\ K_{21A} & \begin{pmatrix} K_{22A} \\ + \\ K_{11B} \end{pmatrix} & K_{12B} & 0 \\ K_{21E} & K_{21B} & \begin{pmatrix} K_{22B} \\ + \\ K_{11C} \\ + \\ K_{22E} \end{pmatrix} & K_{12C} \\ K_{21D} & 0 & K_{21C} & \begin{pmatrix} K_{22C} \\ + \\ K_{22D} \end{pmatrix} \end{bmatrix} \begin{Bmatrix} U_A \\ U_B \\ U_C \\ U_D \end{Bmatrix} = \begin{Bmatrix} P_A + R_A \\ P_B \\ P_C \\ P_D + R_D \end{Bmatrix} \tag{8.19}$$

Each row represents the equations of equilibrium of the corresponding joint. These are in terms of submatrices. Each row is a set of 3 or 6 scalar equations for plane or space frame respectively.

There are interesting features about the locations of the K_{ij} matrices of members in the stiffness matrix of the structure. The first noticeable feature is the occurrence of the sum of the K_{rr} matrices of members at the diagonal locations. Taking A as the location, the contributions $K_{11A}+K_{11D}+K_{11E}$ denote the total forces required at A to cause unit displacements at the same joint. The other elements of the first column, K_{21A}, K_{21E} and K_{21D} represent the forces required at B, C, and D so that all the forces together constitute forces which would be in equilibrium. Figure 8.3 shows the forces due to U_A. The other elements of the first row denote the forces required at A due to unit displacements at B, C and D respectively, namely K_{12A}, K_{12E} and K_{12D}.

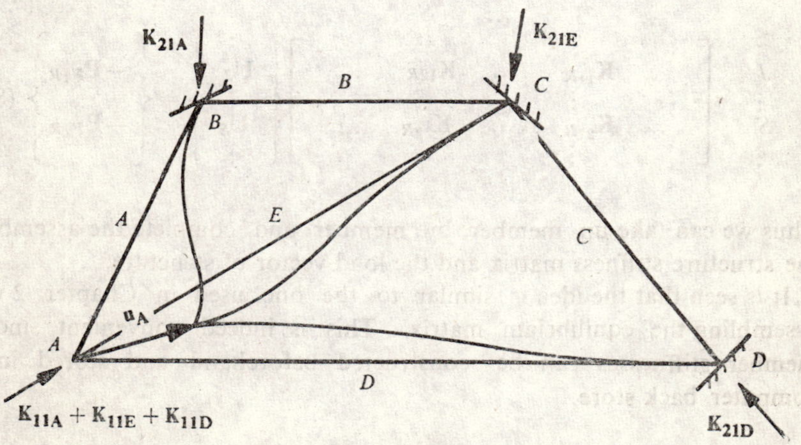

Fig. 8.3.

In the above equations at A and D reactive forces $\mathbf{R}_A$ and $\mathbf{R}_D$ have been shown added to the joint loads. These are the unknown reactive forces since we have assumed joints A and D to be restrained.

Memberwise Assembly

We will now consider the construction of the stiffness of a structural assembly memberwise, a procedure which has great computational advantage. Recall the contents of Sec. 2.7 (Chapter 2) wherein we considered the scheme of assembling the equations of equilibrium memberwise. We will follow here a similar procedure. That such a procedure is possible can be grasped easily. Observe that the displacement method transforms the equation of equilibrium (8.5) which is in terms of internal forces, to that in terms of displacements, through Eqs. (8.6). As was discussed in Chapter 2 the advantage of this conversion is the fact that the equilibrium matrix becomes not only square but also symmetric. Viewed in this angle Eq. (8.7) is a transformation of matrix **A** of Eq. (8.5) to a symmetric one.

Consider a typical member R, whose end joints are J and S. The member end forces affect the equations of equilibrium only of joints J and S. Thus their contributions will be in the Jth and Sth rows. Thus the contributions of the Rth member in the summation (8.18) read :

$$\begin{aligned} \mathbf{K}_{11R}\,\mathbf{U}_J+\mathbf{K}_{12R}\,\mathbf{U}_S &= -\mathbf{P}_{F1R} \\ \mathbf{K}_{21R}\,\mathbf{U}_J+\mathbf{K}_{22R}\,\mathbf{U}_S &= -\mathbf{P}_{F2R} \end{aligned} \qquad (a_5)$$

(Note that we consider the directly applied loads $\mathbf{P}'$ later).

Thus, the scheme according to which the $\mathbf{K}_{rr}$ matrices find place in the structure stiffness matrix is as shown below :

$$\begin{matrix} \\ \vdots \\ J \\ \vdots \\ S \\ \vdots \end{matrix} \begin{matrix} \cdots & J & \cdots & S & \cdots \\ & \vdots & & \vdots & \\ \cdots & \mathbf{K}_{11R} & \cdots & \mathbf{K}_{12R} & \cdots \\ & \vdots & & \vdots & \\ \cdots & \mathbf{K}_{21R} & \cdots & \mathbf{K}_{22R} & \cdots \\ & \vdots & & \vdots & \end{matrix} \begin{Bmatrix} \vdots \\ \mathbf{U}_J \\ \vdots \\ \mathbf{U}_S \\ \vdots \end{Bmatrix} = \begin{Bmatrix} \vdots \\ -\mathbf{P}_{F1R} \\ \vdots \\ -\mathbf{P}_{F2R} \\ \vdots \end{Bmatrix} \qquad (8.20)$$

Thus we can take up member by member and complete the assembly of the structure stiffness matrix and the load vector of structure.

It is seen that the idea is similar to the one used in Chapter 2 while assembling the equilibrium matrix. This is indeed convenient since all member stiffnesses can be constructed beforehand and stored in the computer back store.

Let us consider the problem of Fig. (8.1). The connectivity data is as given below :

Member	*End 1*	*End 2*
A	*A*	*B*
B	*B*	*C*
C	*C*	*D*
D	*A*	*D*
E	*A*	*C*

With the help of this information we may assemble the stiffness matrix **K** as follows. Consider member *A*; it has ends at *A* and *B*; place $\mathbf{K}_{11A}$ at *AA*, $\mathbf{K}_{12A}$ at *AB*, $\mathbf{K}_{21A}$ at *BA*, and $\mathbf{K}_{22A}$ at *BB* locations; take the next member and read the names of the joints at its ends 1 and 2, namely *B* and *C*. Place $\mathbf{K}_{11B}$ at *BB*, $\mathbf{K}_{12B}$ at *BC*, $\mathbf{K}_{21B}$ at *CB*, and $\mathbf{K}_{22B}$ at *CC* locations, and so on. It will be seen at *BB* location there already exists the contribution of member *A*, namely $\mathbf{K}_{22A}$. The contributions at the same locations are to be added. The locations where no contributions occur are treated as being filled by null matrices of appropriate sizes. While programming it is clearing the array first and then posting the contributions.

Assembling the Load Vector

Assembling of the right hand side load vector is relatively simpler. The equivalent end forces $-\mathbf{P}_{F1R}$ and $-\mathbf{P}_{F2R}$ of the *R*th member are to be located at the *J*th and *S*th rows, as shown in Eq. (8.20). After all these are placed, the actual applied forces are to be added. Thus for the frame of Fig. (8.1), the term $\mathbf{P}_B$ is

$$\mathbf{P}_B = -(\mathbf{P}_{F2A} + \mathbf{P}_{F1B}) + \mathbf{P}'_B \tag{a_6}$$

Like stiffness coefficients the loads too get added up.

The scheme shown in (8.20) is in terms of submatrices. Each of the $\mathbf{K}_{ij}$ matrices will be of size equal to, using programming notations, $(NDF \times NDF)$, where *NDF* is the degree of freedom, namely the number of displacement components at a joint, 3 for a plane frame member and 6 for a space frame one. Equation (8.20) written out in full, assuming plane frame case, is actually as shown below :

$$
\begin{array}{cc|c|ccc|c|ccc|c|}
 & & & \multicolumn{3}{c|}{\text{Joint no. } J \text{ (Col. no.)}} & & \multicolumn{3}{c|}{\text{Joint no. } S \text{ (Col. no.)}} & \\
 & & & 3J\text{-}2 & 3J\text{-}1 & 3J & & 3S\text{-}2 & 3S\text{-}1 & 3S & \\
 & & & \vdots & \vdots & \vdots & & \vdots & \vdots & \vdots & \\
 & 3J\text{-}2 & \cdots & * & * & * & \cdots & * & * & * & \cdots \\
J:\ \text{row nos.} & 3J\text{-}1 & \cdots & * & * & * & \cdots & * & * & * & \cdots \\
 & 3J & \cdots & * & * & * & \cdots & * & * & * & \cdots \\
 & & & & & \mathbf{K}_{11} & & \mathbf{K}_{12} & & & \\
 & & & \vdots & \vdots & \vdots & & \vdots & \vdots & \vdots & \\
 & & & & & \mathbf{K}_{21} & & \mathbf{K}_{22} & & & \\
 & 3S\text{-}2 & \cdots & * & * & * & \cdots & * & * & * & \cdots \\
S:\ \text{row nos.} & 3S\text{-}1 & \cdots & * & * & * & \cdots & * & * & * & \cdots \\
 & 3S & \cdots & * & * & * & \cdots & * & * & * & \cdots \\
 & & & \vdots & \vdots & \vdots & & \vdots & \vdots & \vdots &
\end{array}
\left|\begin{array}{c} \vdots \\ u_J \\ v_J \\ \theta_J \\ \vdots \\ u_S \\ v_S \\ \theta_S \end{array}\right|
=
\left\{\begin{array}{c} \vdots \\ -P_{F1} \\ -Q_{F1} \\ -M_{F1} \\ \vdots \\ -P_{F2} \\ -Q_{F2} \\ -M_{F2} \\ \vdots \end{array}\right\}
\begin{array}{c} \\ J \\ \\ \\ S \\ \\ \end{array}
\qquad (8.21)
$$

We will continue to use the notation of submatrices keeping this detail at the background. This detail would be necessary at the time of programming.*

Basic and Direct Stiffness Methods

The reader would perhaps like to see a comparison between the two approaches adopted in assembling the stiffness relations for the structure. The direct stiffness method can also be formally put in the form (8.8) rather than (8.18).

Recall that Eq. (8.8) does imply memberwise assembly, in view of the fact that that the unassembled stiffness matrix **k** is a diagonal one. Thus $\mathbf{A}^t\mathbf{k}\mathbf{A}$ can be written as

*See Subroutine FORMSK given in Apprendix B.

$$\mathbf{A}^t\mathbf{k}\mathbf{A} = \Sigma\ \mathbf{A}_R{}^t\mathbf{k}_R\mathbf{A}_R \tag{a_7}$$

where the summation proceeds over the *numbers of members*. If the number R has its ends 1 and 2 at nodes J and S then the kinematic matrices $\mathbf{A}_{1R}$ and $\mathbf{A}_{2R}$ relate the member deformations $\mathbf{v}_R$ to $\mathbf{U}_J$ and $\mathbf{U}_S$ as,

$$R:\ \begin{bmatrix} \vdots \\ \mathbf{v}_R \\ \vdots \end{bmatrix} = \begin{bmatrix} & \ldots J \ldots & & \ldots S \ldots & \\ & \vdots & & \vdots & \\ \ldots & \mathbf{A}_{1R} & \ldots & \mathbf{A}_{2R} & \ldots \\ & \vdots & & \vdots & \end{bmatrix} \begin{bmatrix} \vdots \\ \mathbf{U}_J \\ \vdots \\ \mathbf{U}_S \\ \vdots \end{bmatrix} \tag{a_8}$$

Thus $\mathbf{A}_R{}^t\ \mathbf{k}_R\ \mathbf{A}_R$ becomes,

$$\begin{matrix} \vdots \\ J \\ \vdots \\ S \\ \vdots \end{matrix} \begin{bmatrix} R \\ \vdots \\ \ldots \mathbf{A}^t{}_{1R} \ldots \\ \vdots \\ \ldots \mathbf{A}^t{}_{2R} \ldots \\ \vdots \end{bmatrix} \begin{bmatrix} \diagdown & & \\ & \mathbf{k}_R & \\ & & \diagdown \end{bmatrix}^{R} \begin{bmatrix} & \ldots J \ldots & & \ldots S \ldots & \\ & \vdots & & \vdots & \\ \ldots & \mathbf{A}_{1R} & \ldots & \mathbf{A}_{2R} & \ldots \\ & \vdots & & \vdots & \end{bmatrix} R$$

$$= \begin{matrix} \vdots \\ J \\ \vdots \\ S \\ \vdots \end{matrix} \begin{bmatrix} & \ldots J \ldots & & \ldots S \ldots & \\ & \vdots & & \vdots & \\ \ldots & \mathbf{A}^t{}_{1R}\mathbf{k}_R\mathbf{A}_{1R} & \ldots & \mathbf{A}^t{}_{1R}\mathbf{k}_R\mathbf{A}_{2R} & \ldots \\ & =\mathbf{K}_{11R} & & =\mathbf{K}_{12R} & \\ & \vdots & & \vdots & \\ \ldots & \mathbf{A}^t{}_{2R}\mathbf{k}_R\mathbf{A}_{1R} & \ldots & \mathbf{A}^t{}_{2R}\mathbf{k}_R\mathbf{A}_{2R} & \ldots \\ & =\mathbf{K}_{21R} & & =\mathbf{K}_{22R} & \\ & \vdots & & \vdots & \end{bmatrix} \tag{a_9}$$

It is thus seen that this is same as the scheme obtained in (8.20). In the direct stiffness method we first compute $\mathbf{K}_{11R}$, $\mathbf{K}_{12R}$, etc. beforehand and directly place them into the structure stiffness matrix.

However a point of difference deserves to be noted. The identity of the two approaches will be lost when the kinematic relations are not in the form (a_8), as when some kinematic assumptions are made. If member deformations cannot be solely determined by the displacements of the end nodes, then the sparseness of the kinematic matrix is lost. Consider the kinematic matrix given by Eq. (a_1) of Example (8.1), in which the assumption of inextensional bending was made. In this, the member of deformations are determined by displacement components which are global, and the kinematic matrices are derived from the overall behaviour of the frame consistent with inextensional bending.

It is usual, in the context of digital computation, to exclude degenerate cases and consider the general kinematics where the member deformations are determined solely by the displacements of member ends. If axial strains are to be made zero for instance, it is usual to place a very large number for axial stiffness which will then give almost zero axial strains.

8.3 NATURE OF STIFFNESS' MATRIX BANDED FORM

We will now show that the srtuctural connectivity data controls the locations of the elements in the structure stiffness matrix. By proper numbering we obtain sparsely populated stiffness matrix. This is a very advantageous feature of the stiffness matrix of the structure unlike the flexibility matrix.

The stiffness matrix such as the one given in Eq. (8.19), is almost fully populated. This is as a result of close interconnection between the joints (Fig. 8.1). Generally we obtain 'banded' stiffness matrices which are denser along the diagonal. Consider the arched structure shown in Fig. 8.4(a).

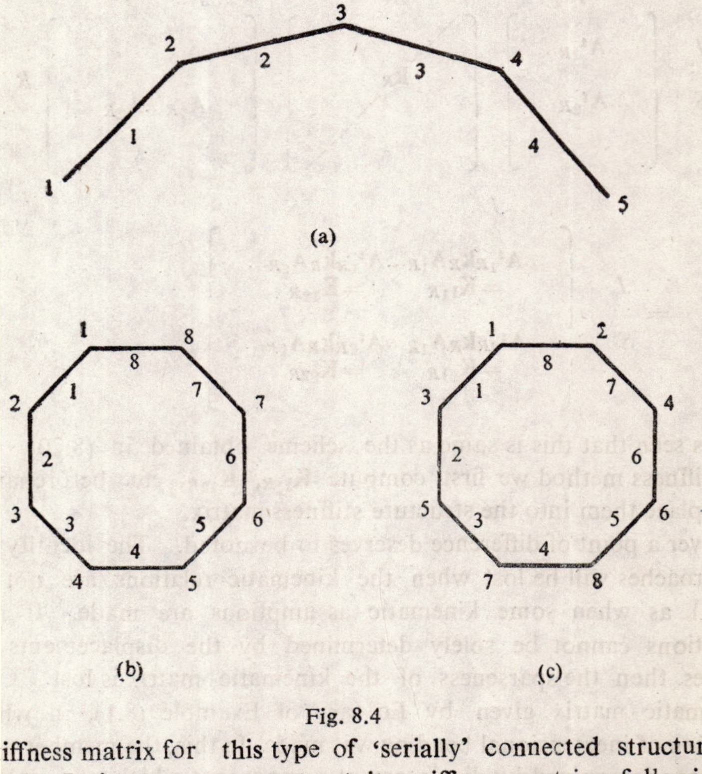

Fig. 8.4

The stiffness matrix for this type of 'serially' connected structure will be banded. It is easy to construct its stiffness matrix following the procedure outlined (Eq. a_1).

Let us suppose that it is a plane frame so that we have 3 degrees of freedom at each joint. Each submatrix then is 3×3. Thus written out in terms of scalar elements this would be a 15×15 matrix as shown in (a_2) :

$$\begin{array}{c} 1 \\ 2 \\ 3 \\ 4 \\ 5 \end{array}\begin{bmatrix} K_1 & (K_{12})_1 & & & \\ (K_{21})_1 & K_2 & (K_{12})_2 & & \\ & (K_{21})_2 & K_3 & (K_{12})_3 & \\ & & (K_{21})_3 & K_4 & (K_{12})_4 \\ & & & (K_{21})_4 & K_5 \end{bmatrix} \begin{Bmatrix} U_1 \\ U_2 \\ U_3 \\ U_4 \\ U_5 \end{Bmatrix} = \begin{Bmatrix} P_1 \\ P_2 \\ P_3 \\ P_4 \\ P_5 \end{Bmatrix} \quad (a_1)$$

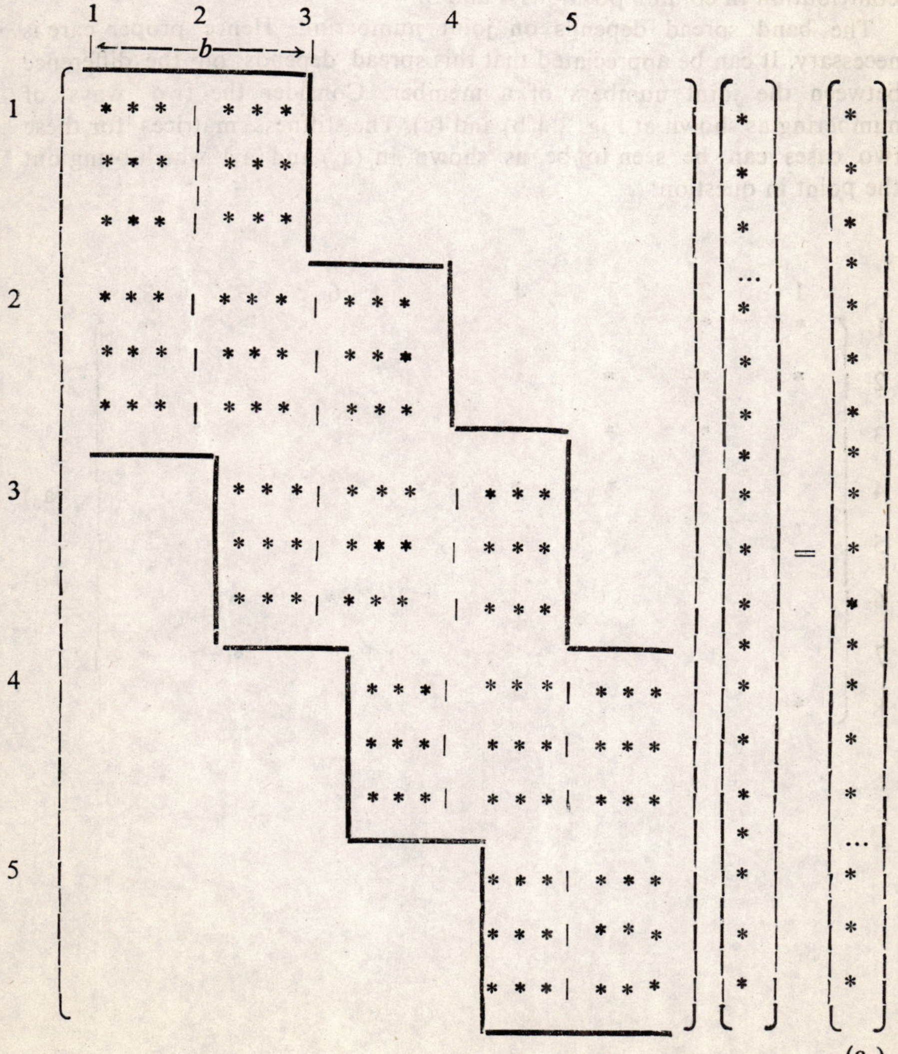

(a_2)

The banded nature is stated in terms of the semiband width b representing the number of elements on one side of the diagonal plus 1 to include the diagonal location itself.

The sparsity can be explained physically also. Recall that elements of each column denote the forces due to unit value of the corresponding displacement. Thus due to $\mathbf{U}_1$ alone we see that there would be forces $(\mathbf{K}_{11})_1\mathbf{U}_1$ at 1 and $(\mathbf{K}_{21})_1\mathbf{U}_1$ at 2 only; since all other joints are held against moments the effect does not 'propagate'. Thus these elements fall in the first and second rows. We can interpret row-wise also. Consider say, the second row, which is the equation of equilibrium $(\mathbf{P}_2)_1+(\mathbf{P}_1)_2 = \mathbf{P}_2$ expressed in terms of displacements. Since the member end forces of members 1 and 2 depend only on the displacements of 1, 2 and 3, there will not be any contribution in column positions 4 and 5.

The band spread depends on joint numbering. Hence proper care is necessary. It can be appreciated that this spread depends on the difference between the joint numbers of a member. Consider the two ways of numbering as shown at Fig. 8.4(b) and (c). The stiffness matrices for these two cases can be seen to be as shown in (a_3) and (a_4), which bring out the point in question:

$$
\begin{array}{c}
 \\ 1 \\ 2 \\ 3 \\ 4 \\ 5 \\ 6 \\ 7 \\ 8
\end{array}
\begin{array}{c}
\begin{array}{cccccccc} 1 & 2 & 3 & 4 & 5 & 6 & 7 & 8 \end{array} \\
\left[\begin{array}{cccccccc}
* & * & & & & & & * \\
* & * & * & & & & & \\
 & * & * & * & & & & \\
 & & * & * & * & & & \\
 & & & * & * & * & & \\
 & & & & * & * & * & \\
 & & & & & * & * & * \\
* & & & & & & * & *
\end{array}\right]
\end{array}
\qquad (a_3)
$$

$$
\begin{array}{c|cccccccc|}
 & 1 & 2 & 3 & 4 & 5 & 6 & 7 & 8 \\
1 & \times & \times & \times & & & & & \\
2 & \times & \times & \cdot & \times & & & & \\
3 & \times & \cdot & \times & \cdot & \times & & & \\
4 & & \times & \cdot & \times & \cdot & \times & & \\
5 & & & \times & \cdot & \times & \cdot & \times & \\
6 & & & & \times & \cdot & \times & \cdot & \times \\
7 & & & & & \times & \cdot & \times & \times \\
8 & & & & & & \times & \times & \times
\end{array}
\qquad (a_4)
$$

In scheme (a_3) the banded scheme is lost. The maximum joint number difference is 7 while as in scheme (a_4) the difference is only 2.

The semiband width can therefore be expressed as

$$b=(d+1)n \qquad (a_5)$$

where d is the maximum difference in joint numbers and n is the degree of freedom. For the arched plane number of Fig. 8.4 the bandwidth as seen from (a_2) is $(1+1)\times 3=6$. Assuming that the frame of Fig. 8.4(b) is a plane frame member, the semiband width of the matrix shown in (a_4) is $(2+1)\times 3=9$.

In view of symmetry of the coefficients we store only the elements above the diagonal and within the semiband width. One convenient from is to store it as a rectangular array as shown in (a_6).

$$
N\updownarrow
\overset{\longleftarrow\, b\, \longrightarrow}{
\begin{bmatrix}
k_{11} & k_{12} & k_{13} & & & \\
\times & k_{22} & k_{23} & k_{24} & & \\
\times & \times & k_{33} & k_{34} & k_{35} & \\
 & \times & \times & k_{44} & k_{45} & k_{46} \\
 & & \times & \times & k_{55} & k_{56} \\
 & & & \times & \times & k_{65}
\end{bmatrix}}
\rightarrow
\overset{\longleftarrow\, b\, \longrightarrow}{
\begin{bmatrix}
k_{11} & k_{12} & k_{13} \\
k_{21} & k_{22} & k_{23} \\
k_{31} & k_{32} & k_{33} \\
k_{41} & k_{42} & k_{43} \\
k_{51} & k_{52} & 0 \\
k_{65} & 0 & 0
\end{bmatrix}}
\qquad (a_6)
$$

Thus if b is the semiband width, and N the number of equations $(b \times N)$ elements are stored. Alternatively we may store elements in a continuous string, (unidimensional array), column (or row) wise :

$$\{k_{11}, k_{12}\, k_{22}, k_{13}, k_{23}\, k_{33}, k_{24}\, k_{34}\, k_{44}, \ldots\ldots\} \qquad (a_7)$$

The addresses of the elements in either case can be related to the original ones*.

Generally the bandwidth varies. We normally work with storage as governed by the *maximum* bandwidth and while solving equations we take note of the variable bandwidth and economise in modification of coefficients during elimination process. We will be discussing this issue in Chapter 10.

8.4 SUPPORT CONDITIONS

Before inverting the stiffness matrix such as in Eq. (8.19) one has to pause and examine if the support conditions have been incorporated so that a unique set of displacements exists. Otherwise the stiffness matrix would be singular. This situation is the same as the one encountered in Chapter 2 while solving the conditions of equilibrium (See Sec. 2.5). The reactive forces were added to the internal forces so that a unique set of internal forces could then be solved for. As already stated the stiffness relations in fact are equations of equilibrium, but expressed in terms of the displacements. Obviously therefore, here too we need to incorporate support conditions.

Consider the equations denoted in (8.19) for the assembly of Fig. 8.1 supported at joints A and B. At these joints there would be reactive forces $\mathbf{R}_A$ and $\mathbf{R}_D$. The general support conditions at A and D can be

$$\mathbf{U}_A = \mathbf{U}^*_A \; ; \; \mathbf{U}_D = \mathbf{U}^*_D \qquad (a_1)$$

where $\mathbf{U}^*_A$ and $\mathbf{U}^*_D$ are prescribed (known) values. When these are set equal to zero we have the fully restrained support conditions. At these joints we have the reactive forces $\mathbf{R}_A$ and $\mathbf{R}_D$ which are unknown and to be determined. Thus we see that in the stiffness relations (8.19), not all the displacement variables are unknowns. Likewise, the right hand load vector is not fully known. Thus we have to separate the known and unknown quantities and solve for the unknowns in terms of the known quantities. Let us write relations (8.19) in general terms as follows :

* See for instance statement no. 35900 of the program of Sec. 10.3 giving the column addresses of elements in the equivalent rectangular array.

$$\begin{bmatrix} K_{AA} & K_{AB} & K_{AC} & K_{AD} \\ K_{BA} & K_{BB} & K_{BC} & K_{BD} \\ K_{CA} & K_{CB} & K_{CC} & K_{CD} \\ K_{DA} & K_{DB} & K_{DC} & K_{DD} \end{bmatrix} \begin{Bmatrix} U_A \\ U_B \\ U_C \\ U_D \end{Bmatrix} = \begin{Bmatrix} P_A + R_A \\ P_B \\ P_C \\ P_D + R_D \end{Bmatrix} \quad (8.22)$$

The above equations can be rearranged so as to separate the displacements into known and unknown ones. This would require *rearrangements of rows and columns of the matrix* and also of the elements of the load vector. However this is not quite convenient in working with computers as the rearrangement of rows and columns and using reduced sizes for matrices and load vectors is cumbersome and entails additional time. An alternative formulation consists in avoiding this by modifying the elements of the stiffness matrix and the load vector so as to incorporate the support conditions and to obtain the unknown reactive forces also.

We can eliminate the variables U^*_A and U^*_D from the given set of equations. Thus transferring these terms to the right hand side, we write,

$$\begin{bmatrix} I & 0 & 0 & 0 \\ 0 & K_{BB} & K_{BC} & 0 \\ 0 & K_{CB} & K_{CC} & 0 \\ 0 & 0 & 0 & I \end{bmatrix} \begin{Bmatrix} U_A \\ U_B \\ U_C \\ U_D \end{Bmatrix} = \begin{Bmatrix} U^*_A \\ P_B - K_{BA}U^*_A - K_{BD}U^*_D \\ P_C - K_{CA}U^*_A - K_{CD}U^*_D \\ U^*_D \end{Bmatrix} \quad (8.23)$$

The coefficients in the off-diagonal terms in the rows corresponding to U^*_A and U^*_D are replaced by zeros and the diagonal elements by unit values. The above equations can be solved with the modified right hand side. In the solution vector one obtains the values $U_A = U^*_A$, $U_D = U^*_D$ and appropriate values for U_B and U_C.

The reactive forces R_A and R_D are obtained from the first and the fourth rows :

$$\begin{aligned} R_A &= -P_A + (K_{AA}U_A + K_{AB}U_B + K_{AC}U_C + K_{AD}U_D) \\ R_D &= -P_D + (K_{DA}U_A + K_{DB}U_B + K_{DC}U_C + K_{DD}U_D) \end{aligned} \quad (8.24)$$

Note that when the prescribed values are all zero, namely when the supports are fully clamped, there is no need of modifying the load terms. In such cases we can assemble the stiffness matrix excluding such foundation joints. Thus if

$$U_A = U_D = 0 \quad (a_2)$$

then we assemble only the portion of the relations stated by Eq. (8.22) :

$$\begin{bmatrix} \mathbf{K}_{BB} & \mathbf{K}_{BC} \\ \mathbf{K}_{CB} & \mathbf{K}_{CC} \end{bmatrix} \begin{Bmatrix} \mathbf{U}_B \\ \mathbf{U}_C \end{Bmatrix} = \begin{Bmatrix} \mathbf{P}_B \\ \mathbf{P}_C \end{Bmatrix} \tag{a$_3$}$$

The modification demonstrated in Eq. (8.23) has been explained in terms of submatrices. It is quite possible that only a few of the components may be prescribed at a joint. In such cases we have to handle the equations in scalar form. In fact the first row of submatrices of (8.23) actually denotes a number of scalar equations, equal to the degree of freedom per joint, namely 3 and 6 for plane and space frame cases respectively. Taking plane frame, the first row reads :

$$\left[\begin{array}{ccc|ccc|ccc|ccc} 1 & 0 & 0 & 0 & 0 & 0 & 0 & 0 & 0 & 0 & 0 & 0 \\ 0 & 1 & 0 & 0 & 0 & 0 & 0 & 0 & 0 & 0 & 0 & 0 \\ 0 & 0 & 1 & 0 & 0 & 0 & 0 & 0 & 0 & 0 & 0 & 0 \\ \hline 0 & 0 & 0 & \times & \times & \times & \times & \times & \times & 0 & 0 & 0 \\ 0 & 0 & 0 & \times & \times & \times & \times & \times & \times & 0 & 0 & 0 \\ 0 & 0 & 0 & \times & \times & \times & \times & \times & \times & 0 & 0 & 0 \\ \hline & \text{etc.} & & & \text{etc.} & & & \text{etc.} & & & \text{etc.} & \end{array}\right] \left\{\begin{array}{c} u_A \\ v_A \\ \theta_A \\ \hline \times \\ \times \\ \times \\ \hline \text{etc.} \end{array}\right\} = \left\{\begin{array}{c} u_A^* \\ v_A^* \\ \theta_A^* \\ \hline \times \\ \times \\ \times \\ \hline \text{etc.} \end{array}\right\} \tag{8.25}$$

If only v_A is prevented at joint A then we alter only the second row and column and the corresponding load terms :

$$\left[\begin{array}{ccc|ccc|ccc|ccc} * & 0 & * & * & * & * & * & * & * & * & * & * \\ 0 & 1 & 0 & 0 & 0 & * & 0 & 0 & 0 & 0 & 0 & 0 \\ * & 0 & * & * & * & * & * & * & * & * & * & * \\ \hline * & 0 & * & * & * & * & * & * & * & * & * & * \\ * & 0 & * & * & * & * & * & * & * & * & * & * \\ * & 0 & * & * & * & * & * & * & * & * & * & * \\ \hline & 0 & & & & & & & & & & \end{array}\right] \left\{\begin{array}{c} \mathbf{u}_A \\ \mathbf{v}_A \\ \theta_A \\ \hline * \\ * \\ * \\ \hline \\ \end{array}\right\} = \left\{\begin{array}{c} P_A - k_{12} v_A^* \\ v_A^* \\ M_A - k_{32} v^*_A \\ \hline * \\ * \\ * \\ \hline \\ \end{array}\right\} \tag{8.26}$$

There is an alternative way which is frequently used. This consists of multiplying the diagonal stiffness by a *large number* and also the prescribed displacement term on the right hand side by this amplified term. Thus consider for illustration the set,

$$\begin{bmatrix} k_{11} & k_{12} & k_{13} \\ k_{21} & k_{22} & k_{23} \\ k_{31} & k_{32} & k_{33} \end{bmatrix} \begin{Bmatrix} u_1(=u^*_1) \\ u_2 \\ u_3 \end{Bmatrix} = \begin{Bmatrix} P_1' \\ P_2 \\ P_3 \end{Bmatrix} \qquad (a_4)$$

This is modified to read as follows :

$$\begin{bmatrix} \alpha k_{11} & k_{12} & k_{13} \\ k_{21} & k_{22} & k_{23} \\ k_{31} & k_{32} & k_{33} \end{bmatrix} \begin{Bmatrix} u_1 \\ u_2 \\ u_3 \end{Bmatrix} = \begin{Bmatrix} \alpha k_{11}\, u_1^* \\ P_2 \\ P_3 \end{Bmatrix} \qquad (8.27)$$

where α is a big number of several orders of magnitude higher than the stiffness. The first equation on dividing by $\alpha\, k_{11}$ gives the value of u_1 very nearly equal to u^*_1 for all practical purposes to within the tolerances accepted in digital computations. No other row modifications are necessary. The alteration of load terms is automatically achieved. The procedure can be given a physical interpretation : multiplying by α is equivalent to connecting a very stiff spring at 1 and at the same time applying a proportionately increased load. The influence of the relatively less stiff structure on the spring is small so that the displacement of the spring is wholly determined by its stiffness and the load on it. We adjust the load also so that we obtain the desired displacement. Thus the first equation can be written as

$$(\alpha\, k_{11})\, u_1 = \alpha k_{11}\, u_1^* - k_{12}\, u_2 - k_{13}\, u_3 \qquad (a_5)$$

and interpreted as the total load on the spring in which the contributions through the stiffnesses of members are negligible. Thus $(\alpha\, k_{11})\, u_1 \approx \alpha\, k_{11}\, u_1^*$ giving $u_1 \approx u_1^*$. In simple parlance, it is equivalent to saying that all the load 'goes' to the heavy spring to produce the desired displacement.

Reactions

We need to calculate the reactive forces according to Eq. (8.24). Observe that in Eq. (8.23) the load term, which includes reactions, is replaced by the prescribed value of the displacement. Therefore, we need to store the right hand side load vectors before placing the prescribed displacements. Observe further that these are precisely the coefficients which have been altered (see Eq. 8·24). Therefore one needs to store these coefficients in the back store to be recalled at the time of calculation of reactions. Alternatively we can modify the equation solving routine to compute the reactions along with the displacement s. We will be illustrating one such procedure in Chapter 10 (Secs. 10.1 and 10.2).

EXAMPLE 8.3

The computer-oriented steps described so far can be applied as an exercise to the frame of previous example. Consider the following loading and support conditions :

Loads : Each of the two members undergo a temperature elongation due to temperature strain $\epsilon°$.

Support Conditions : At support A, a displacement $v_A{}^*=1$ is prescribed. Support C is fully fixed as before :

$$u_C{}^*=v_C{}^*=\theta_C{}^*=0$$

The complete stiffness relations (8.16) are computed for each member by subroutines and stored serially on the back store. $\mathbf{K}_{ii}$ of members are as in Eqs. (a_6) and (a_7) of Example 8.2. The fixing end forces can be computed from Eq. (5.74). For the uniform temperature change, Eq. (a_{26}) of Sec. 5.10 provide the required expressions. Here we are excluding bending and taking only axial elongations. Thus

$$\overline{\mathbf{P}}_{FA}=\{1, 0, 0, -1, 0, 0\}\,(AE\,\epsilon°)_A$$
$$\overline{\mathbf{P}}_{FB}=\{1, 0, 0, -1, 0, 0\}\,(AE\,\epsilon°)_B \qquad (a_1)$$

Transforming these into global coordinates we obtain

$$\overline{\mathbf{P}}_{FA}=\{0.8,\ -0.6,\ 0,\ -0.8,\ 0.6,\ 0\} \qquad (a_2)$$

with $\mathbf{P}_{FB}=\overline{\mathbf{P}}_{FB}$. Presently taking $(AE\,\epsilon°)_A=(AE\epsilon°)_B=200\,EI_o$ we obtain :

$$\mathbf{P}_{FA}=\{160, -120, 0, -160, 120, 0\}\,EI_o$$
$$\mathbf{P}_{FB}=\{200, 0, 0, -200, 0, 0\}\,EI_o \qquad (a_3)$$

We set off a (9×9) matrix and (9×1) column vector for structure stiffness and load vector respectively. When the contributions of the first member are placed, the global equations read as,

$$\begin{bmatrix} \mathbf{K}_{11A} & \mathbf{K}_{12A} & 0 \\ \mathbf{K}_{21A} & \mathbf{K}_{22A} & 0 \\ 0 & 0 & 0 \end{bmatrix} \begin{Bmatrix} \mathbf{u}_A \\ \mathbf{u}_B \\ \mathbf{u}_B \end{Bmatrix} = \begin{Bmatrix} -\mathbf{P}_{F1A} \\ -\mathbf{P}_{F2A} \\ 0 \end{Bmatrix} \qquad (a_4)$$

And when those of second are superimposed,

$$\begin{bmatrix} \mathbf{K}_{11A} & \mathbf{K}_{12A} & 0 \\ \mathbf{K}_{21A} & \mathbf{K}_{22A}+\mathbf{K}_{11B} & \mathbf{K}_{12B} \\ 0 & \mathbf{K}_{21B} & \mathbf{K}_{22B} \end{bmatrix} \begin{Bmatrix} \mathbf{u}_A \\ \mathbf{u}_B \\ \mathbf{u}_C \end{Bmatrix} = \begin{Bmatrix} -\mathbf{P}_{F1A} \\ -(\mathbf{P}_{F2A}+\mathbf{P}_{F1B}) \\ -\mathbf{P}_{F2B} \end{Bmatrix} \qquad (a_5)$$

The nine equations are as in Table (a).

The next step is to modify the above unconstrained set to account for prescribed support conditions, as in Eq. (8.23). Presently we have

$$u^*_A=0\ ;\ v^*_A=1\ ;\ \theta^*_A=0\ ;\ u^*_C=v^*_C=\theta^*_C=0 \qquad (a_6)$$

TABLE (a)—EXAMPLE 8.3

u_A	v_A	θ_A	u_B	v_B	θ_B	u_C	v_C	θ_C	P
26.032	−18.624	1.8	−26.032	18.624	1.8	.	.	.	−160
18.624	15.168	2.4	18.624	−15.168	2.4	.	.	.	120
1.8	2.4	10.0	−1.8	−2.4	5.0	.	.	.	0
−26.032	−18.624	−1.8	76.032	−18.624	−1.8	−50.0	0	0	− 40
18.624	−15.168	−2.4	−18.624	16.668	0.6	0	−1.5	3.0	−120
1.8	2.4	5.0	−1.8	0.6	18.0	0	−3.0	4.0	0
.	.	.	−50.0	0	0	−50.0	0	0	200
.	.	.	0	−1.5	−3.0	0	1.5	−3.0	0
.	.	.	0	3.0	4.0	0	−3.0	8.0	0

Thus we place 1's at the corresponding diagonal locations and place zeros in the corresponding rows and columns. The modified matrix is as in Table (b). The right hand side vector needs modification for non-zero prescribed displacements only. Thus elements of second column multiplied by v^*_A $(=1)$ are to be subtracted from the given load vector for rows corresponding to non-prescribed displacements ; also prescribed values of displacements are to be placed in the load vector at the corresponding rows. The modified vector is as shown in Table (b).

The above equations on solution give :

$$\mathbf{U}=\{0,\ 1.0,\ 0,\ -3.1853,\ -9.845,\ -0.1237,\ 0,\ 0,\ 0\} \qquad (a_7)$$

TABLE (b)—EXAMPLE 8.3

u_A	v_A	θ_A	u_B	v_B	θ_B	u_C	v_C	θ_C	P
1	0	0	0	0	0	0	0	0	0
0	1	0	0	0	0	0	0	0	1
0	0	1	0	0	0	0	0	0	0
0	0	0	76.032	−18.624	−1.8	0	0	0	−58.624
0	0	0	−18.624	16·668	0.6	0	0	0	−104.832
0	0	0	−1.8	0·6	18.0	0	0	0	−2.4
0	0	0	0	0	0	1	0	0	0
0	0	0	0	0	0	0	1	0	0
0	0	0	0	0	0	0	0	1	0

We then obtain the member end forces by substitution into the complete stiffness relations. Thus for instance,

$$\mathbf{P}_{1A}=EI_\circ\begin{bmatrix} 26.032 & -18.624 & 1.8 \\ -18.624 & 15.168 & 2.4 \\ 1.8 & 2.4 & 10.0 \end{bmatrix}\begin{Bmatrix} 0 \\ 1.0 \\ 0 \end{Bmatrix} + EI_\circ\begin{bmatrix} -26.032 & 18{\cdot}624 & 1.8 \\ 18.624 & -15.168 & 2.4 \\ -1.8 & -2.4 & 5.0 \end{bmatrix}\begin{Bmatrix} -3.1853 \\ -9.845 \\ -0.1237 \end{Bmatrix}$$

$$+\begin{Bmatrix} 160.0 \\ -120.0 \\ 0 \end{Bmatrix}EI_\circ=\begin{Bmatrix} 40.7260 \\ -15.1275 \\ 31.1424 \end{Bmatrix}EI_\circ \qquad (a_8)$$

Later in Chapter 10, a solution procedure will be described in which the above steps are accomplished together with triangularization of the coefficient matrix. Instead of actually replacing the coefficients by zeros and ones, we save them to help computing the corresponding reactive forces (see Eq. 8.24).

8.5 FRAMES—PARTICULAR CASES

The most advantageous feature of the direct stiffness method is its generality. The plane frame case was chosen only to detail the steps involved. These steps can be repeated as applied to different kinds of structures. We will be giving below the essential features of some structures to bring out the common forms of the equations of the displacement method.

a. TRANSVERSELY LOADED GRID FRAMEWORKS

The framework is also a plane framework, however it is loaded transversely. In Chapter 2 we discussed this type of structure (see Sec. 2.8 and Fig. 2.12). Equation (2.26) states the equilibrium relations between the member end forces and the independent set of member forces. The latter consists of the torque and the two bending moments. The corresponding deformations are. the torsional deformation ψ and the two end rotations. Thus evoking contragredience, we can obtain from Eq. (2.26) the kinematic relations relating these deformations to the member end displacements :

$$\mathbf{v}=\begin{Bmatrix} \psi \\ \varphi_1 \\ \varphi_2 \end{Bmatrix} = \left\{ \begin{array}{ccc|ccc} -1 & 0 & 0 & 1 & 0 & 0 \\ 0 & 1 & -1/L & 0 & 0 & 1/L \\ 0 & 0 & -1/L & 0 & 1 & 1/L \end{array} \right\} \begin{Bmatrix} \bar{\theta}_{x1} \\ \bar{\theta}_{y1} \\ \bar{w}_1 \\ \bar{\theta}_{x2} \\ \bar{\theta}_{y2} \\ \bar{w}_2 \end{Bmatrix}$$

$$=\bar{\mathbf{A}}_1\,\bar{\mathbf{u}}_1+\bar{\mathbf{A}}_2\,\bar{\mathbf{u}}_2 \;=\bar{\mathbf{A}}\,\bar{\mathbf{u}} \qquad (8.28)$$

The basic stiffness matrix **k** can be easily derived for a member of uniform section :

$$\mathbf{S}=\begin{Bmatrix} T \\ \overline{M}_{y1} \\ \overline{M}_{y2} \end{Bmatrix}=\begin{bmatrix} GJ/L & 0 & 0 \\ 0 & 4EI/L & 2EI/L \\ 0 & 2EI/L & 4EI/L \end{bmatrix}\begin{Bmatrix} \psi \\ \varphi_1 \\ \varphi_2 \end{Bmatrix}=\mathbf{kv} \quad (8.29)$$

In fact this is different from that of a beam only with respect to the axial load term. Here we have GJ/L instead of AE/L as a result of the torque occupying the place of the axial load.

The derivation of the stiffness matrix as per Eq. (5.40) is straightforward. Thus using $\overline{\mathbf{A}}_1$ and $\overline{\mathbf{A}}_2$ matrices of Eq. (8.28), and the above **k** matrix we obtain, finally,

$$\begin{Bmatrix} \overline{M}_{x1} \\ \overline{M}_{y1} \\ \overline{Q}_1 \\ \overline{M}_{x2} \\ \overline{M}_{y2} \\ \overline{Q}_2 \end{Bmatrix}=\left[\begin{array}{ccc|ccc} k_5 & 0 & 0 & -k_5 & 0 & 0 \\ 0 & k_1 & -k_4 & 0 & k_2 & k_4 \\ 0 & -k_4 & k_3 & 0 & -k_4 & -k_3 \\ \hline -k_5 & 0 & 0 & k_5 & 0 & 0 \\ 0 & k_2 & -k_4 & 0 & k_1 & k_4 \\ 0 & k_4 & -k_3 & 0 & k_4 & -k_3 \end{array}\right]\begin{Bmatrix} \overline{\theta}_{x1} \\ \overline{\theta}_{y1} \\ \overline{w}_1 \\ \overline{\theta}_{x2} \\ \overline{\theta}_{y2} \\ \overline{w}_2 \end{Bmatrix} \quad (8.30)$$

where $k_5=GJ/L$ and k_1 to k_4 are same as before; see Eq. (8.12). This is similar to the one in Eq. (8.13), but with rows and columns interchanged as a result of the different order of listing in the vectors concerned.

The derivation of the stiffness matrices in frame coordinates follows the previous pattern. The coordinate transformation matrix **T** in this case is the *same* as for the plane frame case loaded in its own plane (see Sec. 2.8), namely the one given by Eq. (2.12). The $\mathbf{K}_{ij}$ matrices are obtained as $\mathbf{TK}_{ij}\mathbf{T}^t$ as per Eqs. (8.10).

Calculation of the fixing end forces consists of determining the end moments and shears as in beams. The fixing end torques are to be determined following any convenient force or displacement approach.

In practice the analysis of R.C. grids is complicated as a result of the monolithic slab. One way is to include its effect in assessing the moments of inertia of members. The exact procedure would require to treat the slab as an additional structural element. Obviously one has to take recourse to the use of finite element theory in assessing the stiffness of the slab.

b. PIN-JOINTED FRAMES

Analysis of this class of structures simplifies itself as a result of the pin-connections at the joints. The internal force in the member is the axial force S and the member end forces at each end is a set of the three components of the axial force expressed along the frame coordinates. This has been discussed in Sec. 2.6. Equations (2.20) give the member statical matrices $\mathbf{C}_1$ and $\mathbf{C}_2$; to rewrite

$$\mathbf{P}_1=\begin{Bmatrix} P_{x1} \\ P_{y1} \\ P_{z1} \end{Bmatrix}=\begin{Bmatrix} -l \\ -m \\ -n \end{Bmatrix}\{S\}=\mathbf{C}_1\mathbf{S}$$

$$\mathbf{P}_2=\begin{Bmatrix} P_{x2} \\ P_{y2} \\ P_{z2} \end{Bmatrix}=\begin{Bmatrix} l \\ m \\ n \end{Bmatrix}\{S\}=\mathbf{C}_2\mathbf{S} \tag{8.31}$$

with l, m and n as the direction cosines of the axis of the bar. The corresponding kinematic relations are the ones which relate the member elongation to the joint displacements :

$$\begin{aligned}\mathbf{v}=\{e\}&=[-l,\ -m,\ -n,\ l,\ m,\ n]\ \{u_1,\ v_1,\ w_1,\ u_2,\ v_2,\ w_2\}\\ &=\mathbf{A}_1\mathbf{u}_1+\mathbf{A}_2\mathbf{u}_2=\mathbf{A}\mathbf{u}\end{aligned} \tag{8.32}$$

where as usual $\mathbf{A}_1$ and $\mathbf{A}_2$ are transposes of $\mathbf{C}_1$ and $\mathbf{C}_2$ respectively.

The basic member stiffness relation simply relates the axial forces to the axial elongation. Thus

$$\{S\}\equiv\mathbf{S}=\mathbf{k}\mathbf{v}; \tag{8.33}$$

with $\mathbf{k}=[AE/L]$ for a bar of uniform section. The general stiffness matrix is then obtained as per Eq. (8.10). Since in this case the $\mathbf{k}$ matrix is a 1×1 matrix equivalent to a scalar the multiplications are relatively simple. For a pin-ended member we obtain the force-displacement relations

$$\begin{Bmatrix} \mathbf{P}_1 \\ \mathbf{P}_2 \end{Bmatrix} = \begin{bmatrix} \mathbf{K} & -\mathbf{K} \\ -\mathbf{K} & \mathbf{K} \end{bmatrix} \begin{Bmatrix} \mathbf{u}_1 \\ \mathbf{u}_2 \end{Bmatrix} + \begin{Bmatrix} \mathbf{P}_{F1} \\ \mathbf{P}_{F2} \end{Bmatrix} \tag{8.34}$$

where

$$\mathbf{K} = (\mathbf{k}) \begin{bmatrix} l^2 & lm & ln \\ lm & m^2 & mn \\ ln & mn & n^2 \end{bmatrix} \tag{8.34a}$$

Here we have added the fixing end forces to include the common load of temperature and lack of fit. If e° is the free temperature elongation or a lack of fit (too long) the fixing end forces are the ones equivalent to an axial compressive forces of $(AE/L)(e^\circ)$. Thus using (8.31)

$$\mathbf{P}_{F1} = \begin{Bmatrix} l \\ m \\ n \end{Bmatrix} \frac{AE}{L} e^\circ; \quad \mathbf{P}_{F2} = \begin{Bmatrix} -l \\ -m \\ -n \end{Bmatrix} \frac{AE}{L} e^\circ \tag{8.35}$$

We have now the essential relationships for the displacement analysis. The assembly of structure stiffness matrix follows the same pattern. In this case we have to post 3×3 submatrices appearing in Eq. (8.34), and (3×1) load vector in the joint load vector.

c. Space Frames

The procedure of analysing rigid jointed space frames involves no inherent difficulty but is dominated by the sizes of the matrices involved and the total degrees of displacement variables. At each joint (same as at the end of the member) we have six degrees of freedom, both for forces and displacements.

$$\begin{aligned} \mathbf{P}_i &= \{P_{xi} \quad P_{yi} \quad P_{zi} \quad M_{xi} \quad M_{yi} \quad M_{zi}\} \\ \mathbf{u}_i &= \{u_i \quad v_i \quad w_i \quad \theta_{xi} \quad \theta_{yi} \quad \theta_{zi}\} \end{aligned} \quad i=1, 2 \tag{8.36}$$

The vector of independent member forces, S, can be chosen as follows :

$$\mathbf{S} = \{S, \ T, \ \bar{M}_{z1}, \ \bar{M}_{z2}, \ \bar{M}_{y1}, \ M_{y2}\} \tag{8.37}$$

Here S is the axial forces, T the torque, $\bar{M}_{z1}$, $\bar{M}_{z2}$ the bending moments about $\bar{z}$ axis at ends 1 and 2 respectively, and likewise, $\bar{M}_{y1}$ and $\bar{M}_{y2}$ the bending moments about $\bar{y}$ axis at 1 and 2. The coordinate system chosen is a right handed system. Figure (8.5a) shows these forces. It is seen that it is a generalization of the beam model. The $\bar{z}$ and $\bar{y}$ axes are generally the *principal axes* of the section. The corresponding deformations are : an axial elongation *e*, a relative twist ψ between the two cross-sections at ends 1 and 2, $\bar{\varphi}_{z1}$, $\bar{\varphi}_{z2}$, the two end rotations about $\bar{z}$ axis, and, $\bar{\varphi}_{y1}$ and $\bar{\varphi}_{y2}$ similar rotations about $\bar{y}$ axis ; (Figure 8.5 c, d), Thus

$$\mathbf{v} = \{e,\ \psi,\ \bar{\varphi}_{z1},\ \bar{\varphi}_{z2},\ \bar{\varphi}_{y1},\ \bar{\varphi}_{y2}\} \tag{8.38}$$

The basic member stiffness between the forces *S* and these deformations is obtained by considering the beam behaviour in the two principal planes. The simplest set of stiffness relations is the one in which there is no coupling of actions between the forces and deformations resolved in the two orthogonal planes ; such planes are the principal planes of the section. If the section is a thinwalled open section we have to consider the general theory of such sections to include warping stresses. Excluding this case we have the following basic member stiffness :

$$\mathbf{S}=\begin{Bmatrix} S \\ T \\ \bar{M}_{z1} \\ \bar{M}_{z2} \\ \bar{M}_{y1} \\ \bar{M}_{y2} \end{Bmatrix} = \left[\begin{array}{cc|cc|cc} AE/L & 0 & & & & \\ 0 & GJ/L & & & & \\ \hline & & 4EI\bar{z}/L & 2EI\bar{z}/L & & \\ & & 2EI\bar{z}/L & 4EI\bar{z}/L & & \\ \hline & & & & 4EI\bar{y}/L & 2EI\bar{y}/L \\ & & & & 2EI\bar{y}/L & 4EI\bar{y}/L \end{array}\right] \begin{Bmatrix} e \\ \psi \\ \bar{\varphi}_{z1} \\ \bar{\varphi}_{z2} \\ \varphi_{y1} \\ \varphi_{y2} \end{Bmatrix} = \mathbf{k}\,\mathbf{v} \tag{8.39}$$

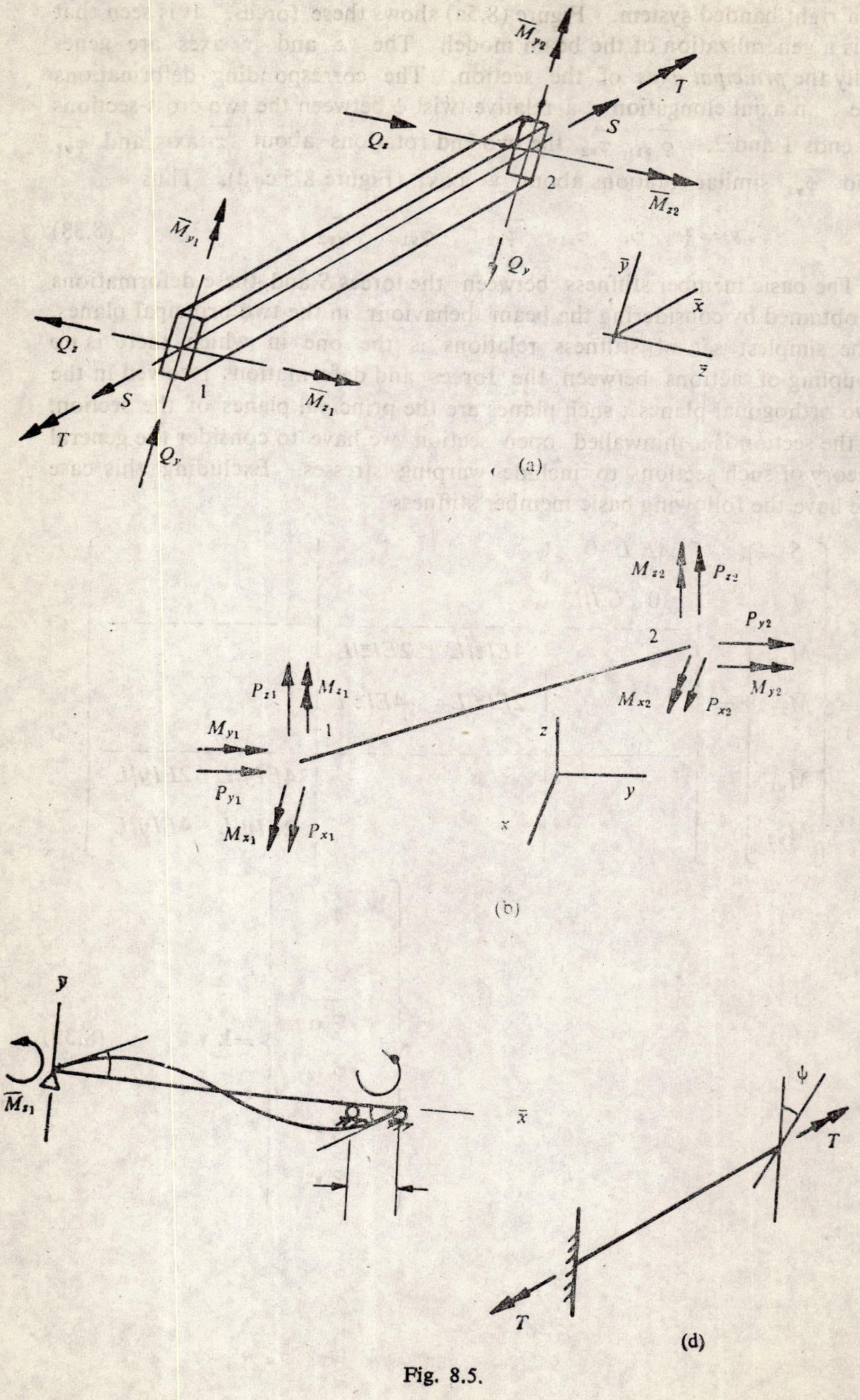

M̄y2
T
S
Qz
2
M̄z2
M̄y1
Qy
ȳ
x̄
z̄
Qz
1
S
M̄z1
T
Qy
(a)
Mz2
Pz2
Py2
2
Mx2
Px2
My2
Pz1
Mz1
1
My1
Py1
Mx1
Px1
z
y
x
(b)
y
M̄z1
x̄
ψ
T
T
(d)

Fig. 8.5.

The next step is to express the member end forces $\bar{P}$ in terms of S and the member deformations v in terms of the member end displacements u. In this case it is easier to write the statical relations and evoke contragredience property to obtain the kinematical relations. Thus we have the relations such as

$$\bar{P}_{x1}=-S\ ;\ \bar{P}_{x2}=S\ ;\ \bar{M}_{x1}=T\ ;\ \bar{M}_{x2}=-T.$$

$$\bar{Q}_{y1}=1/L\,(\bar{M}_{z1}+\bar{M}_{z2})\ \text{etc} \tag{a$_1$}$$

All such relations in matrix form are

$$\left\{\begin{array}{c}\bar{\mathbf{P}}_1\\ \bar{\mathbf{P}}_2\end{array}\right\}=\left[\begin{array}{c}\bar{\mathbf{C}}_1\\ \hline \bar{\mathbf{C}}_2\end{array}\right]\{\mathbf{S}\} \tag{8.40}$$

in which

$$\bar{\mathbf{C}}_1=\left[\begin{array}{cc|cc|cc} -1 & 0 & 0 & 0 & 0 & 0\\ 0 & 0 & 1/L & 1/L & 0 & 0\\ \hline 0 & 0 & 0 & 0 & -1/L & -1/L\\ 0 & -1 & 0 & 0 & 0 & 0\\ \hline 0 & 0 & 0 & 0 & 1 & 0\\ 0 & 0 & 1 & 0 & 0 & 0 \end{array}\right] \tag{8.40a}$$

$$\bar{\mathbf{C}}_2=\left[\begin{array}{cc|cc|cc} 1 & 0 & 0 & 0 & 0 & 0\\ 0 & 0 & -1/L & -1/L & 0 & 0\\ \hline 0 & 0 & 0 & 0 & 1/L & 1/L\\ 0 & 1 & 0 & 0 & 0 & 0\\ \hline 0 & 0 & 0 & 0 & 0 & 1\\ 0 & 0 & 0 & 1 & 0 & 0 \end{array}\right] \tag{8.40b}$$

And the contragredient kinematic relations are,

$$\bar{\mathbf{v}}=[\ \bar{\mathbf{C}}_1^t\quad \bar{\mathbf{C}}_2^t\ ]\{\mathbf{u}\}=[\ \bar{\mathbf{A}}_1\quad \bar{\mathbf{A}}_2]\{\ \mathbf{u}\ \} \tag{8.41}$$

The stiffness matrices $\bar{\mathbf{K}}_{ij}$ are obtained as in Eqs. (8.11), with $\bar{\mathbf{A}}_1$ and $\bar{\mathbf{A}}_2$ from the above equation and the $\mathbf{k}$ matrix from Eq. (8.39). If necessary these can be obtained explicitly by carrying out by hand the required multiplications. It is best to leave it to be carried out within a computer.

Coordinate Transformation

The transformation matrix for a vector quantity in space such as a force or displacement, is a 3×3 matrix containing the direction cosines of the three axes $\bar{x}$, $\bar{y}$ and $\bar{z}$:

$$\mathbf{T}_o = \begin{Bmatrix} \lambda_{\bar{x}x} & \lambda_{\bar{y}x} & \lambda_{\bar{z}x} \\ \lambda_{\bar{x}y} & \lambda_{\bar{y}\bar{y}} & \lambda_{\bar{z}y} \\ \lambda_{\bar{x}z} & \lambda_{\bar{y}z} & \lambda_{\bar{z}z} \end{Bmatrix} \tag{8.42}$$

Here $\lambda_{\bar{x}x}$, $\lambda_{\bar{x}y}$, $\lambda_{\bar{x}z}$ denote the direction cosines of $\bar{x}$ axis. Similarly the elements of the other two columns denote those of $\bar{y}$ and $\bar{z}$ axes. The particular case of this was derived for the plane frame member lying in xy plane. We will give here one convenient method of determining these.* This requires some elementary knowledge of vector products. We can briefly recapitulate what is essentially required as follows :
Let **r** be a vector

$$\mathbf{r} = x\mathbf{i} + y\mathbf{j} + z\mathbf{k} \tag{a$_2$}$$

with **i**, **j** and **k** as unit vectors along x, y and z axis. The modulus of the vector is its length given by $|\mathbf{r}| = (x^2+y^2+z^2)^{1/2}$. Thus $x/|\mathbf{r}|$, $y/|\mathbf{r}|$, $z/|\mathbf{r}|$ represent the direction cosines of the vector. Let l, m, and n be the direction cosines. Since $l^2+m^2+n^2=1$ we can express a *unit* vector as

$$\mathbf{e} = l\mathbf{i} + m\mathbf{j} + n\mathbf{k} \tag{8.43}$$

The vector product $\mathbf{r}_1 \times \mathbf{r}_2$ denotes a vector which is perpendicular to both $\mathbf{r}_1$ and $\mathbf{r}_2$ and is of magnitude equal to the product of their moduli and the sine of the angle between them. This also represents twice the area of the triangle formed by the vectors $\mathbf{r}_1$ and $\mathbf{r}_2$ (Fig. 8.6 b). In particular, if

$$\mathbf{r}_1 = a_1\mathbf{i} + b_1\mathbf{j} + c_1\mathbf{k} \; ; \; \mathbf{r}_2 = a_2\mathbf{i} + b_2\mathbf{j} + c_2\mathbf{k}$$

the vector product can be carried out. Since **i**, **j**, **k** are orthogonal to one another, and therefore $\mathbf{i}\times\mathbf{j} = -\mathbf{j}\times\mathbf{i} = \mathbf{k}$ etc., the vector product is conveniently expressed as

$$\mathbf{r}_3 = \mathbf{r}_1 \times \mathbf{r}_2 = \begin{vmatrix} \mathbf{i} & \mathbf{j} & \mathbf{k} \\ a_1 & b_1 & c_1 \\ a_2 & b_2 & c_2 \end{vmatrix}$$

$$= (b_1c_2 - b_2c_1)\,\mathbf{i} + (c_1a_2 - a_1c_2)\,\mathbf{j} + (a_1b_2 - a_2b_1)\,\mathbf{k}$$

$$= a_3\mathbf{i} + b_3\mathbf{j} + c_3\mathbf{k} \tag{8.44}$$

***This procedure is applicable to area and volume elements where the transformation matrix is obtained in terms of the coordinates of the nodes.**

The magnitude of the vector $\mathbf{r}_3$ is equal to twice the area of triangle formed by the vector $\mathbf{r}_1$ and $\mathbf{r}_2$ (Fig. 8.6b). Thus

$$|\mathbf{r}_3| = 2\Delta = (a_3^2 + b_3^2 + c_3^2)^{1/2} \tag{8.45}$$

Thus the direction cosines of $\mathbf{r}_3$ can be obtained as the components of a unit vector $\mathbf{e}_3$:

$$\mathbf{e}_3 = \mathbf{r}_3 / |\mathbf{r}_3| = \frac{1}{2\Delta}(a_3\mathbf{i} + b_3\mathbf{j} + c_3\mathbf{k}) \tag{8.46}$$

Thus for the space frame member we choose an arbitrary point P in the $\bar{x}\,\bar{y}$ plane, but not lying on $\bar{x}$ axis, so that we have a triangle APB with the two vertices at the ends of the member and the third vertex at P (Fig. 8·6b). The coordinates of the point P are given as additional data. The direction cosines of AB, namely of $\bar{x}$ axis, and the side AP can be calculated. Thus we can express the sides AB and AP as

$$\overline{AB} = a_1\mathbf{i} + b_1\mathbf{j} + c_1\mathbf{k} \tag{8.47}$$

$$\overline{AP} = a_p\mathbf{i} + b_p\mathbf{j} + c_p\mathbf{k} \tag{8.48}$$

where

$$a_1 = x_2 - x_1\ ;\quad b_1 = y_2 - y_1\ ;\quad c_1 = z_2 - z_1$$

$$a_p = x_p - x_1\ ;\quad b_p = y_p - y_1\ ;\quad c_p = z_p - z_1$$

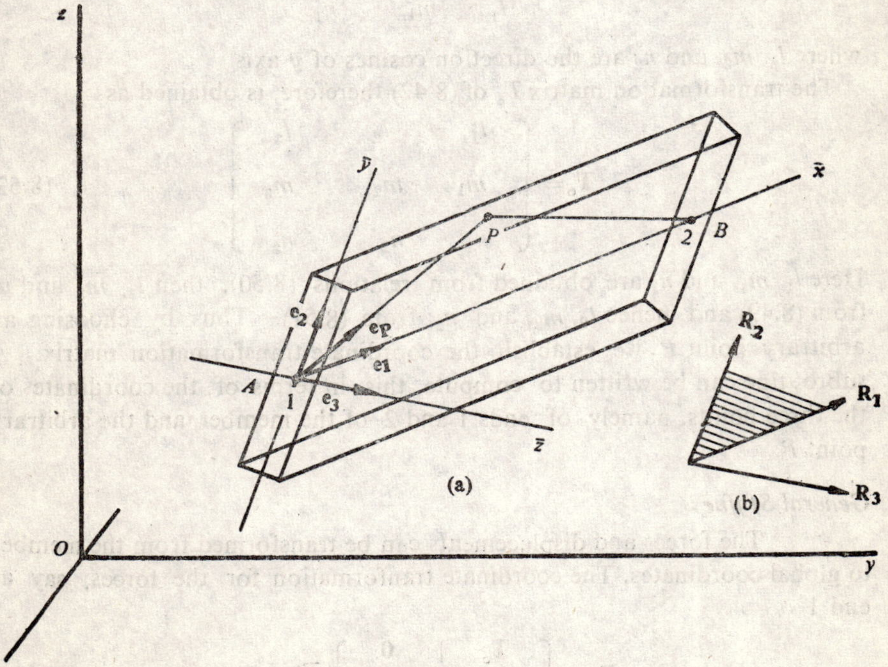

Fig. 8.6

We can perform $\overline{AB} \times \overline{AP}$ giving a vector along $\bar{z}$ axis of magnitude 2Δ where Δ is the area of triangle APB. Thus on dividing by 2Δ we can obtain a unit vector, whose components therefore are the direction cosines of $\bar{z}$. Thus if $\mathbf{e}_3$ is a unit vector along $\bar{z}$, then,

$$\mathbf{e}_3 = \frac{1}{2\Delta} \begin{vmatrix} \mathbf{i} & \mathbf{j} & \mathbf{k} \\ a_1 & b_1 & c_1 \\ a_p & b_p & c_p \end{vmatrix} = l_3\mathbf{i} + m_3\mathbf{j} + n_3\mathbf{k} \tag{8.49}$$

Needless to say, l_3, m_3 and n_3 can be obtained by expanding the determinant. The unit vector $\mathbf{e}_1$ along $\bar{x}$ axis is

$$\mathbf{e}_1 = l_1\mathbf{i} + m_1\mathbf{j} + n_1\mathbf{k}$$

where

$$l_1 = a_1/L\ ;\quad m_1 = b_1/L\ ;\quad n_1 = c_1/L \tag{8.50}$$

We can now perform the vector product $\mathbf{e}_3 \times \mathbf{e}_1$ to obtain $\mathbf{e}_2$, the unit vector along $\bar{y}$ axis :

$$\mathbf{e}_2 = \mathbf{e}_3 \times \mathbf{e}_1 = \begin{vmatrix} \mathbf{i} & \mathbf{j} & \mathbf{k} \\ l_3 & m_3 & n_3 \\ l_1 & m_1 & n_1 \end{vmatrix} = l_2\mathbf{i} + m_2\mathbf{j} + n_2\mathbf{k} \tag{8.51}$$

where l_2, m_2, and n_2 are the direction cosines of $\bar{y}$ axis.

The transformation matrix T_o of (8·42) therefore, is obtained as

$$\mathbf{T}_o = \begin{bmatrix} l_1 & l_2 & l_3 \\ m_1 & m_2 & m_3 \\ n_1 & n_2 & n_3 \end{bmatrix} \tag{8.52}$$

Here l_1, m_1, and n_1 are obtained from relations (8.50), then l_3, m_3 and n_3 from (8.49) and hence l_2, m_2, and n_2 from (8.51). Thus by choosing an arbitrary point P we establish the coordinate transformation matrix. A subroutine can be written to compute this in terms of the coordinates of the three points, namely of ends 1 and 2 of the member and the arbitrary point P.

General Stiffness

The forces and displacements can be transformed from the member to global coordinates. The coordinate tranformation for the forces, say at end 1

$$\underset{(6\times 1)}{\mathbf{P}_1} = \left[\begin{array}{c|c} \mathbf{T}_o & \mathbf{0} \\ \hline \mathbf{0} & \mathbf{T}_o \end{array}\right] \overline{\mathbf{P}}_1 \tag{8.53}$$

with similar expressions for the vector at end 2 and for displacement vectors. Observe that moments and rotations also transform as forces and translations. The stiffness transformation follows the same rule but with the above (6×6) T matrix. The member stiffness matrix would be 12×12 symmetric matrix. The assembly rules and subsequent steps are as in the general procedure outlined in the previous sections.

The reader is advised to grasp the essential common issues described in this section, namely those involved in the analysis of a grid frame work, pinjointed frame work, rigid space frame work etc. This would bring out the essential generality involved in matrix notations describing the basic issues of a method relegating the details of a particular type of structure for later considerations.

8.6 SUBSTRUCTURE ANALYSIS

We will now consider the technique of 'substructuring' which is essentially a generalization of the basic idea of structural analysis. The generalization consists in treating a group of members as an equivalent unit and analysing a structure as an assembly of such units. We will be pursuing this subject again in chapter 10 (Sec. 10.6).

As an elementary application we will derive stiffness and fixing end forccs for a segmental or arched type member such as shown in Fig. 8.7(a). Obviously an arch can be subdivided into a larger number of straight segments. The unit *ACDB* will be analysed as a structure consisting of three members *A*, *B*, *C*. The stiffness and fixing end forces for the member *AB* will be obtained in terms of the stiffnesses and fixing end forces of the three straight members. We will have to determine these in a common (frame) coordinate system so that they can be added on according to the usual rules of the direc. stiffness method. Thus for the frame shown in Fig. 8.7(a), we assemble the following stiffness relations:

$$\left[\begin{array}{cc|cc} \mathbf{K}_{CC} & \mathbf{K}_{CD} & \mathbf{K}_{CA} & \mathbf{0} \\ \mathbf{K}_{DC} & \mathbf{K}_{DD} & \mathbf{0} & \mathbf{K}_{DB} \\ \hline \mathbf{K}_{AC} & \mathbf{0} & \mathbf{K}_{AA} & \mathbf{0} \\ \mathbf{0} & \mathbf{K}_{BD} & \mathbf{0} & \mathbf{K}_{BB} \end{array}\right] \left\{\begin{array}{c} \mathbf{U}_C \\ \mathbf{U}_D \\ \hline \mathbf{U}_A \\ \mathbf{U}_B \end{array}\right\}$$

$$= \left\{\begin{array}{c} \mathbf{P}'_C - \mathbf{P}_{FC} \\ \mathbf{P}'_D - \mathbf{P}_{FD} \\ \hline \mathbf{R}_A + \mathbf{P}'_A - \mathbf{P}_{FA} \\ \mathbf{R}_B + \mathbf{P}'_B - \mathbf{P}_{FB} \end{array}\right\} = \left\{\begin{array}{c} \mathbf{P}_C \\ \mathbf{P}_D \\ \hline \mathbf{P}_A \\ \mathbf{P}_B \end{array}\right\} \tag{8.54}$$

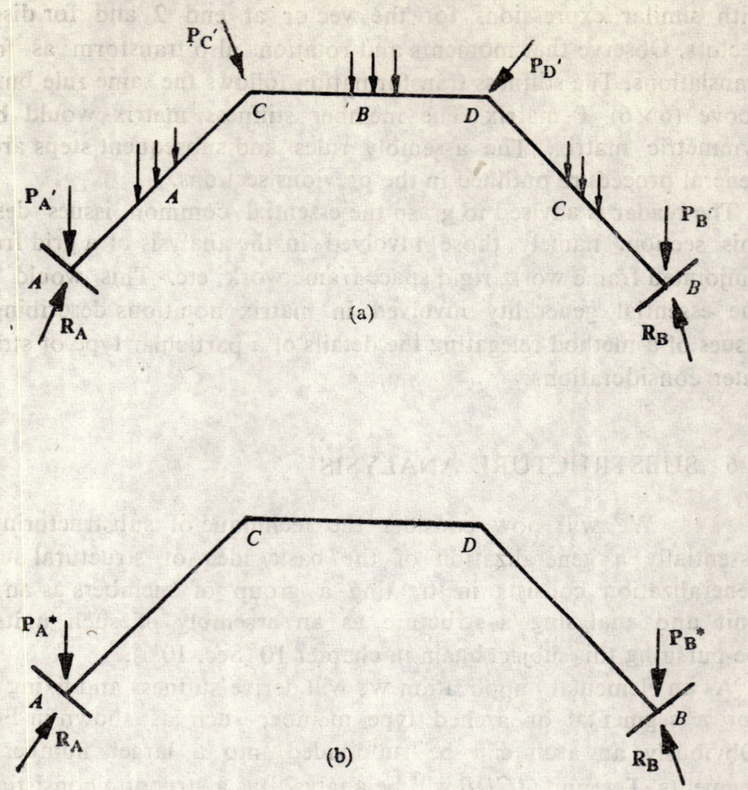

Fig. 8.7

Here $\mathbf{P}'_A$, $\mathbf{P}'_D$, etc. are the loads directly applied to the joints and $\mathbf{R}_A$ and $\mathbf{R}_B$ are the reactive forces at A and B. The joints C and D can be treated as 'internal' joints comparaed to the joints A and B, which can therefore be called 'external' just for the purposes of distinguishing.

We can eliminate the effects of the internal joints and obtain a stiffness relation of the following form:

$$\mathbf{K}^*_E\mathbf{U}_E+\mathbf{P}^*_{FE}=\begin{bmatrix}\mathbf{K}^*_{AA} & \mathbf{K}^*_{AB}\\ \mathbf{K}^*_{BA} & \mathbf{K}^*_{BB}\end{bmatrix}\begin{Bmatrix}\mathbf{U}_A\\ \mathbf{U}_B\end{Bmatrix}+\begin{Bmatrix}\mathbf{P}^*_{FA}\\ \mathbf{P}^*_{FB}\end{Bmatrix}=\begin{Bmatrix}\mathbf{R}_A\\ \mathbf{R}_B\end{Bmatrix}=\mathbf{R}_E \quad (8.55)$$

In such a case we have a single curved member AB, with its associated stiffnessses, and fixing end forces. $\mathbf{R}_E$ are the forces at the ends of the member AB, exerted by the adjoining substructure.

Let us write Eq. (8.54) in the following form:

$$\begin{bmatrix}\mathbf{K}_{II} & \mathbf{K}_{IE}\\ \mathbf{K}_{EI} & \mathbf{K}_{EE}\end{bmatrix}\begin{Bmatrix}\mathbf{U}_I\\ \mathbf{U}_E\end{Bmatrix}=\begin{Bmatrix}\mathbf{P}'_I-\mathbf{P}_{FI}\\ \mathbf{R}_E+\mathbf{P}'_E-\mathbf{P}_{FE}\end{Bmatrix}=\begin{Bmatrix}\mathbf{P}_I\\ \mathbf{R}_E+\mathbf{P}_E\end{Bmatrix} \quad (8.56)$$

Here $\mathbf{U}_E=\{\mathbf{U}_A, \mathbf{U}_B\}$, $\mathbf{U}_I=\{\mathbf{U}_C, \mathbf{U}_D\}$ etc. The subscript E and I stand for 'external' and 'internal' joints respectively. We can eliminate $\mathbf{U}_I$ from the first set of relations:

$$\mathbf{U}_I=\mathbf{K}_{II}^{-1}\{\mathbf{P}_I-\mathbf{K}_{IE}\mathbf{U}_E\} \tag{8.57}$$

Using this in the second set, we obtain

$$\mathbf{K}_{EI}[\mathbf{K}_{II}^{-1}\{\mathbf{P}_I-\mathbf{K}_{IE}\mathbf{U}_E\}]+\mathbf{K}_{EE}\mathbf{U}_E=\mathbf{R}_E+\mathbf{P}_E$$

which therefore gives,

$$[\mathbf{K}_{EE}-\mathbf{K}_{EI}\mathbf{K}_{II}^{-1}\mathbf{K}_{IE}]\,\mathbf{U}_E=\mathbf{R}_E+\{\mathbf{P}_E-\mathbf{K}_{EI}\mathbf{K}_{II}^{-1}\mathbf{P}_I\} \tag{a$_1$}$$

$$=\mathbf{R}_E+\mathbf{P}^*_E$$

This is of the required form (8.55) :

$$\mathbf{K}^*_{EE}\mathbf{U}_E=\mathbf{R}_E+\mathbf{P}^*_E \tag{8.58}$$

where

$$\mathbf{K}^*_{EE}=\mathbf{K}_{EE}-\mathbf{K}_{EI}\mathbf{K}_{II}^{-1}\mathbf{K}_{IE} \tag{8.59}$$

is the modified (condensed) stiffness of the member AB, and,

$$\mathbf{P}^*_E=\mathbf{P}_E-\mathbf{K}_{EI}\mathbf{K}_{II}^{-1}\mathbf{P}_E \tag{8.60}$$

are the equivalent loads as a result of elimination of the effect of the displacements of joint C and D. Relations (8.58) can also be written as

$$\mathbf{K}^*_{EE}\mathbf{U}_E+\mathbf{P}^*_{FE}=\mathbf{R}_E \tag{8.61}$$

where

$$\mathbf{P}^*_{FE}=-\mathbf{P}^*_E \tag{8.62}$$

are the fixing end forces for the member AB (Fig. 8.7b). Observe that $\mathbf{P}^*_E$ depends on applied loads—both distributed and point loads. These are the relations (8.55) which show stiffness and fixing end forces for the member AB with the (internal) joints C and D eliminated.

Let us write Eqs. (8.57) and (8.58) together as follows:

$$\begin{bmatrix} \mathbf{I} & \mathbf{K}_{II}^{-1}\mathbf{K}_{IE} \\ \mathbf{0} & \mathbf{K}^*_{EE} \end{bmatrix}\begin{Bmatrix} \mathbf{U}_I \\ \mathbf{U}_E \end{Bmatrix}=\begin{Bmatrix} \mathbf{K}_{II}^{-1}\mathbf{P}_I \\ \mathbf{R}_E+\mathbf{P}^*_E \end{Bmatrix} \tag{8.63}$$

These are in the familiar form of Gauss elimination. These state that variables $\mathbf{U}_I$ have been eliminated in the system of equations (8.56). The starred quantities are the modified coefficients as a result of subtracting certain multiples of equations corresponding to the variables eliminated. The form (8.63) is the form of Gaussian elimination in which a block of variables are eliminated, instead of a single one. In fact if $\mathbf{K}_{II}$ happens to be a scalar, it is really the usual rule of dividing the row by the pivot, and subtracting the row from each of the rows below that row.

The above discussion illustrates the idea of 'substructuring' as applied to a compound member *AB*, analysed as a structure in which the so-called internal joints (joints *C* and *D*) are eliminated. This becomes an independent process which can be carried out *without* reference to the other joints of the structure having member *AB* as one such member. The joints *C* and *D* would not figure in the general list of joints to be solved by the global structural analysis. We will be explaining this idea further in Chapter 10 as applied to a structure as a whole in which the constituent units are more general than the arched type member discussed so far (see Sec. 10.6).

EXAMPLE 8.4

We will apply the idea discussed in this section to derive the stiffness and fixing end forces for the beam of variable sections shown. Here we will treat this beam as an assembly of two members and the joint *C* as an internal joint to be eliminated. In Example 5.7, the basic stiffness and fixing end forces for the member *A* were derived. We will use the results of the same here:

$$\begin{Bmatrix} M_1 \\ M_2 \end{Bmatrix} = \frac{EI_A}{L_A} \begin{bmatrix} 23.14 & 6.43 \\ 6.43 & 7.29 \end{bmatrix} \begin{Bmatrix} \varphi_1 \\ \varphi_2 \end{Bmatrix} + \begin{Bmatrix} -0.1256 \\ 0.0502 \end{Bmatrix} qL_A^2 \tag{a$_1$}$$

Here I_A is the moment of inertia of the member at its end 2. Let this be the same for the member *B* ; thus $I_A = I_B = I$ say, and further, let $L_A = \lambda L$; $L_B = (1-\lambda) L$; for numerical work we take, $\lambda = 0.4$. Thus eq (a_1) becomes,

$$\begin{Bmatrix} M_1 \\ M_2 \end{Bmatrix} = \frac{EI}{L} \begin{bmatrix} 57.850 & 16.075 \\ 16.075 & 18.225 \end{bmatrix} \begin{Bmatrix} \varphi_1 \\ \varphi_2 \end{Bmatrix} + \begin{Bmatrix} -0.02 \\ 0.00832 \end{Bmatrix} qL^2 \tag{a$_2$}$$

For member *B*, we have the basic stiffness and fixing end forces, given as

$$\begin{Bmatrix} M_1 \\ M_2 \end{Bmatrix} = \frac{EI_B}{L_B} \begin{bmatrix} 4 & 2 \\ 2 & 4 \end{bmatrix} \begin{Bmatrix} \varphi_1 \\ \varphi_2 \end{Bmatrix} + \begin{Bmatrix} -\dfrac{1}{12} \\ \dfrac{1}{12} \end{Bmatrix} qL_B^2$$

$$= \frac{EI}{L} \begin{bmatrix} 6.66 & 3.33 \\ 3.33 & 6.66 \end{bmatrix} \begin{Bmatrix} \varphi_1 \\ \varphi_2 \end{Bmatrix} + \begin{Bmatrix} -0.03 \\ 0.03 \end{Bmatrix} qL^2 \tag{a$_2$}$$

We have to convert these into the general stiffness relations (5.43). Here we will exclude axial load terms. Thus the kinematic relation (3·6) are,

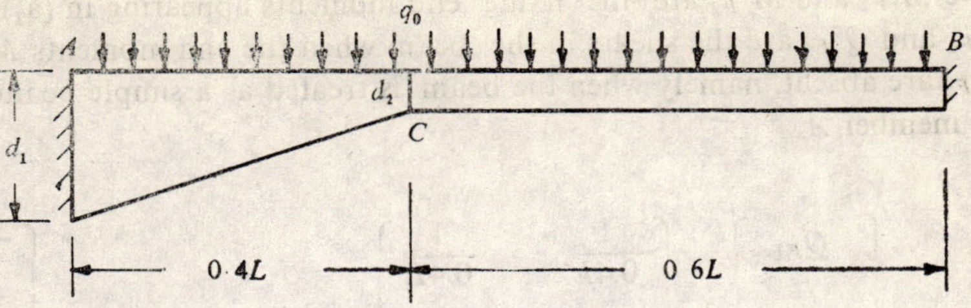

Example 8.4

$$\mathbf{v}=\begin{Bmatrix} \varphi_1 \\ \varphi_2 \end{Bmatrix}=\left[\begin{array}{cc|cc} \frac{1}{L} & 0 & -\frac{1}{L} & 0 \\ \frac{1}{L} & 0 & -\frac{1}{L} & 1 \end{array}\right]\begin{Bmatrix} v_1 \\ \theta_1 \\ \cdots \\ v_2 \\ \theta_2 \end{Bmatrix}=\mathbf{A}_1\mathbf{u}_1+\mathbf{A}_2\mathbf{u}_2 \qquad (a_3)$$

Here the bars on the elements are dropped since no coordinate transformations are involved. Let us write basic stiffness as

$$\mathbf{k}=\begin{bmatrix} a & b \\ b & c \end{bmatrix} \qquad (a_4)$$

And therefore the general stiffness as required by Eq. (5.43) can be derived. Using $\mathbf{A}_1$ and $\mathbf{A}_2$ appearing in Eq. (a_3) above we obtain*

$$\begin{Bmatrix} Q_1 \\ M_1 \\ Q_2 \\ M_2 \end{Bmatrix}=\left[\begin{array}{cc|cc} a_1 & a_2 & -a_1 & a_3 \\ a_2 & a & -a_2 & b \\ \hline -a_1 & -a_2 & a_1 & -a_3 \\ a_3 & b & -a_3 & c \end{array}\right]\begin{Bmatrix} v_1 \\ \theta_1 \\ v_2 \\ \theta_2 \end{Bmatrix}+\begin{Bmatrix} Q_{F1} \\ M_{F1} \\ Q_{F2} \\ M_{F2} \end{Bmatrix} \qquad (a_5)$$

where

$$a_1=(a+2b+c)/L^2; \quad a_2=(a+b)/L; \quad a_3=(b+c)/L \qquad (a_6)$$

Observe that for a beam of uniform section we have $a=c$, $b=a/2$ and therefore $a_2=a_3=6EI/L^2$.

The fixing end forces are

$$\begin{Bmatrix} Q_{F1} \\ M_{F1} \\ Q_{F2} \\ M_{F2} \end{Bmatrix}=\left[\begin{array}{cc} \frac{1}{L} & \frac{1}{L} \\ 1 & 0 \\ \hline -\frac{1}{L} & -\frac{1}{L} \\ 0 & 1 \end{array}\right]\begin{Bmatrix} M'_{F1} \\ M'_{F2} \end{Bmatrix}+\begin{Bmatrix} Q^{\circ}_{F1} \\ 0 \\ \hline Q^{\circ}_{F2} \\ 0 \end{Bmatrix} \qquad (a_7)$$

*See Eq. (a_9) of Sec. 5.7. We could have borrowed from it. However it is best to remember the broad general rules and derive details when required.

Here M'_{F1} and M'_{F2} are the fixing end moments appearing in (a_1) or (a_2); Q°_{F1} and Q°_{F2} are the shears in the beam when the end moments M'_{F1} and M'_{F2} are absent, namely when the beam is treated as a simple beam.* Thus for member A,

$$\mathbf{P}_{FA}=\left\{\begin{array}{c} Q_{F1} \\ M_{F1} \\ \hline Q_{F2} \\ M_{F2} \end{array}\right\}=\left[\begin{array}{cc} \dfrac{1}{0.4L} & \dfrac{1}{0.4L} \\ 1 & 0 \\ \hline -\dfrac{1}{0.4L} & -\dfrac{1}{0.4L} \\ 0 & 1 \end{array}\right]\left\{\begin{array}{c} -0.02 \\ 0.00832 \end{array}\right\}qL^2+\left\{\begin{array}{c} \dfrac{-0.4qL}{2} \\ 0 \\ \hline \dfrac{-0.4qL}{2} \\ 0 \end{array}\right\}$$

or

$$\mathbf{P}_{FA}=\left\{-\frac{0.23}{L},\ -0.02,\ -\frac{0.17}{L},\ 0.00832\right\}qL^2 \qquad (a_8)$$

Similarly, for member B, we obtain,

$$\mathbf{P}_{FB}=\left\{-\frac{0.3}{L},\ -0.03,\ -\frac{0.3}{L},\ 0.03\right\}qL^2 \qquad (a_9)$$

Quantities a_1, a_2 and a_3 (Eq. a_6) of member A are

$$a_1=\frac{57.85+2\times16.075+18.225}{(0.4L)^2}\ \frac{EI}{L}=676.4\left(\frac{EI}{L^3}\right)$$

$$a_2=\frac{57.85+16{\cdot}075}{0.4L}\left(\frac{EI}{L}\right)=184.8\left(\frac{EI}{L^2}\right)$$

$$a_3=\frac{16.075+18.225}{(0.4L)}\ \frac{EI}{L}=85.75\left(\frac{EI}{L^2}\right)$$

Similarly for member B,

$$a_1=55.55\left(\frac{EI}{L^3}\right);\ a_2=16.66\left(\frac{EI}{L^2}\right);\ a_3=3.33\left(\frac{EI}{L^2}\right)$$

The stiffness relations (a_5) can now be constructed. These, dropping multipliers EI/L^3, EI/L^2 etc. for convenience, are as follows:

*Recall Eq. 2.7.

$$\left\{\begin{array}{c} Q_{1A} \\ M_{1A} \\ \hline Q_{2A} \\ M_{2A} \end{array}\right\} = \left[\begin{array}{cc|cc} 676.4 & 184.8 & -676.4 & 85.75 \\ 184.8 & 57.85 & -184.8 & 16.075 \\ \hline -676.4 & -184.8 & 676.4 & -85.75 \\ 85.75 & 16.075 & -85.75 & 18.225 \end{array}\right] \left\{\begin{array}{c} v_1 \\ \theta_1 \\ \hline v_2 \\ \theta_2 \end{array}\right\} + \left\{\begin{array}{c} -0.23 \\ -0.02 \\ \hline -0.17 \\ 0.008 \end{array}\right\} \qquad (a_{12})$$

and

$$\left\{\begin{array}{c} Q_{1B} \\ M_{1B} \\ \hline Q_{2B} \\ M_{2B} \end{array}\right\} = \left[\begin{array}{cc|cc} 55.55 & 16.66 & -55.55 & 16.666 \\ 16.66 & 6.66 & -16.66 & 3.333 \\ \hline -55.55 & -16.66 & 55.55 & -16.66 \\ 16.66 & 3.33 & -16.66 & 6.66 \end{array}\right] \left\{\begin{array}{c} v_1 \\ \theta_1 \\ v_2 \\ \theta_2 \end{array}\right\} + \left\{\begin{array}{c} -0.3 \\ -0.3 \\ \hline -0.3 \\ 0.03 \end{array}\right\} \qquad (a_{13})$$

These are in the general form,

$$\left\{\begin{array}{c} \mathbf{P}_1 \\ \mathbf{P}_2 \end{array}\right\} = \left[\begin{array}{cc} \mathbf{K}_{11} & \mathbf{K}_{12} \\ \mathbf{K}_{22} & \mathbf{K}_{22} \end{array}\right] \left\{\begin{array}{c} \mathbf{U}_1 \\ \mathbf{U}_2 \end{array}\right\} + \left\{\begin{array}{c} \mathbf{P}_{F1} \\ \mathbf{P}_{F2} \end{array}\right\} \qquad (a_{14})$$

ready for the construction of the stiffness matrix for the structure. The assembled stiffness matrix as per usual rules of the direct stiffness method, for the beam *ACB* is

$$\left[\begin{array}{c|cc} \mathbf{K}_{22A}+\mathbf{K}_{11B} & \mathbf{K}_{21A} & \mathbf{K}_{12B} \\ \hline \mathbf{K}_{12A} & \mathbf{K}_{11A} & 0 \\ \mathbf{K}_{21B} & 0 & \mathbf{K}_{22B} \end{array}\right] \left\{\begin{array}{c} \mathbf{U}_C \\ \mathbf{U}_A \\ \mathbf{U}_B \end{array}\right\} = \left\{\begin{array}{c} \mathbf{P}'_C-(\mathbf{P}_{F2A}+\mathbf{P}_{F1B}) \\ \mathbf{R}_A+\mathbf{P}'_A-\mathbf{P}_{F1A} \\ \mathbf{R}_B+\mathbf{P}'_B-\mathbf{P}_{F2B} \end{array}\right\} \qquad (a_{15})$$

Here $\mathbf{P}'_A$, $\mathbf{P}'_B$ etc. are the applied point loads. $\mathbf{R}_A$ and $\mathbf{R}_B$ are the member end forces at *A* and *B*, which are the forces exerted by the adjoining member for the equilibrium of the member. We will take here

$$\mathbf{P'}_A=\mathbf{P'}_B=\mathbf{P'}_C=0$$

which means that there are no point loads applied.

The above relations written with the help of the elements of the matrices appearing in (a_{12}) and (a_{13}) become,

$$\left[\begin{array}{cc|cccc} 731.95 & -69.08 & -676.4 & -184.8 & -55.55 & 16.66 \\ -69.08 & 24.89 & 85.75 & 16.075 & -16.66 & 3.33 \\ \hline -676.4 & 85.75 & 676.4 & 184.8 & 0 & 0 \\ -184.8 & 16.075 & 184.8 & 57.85 & 0 & 0 \\ -55.55 & -16.66 & 0 & 0 & 55.55 & -16.66 \\ 16.66 & 3.33 & 0 & 0 & -16.66 & 6.66 \end{array}\right] \left\{\begin{array}{c} \mathbf{U}_C \\ \hline \mathbf{U}_A \\ \hline \mathbf{U}_B \end{array}\right\}$$

$$=\left\{\begin{array}{c} 0.47 \\ 0.022 \\ \hline 0.23+R_1 \\ 0.02+R_2 \\ 0.3+R_3 \\ -0.03+R_4 \end{array}\right\} \qquad (a_{16})$$

These equations are partitioned as follows :

$$\begin{bmatrix} \mathbf{K}_{II} & \mathbf{K}_{IE} \\ \mathbf{K}_{EI} & \mathbf{K}_{EE} \end{bmatrix} \begin{Bmatrix} \mathbf{U}_I \\ \mathbf{U}_E \end{Bmatrix} = \begin{Bmatrix} \mathbf{P}_I \\ \mathbf{P}_E+\mathbf{R}_E \end{Bmatrix} \qquad (a_{17})$$

Here $\mathbf{U}_I\equiv\mathbf{U}_C$; $\mathbf{U}_E=\{\mathbf{U}_A,\ \mathbf{U}_B\}$ etc. The internal joint $\mathbf{U}_I$ can now be eliminated and the reduced set (8.58) obtained. Presently,

$$\mathbf{K}_{II}^{-1}=\begin{bmatrix} 731.95 & -69.08 \\ -69.08 & 24.89 \end{bmatrix}^{-1}=\begin{bmatrix} 0.00185 & .005137 \\ 0.005137 & 0.0544 \end{bmatrix} \qquad (a_{18})$$

$\mathbf{K}_{II}^{-1}\ \mathbf{K}_{IE}=$

$$\begin{bmatrix} 0.00185 & 0.005137 \\ 0.005137 & 0.0544 \end{bmatrix} \begin{bmatrix} -676.4 & -184.8 & -55.55 & 16.66 \\ 85.75 & 16.075 & 16.66 & 3.33 \end{bmatrix}$$

$$=\begin{bmatrix} -0.8108 & -0.2593 & -0.1883 & 0.0479 \\ 1.1901 & -0.0748 & -1.1921 & 0.2671 \end{bmatrix} \qquad (a_{19})$$

Premultiplying the above by $\mathbf{K}_{EI}$ (which is $\mathbf{K}_{IE}^t$) we obtain

$$\mathbf{K}_{EI}\,\mathbf{K}_{II}^{-1}\,\mathbf{K}_{IE}=\begin{bmatrix} 651.176 & 168.976 & 25.143 & -9.52 \\ 168.96 & 46.716 & 15.634 & -4.567 \\ 25.209 & 15.74 & 30.329 & -7.1155 \\ -9.546 & -4.571 & -7.112 & 1.689 \end{bmatrix} \quad (a_{20})$$

Thus the modified stiffness $\mathbf{K}_{EE}^*$, as per (8.59) is

$$\mathbf{K}_{EE}^*=\left[\begin{array}{cc|cc} 25.224 & 15.824 & -25.143 & 9.542 \\ 15.84 & 11.134 & -15.634 & 4.569 \\ \hline -25.375 & -15.74 & 25.375 & -9.542 \\ 9.542 & 4.569 & -9.542 & 4.977 \end{array}\right] \quad (a_{21})$$

Similarly the modified equivalent load term, as per Eq. (8.60), reduces to

$$\mathbf{P}^*_F=\begin{Bmatrix} 0.585 \\ 0.1435 \\ 0.415 \\ -0.0584 \end{Bmatrix} \quad (a_{22})$$

Thus the required modified stiffness relations (Eq. 8.61) for member AB, (after introducing the multipliers EI/L^3, etc.) are as follows :

$$\begin{Bmatrix} Q_A \\ M_A \\ \hline Q_B \\ M_B \end{Bmatrix}=\frac{EI}{L}\left[\begin{array}{cc|cc} 25.375/L^2 & 15.74/L & -25.375/L^2 & 9.542/L \\ 15.74/L & 11.134 & -15.74/L & 4.569 \\ \hline -25.375/L^2 & -15.74/L & 25.375/L^2 & -9.542/L \\ 9.542/L & 4.569 & 9.542/L & 4.977 \end{array}\right]\begin{Bmatrix} v_A \\ \theta_A \\ \hline v_A \\ \theta_B \end{Bmatrix}+\begin{Bmatrix} -0.585 \\ -0.1435/L \\ -0.415 \\ 0.0584/L \end{Bmatrix} qL$$

Here, some of the terms have been averaged out, since there are some rounding errors in the terms of (a_{20}) and (a_{21}), both of which should show perfect symmetry. Observe that relations (a_6) *should hold here also* for the member AB as a whole in which $a=11.134$, $b=4.569$ and $c=4.977$.

8.7 STATICO–GEOMETRIC ANALOGY

We will now consider an analogy which can be observed in the structure of the equations of the force and displacement methods, a matter of some academic interest.

The set of (internal) displacements $\mathbf{U}_I$ of the previous section are sometimes termed 'inactive' variables, and are eliminated *before* the structural assembly. The word 'kinematic redundancy' as introduced by Argyris [7] can also be used. There is a close analogy between the equations of the force and the displacement method, and these definitions were introduced in this context.

Consider the particular case of the above discussion, in which we concentrate on the stiffness of the member of Fig. 8.7 with end forces $\mathbf{R}_A$ and $\mathbf{R}_B$ alone without any distributed loads or point loads on the member. In Eq. (8.56) we have $\mathbf{P}\equiv\mathbf{0}$. Thus Eq. (8.56) are

$$\begin{bmatrix} \mathbf{K}_{II} & \mathbf{K}_{IE} \\ \mathbf{K}_{EI} & \mathbf{K}_{EE} \end{bmatrix} \begin{Bmatrix} \mathbf{U}_I \\ \mathbf{U}_E \end{Bmatrix} = \begin{Bmatrix} \mathbf{0} \\ \mathbf{P}_E \end{Bmatrix} \tag{8.64}$$

Here $\mathbf{R}_A$ are $\mathbf{R}_B$ are member end forces of the curved member *AB*. Elimination of $\mathbf{U}_I$ only modifies $\mathbf{K}_{EE}$ to $\mathbf{K}_{EE}^*$ as in Eq. (8.59). The right hand load term remains unaffected on account of the fact that $\mathbf{P}_I\equiv\mathbf{0}$. (See Eq. (8.60), Thus we have the relations,

$$\mathbf{K}^*_{EE}\,\mathbf{U}_E=\mathbf{P}_E \tag{8.65}$$

The stiffness matrix $\mathbf{K}^*_{EE}$ is thc so called 'condensed' stiffness matrix. Since the load $\mathbf{P}_I\equiv\mathbf{0}$, the corresponding variables $\mathbf{U}_I$ are in a way superfluous or say 'redundant'. The process involved in obtaining $\mathbf{K}^*$ is sometimes called 'condensation' of stiffness matrix.

To consider the analogy with the force method, we begin by considering the expression for member deformations in terms of joint displacements. Presently for the frame of Fig. 8.7.

$$\mathbf{V}=\begin{Bmatrix} \mathbf{v}_A \\ \mathbf{v}_B \\ \mathbf{v}_C \end{Bmatrix}=\left[\begin{array}{cc|cc} \mathbf{A}_{2A} & \mathbf{0} & \mathbf{A}_{1A} & \mathbf{O} \\ \mathbf{A}_{1B} & \mathbf{A}_{2B} & \mathbf{O} & \mathbf{O} \\ \mathbf{O} & \mathbf{A}_{1C} & \mathbf{O} & \mathbf{A}_{2C} \end{array}\right] \begin{Bmatrix} \mathbf{U}_C \\ \mathbf{U}_D \\ \hline \mathbf{U}_A \\ \mathbf{U}_B \end{Bmatrix}=\mathbf{AU} \tag{8.66}$$

We can partition this as shown, and write

$$\mathbf{V}=[\mathbf{A}_I \,\vdots\, \mathbf{A}_E]\begin{Bmatrix} \mathbf{U}_I \\ \hline \mathbf{U}_E \end{Bmatrix}=\mathbf{A}_I\,\mathbf{U}_I+\mathbf{A}_E\,\mathbf{U}_E \tag{8.67}$$

This expression is analogous to the following expression used in force method : (See Eq. 7.1)

$$\mathbf{S} = \mathbf{bR}+\mathbf{BP} = [\mathbf{b}\ \ \mathbf{B}]\begin{Bmatrix}\mathbf{R}\\ \mathbf{P}\end{Bmatrix} \tag{8.68}$$

Here the term **BP** is written after the term **bR** to match with the quantities in Eq. (8.67). The stiffness matrix is the same as the one appearing in Eq. (8.64). Thus

$$\mathbf{A}^t\mathbf{KA} = \begin{bmatrix}\mathbf{A}_I{}^t\\ \mathbf{A}_E{}^t\end{bmatrix}\mathbf{K}[\mathbf{A}_I\ \ \mathbf{A}_E] = \begin{bmatrix}\mathbf{A}_I{}^t\,\mathbf{KA}_I & \mathbf{A}_I{}^t\,\mathbf{KA}_E\\ \mathbf{A}_E{}^t\,\mathbf{KA}_I & \mathbf{A}_E{}^t\,\mathbf{KA}_E\end{bmatrix} = \begin{bmatrix}\mathbf{K}_{II} & \mathbf{K}_{IE}\\ \mathbf{K}_{EI} & \mathbf{K}_{EE}\end{bmatrix} \tag{8.69}$$

Thus we will write Eq. (8.64) as

$$\begin{bmatrix}\mathbf{A}_I{}^t\,\mathbf{K}\,\mathbf{A}_I & \mathbf{A}_I{}^t\,\mathbf{K}\,\mathbf{A}_E\\ \mathbf{A}_E{}^t\,\mathbf{K}\,\mathbf{A}_I & \mathbf{A}_E{}^t\,\mathbf{K}\,\mathbf{A}_E\end{bmatrix}\begin{Bmatrix}\mathbf{U}_I\\ \mathbf{U}_E\end{Bmatrix} = \begin{Bmatrix}\mathbf{0}\\ \mathbf{P}_E\end{Bmatrix} \tag{8.70}$$

Now let us consider the flexibility relations corresponding to the internal forces taken in the form (8.68). Let the displacements corresponding to **R** be $\mathbf{U}_R$, and those due to **P** be say $\mathbf{U}_P$. Recall that **R** constitute *force pairs* and the generalized displacement for this system is zero since the member ends at the forcepairs are continuous. By virtual work considerations, Eq. (7.5), we obtain $\mathbf{U}_R=\mathbf{b}^t\,\mathbf{f}\,\mathbf{S}$ and $\mathbf{U}_P=\mathbf{B}^t\,\mathbf{f}\,\mathbf{S}$ with **f** as the vector of member flexibilities. Thus

$$\begin{Bmatrix}\mathbf{b}^t\\ \mathbf{B}^t\end{Bmatrix}\mathbf{f}\ \mathbf{S} = \begin{Bmatrix}\mathbf{0}\\ \mathbf{U}_P\end{Bmatrix}$$

This, in view of (8.68) becomes

$$\begin{Bmatrix}\mathbf{b}^t\\ \mathbf{B}^t\end{Bmatrix}\mathbf{f}\ [\mathbf{b}\ \ \mathbf{B}]\begin{Bmatrix}\mathbf{R}\\ \mathbf{P}\end{Bmatrix} = \begin{Bmatrix}\mathbf{0}\\ \mathbf{U}_P\end{Bmatrix}$$

or

$$\begin{bmatrix}\mathbf{b}^t\,\mathbf{f}\,\mathbf{b} & \mathbf{b}^t\,\mathbf{f}\,\mathbf{B}\\ \mathbf{B}^t\,\mathbf{f}\,\mathbf{b} & \mathbf{B}^t\,\mathbf{f}\,\mathbf{B}\end{bmatrix}\begin{Bmatrix}\mathbf{R}\\ \mathbf{P}\end{Bmatrix} = \begin{Bmatrix}\mathbf{0}\\ \mathbf{U}_P\end{Bmatrix} \tag{8.71}$$

This set is analogous to the set (8.70). Elimination of the internal nodes $\mathbf{U}_I$ in (8.70) and of **R** in (8.71) are analogous. In the force method the displacements corresponding to the redundants are zero, and therefore to invoke an analogy in the displacement method the displacement variables whose corresponding forces are zero, can be termed 'kinematic redundants'. They are in a way 'redundant' displacement variables which are eliminated just as the redundants force variables are introduced and eliminated.

There are other similar analogies which can be helpful. Recall the use of the word kinematically determinate (or basic) system to the first stage of analysis in the displacement method in which the effects of distributed loads on members are replaced by equivalent loads at joints (see Sec. 1.5). This is analogous to the statically determinate (basic) system, where the redundants are set equal to zero, and displacements of releases are computed due to loads alone. Reference [7] gives a 'dictionary' of analogies between force and displacement methods some of which are useful.

8.8 AXIAL LOAD EFFECTS

In tall buildings the effect of axial loads in columns assumes importance as they influence the load carrying capacity of the frame. There are two essential problems in structures of this type in which effect of axial loads can be significant:

(a) Beam-column effect

(b) Critical load estimate.

The first is the usual *equilibrium problem* in which the internal forces are to be determined by taking the effect of axial force on the stiffness of the member. The approach is to modify the stiffness for axial load and to employ the displacement method. However except in a determinate frame the axial forces are not known *a priori*. An iterative procedure is therefore used to estimate the axial force and reestimate it by a subsequent analysis performed on the basis of the assumed axial forces.

The second problem is the estimation of the *critical* (*buckling*) load. The magnitude of the axial forces at which the given frame can become unstable is required to be determined. At the instant of instability the frame looses its stiffness. Thus the problem becomes one of investigating the condition at which the frame as a whole looses stiffness. This loss of stiffness arises on account of any or some of its members loosing the stiffness. We will briefly discuss these issues in this section.

a. Stiffness of Beam

Consider the beam shown in Fig. 8.8. The stiffness of this was derived in Example 5.2 using the governing differential equation. The contents of these may be recalled here, in particular the feature of considering the equilibrium of a differential element in a deformed configuration,* and obtaining Eq. (a). The stiffness relations (a_5) were derived using stability functions (a_{12}). To rewrite, with slightly different form for the stability functions (a_{12}), (a_{13}),

$$\begin{Bmatrix} M_1 \\ M_2 \end{Bmatrix} = \begin{bmatrix} \frac{4EI}{L}\psi_1 & \frac{2EI}{L}\psi_2 \\ \frac{2EI}{L}\psi_2 & \frac{4EI}{L}\psi_1 \end{bmatrix} \begin{Bmatrix} \varphi_1 \\ \varphi_2 \end{Bmatrix} \tag{8.72}$$

*Recall the elementary theory of Euler buckling load analysis of a column where we derive the differential equation in deformed condition of the column.

with

$$\psi_1 = \frac{\gamma}{4}\ \frac{1-2\gamma \cot 2\gamma}{\tan \gamma-\gamma}\ ;\ \psi_2 = \frac{\gamma}{2}\ \frac{2\gamma \operatorname{cosec} 2\gamma-1}{\tan \gamma-\gamma} \tag{8.73}$$

$$\gamma = \frac{\beta L}{2} = \sqrt{\frac{PL^2}{4EI}} = \frac{\pi}{2}\sqrt{\frac{P}{P_E}}\ ;\ P_E = \frac{\pi^2 EI}{L^2}$$

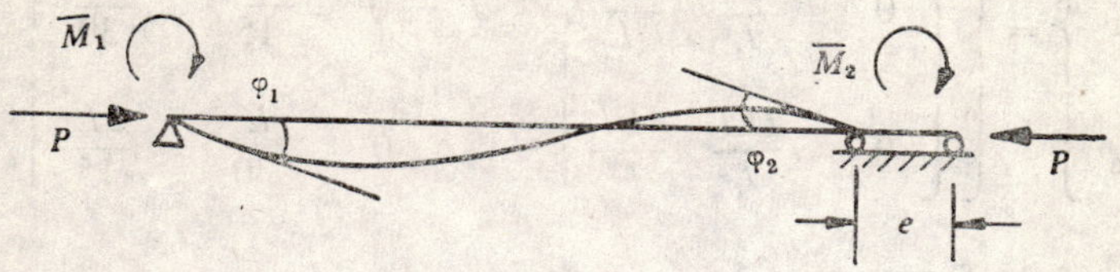

Fig. 8.8

Note that here P is the axial compressive force ; in other words $P=-S$. (When the force S is positive γ is imaginary and the functions will be as in Eq. (a_{15}) of Example 5.2.)

Basic Stiffness

The complete basic stiffness relation for the beam column is taken as

$$\mathbf{S} = \begin{Bmatrix} S \\ M_1 \\ M_2 \end{Bmatrix} = \begin{bmatrix} \frac{AE}{L} & 0 & 0 \\ 0 & \frac{4EI}{L}\psi_1 & \frac{2EI}{L}\psi_2 \\ 0 & \frac{2EI}{L}\psi_1 & \frac{4EI}{L}\psi_2 \end{bmatrix} \begin{Bmatrix} e \\ \varphi_1 \\ \varphi_2 \end{Bmatrix} = \mathbf{kv} \tag{8.74}$$

Here the stiffness relation for the axial load is left unchanged. Flexure however also affects its stiffness relation. But it would be necessary to incorporate the effect only in a proper large displacement analysis. The critical load analysis only investigates the stability of the undeformed configuration of the structure as a limit of the applicability of small displacements.

In the analysis of frames we can sometimes approximate the stability functions ψ_1 and ψ_2. For smaller values of γ we can approximate these as follows:

$$\psi_1 = 1-\frac{PL^2}{30\,EI} = 1+\frac{SL^2}{30\,EI}$$

$$\psi_2 = 1+\frac{PL^2}{60\,EI} = 1-\frac{SL^2}{60\,EI} \tag{8.75}$$

It can be shown that this approximation corresponds to a cubic displacement shape for the bending. Substituting these in (8.74) we can now write,

$$\begin{Bmatrix} S \\ M_1 \\ M_3 \end{Bmatrix} = \left[\begin{bmatrix} \frac{AE}{L} & 0 & 0 \\ 0 & \frac{4EL}{L} & \frac{2EI}{L} \\ 0 & \frac{2EI}{L} & \frac{4EI}{L} \end{bmatrix} + P \begin{bmatrix} 0 & 0 & 0 \\ 0 & -\frac{2L}{15} & \frac{L}{30} \\ 0 & \frac{L}{30} & -\frac{2L}{15} \end{bmatrix} \right] \begin{Bmatrix} e \\ \varphi_1 \\ \varphi_2 \end{Bmatrix} \tag{8·76}$$

This is of the form

$$\mathbf{S} = [\mathbf{k}_0 + \mathbf{k}_G]\mathbf{v}$$

General Stiffness

To obtain the general stiffness between $\overline{\mathbf{P}}$ and $\overline{\mathbf{u}}$ of the member we first obtain the kinematic matrix $\overline{\mathbf{A}}$, Eq. (3.3) ; to now rewrite, as $\overline{\mathbf{A}}_0$,

$$\mathbf{e} = \begin{Bmatrix} e \\ \varphi_1 \\ \varphi_2 \end{Bmatrix} = \left[\begin{array}{ccc|ccc} -1 & 0 & 0 & 1 & 0 & 0 \\ 0 & \frac{1}{L} & 1 & 0 & -\frac{1}{L} & 0 \\ 0 & \frac{1}{L} & 1 & 0 & -\frac{1}{L} & 1 \end{array} \right] \begin{Bmatrix} \overline{u}_1 \\ \overline{v}_1 \\ \overline{\theta}_1 \\ \cdots\cdots \\ \overline{u}_2 \\ \overline{v}_2 \\ \overline{\theta}_2 \end{Bmatrix}$$

$$= \overline{\mathbf{A}}_0 \ \overline{\mathbf{u}} \tag{8.77}$$

In obtaining the equation of equilibrium we take into account the *additional effect* of the relative transverse deflection ($\overline{v}_2 - \overline{v}_1$). There would be additional shear forces to equilibrate the moment caused. From Fig. 8.9* we obtain the additional shear forces (upward at 1 and downward at 2),

$$\frac{\overline{P}_2(\overline{v}_2 - \overline{v}_1)}{L} = S\frac{\overline{v}_2 - \overline{v}_1}{L} = -P\frac{\overline{v}_2 - \overline{v}_1}{L} \tag{a$_1$}$$

*Note that in Fig. 8.9, $\overline{P}_2$ is shown. Recall $\overline{P}_2 = S = -P$.

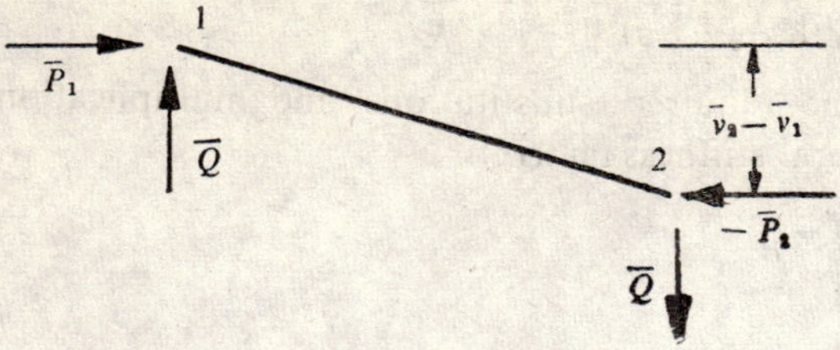

Fig. 8. 9

Thus we write the transformations

$$
\begin{Bmatrix} \bar{P}_1 \\ \bar{Q}_1 \\ \bar{M}_1 \\ \hline \bar{P}_2 \\ \bar{Q}_2 \\ \bar{M}_2 \end{Bmatrix} = \begin{bmatrix} -1 & 0 & 0 \\ 0 & \frac{1}{L} & \frac{1}{L} \\ 0 & 1 & 0 \\ \hline 1 & 0 & 0 \\ 0 & -\frac{1}{L} & -\frac{1}{L} \\ 0 & 0 & 1 \end{bmatrix} \begin{Bmatrix} S \\ M_1 \\ M_2 \end{Bmatrix}
$$

$$
+ \begin{bmatrix} 0 & 0 & 0 & 0 & 0 & 0 \\ 0 & -\frac{P}{L} & 0 & 0 & \frac{P}{L} & 0 \\ 0 & 0 & 0 & 0 & 0 & 0 \\ \hline 0 & 0 & 0 & 0 & 0 & 0 \\ 0 & \frac{P}{L} & 0 & 0 & -\frac{P}{L} & 0 \\ 0 & 0 & 0 & 0 & 0 & 0 \end{bmatrix} \begin{Bmatrix} \bar{u}_1 \\ \bar{v}_1 \\ \bar{\theta}_1 \\ \bar{u}_2 \\ \bar{v}_2 \\ \bar{\theta}_2 \end{Bmatrix} \tag{8.78}
$$

This is of the form $\qquad \bar{\mathbf{P}} = \bar{\mathbf{A}}_0{}^t\,\mathbf{S} + \bar{\mathbf{K}}_G\ \bar{\mathbf{u}}$

Here $\bar{\mathbf{K}}_G$ is the additional matrix which depends on internal force S. This is termed *initial stress matrix or geometric matrix*. The matrix $\mathbf{k}_G$ appearing in Eq. (8.76) also belongs to this class of matrices. Thus combining (8.74, 8.77 and 8.78) we get

$$\overline{\mathbf{P}} = (\overline{\mathbf{A}}_0{}^t\ \mathbf{k}\ \mathbf{A}_0 + \overline{\mathbf{K}}_G)\ \overline{\mathbf{u}} = \overline{\mathbf{K}}\ \overline{\mathbf{u}} \tag{8.79}$$

On carrying out the required substitutions and multiplications we obtain the following general stiffness matrix:

$$\overline{\mathrm{P}} = \left\{ \begin{array}{c} \overline{P}_1 \\ \overline{Q}_1 \\ \overline{M}_1 \\ \cdots\cdots\cdots \\ \overline{P}_2 \\ \overline{Q}_2 \\ \overline{M}_2 \end{array} \right\}$$

$$= \left[\begin{array}{ccc|ccc} \frac{AE}{L} & 0 & 0 & -\frac{AE}{L} & 0 & 0 \\ & \frac{12EI}{L^3}\gamma_1 & \frac{6EI}{L^2}\gamma_2 & 0 & -\frac{12EI}{L^3}\gamma_1 & \frac{6EI}{L^2}\gamma_2 \\ & & \frac{4EI}{L}\gamma_3 & 0 & -\frac{6EI}{L^2}\gamma_2 & \frac{2EI}{L}\gamma_4 \\ \hline & & & \frac{AL}{L} & 0 & 0 \\ & Sym. & & & \frac{12EI}{L^3}\gamma & -\frac{6EI}{L^2}\gamma_2 \\ & & & & & \frac{4EI}{L}\gamma_3 \end{array} \right] \left\{ \begin{array}{c} \overline{u}_1 \\ \overline{v}_1 \\ \overline{\theta}_1 \\ \overline{u}_2 \\ \overline{v}_2 \\ \overline{\theta}_2 \end{array} \right\} \tag{8.80}$$

where

$$\gamma_1 = \gamma_2\ \gamma \cot\gamma\ ; \quad \gamma_2 = \tfrac{1}{3}\frac{\gamma^2}{1-\gamma\cot\gamma} \tag{8.81}$$

$$\gamma_3 = \tfrac{3}{4}\gamma_2 + \tfrac{1}{4}\gamma\cot\gamma\ ; \quad \gamma_4 = \tfrac{3}{4}\gamma_2 - \tfrac{1}{2}\gamma\cot\gamma$$

Some algebra is involved in arriving at these forms. Terms containing

2γ are replaced by those involving γ alone. In terms of $k_{11} = \dfrac{4EI}{L}\,\psi_1$; $k_{12} = \dfrac{2EI}{L}\,\psi_2$ we have,

$$\begin{aligned}
\frac{12EI}{L^3}\,\gamma_1 &= \frac{2k_{11}+k_{12}}{L^2} - \frac{P}{L} = \frac{4EI}{L^3}\,(2\psi_1+\psi_2-\gamma^2)\\
\frac{6EI}{L^2}\,\gamma_2 &= \frac{k_{11}+k_{12}}{L} = \frac{2EI}{L^2}\,(2\psi_1+\psi_2)\\
\gamma_3 = \psi_1 &\equiv \frac{\gamma}{4}\;\frac{1+\gamma\;\tan\gamma-\gamma\cot\gamma}{\tan\gamma-\gamma} \qquad (8.82)\\
\gamma_4 = \psi_2 &= \frac{\gamma}{2}\;\frac{\gamma\,(\tan\gamma+\cot\gamma)-1}{\tan\gamma-\gamma}
\end{aligned}$$

Stability functions (8.81) are the ones used by Livesely [6] and Merchant [9].

Instead of the exact stiffness if we use the one in (8.76) for **k** we obtain, for small values of γ, the following expressions for γ_1, γ_2, γ_3 and γ_4 :

$$\begin{aligned}
\gamma_1 &= \left(1-\frac{PL^2}{10EI}\right) = 1-\left(\frac{2}{5}\right)\gamma^2\\
\gamma_2 &= \left(1-\frac{PL^2}{60EI}\right) = 1-\left(\frac{1}{15}\right)\gamma^2 \qquad (8.83)\\
\gamma_3 &= \left(1-\frac{PL^2}{30EI}\right) = 1-\left(\frac{2}{15}\right)\gamma^2\\
\gamma_4 &= \left(1+\frac{PL^2}{60EI}\right) = 1+\left(\frac{1}{15}\right)\gamma^2
\end{aligned}$$

In this case we can separate out into the ordinary stiffness matrix $\overline{\mathbf{K}}_0$ and geometric matrix $\overline{\mathbf{K}}_G$

$$\overline{\mathbf{P}} = (\mathbf{K}_0 + \overline{\mathbf{K}}_G)\,\overline{\mathbf{u}} = \overline{\mathbf{K}}\,\overline{\mathbf{u}} \qquad (8.84)$$

Here $\overline{\mathbf{K}}_0$ is the usual matrix whose elements are as in Eq. (5.45) or what is equivalent, as in Eq. (8.80) in which γ_1, γ_2 etc. are treated as unity. The general geometric matrix $\overline{\mathbf{K}}_G$ is as given below :

$$\overline{\mathbf{K}}_G = \begin{bmatrix}
0 & 0 & 0 & 0 & 0 & 0\\
 & -\dfrac{6}{5L} & -\dfrac{1}{10} & 0 & \dfrac{6}{5L} & -\dfrac{1}{10}\\
 & & -\dfrac{2L}{15} & 0 & \dfrac{1}{10} & +\dfrac{L}{30}\\
 & & & 0 & 0 & 0\\
 & & Sym. & & -\dfrac{6}{5L} & \dfrac{1}{10}\\
 & & & & & -\dfrac{2L}{15}
\end{bmatrix} P \qquad (8.85)$$

Coordinate transformation

The general stiffness matrix in frame coordinates follows the same transformation, $\mathbf{T}\ \overline{\mathbf{K}}\ \mathbf{T}^t$

$$\mathbf{P} = \mathbf{K}\mathbf{U} = (\mathbf{K}_0 + \mathbf{K}_G)\,\mathbf{U} \tag{8.86}$$

with

$$\mathbf{K} = \mathbf{T}\overline{\mathbf{K}}\ \mathbf{T}^t\ ;\ \mathbf{K}_0 = \mathbf{T}\,\overline{\mathbf{K}}_0\,\mathbf{T}^t\ ;\ \mathbf{K}_G = \mathbf{T}\ \overline{\mathbf{K}}_G\,\mathbf{T}^t \tag{8.87}$$

b. ANALYSIS WITH BEAM COLUMN EFFECTS

The equations of equilibrium for the structure as a whole assembled according to the usual method outlined in Chapter 8, will be of the form

$$\mathbf{KU} = \mathbf{P} \tag{8.88}$$

Since, the axial forces in the members are unknown, we start with an approximate guess of their magnitudes and improve it by iteration. Thus assembling the ordinary stiffness matrix $\mathbf{K}_0$ an initial set $\mathbf{U}_0$ are obtained from

$$\mathbf{K}_0\mathbf{U}_0 = \mathbf{P} \tag{a$_1$}$$

With these, the displacements and the internal forces $\mathbf{S}_0$ are determined. The stiffness matrices of individual members are revised incorporating the axial load effects, and the structure stiffness matrix is reassembled. A revised calculation of the displacements is made from

$$\mathbf{K}_1\mathbf{U}_1 = \mathbf{P} \tag{a$_2$}$$

and the resulting axial forces are recalculated. If these are still found to be different from the assumed ones, beyond a prescribed tolerable limit, the stiffness matrix is revised and the process of calculating the new displacements is repeated. Generally the process converges with two or three iterations. We will be explaining this in an example below.

EXAMPLE 8.5

We will analyse the frame of Example 7.3, considering now the effects of axial loads on the stiffness. We will take $EI_0 = 30$ kN m^2.

Thus (see Example 8.2 Table (a)),

$$EI_A = 12.5\,EI_0 = 12{\cdot}5 \times 30 = 375 \text{ kN m}^2$$
$$EI_B = 8.0 \times 30 = 240 \text{ kN m}^2 \tag{a$_1$}$$

The first cycle of iteration is the ordinary elastic analysis. Thus from Example 8.2, we have, (see Eqs. (a_{10}), (a_{11})),

$$P_{1A} = 87.78 \text{ kN}; \quad Q_{1A} = -82.68 \text{ kN (upwards)}$$
$$P_{2B} = -91.38 \text{ kN (to the right)}. \tag{a$_2$}$$

These on resolution give the axial forces

$$S_A = -(87.7 \times 0.8 + 82.68 \times 0.6) = -119.83 \text{ kN}$$
$$S_B = -91.38 \text{ kN} \tag{a$_3$}$$

These are compressive, and therefore the axial compressive forces are

$$P_A = -S_A = 119.83 \text{ kN}; \quad P_B = 91.38 \text{ kN} \tag{a$_4$}$$

The Euler loads $P_E = \pi^2 EI/L^2$, for these members are

$$P_{EA} = \frac{\pi^2 \times 375}{5^2} = 148.04 \text{ kN}; \quad P_{EB} = \frac{\pi^2 \times 240}{4^2}$$
$$= 148.04 \text{ kN} \tag{a$_5$}$$

(These are same since $\frac{L}{r}$ ratio for both is the same, 20). Thus

$$\left(\frac{P}{P_E}\right)_A = \frac{119.83}{148.04} = 0.809; \quad \left(\frac{P}{P_E}\right)_B = \frac{91.38}{148.04}$$
$$= 0.617 \tag{a$_6$}$$

We can now compute the stability functions using Eqs. (8.81).

For member A:

$$\gamma = \frac{\pi}{2}\sqrt{0.809} = 1.413; \quad \cot\gamma = 0.1593$$

$$\gamma \cot\gamma = 1.413 \times 0.1593 = 0.2251$$
$$\gamma_2 = (1.413)^2/3\,(1-0.2251) = 0.8588 \tag{a$_7$}$$
$$\gamma_1 = 0.8588 \times 0.2251 = 0.1931$$
$$\gamma_3 = 0.75 \times 0.8588 + 0.25 \times 0.2251 = 0.7003$$

Similarly for member B:

$$\gamma = \frac{\pi}{2}\sqrt{0.617} = 1.234; \quad \gamma\cot\gamma = 0.4321$$

$$\gamma_2 = (1.234)^2/3\,(1-0.4321) = 0.8937 \tag{a$_8$}$$
$$\gamma_1 = 0.8937 \times 0.4321 = 0.3861$$
$$\gamma_3 = 0.75 \times 0.8937 + 0.25 \times 0.4321 = 0.7783$$

The stiffness coefficients, as per Eqs. (8.80) are according to notations of (8.12), for member A,

$$k_5 = 1200.0; \quad k_1\gamma_3 = 300 \times 0.7003 = 210.09$$
$$k_3\gamma_1 = 36 \times 0.1931 = 6.95$$
$$k_4\gamma_2 = 90 \times 0.8588 = 77.29 \tag{a$_9$}$$

And for member B,

$$k_5 = 1500.0;\ k_1\Upsilon_3 = 240 \times 0.7783 = 168.07$$
$$k_3\Upsilon_1 = 45 \times 0.3861 = 17.37 \qquad (a_{10})$$
$$k_4\Upsilon_2 = 90 \times 0.8937 = 80.43$$

We have to convert the above stiffnesses to those in frame coordinates. Thus for member A, (see Eqs. (a_6) of Example 8.2)

$$k_5c^2+k_3s^2 = 1200\times0.8^2+6.95\times0.6^2 = 770.5$$
$$(k_5-k_3)cs = (1200-6.95)(0.8)(-0.6) = -572.66$$
$$k_3c^2+k_5s^2 = 6.95\times0.64+1200\times0.36 = 436.448$$
$$k_4s = 77.29\times(-0.6) = -46.374 \qquad (a_{11})$$
$$k_4c = 77.29\times0.8 = 61.832.$$

There is no need of coordinate transformation for member B. Thus the assembled stiffness relations as per Eq. (a_4) of Example 8.2, are as follows :

$$\begin{bmatrix} 2270.5 & -572.66 & -46.374 \\ -572.66 & 453.818 & 18.598 \\ -46.374 & 18.598 & 378.16 \end{bmatrix} \begin{Bmatrix} u_B \\ v_B \\ \theta_B \end{Bmatrix} = \begin{Bmatrix} 3.6 \\ 88.8 \\ -29.0 \end{Bmatrix} \qquad (a_{12})$$

These can now be compared with Eq. (a_8) of Example 8.2, by noting that $EI_0=30$. Thus with this value of EI_0 Eq. (a_8) of Example 8.2 is

$$\begin{bmatrix} 2280.96 & -558.72 & -54.00 \\ -558.72 & 500.04 & 18.0 \\ -54.0 & 18.0 & 540.0 \end{bmatrix} \begin{Bmatrix} u_B \\ v_B \\ \theta_B \end{Bmatrix} = \begin{Bmatrix} 3.6 \\ 88.8 \\ -29.0 \end{Bmatrix} \qquad (a_{13})$$

We see that the stiffness coefficients are in general smaller in magnitude, with considerable decrease in the rotational stiffness of the joint.

On solution of Eq. (a_{12}) we obtain

$$\mathbf{u} = \{0.07349,\ 0.29178,\ -0.08203\}$$
$$= \frac{1}{EI_0}\left\{2.2047,\quad 8.7534,\quad 2.4609\right\} \qquad (a_{14})$$

These are an improvement to the solution given by Eq. (a_9) of Example 8.2. The member end forces are

$$\mathbf{P}_{2A} = \begin{bmatrix} 770.5 & -572.66 & -46.374 \\ -572.66 & +36.448 & -61.832 \\ -46.374 & -61.832 & 210.09 \end{bmatrix} \begin{Bmatrix} 0.07349 \\ 0.29178 \\ -0.08203 \end{Bmatrix} = \begin{Bmatrix} -106.66 \\ 90.33 \\ -38.681 \end{Bmatrix}$$
$$(a_{15})$$

$$\mathbf{P}_{1B}=\begin{bmatrix} 1500 & 0 & 0 \\ 0 & 17.37 & 80.43 \\ 0 & 80.43 & 17.37 \end{bmatrix}\begin{Bmatrix} 0.07349 \\ 0.29178 \\ -0.08203 \end{Bmatrix}=\begin{Bmatrix} 110.235 \\ -1.528 \\ 9.68 \end{Bmatrix} \qquad (a_{16})$$

Thus the axial compressive forces, determined as in Eq. (a_3) are

$$P_A = 110.66\times 0.8+90.33\times 0.6 = 142.726 \text{ kN}$$
$$P_B = 110.235 \text{ kN}. \qquad (a_{17})$$

Second Iteration

We now repeat the above steps using the axial forces of Eq. (a_{17})

For member A,

$$\frac{P}{P_E}=\frac{142.726}{148.04}=0.9641 \ ; \quad \gamma=\frac{\pi}{2}\sqrt{0.9641} = 1.542.$$
$$\gamma_1=0.03638 \ ; \quad \gamma_2=0.8289 \ ; \quad \gamma_3=0.6327 \qquad (a_{18})$$

For member B

$$\frac{P}{P_E}=0.745 \ ; \quad \gamma=1.3558$$
$$\gamma_1=0.2568 \ ; \quad \gamma_2=0.8675 \ ; \quad \gamma_3=0.7247 \qquad (a_{19})$$

The above calculations are straight forward and a computer subroutine can be written to compute these for a given value of (P/P_E). The stiffness coefficients are

$$k_{5A}=1200 \ ; \quad k_{4A}=90\times 0.8289=74.421$$
$$k_{3A}=36\times 0.03638=1.31 \ ; \quad k_1=300\times 0.6327=189.81.$$
$$k_{5B}=1500 \ ; \quad k_{4B}=90\times 0.8675=78.075 \qquad (a_{20})$$
$$k_{3B}=45\times 0.2568=11.556 \ ; \quad k_{1B}=240\times 0.7247=173.93.$$

The stiffnesses of member A, in global coordinates are

$$k_5c^2+k_3s^2=768.471 \ ; \quad k_3c^2+k_5s^2=432.84 \qquad (a_{21})$$
$$(k_5-k_3)cs=-575.371 \ ; \quad k_4s=-44.65 \ ; \quad k_4c=59.537$$

And the assembled stiffness relations are as follows :

$$\begin{bmatrix} 2268.47 & -575.37 & -44.65 \\ -575.37 & 444.39 & 18.538 \\ -44.65 & 18.538 & 363.74 \end{bmatrix}\begin{Bmatrix} u_B \\ v_B \\ \theta_B \end{Bmatrix}=\begin{Bmatrix} 3.6 \\ 88.8 \\ -29.0 \end{Bmatrix} \qquad (a_{22})$$

On solution, we obtain

$$\mathbf{u}=\{0.07665, \quad 0.30263, \quad -0.08547\} \qquad (a_{23})$$

which represent an improvement of the first solution obtained in Eq. (a_{14}). The member forces are

$$P_{2A}=\begin{bmatrix} 768.471 & -575.371 & -44.65 \\ -575.371 & 432.84 & -59.537 \\ -44.65 & -59.537 & 189.81 \end{bmatrix}\begin{Bmatrix} 0.07665 \\ 0.30263 \\ -0.08547 \end{Bmatrix} \quad (a_{24})$$

$$=\begin{Bmatrix} -110.40 \\ 91.97 \\ -37.663 \end{Bmatrix} \quad (a_{25})$$

$$P_{1B}=\begin{bmatrix} 1500 & 0 & 0 \\ 0 & 11.556 & 78.075 \\ 0 & 78.075 & 173.93 \end{bmatrix}\begin{Bmatrix} 0.07665 \\ 0.30263 \\ -0.08547 \end{Bmatrix}=\begin{Bmatrix} 114.975 \\ -3.176 \\ 8.762 \end{Bmatrix} \quad (a_{26})$$

Thus the axial forces are

$$P_A = 111.4\times0.8+91.97\times0.6 = 144.762$$
$$P_B = 114.975 \quad (a_{27})$$

Third Iteration

Comparing the trend of the changes (Eqs. a_4, a_{17}, a_{27})) we assume for the present iteration

$$P_A = 146.0\ ; \quad P_B \doteq 117.0 \quad (a_{28})$$

rather than the actual forces of Eq. (a_{27}). Such 'over estimates' are common in iterative procedures. For these values

$$\left(\frac{P}{P_E}\right)_A=\frac{146.0}{148.04}=0.986\ ; \quad \left(\frac{P}{P_E}\right)_B=\frac{117.0}{148.04} = 0.7903 \quad (a_{29})$$

For member A:

$$\gamma=1.56\ ; \quad \gamma_1=0.01397, \quad \gamma_2=0.8251$$
$$\gamma_3=0.62306 \quad (a_{30})$$

and for member B:

$$\gamma=1.3964\ ; \quad \gamma_1=0.2120\ ; \quad \gamma_2=0.8620 \quad (a_{31})$$
$$\gamma_3=0.7080$$

The assembled structure stiffness relations become,

$$\begin{bmatrix} 2268.18 & -575.76 & -44.556 \\ -575.76 & 441.94 & 18.17 \\ -44.556 & 18.17 & 356.84 \end{bmatrix} \begin{Bmatrix} u_B \\ v_B \\ \theta_B \end{Bmatrix} = \begin{Bmatrix} 3.6 \\ 88.8 \\ -29.0 \end{Bmatrix} \qquad (a_{32})$$

which can be compared with Eq. (a_{22}). The solution yields

$$\mathbf{u} = \{0.07738, \quad 0.30532, \quad -0.08715\}, \qquad (a_{33})$$

The member forces,

$$\begin{aligned} P_{2A} &= \{-112.46, \quad 92.64, \quad -37.87\} \\ P_{1B} &= \{116.07, \quad -3.84, \quad 8.88\} \end{aligned} \qquad (a_{34})$$

and the axial compressive forces,

$$\begin{aligned} P_A &= 112.46 \times 0.8 + 92.64 \times 0.6 = 145.55 \text{ kN} \\ P_B &= 116.07 \text{ kN} \end{aligned} \qquad (a_{35})$$

These forces are seen to agree with the forces assumed (Eq. (a_{28})) to within acceptable tolerance. Hence the iterations can be considered to have converged to the true answers, given by Eqs. (a_{33}) and (a_{34}).

c. CRITICAL LOAD ANALYSIS

The stiffness of a member becomes zero at certain values of the axial load. This is the critical load for the member. It can be seen from Eq. (8.83) that γ_1 becomes zero at $P = 10\,EI/L^2$, and γ_2 at $60\,EI/L^2$ etc. For a frame as a whole the stiffness matrix becomes singular when the critical load value is reached. The determinant of the stiffness matrix becomes zero at this value of the load.

Suppose that in a frame such as shown in Fig 8.10 (a) the *axial* loads are of a given magnitude

$$\mathbf{P}_a = \{P_A, P_B, P_C, \ldots\} \qquad (a_1)$$

The critical load analysis concerns with the problem of determining a critical load factor λ_c such that for

$$\lambda_c\,\mathbf{P}_a = \{\lambda_c P_A, \lambda_c P_B \cdots\} \qquad (a_2)$$

the stiffness of the structure becomes zero. In general, when the load factor is lesser than λ_c, (loads lesser than critical loads), we can solve for the displacements $\Delta\mathbf{u}$ for a given $\Delta\mathbf{Q}$ from Eq. (8.88):

$$[\mathbf{K}(\lambda)]\,\Delta\mathbf{u} = \Delta\mathbf{Q} \qquad (8.89)$$

Here $\Delta\mathbf{Q}$ can be considered as a set of small *disturbing* forces. The determinant of the matrix at this load level will be nonzero. But when the axial load level is increased to λ_c, the stiffness of the structure becomes zero; therefore

$$[\mathbf{K}(\lambda_c)]\ \Delta\,\mathbf{u} = \mathbf{0} \tag{8.90}$$

The problem therefore reduces to the one of solving for the lowest critical load value λ_c such that

$$\det\,[\mathbf{K}(\lambda_c)] = 0 \tag{8.91}$$

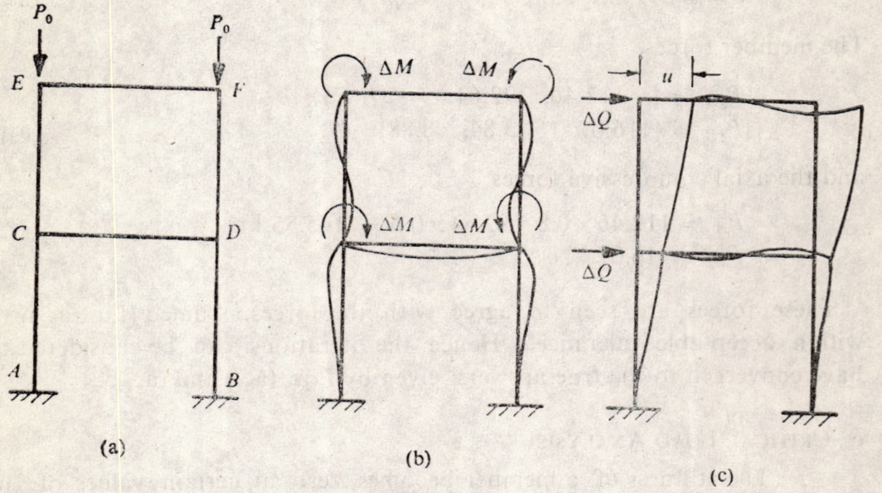

Fig. 8.10

Thus there are two possible approaches to determine the lowest value of λ_c. The first procedure is to test the value of the determinant to see if it is zero. There are standard computer procedures for evaluating the determinant or testing for the condition when it becomes zero.

The alternative procedure is to use (8.89) and obtain λ_c by an extrapolation procedure. $\Delta\mathbf{Q}$ would be a set of small disturbing forces taken on the basis of investigating a particular type of buckling mode. Thus for the frame of Fig. 8.10 (a) the forces shown in (b) correspond to a symmetric mode, while the forces shown at (c) correspond to the sway mode. For a given (small) set of disturbing forces the displacements $\Delta\mathbf{u}$ are computed for increasing values of λ. A plot of a typical displacement versus the load factor λ is obtained, by the series of calculations based on increasing λ. Such a plot would be as in Fig. 8.11. It would be advantageous to plot $(\Delta Q/u)$, the stiffness, versus λ and to obtain λ_c from the point where the curve meets the λ axis on extrapolation.

The second procedure is evidently easier to be incorporated in an elastostatic program. Thus if we have a program for elastic analysis we can modify it to be rerun with different stiffness matrices depending on the load factor λ. This procedure also provides the buckling mode. In many practical cases it is not very difficult to guess the possible mode shapes and the associated disturbing forces. The first procedure however requires a special algorithm to test the determinant of a matrix for positive nonzero values and what is important, to work back the buckling mode shapes.

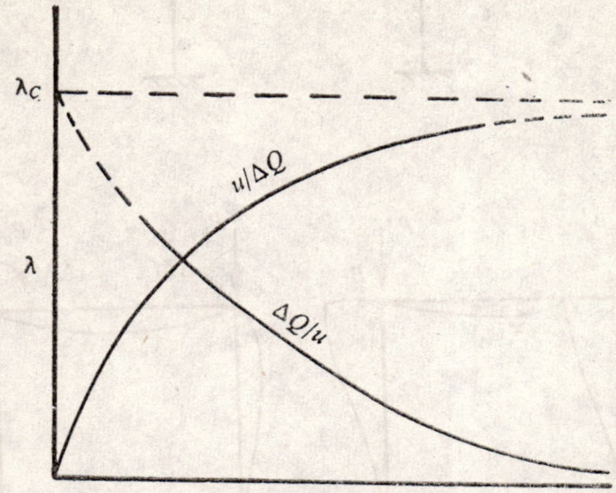

Fig. 8·11

EXAMPLE 8.6

Consider the portal frame shown in (a). This has six degrees of freedom, 3 at each of the joints A and B. Thus this frame has in general six buckling modes. We will simplify this by assuming that (a) axial deformations in members are neglected and (b) the compressive forces in the columns will always be P_o ; in other words the changes in the axial loads due to the different end moments of the beam are neglected. In such a case we will have three mode shapes corresponding to θ_A, θ_B and $u_A = u_B = u$.

We will further take, for the purpose of illustrating numerical example, that

$$\frac{EI_A}{L_A} = \frac{EI_B}{L_B} = \frac{EI}{L} \tag{a_1}$$

so that we have, in the case of antisymmetrical mode, (Fig. (b))

$$\theta_B = \theta_A \tag{a_2}$$

and in the case of symmetrical mode, (Fig. (c)),

$$\theta_B = -\theta_A \ ; \ u = 0 \tag{a_3}$$

We thus have only two buckling mode shapes as shown in Figs. (b) and (c). The first is the 'sway' mode and second can be called, for want of a better word, 'nonsway' mode.

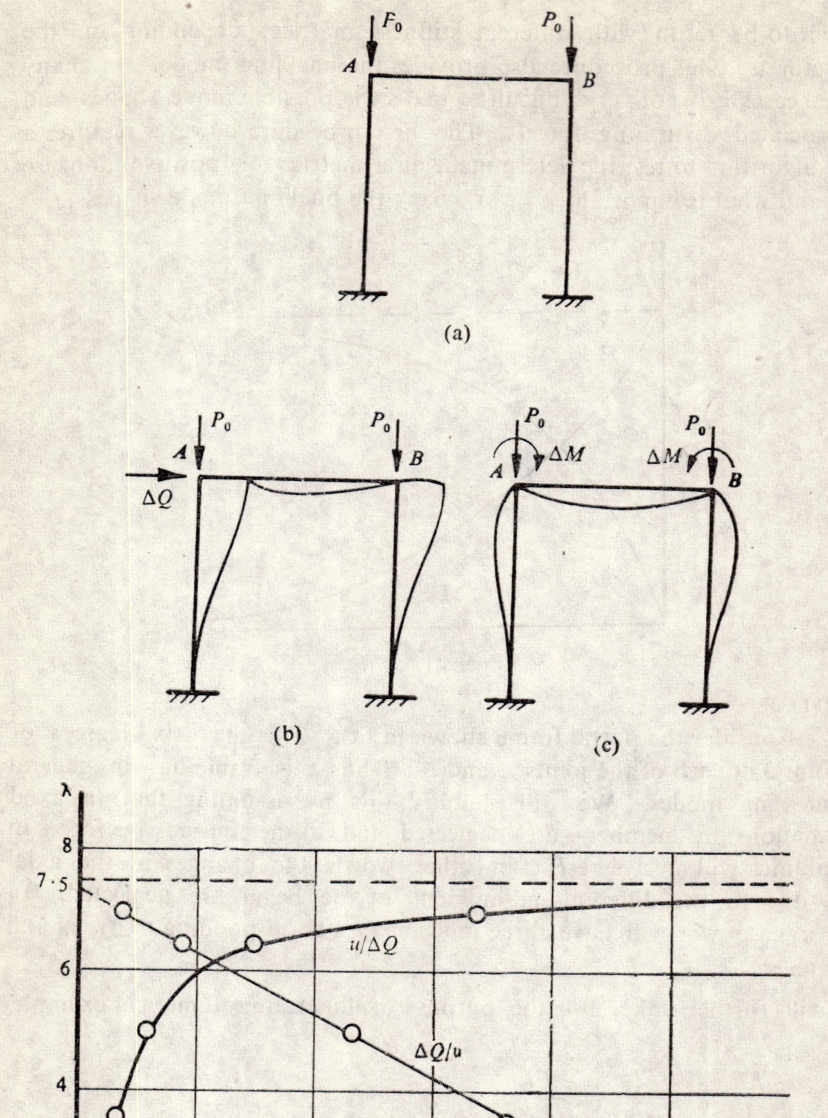

Fig. Example 8.6

Consider the assembly of the structure stiffness matrix. We have, for member A, (column)

$$\mathbf{K}_{11A}=\begin{bmatrix} k_3 & 0 & -k_4 \\ 0 & k_5 & 0 \\ -k_4 & 0 & k_1 \end{bmatrix} \tag{a_4}$$

where (See Eq. (8.12))

$$k_1=\frac{4EI}{L}\ ;\quad k_2=\frac{2EI}{L} \tag{a_5}$$

$$k_3=\frac{12EI}{L^3}\ ;\quad k_4=\frac{6EI}{L^2} \tag{a_6}$$

The above matrix is obtained by using $\mathbf{K}_{11}$ from Table 8.1 with cos $\alpha=1.0$; sin $\alpha=0.0$. The stiffness matrix is

$$\begin{bmatrix} \mathbf{K}_{11A}+\mathbf{K}_{11B} & \mathbf{K}_{12B} \\ \mathbf{K}^t_{12B} & \mathbf{K}_{11A}+\mathbf{K}_{22B} \end{bmatrix}\begin{Bmatrix} \mathbf{u}_A \\ \mathbf{u}_B \end{Bmatrix}=\begin{Bmatrix} \mathbf{P}_A \\ \mathbf{P}_B \end{Bmatrix} \tag{a_7}$$

Using $\mathbf{K}_{11A}$ given in (a_4) above, and $\mathbf{K}_{11B}$, $\mathbf{K}_{12B}$ etc. as given in Eq. (8.13), we can write the above relations in full as follows :

$$\left[\begin{array}{ccc|ccc} k_{3A}+k_{5B} & 0 & -k_{4A} & -k_{5B} & 0 & 0 \\ 0 & k_{5A}+k_{3B} & k_{4B} & 0 & -k_{3B} & k_{4B} \\ -k_{4A} & k_{4B} & k_{1A}+k_{1B} & 0 & -k_{4B} & k_{2B} \\ \hline -k_{5B} & 0 & 0 & k_{3A}+k_{5B} & 0 & -k_{4A} \\ 0 & -k_{3B} & -k_{4B} & 0 & k_{5A}+k_{3B} & -k_{4B} \\ 0 & k_{4B} & k_{2B} & -k_{4A} & -k_{4B} & k_{1A}+k_{1B} \end{array}\right]$$

$$\begin{Bmatrix} u_A \\ v_A \\ \theta_A \\ u_B \\ v_B \\ \theta_B \end{Bmatrix}=\begin{Bmatrix} P_A \\ Q_A \\ M_A \\ P_B \\ Q_B \\ M_B \end{Bmatrix} \tag{a_8}$$

We have in view of inextensional deformations for members

$$v_A=v_B=0 \tag{a_9}$$

$$u_A=u_B=u \tag{a_{10}}$$

We therefore exclude the second and fifth rows and columns, and add the fourth row to the first, to obtain

$$\mathbf{KU} = \begin{bmatrix} 2k_{3A} - k_{4A} & & -k_{4A} \\ -k_{4A} & k_{1A}+k_{1B} & k_{2B} \\ -k_{4A} & k_{2B} & k_{1A}+k_{1B} \end{bmatrix} \begin{Bmatrix} u \\ \theta_A \\ \theta_B \end{Bmatrix} = \begin{Bmatrix} P_B+P_C \\ M_B \\ M_C \end{Bmatrix} = \mathbf{P} \tag{a$_{11}$}$$

The first equation is the so called sway equation and the other two are the moment equilibrium equations.

We will, as said before, take the axial compressive load in the columns to always remain equal to P_0. Thus

$$k_{3A} = \frac{12EI}{L^3}\gamma_1\ ; \quad K_{4A} = \frac{6EI}{L^2}\gamma_2$$

$$k_{1A} = \frac{4EI}{L}\gamma_3 \tag{a$_{12}$}$$

with $\gamma = \sqrt{P_0L^2/4EI}$. We *ignore* the small possible axial force in the beam and take only the ordinary stiffnesses $k_3 = 12\ EI/L^3$ etc.

Let us now consider the possible buckling loads. The homogeneous set giving these is

$$\mathbf{KU} = \mathbf{0} \tag{a$_{13}$}$$

with **K** of Eq. (a$_{11}$). One way of solving the problem will be to determine the magnitudes of γ such that

$$\det |\mathbf{K}| = 0 \tag{a$_{14}$}$$

and for each value of γ to determine **U**. The above determinantal equation is in general a transcendental equation in γ :

$$f(\gamma) = 0 \tag{a$_{15}$}$$

This can be explicitly obtained for simple problems such as the present one. In general however the determinant has to be evaluated for a series of increasing values of γ and tested if it is zero. Presently we will follow an equivalent process in which we consider a nonhomogeneous set

$$\mathbf{KU} = \Delta\mathbf{Q} \tag{a$_{16}$}$$

instead of Eq. (a$_{13}$) where ΔQ is a predetermined set of disturbing forces. The value of **U** is evaluated for a series of increasing values of γ, such that **U** tend to very large values. The limiting value of γ when these displacements tend to become infinite is the buckling load corresponding to the assumed set of disturbing forces. By assuming different type of disturbing forces, such as those which can lead to sway or nonsway modes, we investigate the different buckling mode shapes and the associated buckling loads. This procedure has a physical appeal and is practical in the sense that we examine the modes of buckling we consider are likely.

(*a*) *Sway Mode*

Let us apply a set of disturbing forces

$$\Delta \mathbf{Q} = \{1, \quad 0, \quad 0\}\Delta Q \qquad (a_{17})$$

as shown in Fig. (b). We use this vector on the right hand side of Eq. (a_{16}).

We have, here and likewise, in any problem, made a first choice for γ and a guiding factor for increasing this in steps. Presently the lowest value of buckling load is $\pi^2 EI/4L^2$ which corresponds to the case where the beam is ignored and the column is a free cantilever. The other extreme is when the beam is rigid, in which case the effective length is $L/2$ and the buckling load is four times. Thus the actual buckling load can be between $\pi^2 EI/L^2$ and $\pi^2 EI/4L^2$.

Let us take $P_0=100$ and determine λ_c such that $\lambda_c P_0$ will be the buckling load. We will start with $\lambda=2.5$ and increment it gradually by observing the trend in the increase of displacements. We have, for the given frame,

$$\pi^2 EI/L^2 = \frac{\pi^2 \times 10{,}000}{10^2} = 986.96 \text{ kN}.$$

$$\text{Thus } \gamma = \frac{\pi}{2}\sqrt{\frac{2.5\times 100}{986.96}} = 0.7905 \qquad (a_{18})$$

Proceeding as required by Eq. (8.81), we obtain $\gamma_1=0.7487$; $\gamma_2=0.9570$; $\gamma_3=0.9133$ (a_{19})

$$k_{1A} = \frac{4\times 10{,}000}{10}\times 0.9133 = 3653.2$$

$$k_{3A} = \frac{12\times 10{,}000}{10^3}\times 0.7487 = 89.844 \qquad (a_{20})$$

$$k_{4A} = \frac{6\times 10{,}000}{10^2}\times 0.9590 = 574.2$$

$$k_{1B} = \frac{4\times 10{,}000}{40} = 4000.0$$

$$k_{2B} = 2000.0$$

And therefore the stiffness relations (a_{16}), with **K** of Eq.(a_{11}) and $\Delta \mathbf{Q}$ of Eq. (a_{17}) become

$$\mathbf{KU} = \begin{bmatrix} 178.688 & -574.2 & -574.2 \\ -574.2 & 7653.2 & 2000.0 \\ -574.2 & 2000.0 & 7653.2 \end{bmatrix} \begin{Bmatrix} u \\ \theta_A \\ \theta_B \end{Bmatrix} = \begin{Bmatrix} 1.0 \\ 0 \\ 0 \end{Bmatrix} \Delta Q \qquad (a_{21})$$

These on solution give

$$\mathbf{U} = \{9.0584, \quad 0.5387, \quad 0.5387\} \times 10^{-3}\, \Delta \mathbf{Q} \qquad (a_{22})$$

Observe that in view of equal stiffnesses of columns we always obtain $\theta_A=\theta_B$.

Let now λ be increased to 3.5 giving a load 3.5×100=350.0 kN. We have for this load

$$\gamma = \frac{\pi}{2}\sqrt{\frac{350}{986.96}} = 0.9354\ ;$$

$$\gamma_1 = 0.6484\ ;\ \gamma_2 = 0.9399\ ;\ \gamma_3 = 0.8773$$

$$k_{1A} = 4000 \times 0.8773 = 3309.2 \qquad (a_{23})$$

$$k_{3A} = 120 \times 0.6484 = 77.808$$

$$k_{4A} = 600 \times 0.9399 = 563.94$$

with k_{1B}, k_{2B} remaining the same. Thus the stiffness relations obtained would be

$$\begin{bmatrix} 155.616 & -563.94 & -563.94 \\ -563.94 & 7309.2 & 2000.0 \\ -563.9 & 2000.0 & 7309.2 \end{bmatrix} \begin{Bmatrix} u \\ \theta_A \\ \theta_B \end{Bmatrix} = \begin{Bmatrix} 1.0 \\ 0 \\ 0 \end{Bmatrix} \Delta Q \qquad (a_{24})$$

These on solution yield

$$U = \{11.459,\ \ 0.694,\ \ 0.694\} \times 10^{-3}\ \Delta Q \qquad (a_{25})$$

The next value of λ chosen is 5.0 since it can be felt that the increases are almost linear. Thus, for $\lambda P_0 = 5.0 \times 100$ we obtain =1.118, and proceeding as before,

$$\begin{bmatrix} 119.22 & -548.1 & -548.1 \\ -548.1 & 7284.4 & 2000.0 \\ -548.1 & 2000.0 & 7284.4 \end{bmatrix} \begin{Bmatrix} u \\ \theta_A \\ \theta_B \end{Bmatrix} = \begin{Bmatrix} 1.0 \\ 0 \\ 0 \end{Bmatrix} \Delta Q \qquad (a_{26})$$

These on solution give,

$$U = \{18.34,\ 1.083,\ 1.083\} \times 10^{-3}\ \Delta Q \qquad (a_{27})$$

We can check the *rate of increase* of deformations. The rate of increment in u for the load ranges 250 to 350 and 350 to 500 are seen to be

$$\frac{11.459-9.0584}{350-250} = .02 \qquad (a_{28})$$

$$\frac{18.3-11.3459}{500-350} = .045$$

This shows that the rate has almost doubled. In general, a computer code can be drawn to check the rate of increases in displacements and a convenient guide can be to limit the successive increases to within certain limits.

Let us choose $\lambda=6.5$. The stiffness relations for this load level will be found to be,

$$\begin{bmatrix} 82.784 & -531.78 & -531.78 \\ -531.78 & 7047.8 & 2000.0 \\ -531.78 & 2000.0 & 7047.8 \end{bmatrix} \begin{Bmatrix} u \\ \theta_A \\ \theta_A \end{Bmatrix} = \begin{Bmatrix} 1.0 \\ 0 \\ 0 \end{Bmatrix} \Delta Q \qquad (a_{29})$$

These have a solution

$$U = \{49.17,\ 2.89,\ 2.89\}\ 10^{-3}\Delta Q \qquad (a_{30})$$

It is seen that the increase in u has been at a rate of $\left(\dfrac{49.17-18.34}{650-500}\right)$ $=0.2$, which is five times the rate in the previous step. With $\lambda=7.0$ we obtain

$$\begin{bmatrix} 70.488 & -526.2 & -526.2 \\ -526.2 & 6965.9 & 2000.0 \\ -526.2 & 2000.0 & 6965.9 \end{bmatrix} \begin{Bmatrix} u \\ \theta_A \\ \theta_B \end{Bmatrix} = \begin{Bmatrix} 1.0 \\ 0 \\ 0 \end{Bmatrix} \Delta Q \qquad (a_{31})$$

with a solution

$$U = \{114.62,\ 6.727,\ 6.727,\} \times 10^{-3}\ \Delta Q \qquad (a_{32})$$

Since we have sufficient sets we can extrapolate the results. Fig (d) shows the plots of $\Delta Q/u$ and $u/\Delta Q$ against λ. The latter is the familiar load-displacement curve. The former which is a plot of the reciprocal of u against the load is more practical. It is seen that $\lambda=7.5$ would make $(\Delta Q/u)$ tend to zero which is equivalent to u tending to infinity. A computer subroutine can be written to extrapolate the values and obtain the limiting value of λ.

Having obtained $\lambda=7.5$ we assemble the stiffness relations and check. Presently for $\lambda=7.5$ we obtain

$$\begin{bmatrix} 57.6 & -515.52 & 515.52 \\ -515.52 & 6856.8 & 2000.0 \\ -515.52 & 2000.0 & 6856.6 \end{bmatrix} \begin{Bmatrix} u \\ \theta_A \\ \theta_B \end{Bmatrix} = \begin{Bmatrix} 1.0 \\ 0 \\ 0 \end{Bmatrix} \Delta Q \qquad (a_{33})$$

The solution obtained is

$$U = \{-41.867,\ -24.155,\ -24.155\} \times 10^{-3}\ \Delta Q \qquad (a_{34})$$

which shows that the signs of the displacements have changed. This indicates that the determinant of the stiffness matrix has changed sign, showing that it has crossed zero.

The above example demonstrates a method which can be seen to be a general one applicable to frames. The point to be noted is that the set of disturbing forces have to be chosen, which is expected to excite the frame into a buckling mode anticipated, such as a sway force here. Thus we can examine as many modes as we may intuitively feel are significant.

Symmetrical Mode

Let us now apply a set of disturbing forces

$$\Delta Q = \{O \quad \Delta M \quad -\Delta M\} \tag{a$_{35}$}$$

These are as shown in Fig. (c). The starting point for λ would be to consider the beam to be of negligible moment of inertia, in which case the column is fixed at one end and pinned at the other. This is a well known case having a buckling load of 20.193 EI/L^2 or $P/P_E=2.046$, The upper limit is when the beam is infinitely rigid; in such a case $P = 4\pi^2 EI/L^2 = 39.47\ (EI/L^2)$. Thus we can expect a value between 20.193 (EI/L^2) and 39.47 (EI/L^2)

Thus take $\lambda=2.0$, giving $\gamma=2.2360$

$$\gamma_1 = -1.0591; \quad \gamma_2 = 0.60507; \quad \gamma_3 = 0.015 \tag{a$_{36}$}$$

The stiffness coefficients are

$$\begin{aligned} k_{1A} &= 4000.0 \times 0.015 = 60.0 \\ k_{3A} &= 120 \times (-10591) = -127.09 \\ k_{4A} &= 600 \times 0.60507 = 363.04. \end{aligned} \tag{a$_{37}$}$$

And the stiffness relation becomes,

$$\begin{bmatrix} -254.18 & -363.04 & -363.04 \\ -363.04 & 4060.0 & 2000.0 \\ -363.04 & 2000.0 & 4060.0 \end{bmatrix} \begin{Bmatrix} u \\ \theta_A \\ \theta_B \end{Bmatrix} = \begin{Bmatrix} 0 \\ 1.0 \\ -1.0 \end{Bmatrix} \Delta M \tag{a$_{38}$}$$

These have a solution

$$U = \{0, 0.485, -0.485\}\ 10^{-3} \Delta M \tag{a$_{39}$}$$

As expected, the lateral sway displacement turns out to be zero and $\theta_B = -\theta_A$. This defines the symmetrical mode shape.

We can continue the solution as in the previous case. We give below the different trials:

$$\lambda=2.2,\ U=\{0, \quad 0.7027, \ -0.7097\} \times 10^{-3}\ \Delta M \tag{a$_{40}$}$$

$$\lambda=2.4,\ U=\{0, \quad 1.780, \ -1.780\} \times 10^{-3}\ \Delta M \tag{a$_{41}$}$$

$$\lambda=2.5,\ U=\{0, \ 10.9, \ -10.9\} \times 10^{-3}\ \Delta M \tag{a$_{42}$}$$

$$\lambda=2.525,\ U=\{0, -29.2, +28.2\} \times 10^{-3}\ \Delta M$$

A plot similar to the one shown in Fig. (d) indicates that $\lambda=2.5$ is the critical load factor.

8.9 PLASTIC ANALYSIS OF FRAMES*

We will briefly demonstrate the approach of the method of successive elastic solutions as applied to frames under gradually increasing

* It is presumed that the reader has some elementary ideas about plastic theory and the concept of plastic hinges from their structural design courses.

loads causing plastic flows at certain locations of the frame. The engineering theory reduces the problem to the one of determining a discrete number of unknowns, namely a certain number of plastic deformation parameters. Such a discretization has been rendered possible by the use of 'plastic hinge' idealization of the zones of plastic deformation. The structural analysis problem will be to determine in addition to the usual displacement unknowns, a certain number of plastic hinge rotations and in turn from these the internal force distributions within the frame. We will show how by modifying the stiffnesses of members we can apply the elastic analysis to obtain the elastoplastic behaviour. Such a process repeated with increments of loads will determine finally the *collapse load* of the frame. The reader might have been exposed to the simple procedures of a *direct* determination of the collapse load of a frame. The direct determination of this load without having to trace the complete history of the response of the structure is of interest in design. This knowledge alone is adequate for steel frames which possess considerable ductility. However in reinforced concrete frames the extent of plastic deformations is limited and therefore the estimate of deformations becomes an additional necessity. In such a situation one has to perform an *elastoplastic analysis* keeping the plastic deformations to within permissible limits.

a. Plastic Hinge

This is based on the idealization of the moment-curvature relation. This relationship depends in general on the stress strain curve and the properties of the section. The general response is as shown by the curve OD of Fig. 8.12(a). The initial portion is a straight line and the relationship is $\chi = M/EI$ where EI denotes the flexural rigidity of the section. As the moment is increased elastoplastic behaviour sets in. Extent of plasticity varies along the depth of the cross-section. As the moment is increased the element becomes more and more flexible. The broken (bilinear) line OAB approximately sums up the behaviour. The slope of the line AB can be denoted by EI_t and called *reduced flexural stiffness*. This is several times smaller than the elastic stiffness EI. If we neglect this altogether we then obtain an *idealized* behaviour as shown by line OAC. This is equivalent to assuming that the flexural stiffness becomes suddenly zero when the moment reaches a limiting value M_P. This is the maximum moment carrying capacity of the section, beyond which its stiffness is (suddenly) zero, or in other words its flexibility is infinite. Since χ denotes the ratio $d\psi/dx$, which is the ratio of the relative angular rotation of adjacent cross-sections dx apart, infinite flexibility means very large angular deformations between two cross-sections small distance apart. This is the situation of a hinge. Thus we assume that when at a section the moment has reached the maximum capacity of the section, a small zone near that section behaves as a hinge. If after loading to the point E the moment is removed the unloading is elastic as represented by the line EF parallel to OA. The curvature denoted by OF is

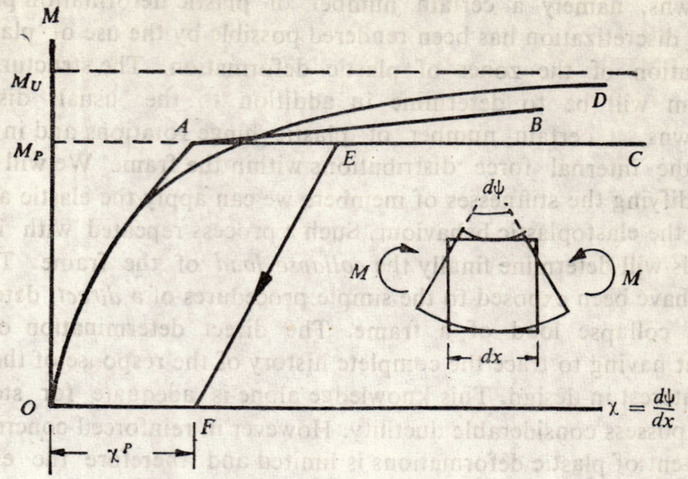

(a)

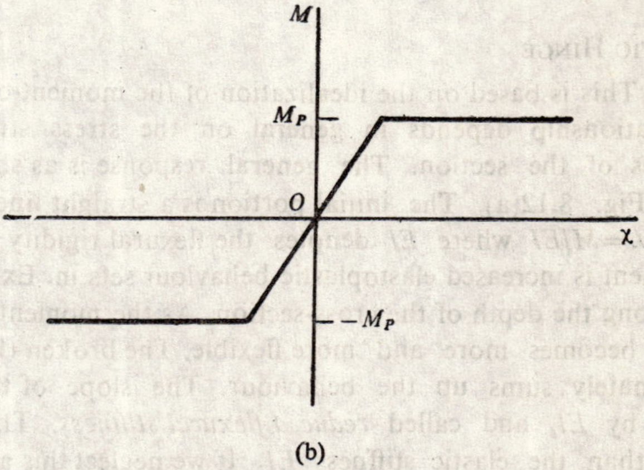

(b)

Fig. 8.12

the plastic deformation which is irrecoverable. This is the permanent set at the section.

b. Structural Analysis

The analysis proceeds by considering loads gradually increased in steps. The stiffness of a member is modified when at any of its ends a plastic hinge forms. This in turn alters the stiffness of the structure. When the loads are incremented gradually in steps new hinges develop and the plastic deformations (rotations) occur at the previously developed hinges.

The stiffness of the structure goes on diminishing until finally it becomes zero as a result of the stucture having turned into a mechanism. Figure 8.13(e) shows the load-displacement relation for the beam shown in Fig. 8.13(a). Curve OA denotes the elastic behaviour of the beam upto a load level W_1. This happens till the most heavily stressed section, section C, reaches its capacity. (Fig. 8.13(b)). For further load increments the beam behaves as a beam with hinge at C. Thus the stiffness of the member BC can be changed to take this into account and the structure stiffness recomputed. An additional load $\lambda_1 W$ is assumed and the beam of the modified stiffness analysed elastically. The increments in the moments are added to the moments due to the load W_1 and the total moments obtained. The scaling factor λ_1 is so determined that the most severely loaded section reaches its moment carrying capacity. In this example it would at a section near the load* as shown in Fig. 8.13(c). We have thus

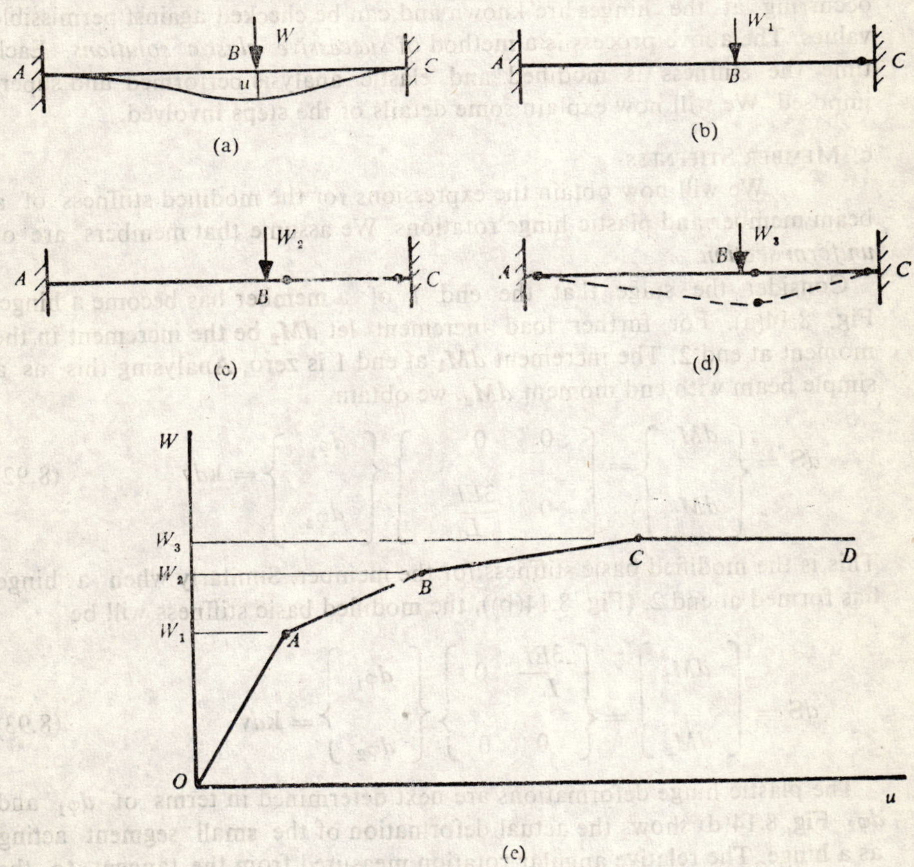

Fig. 8.13

* We have assumed that the hinge is to a little right of the load. It is at B of the member BC.

traced the line AB of Fig. 8.13(e). The load level is now $W_1+\lambda_1 W=W_2$. A next increment $\lambda_2 W$ is taken and the beam analysed with its stiffness computed on the modified situation of two hinges. The scaling factor λ_2 is again determined so that the third hinge forms at section A. (Fig. 8.13(d)). The total load is $W_3=W_2+\lambda_2 W$ giving the portion BC of the load displacement curve. If one modifies the stiffnesses of members again and reassembles the stiffness matrix for the structure one finds that its determinant is zero thus showing that the structure has lost its stiffness for further load increments. Observe that slopes of lines OA, AB, BC etc denote the structure stiffness. These go on decreasing as the hinges go on developing in the structure. Physically. the final stage means that the hinges have converted the structure into a mechanism which obviously has zero stiffness. We have here in described the incremental method of tracing the response of a structure till collapse load. At any stage the plastic deformations occurring at the hinges are known and can be checked against permissible values. The above process is a method of *successive elastic solutions.* Each time the stiffness is modified and elastic analysis performed and superimposed. We will now explain some details of the steps involved.

c. MEMBER STIFFNESS

We will now obtain the expressions for the modified stiffness of a beam member and plastic hinge rotations. We assume that members are of *uniform section.*

Consider the stage that the end 1 of a member has become a hinge, Fig. 8.14(a). For further load increment let dM_2 be the increment in the moment at end 2. The increment dM_1 at end 1 is zero. Analysing this as a simple beam with end moment dM_2, we obtain

$$d\mathbf{S}=\begin{Bmatrix} dM_1 \\ dM_2 \end{Bmatrix}=\begin{bmatrix} 0 & 0 \\ 0 & \dfrac{3EI}{L} \end{bmatrix}\begin{Bmatrix} d\varphi_1 \\ d\varphi_2 \end{Bmatrix}=\mathbf{k}d\mathbf{v} \tag{8.92}$$

This is the modified basic stiffness for the member. Similarly when a hinge has formed at end 2, (Fig. 8.14(b)), the modified basic stiffness will be

$$d\mathbf{S}=\begin{bmatrix} dM_1 \\ dM_2 \end{bmatrix}=\begin{Bmatrix} \dfrac{3EI}{L} & 0 \\ 0 & 0 \end{Bmatrix}\begin{Bmatrix} d\varphi_1 \\ d\varphi_2 \end{Bmatrix}=\mathbf{k}d\mathbf{v} \tag{8.93}$$

The plastic hinge deformations are next determined in terms of $d\varphi_1$ and $d\varphi_2$. Fig. 8.14(d) shows the actual deformation of the small segment acting as a hinge. The relative angular rotation measured from the tangent to the member side of the hinge to the tangent to the joint side, is the plastic deformation. Consider a hinge at end 1 (Fig. 8.14(a)). Let the cross-section attached to the joint rotate by an angle $d\varphi_1$. The rotation of the member end 1 due to the applied increment dM_2 at end 2 is $d\varphi_1'=\frac{1}{2}\,d\varphi_2=(L/6EI)\,dM_2$.

The plastic deformation at end 1 is therefore $d\psi_1=d\varphi_1+\frac{1}{2}d\varphi_2$. Since the end 2 is elastic $d\psi_2=0$. Similarly when the hinge is at end 2 and the moment dM_1 is applied at end 1 we obtain, (Fig. 8.14(b)), $d\psi_2=d\varphi_2+\frac{1}{2}\,d\varphi_1$ and

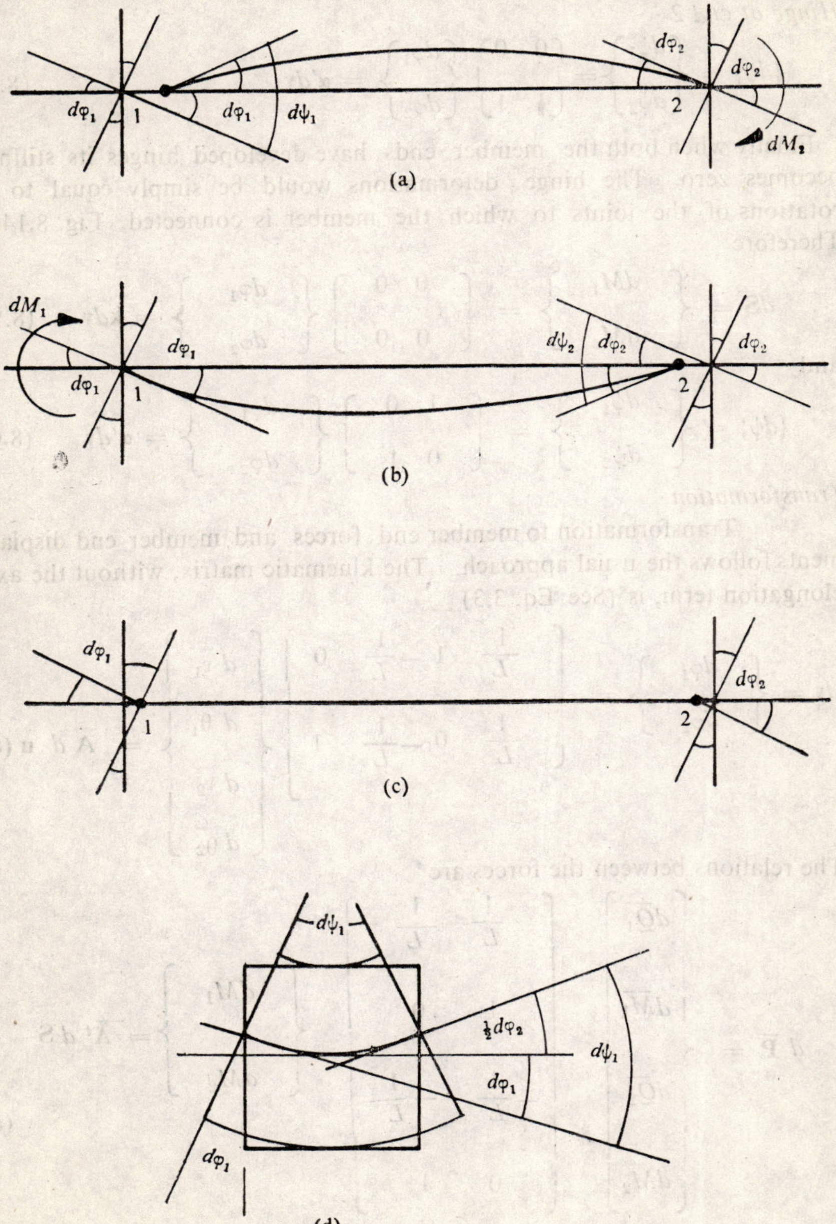

Fig. 8.14

$d\psi_1=0$. We therefore write the following relations for plastic hinge deformations :

Hinge at end 1

$$\{d\psi\} = \begin{Bmatrix} d\psi_1 \\ d\psi_2 \end{Bmatrix} = \begin{bmatrix} 1 & \frac{1}{2} \\ 0 & 0 \end{bmatrix} \begin{Bmatrix} d\varphi_1 \\ d\varphi_2 \end{Bmatrix} = \boldsymbol{\alpha}' d\mathbf{v} \tag{8·94}$$

Hinge at end 2

$$\{d\psi\} = \begin{Bmatrix} d\psi_1 \\ d\psi_2 \end{Bmatrix} = \begin{bmatrix} 0 & 0 \\ \frac{1}{2} & 1 \end{bmatrix} \begin{Bmatrix} d\varphi_1 \\ d\varphi_2 \end{Bmatrix} = \boldsymbol{\alpha}' d\mathbf{v} \tag{8.95}$$

Finally when both the member ends have developed hinges its stiffness becomes zero. The hinge deformations would be simply equal to the rotations of the joints to which the member is connected, Fig. 8.14(c). Therefore

$$d\mathbf{S} = \begin{Bmatrix} dM_1 \\ dM_2 \end{Bmatrix} = \begin{bmatrix} 0 & 0 \\ 0 & 0 \end{bmatrix} \begin{Bmatrix} d\varphi_1 \\ d\varphi_2 \end{Bmatrix} = \mathbf{k} d\mathbf{v} \tag{8.96}$$

and

$$\{d\psi\} = \begin{Bmatrix} d\psi_1 \\ d\psi_2 \end{Bmatrix} = \begin{bmatrix} 1 & 0 \\ 0 & 1 \end{bmatrix} \begin{Bmatrix} d\varphi_1 \\ d\varphi_2 \end{Bmatrix} = \boldsymbol{\alpha}' d\mathbf{v} \tag{8.97}$$

Transformation

Transformation to member end forces and member end displacements follows the usual approach. The kinematic matrix, without the axial elongation term, is (See Eq. 3.3)

$$d\mathbf{v} = \begin{Bmatrix} d\varphi_1 \\ d\varphi_2 \end{Bmatrix} = \begin{bmatrix} \frac{1}{L} & 1-\frac{1}{L} & 0 \\ \frac{1}{L} & 0-\frac{1}{L} & 1 \end{bmatrix} \begin{Bmatrix} d\,\overline{v}_1 \\ d\,\overline{\theta}_1 \\ d\,\overline{v}_2 \\ d\,\overline{\theta}_2 \end{Bmatrix} = \overline{\mathbf{A}}\, d\,\overline{\mathbf{u}} \tag{a$_1$}$$

The relations between the forces are

$$d\,\overline{\mathbf{P}} = \begin{Bmatrix} d\overline{Q_1} \\ d\overline{M_1} \\ d\overline{Q_2} \\ d\overline{M_2} \end{Bmatrix} \begin{bmatrix} \frac{1}{L} & \frac{1}{L} \\ 1 & 0 \\ -\frac{1}{L} & -\frac{1}{L} \\ 0 & 1 \end{bmatrix} \begin{Bmatrix} dM_1 \\ dM_2 \end{Bmatrix} = \overline{\mathbf{A}}^t\, d\,\mathbf{S} \tag{a$_2$}$$

We therefore obtain

$$d\,\overline{\mathbf{P}} = [\overline{\mathbf{A}}^t \quad \mathbf{K}\,\overline{\mathbf{A}}]\; d\,\overline{\mathbf{u}} = \overline{\mathbf{K}}\; d\,\overline{\mathbf{u}} \tag{8.98}$$

$$d\,\overline{\psi} = \alpha'\,\overline{\mathbf{A}}\; d\,\overline{\mathbf{u}} = \overline{\alpha}\; d\,\overline{\mathbf{u}} \tag{8.99}$$

The relations for the different cases are as follows:

Hinge at end 1

$$d\,\overline{\mathbf{P}} = \begin{Bmatrix} d\,\overline{Q}_1 \\ d\overline{M}_1 \\ d\,\overline{Q}_2 \\ d\overline{M}_2 \end{Bmatrix} = \begin{bmatrix} k_3' & 0 & -k_3' & k_4' \\ 0 & 0 & 0 & 0 \\ -k_3' & 0 & k_3' & -k_4' \\ k_4' & 0 & -k_4' & k_1' \end{bmatrix} \begin{Bmatrix} d\,\overline{v}_1 \\ d\,\overline{\theta}_1 \\ d\,\overline{v}_2 \\ d\,\overline{\theta}_2 \end{Bmatrix} = \overline{\mathbf{K}}\,d\,\overline{\mathbf{u}} \tag{8.100}$$

$$\{d\psi\} = \begin{Bmatrix} d\psi_1 \\ d\psi_2 \end{Bmatrix} = \begin{bmatrix} \frac{3}{2L} & 1 & -\frac{3}{2L} & \frac{1}{2} \\ 0 & 0 & 0 & 0 \end{bmatrix} [\,d\,\overline{\mathbf{u}}\,] = \overline{\alpha}\,d\,\overline{\mathbf{u}} \tag{8.101}$$

Hinge at end 2

$$\overline{\mathbf{K}} = \begin{bmatrix} k_3' & k_4' & -k_3' & 0 \\ k_4' & k_1' & -k_4' & 0 \\ -k_3' & -k_4' & k_3' & 0 \\ 0 & 0 & 0 & 0 \end{bmatrix} \tag{8.102}$$

$$\overline{\alpha} = \begin{bmatrix} 0 & 0 & 0 & 0 \\ \frac{3}{2L} & \frac{1}{2} & -\frac{3}{2L} & 1 \end{bmatrix} \tag{8.103}$$

Hinges at both ends

$$\mathbf{K} \equiv 0 \tag{8.104}$$

$$\overline{\alpha} = \begin{bmatrix} \frac{1}{L} & 1 & -\frac{1}{L} & 0 \\ \frac{1}{L} & 0 & -\frac{1}{L} & 0 \end{bmatrix} \tag{8.105}$$

Here,

$$k_1' = \frac{3EI}{L} \quad k_3' = \frac{3EI}{L^3} \quad k_4' = \frac{3EI}{L^2} \tag{8.106}$$

We have thus obtained the modified stiffness relations and also the expressions for plastic deformations in terms of the displacements. We have introduced the differential quantities to remind that these are *increments* to be added to the previous values. These will, therefore, be called *incremental* relations.

The above matrices can be further transformed for *coordinate transformation* to handle members oriented arbitrarily in the plane of the frame.

The incremental approach outlined in para *c* will be explained with respect to an example below. The loading would be incremented by multiplying the loads by a factor λ and determining it so that one hinge is formed at a time.

EXAMPLE 8.7

Consider the beam with the off central load W as shown in Fig. 8.13(a). This would be treated as a beam with two members A and B. We will incorporate the following support conditions and the continuity at B:

$$\theta_{1A} = v_{1A} = 0\ ;\ \theta_{2B} = v_{2B} = 0 \tag{a$_1$}$$

$$\theta_{2A} = \theta_{1B} = \theta_B\ ;\ v_{3B} = v_{1B} = v_B \tag{a$_2$}$$

Consider now the initial elastic analysis. Let dW_1 be the load. Here, and in what follows we drop the bars on top for convenience as no coordinate transformations are involved. Therefore, for member A, noting that length is 2 L

$$d\mathbf{P}_{2A} = \begin{Bmatrix} dQ_{2A} \\ dM_{2A} \end{Bmatrix} = \begin{bmatrix} \dfrac{3EI}{2L^3} & -\dfrac{3EI}{2L^2} \\ -\dfrac{3EI}{2L^2} & \dfrac{2EI}{L} \end{bmatrix} \begin{Bmatrix} dv_B \\ d\theta_B \end{Bmatrix} \tag{a$_3$}$$

$$d\mathbf{P}_{1B} = \begin{Bmatrix} dQ_{1B} \\ dM_{1B} \end{Bmatrix} = \begin{bmatrix} \dfrac{12EI}{L^3} & \dfrac{6EI}{L^2} \\ \dfrac{6EI}{L^2} & \dfrac{4EI}{L} \end{bmatrix} \begin{Bmatrix} dv_B \\ d\theta_B \end{Bmatrix} \tag{a$_4$}$$

The equation of equilibrium at B is simply

$$d\mathbf{P}_{2A} + d\mathbf{P}_{1B} = \{dW_1 \quad 0\} = d\mathbf{P}_1 \tag{a$_5$}$$

which therefore becomes

$$\mathbf{K}d\mathbf{u}_1 = \begin{bmatrix} \dfrac{27}{2}\dfrac{EI}{L^2} & \dfrac{9}{2}\dfrac{EI}{L^2} \\ \dfrac{9}{2}\dfrac{EI}{L^2} & \dfrac{6EI}{L} \end{bmatrix} \begin{Bmatrix} dv_B \\ d\theta_B \end{Bmatrix} = \begin{Bmatrix} dW_1 \\ 0 \end{Bmatrix} = d\mathbf{P}_1 \tag{a$_6$}$$

This on solution gives

$$
d\mathbf{u}_1 = \begin{bmatrix} 0.09876\dfrac{L^3}{EI} & -0.07407\dfrac{L^2}{EI} \\ -0.07407\dfrac{L^2}{EI} & 0.2222\dfrac{L}{EI} \end{bmatrix} \begin{Bmatrix} dW_1 \\ 0 \end{Bmatrix}
$$

$$
= \begin{Bmatrix} -0.09876\dfrac{L^3}{EI} \\ -0.07407\dfrac{L^2}{EI} \end{Bmatrix} dW_1 \qquad (a_7)
$$

The end moments are,

$$
d\mathbf{M} = \begin{Bmatrix} dM_{1A} \\ dM_{2A} \\ dM_{1B} \\ dM_{2B} \end{Bmatrix} = \begin{bmatrix} -\dfrac{3}{2}\dfrac{EI}{L^2} & \dfrac{EI}{L} \\ -\dfrac{3}{2}\dfrac{EI}{L^2} & \dfrac{2EI}{L} \\ \dfrac{6EI}{L^2} & \dfrac{4EI}{L} \\ \dfrac{6EI}{L^2} & \dfrac{2EI}{L} \end{bmatrix} \begin{Bmatrix} 0.09876\dfrac{L^3}{EI} \\ -0.07407\dfrac{L^2}{EI} \end{Bmatrix} dW_1
$$

$$
= \begin{Bmatrix} -0.2222 \\ -0.2963 \\ 0.2963 \\ 0.4444 \end{Bmatrix} dW_1 L \qquad (a_8)
$$

It is seen that M_{2B} is largest governing the lowest value of dW_1. Thus if M_P is the maximum moment carrying capacity then

$$
0.4444\, dW_1 L = M_P \qquad (a_9)
$$

giving $dW_1 = 2.25\,(M_P/L)$. Therefore, substituting back, we obtain

$$
\mathbf{u}_1 \equiv d\mathbf{u}_1 = \begin{Bmatrix} 0.2222\dfrac{L^2}{EI} \\ -0.1666\dfrac{L}{EI} \end{Bmatrix} M_P \qquad (a_{10})
$$

$$
\mathbf{M}_1 \equiv d\mathbf{M}_1 = \{-0.5 \quad -0.6666 \quad +0.6666 \quad 1.0\} M_P \qquad (a_{11})
$$

We now consider the next increment dW_2 to be applied. For this increment of load the stiffnesses of the members affected by plasticity are altered and reanalysed. Presently we need to change the stiffness of member B for hinge developed at end 2. Thus we use **K** of Eq (8.102) for member B, and ordinary elastic matrix for member A. For convenience, we adopt $L = EI = 1$.

$$\mathbf{K}_A=\begin{bmatrix} 1.5 & 1.5 & -1.5 & 1.5 \\ 1.5 & 2 & -1.5 & 1 \\ -1.5 & -1.5 & 1.5 & -1.5 \\ 1.5 & 1 & -1.5 & 2 \end{bmatrix}; \quad \mathbf{K}_B=\begin{bmatrix} 3 & 3 & -3 & 0 \\ 3 & 3 & -3 & 0 \\ -3 & -3 & 3 & 0 \\ 0 & 0 & 0 & 0 \end{bmatrix} \qquad (a_{12})$$

The structure stiffness equations are

$$\begin{bmatrix} 4.5 & 1.5 \\ 1.5 & 5 \end{bmatrix}\begin{Bmatrix} dv_B \\ d\theta_B \end{Bmatrix}=\begin{Bmatrix} dW_2 \\ 0 \end{Bmatrix}=d\mathbf{P}_3 \qquad (a_{13})$$

These on solution give

$$d\mathbf{u}_2=\{0.2469 \quad -0.07407\}dW_2 \qquad (a_{14})$$

The end moments obtainable from Eq. (a_{12}) are

$$\begin{Bmatrix} dM_{1A} \\ dM_{2A} \\ dM_{1B} \\ dM_{2B} \end{Bmatrix}=\begin{bmatrix} -1.5 & 1 \\ -1.5 & 2 \\ 3 & 3 \\ 0 & 0 \end{bmatrix}\begin{Bmatrix} 0.2469 \\ -0.07407 \end{Bmatrix}dW_2=\begin{Bmatrix} -0.4444 \\ -0.5185 \\ +0.5185 \\ 0 \end{Bmatrix}dW_2 \qquad (a_{15})$$

These when superimposed on the moments given by Eq. (a_{11}) determine the total moments. We see that M_{2A} is always equal in magnitude and opposite in sign to M_{1B} since the applied moment at B is zero. Thus the next heavily stressed sections are these sections which reach plasticity simultaneously. We can determine the magnitude dW_2 by equating the total moment to M_P or $-M_P$ for member end moments M_{1B} and M_{2A} respectively.

$$\pm 0.6666\, M_P \pm 0.5185\, dW_2 = \pm M_P \qquad (a_{16})$$

This gives $dW_2= 0.6428\, M_P$. Thus the increments are

$$d\mathbf{u}_2 = \{0.1587 \quad -0.04761\}\, M_P$$
$$d\mathbf{M} = \{-0.2856 \quad -0.333 \quad +0.333 \quad 0\}\, M_P \qquad (a_{17})$$

Thus the total load and the corresponding values of displacements and moments are

$$W_2=(2.25+0.6428)M_P=2.8928\, M_P \qquad (a_{18})$$
$$\mathbf{u}_2=\{0.3809, \quad -0.2143\}M_P \qquad (a_{19})$$
$$\mathbf{M}_2=\{-0.7856 \quad -1.0, \quad 1.0, \quad 1.0\}M_P \qquad (a_{20})$$

Hinge rotation

We can now calculate the plastic hinge rotation which would have taken place at end 2 of member B during the increment dW_2. We determine this using Eq. (8.103)

$$\begin{Bmatrix} d\psi_1 \\ d\psi_2 \end{Bmatrix} = \begin{bmatrix} 0 & 0 & 0 & 0 \\ \frac{3}{2} & \frac{1}{2} & -\frac{3}{2} & 1 \end{bmatrix} \begin{Bmatrix} 0.1587 \\ -0.04761 \\ 0 \\ 0 \end{Bmatrix} M_P = \begin{Bmatrix} 0 \\ 0.2142 \end{Bmatrix} M_P \qquad (a_{21})$$

Note that the multiplier for this quantity is $\left(\frac{L}{EI}\right)$, presently taken as unity.

Consider now the next load increment dW_3. We will assume that section at end 1 of member B alone has become plastic. The situation which would arise by assuming both this and the section at end 2 of member A show plastic deformation for further load increments will be dealt with after the first assumption. The first assumption is not a difficult one to visualize. We can depart from the ideal situation of uniform moment carrying capacity of the beam and imagine that the end 2 of A is only a little too stronger than end 1 of member B or vice versa. Pending further consideration of this matter, let us pursue the calculations based on the first assumption.

Consider then the increment of load dW_3. The member B now has hinges at both its ends. Thus the stiffness $\mathbf{K}_B \equiv \mathbf{0}$, and since both the ends of member A are elastic we have $\mathbf{K}_A$ as in Eq. (a_{12}). Thus the total structure stiffness relations are

$$\begin{bmatrix} 1.5 & -1.5 \\ -1.5 & 2 \end{bmatrix} \begin{Bmatrix} dv_B \\ d\theta_B \end{Bmatrix} = \begin{Bmatrix} dW_3 \\ 0 \end{Bmatrix} \qquad (a_{22})$$

This shows that the structure has still the stiffness, namely that of a member A acting as cantilever.

The above relation gives

$$d\mathbf{u}_3 = \{2.666 \quad 2\} dW_3 \qquad (a_{23})$$

Therefore using stiffness of A as from Eq. (a_{12})

$$d\mathbf{M}_3 = \{-2, \ 0, \ 0 \ \ 0\} dW_3 \qquad (a_{24})$$

These moments when added to those in Eq. (a_{20}) determine the magnitude of dW_3 such that the moment M_{1A} reaches its limit equal to $-M_P$. Therefore

$$-0.7856\, M_P - 2dW_3 = -M_P \qquad (a_{25})$$

which gives $dW_3 = 0.1072\, M_P$. The corresponding quantities are therefore as follows :

$$d\mathbf{u}_3 = \{0.2858 \quad 0.2144\} M_P \qquad (a_{26})$$

$$d\mathbf{M}_3 = \{-0.2144, \ 0, \ 0, \ 0\} M_P \qquad (a_{27})$$

The total aggregated values therefore are

$$W_3 = (2.8928 + 0.1072) M_P = 3.0\, M_P \qquad (a_{28})$$

$$\mathbf{u}_3 = \{06667, \quad 0\}M_P \qquad (a_{29})$$

$$\mathbf{M}_3 = \{-1.0, \quad -1.0, \quad 1.0, \quad 1.0\}M_P \qquad (a_{30})$$

We again calculate the plastic hinge rotations due to $dW_3=0.1072\ M_P$. Recall that we have assumed a hinge at end 1 of B, and a hinge had earlier formed at end 2 due to dW_2. Thus $d\psi_{1B}$ and $d\psi_{2B}$ due to dW_3 to be obtained from Eq. (8.105) are

$$\begin{Bmatrix} d\psi_{1B} \\ d\psi_{2B} \end{Bmatrix} = \begin{bmatrix} 1 & 1 & -1 & 0 \\ 1 & 0 & -1 & 1 \end{bmatrix} \begin{Bmatrix} 0.2858 \\ 0.2144 \\ 0 \\ 0 \end{Bmatrix} M_P = \begin{Bmatrix} 0.5 \\ 0.2858 \end{Bmatrix} M_P \qquad (a_{31})$$

Thus the total hinge rotation at $W_3=3.0\ M_P$ obtained by superposing the above on those of given in Eq. (a_{21}) are,

$$\begin{Bmatrix} \psi_1 \\ \psi_2 \end{Bmatrix} = \begin{Bmatrix} 0+0.5 \\ 0.2142+0.2858 \end{Bmatrix} M_P = \begin{Bmatrix} 0.5 \\ 0.5 \end{Bmatrix} M_P \qquad (a_{32})$$

Consider the trend of the present scheme of calculations. Let us now evaluate the present structure stiffness for further load increments. Modify the stiffness of member A for a hinge at end 1 as per Eq. (8.100). Since $K_B=0$, we have the structure stiffness relations

$$\begin{bmatrix} \frac{3}{8} & -\frac{3}{8} \\ -\frac{3}{8} & \frac{3}{2} \end{bmatrix} \begin{Bmatrix} dv_B \\ d\theta_B \end{Bmatrix} = \begin{Bmatrix} dW_4 \\ 0 \end{Bmatrix} \qquad (a_{33})$$

The determinant of the matrix is zero showing that the structure has lost its stiffness for further loads. This represents the collapse load condition, calling for the stop of the calculations. The loss of stiffness of the structure is as a result of sufficient plastic hinges converting it into a mechanism. Since there are no increases in bending moments at member ends they remain straight between the hinges as shown. The *limit equilibrium* methods depend on the kinematics of such a mechanism to determine the collapse load *directly*. Fig. 8.13(e) shows the progressive loss of stiffness of the structure. At C the stiffness becomes zero.

Consider now the second assumption of assuming plastic hinges on either side of the joint B, after the load increment dW_2. If we assume that $d\psi_{2A}$, $d\psi_{1B}$ and $d\psi_{2B}$ occur for load increment dW_3, let us then modify the stiffness of member A according to Eq. (8.102) and of member B according to Eq. (8.104). Thus,

$$\mathbf{K}_A = \begin{bmatrix} \frac{3}{8} & \frac{3}{4} & \frac{3}{8} & 0 \\ \frac{3}{4} & \frac{3}{2} & -\frac{3}{4} & 0 \\ -\frac{3}{8} & \frac{3}{4} & -\frac{3}{8} & 0 \\ 0 & 0 & 0 & 0 \end{bmatrix}; \quad \mathbf{K}_B \equiv \mathbf{0} \tag{a$_{34}$}$$

Therefore the structure stiffness relations are

$$\begin{bmatrix} \frac{3}{8} & 0 \\ 0 & 0 \end{bmatrix} \begin{Bmatrix} dv_B \\ d\theta_B \end{Bmatrix} = \begin{Bmatrix} dW_3 \\ 0 \end{Bmatrix} \tag{a$_{35}$}$$

This again shows that the determinant is zero. Also, this shows that the structure has no stiffness against a moment applied about the joint B. The loss of structure stiffness represented by the vanishing of the determinant of Eq. (a$_{35}$) is as a result of the 'joint mechanism' in which there is no rotational stiffness of members meeting at the joint as a result of plasticity which can develop at their ends. This is a 'local' collapse and has to be excluded.

EXAMPLE 8.8

Consider the frame shown in Fig. (a). We will take the member end forces for the column as shown in (b). Presently we have two unknowns v_B and θ_B at B ; We will assume uniform EI for members and further that plastic hinge forms at the top end of the column rather than at the beam end.

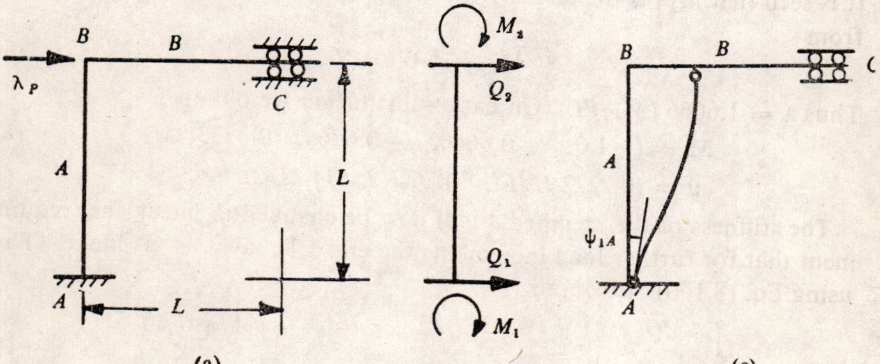

(a) (b) (c)

Fig. Example 8.8

The elastic stiffness relations for the column, obtainable from Eq. (5.45) are, (once again dropping the bars on top).

$$\begin{Bmatrix} Q_1 \\ M_1 \\ Q_2 \\ M_2 \end{Bmatrix} = \begin{bmatrix} -12EI/L^3 & 6EI/L^2 \\ -6EI/L^2 & 2EI/L \\ 12EI/L^3 & -6EI/L^2 \\ -6EI/L^2 & 4EI/L \end{bmatrix} \begin{Bmatrix} v_B \\ \theta_B \end{Bmatrix} \tag{a$_1$}$$

Similarly for beam B

$$\{M_{1B} \quad M_{2B}\} = \{4EI/L \quad 2EI/L\}\ \theta_B \tag{a$_2$}$$

The equations of equilibrium are

$$Q_{2A} = \lambda P\ ; \quad M_{2A} + M_{1B} = 0 \tag{a$_3$}$$

These in terms of the displacements give

$$\mathbf{KU} = \begin{bmatrix} 12EI/L^3 & -6EI/L^2 \\ -6EI/L^2 & 8EI/L \end{bmatrix} \begin{Bmatrix} v_B \\ \theta_B \end{Bmatrix} = \lambda \begin{Bmatrix} P \\ 0 \end{Bmatrix} \tag{a$_4$}$$

On solving,

$$\begin{Bmatrix} v_B, \theta_B \end{Bmatrix} = \begin{Bmatrix} \frac{2}{15}\frac{L^3}{EI} & \frac{1}{10}\frac{L^2}{EI} \end{Bmatrix} \lambda P \tag{a$_5$}$$

These determine moments

$$\mathbf{M} = \begin{Bmatrix} M_{1A} \\ M_{2A} \\ M_{1B} \\ M_{2B} \end{Bmatrix} = \begin{bmatrix} -6EI/L^2 & 2EI/L \\ -6EI/L^2 & 4EI/L \\ 0 & 4EI/L \\ 0 & 2EI/L \end{bmatrix} \begin{Bmatrix} \frac{2}{15}\frac{L^3}{EI} \\ \frac{1}{10}\frac{L^2}{EI} \end{Bmatrix} (\lambda P) = \begin{Bmatrix} -0.6 \\ -0.4 \\ 0.4 \\ 0.2 \end{Bmatrix} \lambda PL \tag{a$_6$}$$

It is seen that M_{1A} is the largest; the value of load factor λ is obtained from

$$-M_P = -0.6\ \lambda PL \tag{a$_7$}$$

Thus $\lambda = 1.6666\ (M_P/PL)$. On back substitution we obtain

$$\mathbf{M} = \{-1.0, \quad -0.6666, \quad +0.6666, \quad 0.3333\}\ M_P \tag{a$_8$}$$

$$\mathbf{u} = \{0.2222\ L^2/EI, \quad 0.2666\ L/EI\}\ M_P.$$

The stiffness of the member A will now be changed to meet the requirement that for further load increments the end 1 behaves as a hinge. Thus using Eq. (8.100), we obtain

$$\begin{Bmatrix} dQ_{2A} \\ dM_{2A} \end{Bmatrix} = \begin{bmatrix} 3EI/L^3 & -3EI/L^2 \\ -3EI/L^2 & 3EI/L \end{bmatrix} \begin{Bmatrix} dv_B \\ d\theta_B \end{Bmatrix} \tag{a$_9$}$$

The assembled stiffness relations stating Eqs. (a$_3$) are

$$\begin{bmatrix} 3EI/L^3 & -3EI/L^2 \\ -3EI/L^2 & 7EI/L \end{bmatrix} \begin{Bmatrix} dv_B \\ d\theta_B \end{Bmatrix} = \begin{Bmatrix} P \\ 0 \end{Bmatrix} d\lambda \tag{a$_{10}$}$$

On solution these yield

$$\mathbf{du} = \{0.5883\ L^3/EI \quad 0.25\ L^2/EI\}\ d\lambda P \tag{a$_{11}$}$$

And on back substitution, the moments would be

$$\begin{Bmatrix} dM_{1A} \\ dM_{2A} \\ dM_{1B} \\ dM_{2B} \end{Bmatrix} = \begin{bmatrix} 0 & 0 \\ -3EI/L^2 & 3EI/L \\ 0 & 4EI/L \\ 0 & 2EI/L \end{bmatrix} \begin{Bmatrix} 0.5883\ L^3/EI \\ 0.25L^2/EI \end{Bmatrix} d\lambda P = \begin{Bmatrix} 0 \\ -1.0 \\ 1.0 \\ 0.5 \end{Bmatrix} d\lambda PL \tag{a$_{12}$}$$

The value of $d\lambda$ so as to cause yielding at end 2 of column is obtained from

$$-\tfrac{2}{3}M_P + (-1.0)\ d\lambda PL = -M_P$$

This gives $d\lambda = \frac{1}{3}(M_P/PL)$. Thus Eq. (a$_{12}$) give

$$\begin{aligned} d\mathbf{M} &= \{0 \quad -\tfrac{1}{3} \quad -\tfrac{1}{3} \quad 0\}\ M_P \\ \mathbf{du} &= \{0.1944\ L^2/EI,\ 0.0833\ L/EI\}\ M_P \end{aligned} \tag{a$_{13}$}$$

Thus the total values are

$$\begin{aligned} \lambda &= 1.6666 + \tfrac{1}{3} = 2.0 \\ \mathbf{M} &= \{-1.0,\ -1.0,\ 1.0 \quad 0.5\}\ M_P \\ \mathbf{u} &= \{0.4166\ L^3/EI, \quad 0.35\ L^2/EI\}\ M_P \end{aligned} \tag{a$_{14}$}$$

The increment in the plastic hinge rotation at end 1 during the above load increment is to be obtained using (8.101). Thus

$$d\psi_1 = [-\tfrac{3}{2}L \quad \tfrac{1}{2}] \begin{Bmatrix} 0.1944\ L^3/EI \\ 0.0833\ L/EI \end{Bmatrix} M_P/L = 0.25\ (M_PL/EI) \tag{a$_{15}$}$$

The reader can verify that the structure stiffness with hinges at both ends of column A becomes zero indicating that the structure has reached the collapse load (Fig. (c)).

d. The Incremental Procedure

The above examples have demonstrated the essential principles involved. It is seen that the procedure consists in applying the load in increments. In simple problems the procedure consists of scaling down the load increment so that only one plastic hinge forms. For bigger problems procedures overcoming this limitation are however possible.

The procedure indicated in reference [30] is slightly different from the one presented here. An augmented vector consisting of the displacement variables and the plastic hinge rotations is considered and a corresponding augmented matrix of stiffness coefficients is obtained. The coefficients are those which state the equations of equilibrium corresponding to the displacement variables, and the 'yield conditions' corresponding to the locations where plastic hinges are formed. The stiffnesses of members are *not* modified as done here but the handling of the augmented matrix is equivalent to the procedure of modifying the stiffnesses of member first and determining the hinge rotations next as is adopted here.

There is a lot to be said about the details which would be necessary in handling a bigger frame. Reference [30] is a reference where one can take note of such matters.

EXERCISES

8.1 Use the basic stiffness method in analysing the frame of Ex. 2.13. Take the data as in Ex. 7.3 and compare the answers obtained for the case when the axial flexibilities of members were ignored. (Observe that the basic stiffness method involves the same steps as in the classical slope deflection equation method but with equations of equilibrium obtained by evoking the principle of virtual work.)

8.2 The beam of the portal frame shown is rigid. Construct the stiffness relation between P and the deflection (in its direction) following the basic stiffness approach. The beam is continuous over the columns. Consider only axial deformations of the columns; how would the analysis extend itself if the columns are monolithic with the beam?

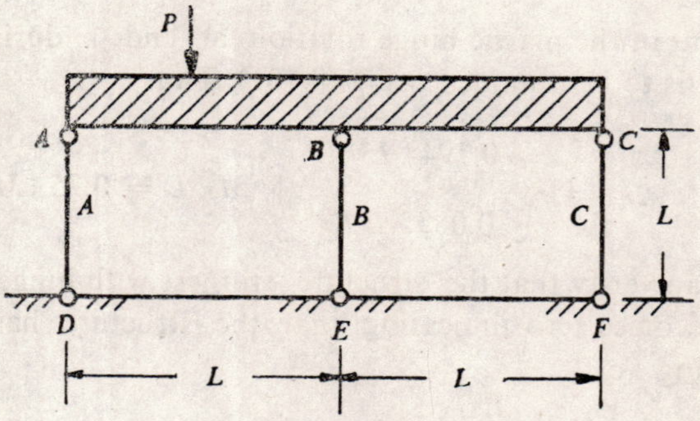

Fig. Ex. 8.2

8.3 For the problem of Ex. 8.1 obtain the stiffness relations by following the direct stiffness approach. In the kinematic relations introduce the conditions that member elongations are zero. By a pure transformation derive the stiffness relations obtained in Ex. 8.1 Take help of Sec. 3.3.

8.4 Solve Ex. 8.1 on the lines of Example 8.2. Exclude the support joints from the beginning; compare the answers obtained in Ex. 7.3.

8.5 Assemble the stiffness matrix for the frame of Example 8.2 including the foundation joints following the memberwise assembly. Modify the same for the prescribed support displacements. Assume that all the support displacements can assume prescribed values. Solve for unit displacements imposed at support A. Use modification suggested in Eq. (8.23) or in Eq. (a_5) of Sec. 8.4.

8.6 Determine the band width for the frame shown treating (a) the frame to be a plane frame and (b) a space frame.

8.7 Solve the equations of Ex. 8.5 adopting the procedure of Sec. 10.1, Chapter 10.

8.8 Analyse the grid frame of Ex. 7.4 by the direct stiffness method using stiffnesses of Eq. (8.30).

8.9 Repeat Ex. 8.4 for temperature elongations in all the members. Assume suitable numerical magnitudes if necessary.

8.10 Analyse the frame shown considering the effects of axial load in the column by taking P such that it is 80 percent of the Euler load for the column assumed as a free cantilever.

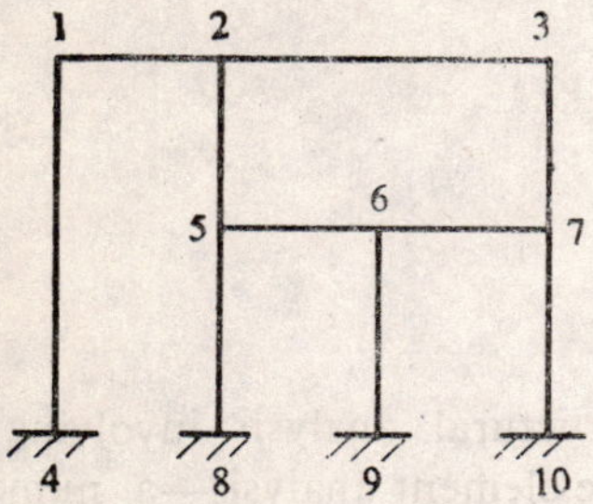

Fig. Ex. 8.6

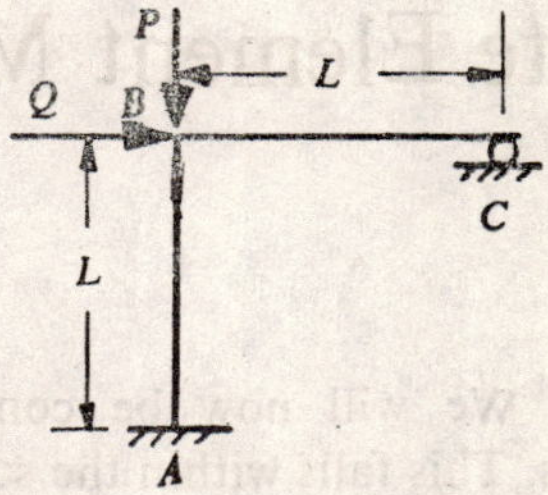

Fig. Ex. 8.10

8.11 Determine the sway mode critical load for the frame shown. Take EI and AE to be the same for all members.

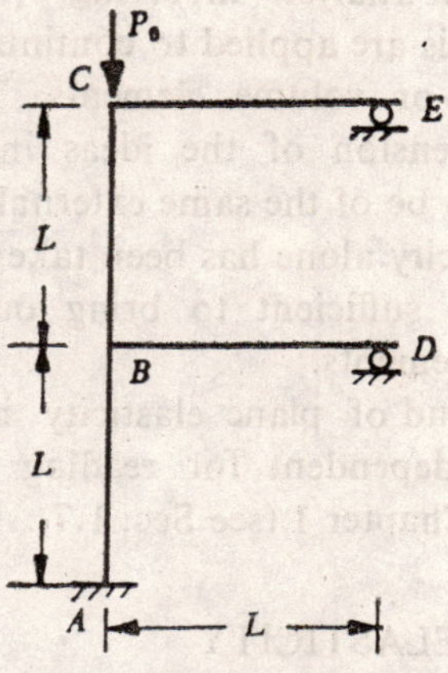

Fig. Ex. 8.11

8.12 Determine the collapse load for the portal frame shown in Fig. Ex. 8.10 assuming that load Q only acts. Take members of uniform section.

8.13 Determine the collapse load for the portal frame of Fig. Ex. 8.11 under only horizontal loads Q acting at B and C.

Chapter 9

Finite Element Method

We will now be considering structural analysis involving area elements. This falls within the scope of finite element analysis—a numerical procedure of analysing continuum problems. This is now a well developed procedure that has shown its power and versatility.

The object of this chapter is very limited. The intention is to show how the basic ideas of structural analysis involving : (a) element analysis, and (b) global assembly analysis are applied to continuum problems considered as an assembly of area or volume elements. The reader will find this chapter as a logical extension of the ideas introduced so far. Many equations will be found to be of the same external format.

The case of plane elasticity alone has been taken as an illustration. This one case is in fact just sufficient to bring out the essential aspects of analysis involving finite elements.

The necessary background of plane elasticity has been first introduced to make the chapter independent for reading. The reader may recall the introduction outlined in Chapter 1 (see Sec. 1.7).

9.1 EQUATIONS OF ELASTICITY

We will be taking the case of two dimensional system to illustrate the application of the finite element discretization.

A two dimensional system is a plane system. The physical variables are functions of the two coordinates x, y in the plane of the system. The behaviour in the perpendicular direction is known. Two types of stress distributions usually arise in engineering practice. These are termed 'plane stress' and 'plane strain' cases. The former is applicable to thin plate like bodies and the latter to bodies which are infinitely long in a direction perpendicular to the plane. Figure 9.1 shows the case of a deep beam and

Fig. 9.2, the case of a dam. These represent the two cases respectively. In the former, because of the small thickness the stress σ_z in the z direction is taken to be zero. In the latter case, a slice such as shown in Fig. 9.2(b) is held by the adjacent slices so that the strain ϵ_z in the z direction is taken as zero. Since the body is sufficiently long this assumption is justified. One has to examine if the cross section at ends can be considered to be in a condition of plane strain depending on whether or not the abutments of the dam can exert restraining forces to prevent lateral strains. If it is free, a small slice at ends can be considered to be in a condition of plane stress.

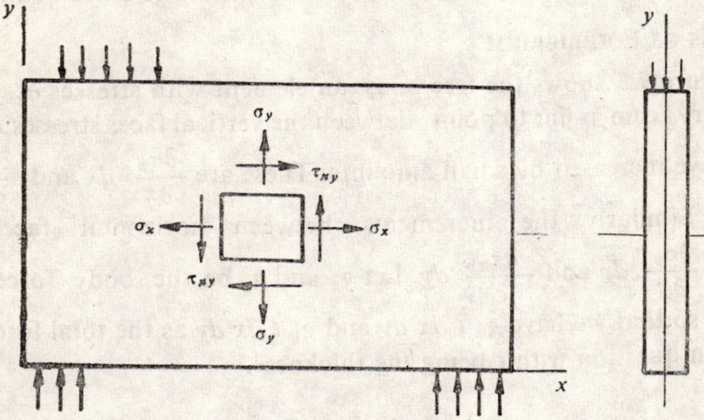

Fig. 9.1

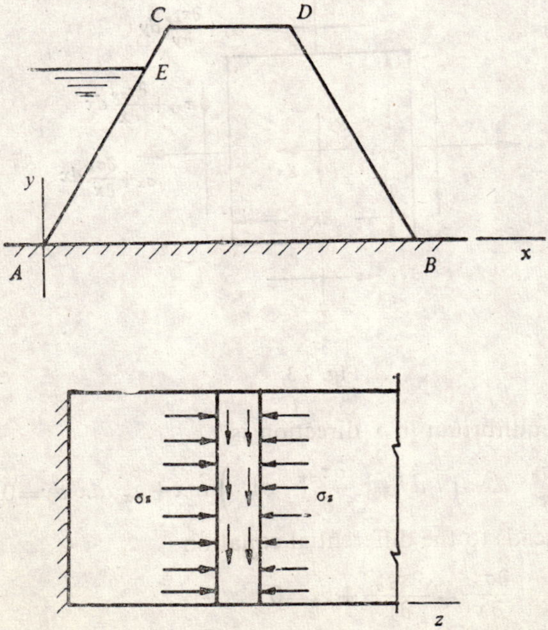

Fig. 9.2

The equations of elasticity are essentially the same as the ones considered in structural analysis but in different form. The equations to be handled are, again, those of (a) equilibrium, (b) compatibility (kinematics) and (c) elasticity. We have here too, the displacement and the force methods, which in the parlance of continuum mechanics are the methods of displacement and stress functions. Instead of working with forces and deformations we deal in terms of stresses and strains. In fact viewed in a reversed sense the structural analysis is an approximate analysis of solids in which kinematic assumptions are introduced *a priori* and corresponding stress resultants (moments, shear forces, etc.) and deformations (rotations, elongations, etc.) are defined and analysis formulated in terms of these quantities.

a. EQUATIONS OF EQUILIBRIUM

Figure 9.3 shows the free body an element with stresses σ_x, σ_y and τ_{xy} which vary from point to point. Between the vertical faces stresses σ_x and τ_{xy} would have increased by small amounts. These are $\frac{\partial \sigma_x}{\partial x}\,dx$ and $\frac{\partial \tau_{xy}}{\partial x}\,dx$ respectively. Similarly the increments between horizontal faces are respectively $\frac{\partial \sigma_y}{\partial y}\,dy$ and $\frac{\partial \tau_{xy}}{\partial y}\,dy$. Let g_x and g_y be the body forces per unit volume, so that we have $g_x\,t\,dx\,dy$ and $g_y\,t\,dx\,dy$ as the total forces for the element in question with t being the thickness.

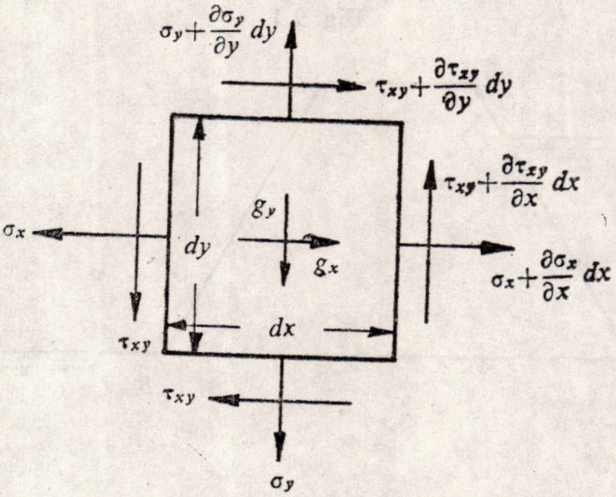

Fig. 9.3

The equation of equlibrium in x direction is

$$\left(\frac{\partial \sigma_x}{\partial x}\,dx\right)t\,dy+\left(\frac{\partial \tau_{xy}}{\partial y}\,dy\right)t dx+g_x\,t dx\,dy=0$$

This in the limit leads to the differential equation

$$\frac{\partial \sigma_x}{\partial x}+\frac{\partial \tau_{xy}}{\partial y}+g_x=0$$

This is the condition for the rates of changes in σ_x and τ_{xy}. Recall the engineering theory of beams in which the shear stress is determined in terms of the change in the bending stresses between two adjacent cross sections. The above equation, say with $g_x \equiv 0$ implies that change in τ_{xy} between two horizontal sections balances the change in direct stresses between two vertical sections.

Thus the two equations of equilibrium are

$$\frac{\partial \sigma_x}{\partial x} + \frac{\partial \tau_{xy}}{\partial y} + g_x = 0$$

$$\frac{\partial \tau_{xy}}{\partial x} + \frac{\partial \sigma_y}{\partial y} + g_y = 0 \tag{9.1}$$

b. KINEMATICAL EQUATIONS

The displacement of a point (Fig. 9.4) is characterized by its components u, v, along x and y respectively. The displaced point 1 has new coordinates $(x+u,\ y+v)$. A neighbouring point 2, $(x+dx,\ y)$ gets displaced to

$$\left(x+dx+u+\frac{\partial u}{\partial x}dx,\ y+v+\frac{\partial v}{\partial x}\ dx \right)$$

Similarly the neighbouring point 4, $(x,\ y+dy)$ gets displaced to

$$\left(x+u+\frac{\partial u}{\partial y}\ dy,\ y+dy+v+\frac{\partial v}{\partial y}\ dy \right)$$

The elongation of the horizontal segment dx (segment 1-2) can be obtained from the geometry of sketch (b) :

$$L'^2 = \left(dx + \frac{\partial u}{\partial x} dx \right)^2 + \left(\frac{\partial v}{\partial x}\ dx \right)^2$$

where L' is the elongated length of the segment dx. If ϵ_x is the strain then $L' = dx + \epsilon_x\ dx = dx(1+\epsilon_x)$.

Thus

$$(dx(1+\epsilon_x))^2 = dx^2 + \left(\frac{\partial u}{\partial x}\right)^2 dx^2 + 2\ \frac{\partial u}{\partial x}\ dx^2 + \left(\frac{\partial v}{\partial x}\right)^2 dx^2$$

or

$$1+\epsilon_x^2+2\epsilon_x = 1 + 2\ \frac{\partial u}{\partial x} + \left(\frac{\partial u}{\partial x}\right)^2 + \left(\frac{\partial v}{\partial x}\right)^2 \tag{a$_1$}$$

Strains being *small* ϵ_x^2 term can be neglected in comparison to ϵ_x. Thus we obtain

$$\epsilon_x = \frac{\partial u}{\partial x} + \frac{1}{2}\left(\left(\frac{\partial u}{\partial x}\right)^2 + \left(\frac{\partial v}{\partial x}\right)^2\right) \tag{9.2}$$

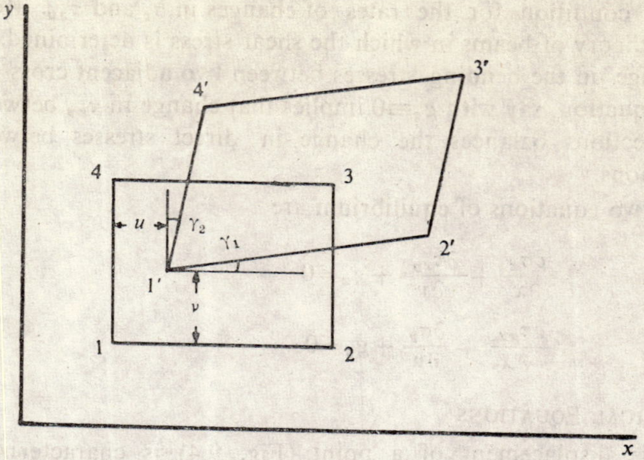
y
x
1
2
3
4
1'
2'
3'
4'
u
v
γ2
γ1

(a)

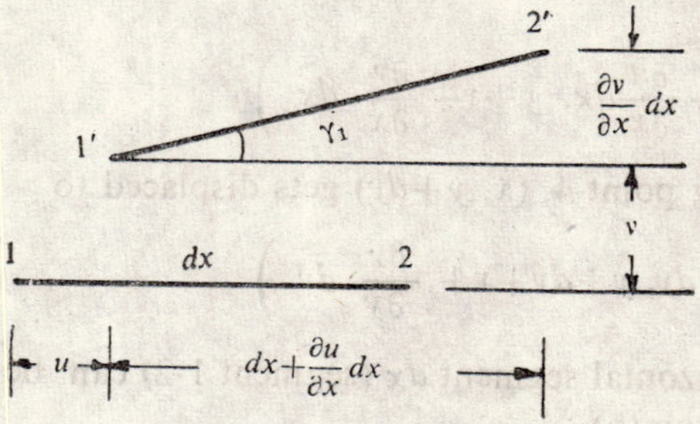
1'
2'
γ1
∂v/∂x dx
v
1
2
dx
u
dx+∂u/∂x dx

(b)

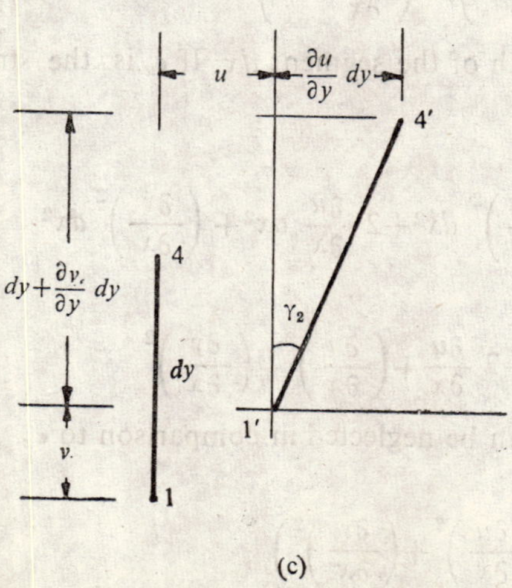
u
∂u/∂y dy
4'
4
dy+∂v/∂y dy
dy
γ2
1'
v
1

(c)

Fig. 9.4

These are the (nonlinear) kinematic relations relating the strain to the displacements. If further we consider that *displacements* are also *small* then the nonlinear terms can be neglected and we obtain $\epsilon_x = \frac{\partial u}{\partial x}$, a relation immediately appealing. Similarly we obtain the strain ϵ_y in the segment dy as,

$$\epsilon_y = \frac{\partial v}{\partial y} + \frac{1}{2}\left(\left(\frac{\partial v}{\partial y}\right)^2 + \left(\frac{\partial u}{\partial y}\right)^2\right) \tag{9.3}$$

The shearing strain is characterized by the change in the angle between segments dx and dy. We define

$$\gamma_{xy} = \gamma_1 + \gamma_2 \tag{a$_2$}$$

where γ_1 and γ_2 are the rotations of segments dx and dy. These can be obtained from sketches (b) and (c) of Fig. 9.4.

$$\gamma_1 = \left(\frac{\partial v}{\partial x}\, dx\right) \Big/ \left(dx + \frac{\partial u}{\partial x}\, dx\right)$$

$$\gamma_2 = \left(\frac{\partial v}{\partial y}\, dy\right) \Big/ \left(dy + \frac{\partial v}{\partial y}\, dy\right)$$

On adding these and neglecting terms higher than $dx\, dy$, we obtain

$$\gamma_{xy} = \gamma_1 + \gamma_2 = \frac{\partial v}{\partial x} + \frac{\partial u}{\partial y} + \frac{\partial u}{\partial x}\,\frac{\partial u}{\partial y} + \frac{\partial v}{\partial x}\,\frac{\partial v}{\partial y} \tag{9.4}$$

For *small displacement* theory the linear kinematic relations therefore, are,

$$\epsilon_x = \frac{\partial u}{\partial x}\ ;\ \epsilon_y = \frac{\partial v}{\partial y}\ ;\ \gamma_{xy} = \frac{\partial v}{\partial x} + \frac{\partial u}{\partial y} \tag{9.5}$$

Compatibility Conditions

Equations (9.5) constitute three equations in terms of two variables u and v. Thus ϵ_x, ϵ_y and γ_{xy} have to satisfy a certain relationship between them. This in physical terms means that the definition of strain components introduced have to satisfy certain relationships between them. These are the continuity conditions. Differentiating the first twice with respect to y, the second twice with respect to x and the third once with x and once with y, shows that following relation must hold:

$$\frac{\partial^2 \epsilon_x}{\partial y^2} + \frac{\partial^2 \epsilon_y}{\partial x^2} = \frac{\partial^2 \gamma_{xy}}{\partial x \partial y} \tag{9.6}$$

If for instance we assume a strain distribution, which is what we frequently do, we have to assume smooth enough functions so that the above relation is satisfied. If this is disregarded we only have small discontinuities in the deformation pattern. Consider the case wherein ϵ_x, ϵ_y and γ_{xy} are assumed to be *linear* functions of x and y, a case commonly adopted. It is seen that this satisfies the compatibility of strains. So also is the case of *constant* strains.

The above condition of compatibility is a generalization of the one used in bar frameworks wherein we talk of member elongations being consistent with joint displacements.

c. ELASTICITY EQUATIONS

For an isotropic body two constants, the Young's Modulus and Poisson's ratio define the elastic relationships. The Hooke's laws give,

$$\boldsymbol{\epsilon}=\begin{Bmatrix}\epsilon_x\\ \epsilon_y\\ \epsilon_z\\ \gamma_{xy}\end{Bmatrix}=1/E\begin{bmatrix}1 & -\nu & -\nu & 0\\ -\nu & 1 & -\nu & 0\\ -\nu & -\nu & 1 & 0\\ 0 & 0 & 0 & 2(1+\nu)\end{bmatrix}\begin{Bmatrix}\sigma_x\\ \sigma_y\\ \sigma_z\\ \tau_{xy}\end{Bmatrix}=\mathbf{D}^{-1}\boldsymbol{\sigma} \quad (a_3)$$

From these we can obtain the elasticity relation for plane stress and plain strain. For plane stress $\sigma_z=0$, and therefore

$$\boldsymbol{\epsilon}=\begin{Bmatrix}\epsilon_x\\ \epsilon_y\\ \gamma_{xy}\end{Bmatrix}=1/E\begin{bmatrix}1 & -\nu & 0\\ -\nu & 1 & 0\\ 0 & 0 & 2(1+\nu)\end{bmatrix}\begin{Bmatrix}\sigma_x\\ \sigma_y\\ \tau_{xy}\end{Bmatrix}=\mathbf{D}^{-1}\boldsymbol{\sigma} \quad (9.7)$$

Here $\mathbf{D}^{-1}$ is the flexibility matrix. On inversion we obtain

$$\boldsymbol{\sigma}=\begin{Bmatrix}\sigma_x\\ \sigma_y\\ \tau_{xy}\end{Bmatrix}=\frac{E}{1-\nu^2}\begin{bmatrix}1 & \nu & 0\\ \nu & 1 & 0\\ 0 & 0 & (1-\nu)/2\end{bmatrix}\begin{Bmatrix}\epsilon_x\\ \epsilon_y\\ \gamma_{xy}\end{Bmatrix}=\mathbf{D}\,\boldsymbol{\epsilon} \quad (9.8)$$

Here **D** is the matrix of elastic constants.

For the plain strain case we have $\epsilon_z=0$ giving $\sigma_z=\nu\ (\sigma_x+\sigma_y)$, and therefore,

$$\boldsymbol{\epsilon}=\begin{Bmatrix}\epsilon_x\\ \epsilon_y\\ \gamma_{xy}\end{Bmatrix}=(1+\nu)/E\begin{bmatrix}1-\nu & -\nu & 0\\ -\nu & 1-\nu & 0\\ 0 & 0 & 2\end{bmatrix}\begin{Bmatrix}\sigma_x\\ \sigma_y\\ \tau_{xy}\end{Bmatrix}=\mathbf{D}^{-1}\boldsymbol{\sigma} \quad (9.9)$$

This on inversion gives,

$$\boldsymbol{\sigma}=\begin{Bmatrix}\sigma_x\\ \sigma_y\\ \tau_{xy}\end{Bmatrix}=\frac{E}{(1+\nu)(1-2\nu)}\begin{bmatrix}1-\nu & \nu & 0\\ \nu & 1-\nu & 0\\ 0 & 0 & (1-2\nu)/2\end{bmatrix}=\mathbf{D}\,\boldsymbol{\epsilon} \quad (9.10)$$

It is observed therefore that the elasticity matrices for the two cases are similar. In general we define suitable elasticity matrices depending on the material properties, such as orthotropy, etc.

d. Boundary Conditions

The stress and displacement distributions have also to satisfy the boundary conditions. The plane strip would be supported along a certain part of the contour (Figs 9.1, 2). At the supported contour the displacements have to be the same as required by the support condition *s*. For the cross section of the dam shown in Fig. 9.2 the displacement pattern to be obtained should have zero values along the base. In continuum problems we have to ensure that the stresses at the loaded part of the boundary should be equal to the applied load intensities. For example along the edge *AE* of the dam (Fig 9.2) the stresses should agree with the magnitudes of loads actually acting at the boundary.

Figure 9.5 shows the stresses p_x and p_y acting along the boundary. The stresses σ_x, σ_y and τ_{xy} are required to equilibrate the boundary stresses. Therefore

$$\sigma_x(dy)+\tau_{xy}\,(-dx)=p_x\;ds$$
$$\sigma_y(-dx)+\tau_{yx}\,(dy)=p_y\;ds$$

Dividing throughout by *ds* we write this as

$$l\tau_x+m\tau_{xy}=p_x \qquad (9.11)$$
$$m\sigma_y+l\tau_{xy}=p_y$$

where $l=dy/dx$ and $m=-\,dx/ds$ are the direction cosines of the outward normal of the boundary. Observe that dx is negative for a ds considered positive when taken counter-clockwise.

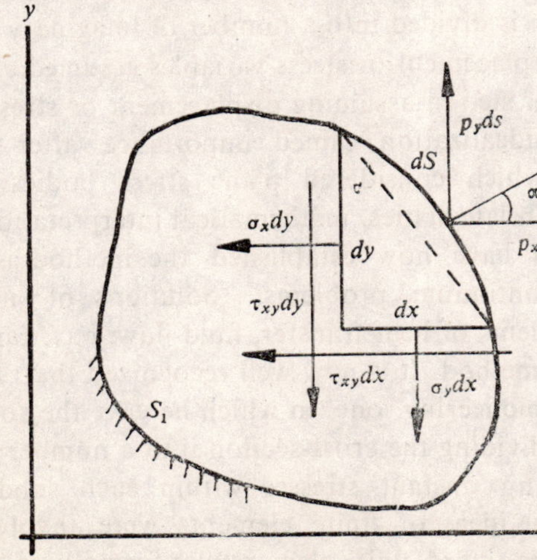

Fig. 9.5

e. SOLUTION OF EQUATIONS

The classical solutions construct differential equations expressed in terms of either displacement or stress functions, and provide solution for a given set of loads and boundary conditions. Alternatively an energy functional is formulated and the solution is sought as an extremum value of such a functional. Closed form solutions are possible only for cases of simple boundaries and loads.

Numerical solutions have always been put forward from time to time. The two chief approaches can be grouped as

(a) finite difference solutions,

(b) solutions based on assumed displacement or stress distributions.

The above procedures discretize the problem and give either of the set of simultaneous equations.

$$\mathbf{Ka}=\mathbf{b}$$
$$\mathbf{Fc}=\mathbf{d} \tag{9.12}$$

In the first case $\mathbf{a}=\{a_1, a_2,......\}$ is a vector of the displacement variables, and $\mathbf{b}$ is a vector of the corresponding generalised loads. $\mathbf{K}$ is the coefficient matrix. In the second case $\mathbf{c}$ is the vector of stress variables and $\mathbf{d}$ is the vector of the associated strain variables. $\mathbf{F}$ is a similar coefficient matrix. $\mathbf{K}$ and $\mathbf{F}$ are analogous to stiffness and flexibility matrices.

The finite difference method replaces the derivatives by their finite difference approximations. The procedures of type (b) are more versatile. The finite element formulation belongs to this class. The early applications of this principle were demonstrated by Ritz [22] and Galerkin [23]. In such procedures the displacement or stress variables are assumed over the *whole* domain of the problem in terms of a finite number of parameters, and Eqs, (9.12) are derived using principles of virtual work or alternatively energy minimization criteria. The finite element technique is a recent development in which the whole region is divided into a number of imaginary small regions (finite domains) and displacement or stress variables assumed *only within* the region. This is an easier step in assuming displacement or stress patterns.

The finite element idealization gained importance after the notable publications [7, 20] which considered plane stress bodies as structural assembly of elements. In late sixties, mathematical interpretations began to be put forward which have now established the method as a numerical method of solving continuum problems. Solutions of non-structural problems, such as problems of heat transfer, fluid flow, etc., came under the scope of finite element method. It is now well recognized that the work of Courant [24] is the pioneering one in which he gave the solution to the problem of torsion by dividing the cross-section into a number of triangular subdomains and assuming constant stresses within each subdomain. It is also now known that the ideas of finite elements were involved in some works of Leibnitz. The word finite element was formally introduced later in engineering literature.

We will pursue here the structural idea, namely the plane stress or strain domain as an assembly of area elements. The mathematical idea, which considers finite element method as a tool to provide a discretized solution to continuum problems, though more effective and general, will not be pursued here. The object of this chapter is to generalize the ideas of displacement method to assemblies involving area or volume elements.

9.2 OUTLINE OF FINITE ELEMENT METHOD

Consider the deep beam shown in Fig. 9.6(a) as a typical example. The first step is to imagine that the body is subdivided into a number of

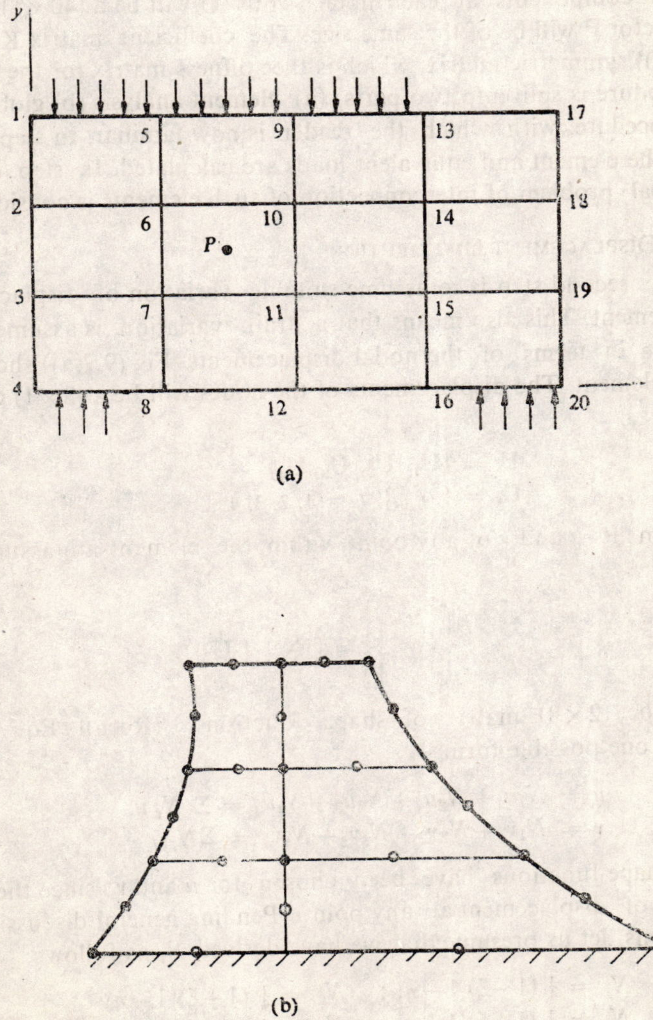

Fig. 9.6

convenient finite elements. It is expected that we repeat the problem few times, by subdividing it into a larger number of smaller size elements and make sure that the solution converges, namely that displacements and stresses converge to some definite values. Consider that it is divided into 12 rectangular elements. Fig 9.6(b) shows another example which employs curvilinear elements, better suited to fit curved boundaries.

The displacement method arrives at a set of simultaneous equations

$$\mathbf{KU}=\mathbf{P} \tag{9.13}$$

Here **U** will be the vector of nodal displacements and **P** the vector of nodal loads. For the example of deep beam, we have 20 nodes, with two displacement components at each node. Thus **U** will be a (40×1) vector. The load vector **P** will be of the same size. The coefficient matrix **K** would be a (40×40) symmetric matrix, which is the stiffness matrix for the beam.

The procedure is split into two parts: (a) element analysis (b) global analysis, a procedure with which the reader is now familiar. In step (a) the stiffness of the element and equivalent loads are calculated. In step (b) the overall (global) problem of interconnection of such elements is considered.

a. Assumed Displacement Distributions

The second step is to assume suitable variation of displacements within an element. This also means that a strain variation is assumed. We assume these in terms of the nodal displacements. Fig (9.7(a)) shows the rectangular element. The displacements of the nodes will be a (8×1) column vector.

$$\begin{aligned}\mathbf{U} &= \{\mathbf{U}_1, \mathbf{U}_2, \mathbf{U}_3, \mathbf{U}_4\}\\ \mathbf{U}_i &= \{u_i\ v_i\},\ i = 1, 2, 3, 4\end{aligned} \tag{a$_1$}$$

The displacements u and v of any point within the element are assumed in terms of **U**:

$$\begin{Bmatrix} u \\ v \end{Bmatrix} = [\,\mathbf{N}\,]\,\{\,\mathbf{U}\,\} \tag{a$_2$}$$

Here **N** is the (2×8) matrix of shape functions. (Recall Eq. (5.55)) Following is one possible form:

$$\begin{aligned} u &= N_1u_1+N_2u_2+N_3u_3+N_4u_4 = \Sigma\, N_i\, u_i \\ v &= N_1v_1+N_2v_2+N_3v_2+N_4v_4 = \Sigma N_i v_i \end{aligned} \tag{a$_3$}$$

Here same shape functions have been chosen for u and v, since these are components of displacement at any point. Pending general discussion on shape functions, let us presume that we have derived $\mathbf{N}_i$ as follows:

$$\begin{aligned} N_1 &= \tfrac{1}{4}\,(1-\xi)(1-\eta)\ ; \quad N_2 = \tfrac{1}{4}\,(1+\xi)(1-\eta) \\ N_3 &= \tfrac{1}{4}\,(1+\xi)(1+\eta)\ ; \quad N_4 = \tfrac{1}{4}\,(1-\xi)(1+\eta) \end{aligned} \tag{9.14}$$

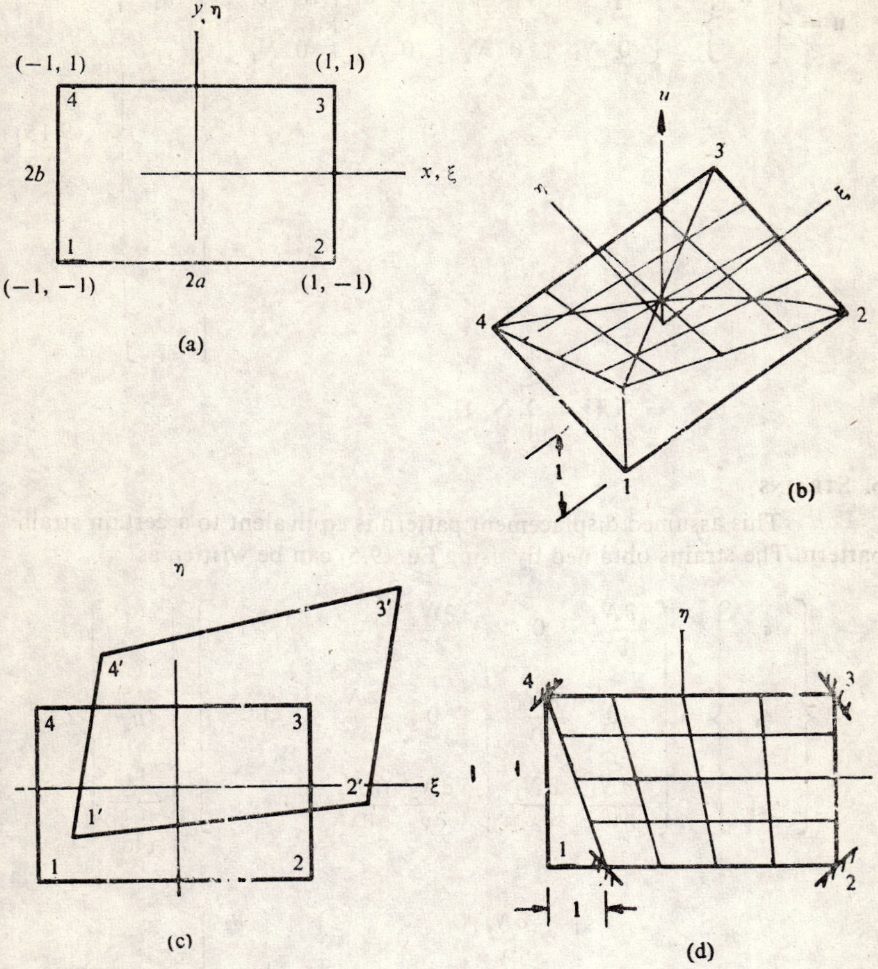

Fig. 9.7

with $\xi = x/a$ and $\eta = y/b$.

Observe that nodes 1, 2, 3 and 4 have coordinates (−1, −1), (1, −1), (1, 1) and (−1, 1) respectively. Shape functions N_i have the same physical meaning as introduced in Sec. 5.8 (See Eq. (5.55)). N_i represents $u(\xi, \eta)$ due to $u_i = 1$ alone. $N_i(\xi_i, \eta_i) = 1$ and zero at other three nodes. It is seen that N_i vary linearly along lines ξ = constant or η = constant. Figure 9.7(b) shows N_1 plotted perpendicular to the plane. Figure 9.7(c) shows the general deformed shape.

Equation (a_2) can now be written as

$$\mathbf{u} = \begin{Bmatrix} u \\ v \end{Bmatrix} = \left[\begin{array}{cc|cc|cc|cc} N_1 & 0 & N_2 & 0 & N_3 & 0 & N_4 & 0 \\ 0 & N_1 & 0 & N_2 & 0 & N_3 & 0 & N_4 \end{array}\right] \begin{Bmatrix} u_1 \\ v_1 \\ \hline \cdot \\ \cdot \\ \cdot \\ \hline u_4 \\ v_4 \end{Bmatrix} \tag{9.15}$$

$$= \mathbf{N}\,\mathbf{U} = \sum_1^4 \mathbf{N}_i\,\mathbf{U}_i$$

b. STRAINS

This assumed displacement pattern is equivalent to a certain strain pattern. The strains obtained by using Eq. (9.5) can be written as

$$\begin{Bmatrix} \epsilon_x \\ \epsilon_y \\ \gamma_{xy} \end{Bmatrix} = \left[\begin{array}{cc|cc|c} \dfrac{\partial N_1}{\partial x} & 0 & \dfrac{\partial N_2}{\partial x} & 0 & \\ 0 & \dfrac{\partial N_1}{\partial y} & 0 & \dfrac{\partial N_2}{\partial y} & \text{etc.} \\ \dfrac{\partial N_1}{\partial y} & \dfrac{\partial N_1}{\partial x} & \dfrac{\partial N_2}{\partial y} & \dfrac{\partial N_2}{\partial x} & \end{array}\right] \begin{Bmatrix} u_1 \\ v_1 \\ \hline u_2 \\ v_2 \\ \hline \text{etc.} \end{Bmatrix}$$

$$= \sum_1^4 \begin{bmatrix} \dfrac{\partial N_i}{\partial x} & 0 \\ 0 & \dfrac{\partial N_i}{\partial y} \\ \dfrac{\partial N_i}{dy} & \dfrac{\partial N_i}{\partial x} \end{bmatrix} \begin{Bmatrix} u_i \\ v_i \end{Bmatrix}$$

Or

$$\boldsymbol{\epsilon} = \Sigma\, \mathbf{A}_i\, \mathbf{u}_i = \mathbf{AU} \tag{9.16}$$

This is the familiar form of kinematics. The kinematic matrix **A** written in full is as follows:

$$\mathbf{A} = \left[\begin{array}{cc|cc|cc|cc} -\frac{1-\eta}{4a} & 0 & \frac{1-\eta}{4a} & 0 & \frac{1+\eta}{4a} & 0 & -\frac{1+\eta}{4a} & 0 \\ 0 & -\frac{1-\xi}{4b} & 0 & -\frac{1+\xi}{4b} & 0 & \frac{1+\xi}{4b} & 0 & \frac{1-\xi}{4b} \\ -\frac{1-\xi}{4b} & -\frac{1-\eta}{4a} & -\frac{1+\xi}{4b} & \frac{1-\eta}{4a} & \frac{1+\xi}{4b} & \frac{1-\eta}{4a} & \frac{1-\xi}{4b} & -\frac{1+\eta}{4a} \end{array}\right] \tag{a_4}$$

Observe that in this case strains are linear functions of x and y. Therefore these satisfy the compatibility equations (9.6). Further, if the strains within the element tend to constant values in the limit when the size of the element is reduced, convergence is assured. This criterion, which has a physical appeal, can here be verified. The component ϵ_x is seen to be

$$\epsilon_x = \frac{1-\eta}{4a}(u_2-u_1)+\frac{1+\eta}{4a}(u_3-u_4) \tag{a_5}$$

In the limit when $\eta \to 0$, ϵ_x approaches the constant value

$$\epsilon_x \to \frac{1}{2}\left(\frac{u_2-u_1}{2a}+\frac{u_3-u_4}{2a}\right) \tag{a_6}$$

This is seen to be the average of the strains of edges 12 and 34. It may be similarly verified that other components also tend to constant values. Observe that due to u_1 alone we have

$$\epsilon_x = -\frac{1-\eta}{4a}u_1\ ;\ \epsilon_y = 0;\ \gamma_{xy} = -\frac{1-\xi}{4b}u_1 \tag{a_7}$$

The deformed shape is shown in Fig. 9.7(d). Observe that along edge 12 ($\eta=-1$) we have $\epsilon_x=-u_1/2u$; this uniformly decreases to zero at edge 34 ($\eta=1$). $\epsilon_y=0$, is obviously an approximation. γ_{xy} also is seen to be maximum at node 1 ($\xi=-1$) and equal to $-u_1/2b$, which is the angle of edge 41 with the vertical. It is seen that shearing strain varies only with ξ. In fact we can visualize *a priori* such simple states of deformation based on engineering judgement and arrive at the corresponding equations for u and v by integrating the strain quantities.

Let us now take a look at *interelement compatibility*. Along edge 34, corresponding to $\eta = 1$, we have from Eq. (9.15)

$$u_{3-4} = \frac{1+\xi}{2}u_3+\frac{1-\xi}{2}u_4\ ;\ v_{3-4} = \frac{1+\xi}{2}u_3+\frac{1-\xi}{2}v_4 \tag{a_8}$$

Similarly along edge 12, $\eta = -1$

$$u_{1-2} = \frac{1-\xi}{2}u_1+\frac{1+\xi}{2}u_2\ ;\ v_{1-2} = \frac{1-\xi}{2}v_1+\frac{1+\xi}{2}v_2 \tag{a_9}$$

Relations such as these show that the displacements along an edge are determined solely by the nodes lying on that edge. This ensures interelement compatibility between the adjacent element edges as shown in Fig. 9.8.

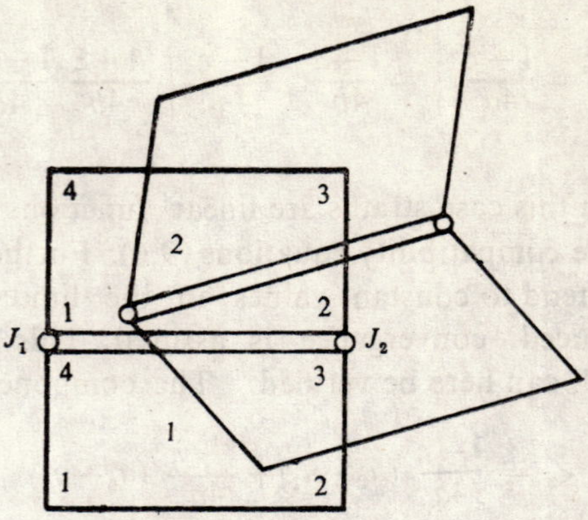

Fig. 9.8

Edge 43 of element 1 and edge 12 of element 2 will have common nodes (J_1 and J_2). The global analysis would ensure that $(u_4)_1 = (u_1)_2 = u_{J_1}$ and $(u_3)_1 = (u_2)_2 = u_{J_2}$. Therefore it follows that after deformation the element edges undergo identical displacements thus ensuring continuity between the elements. If the chosen displacement pattern does not ensure interelement continuity we term such shape functions as 'non-conforming'. In such a case there would be discontinuities or 'gaps' between the edges.

Here then we summarise below the two essential criteria to be satisfied by the assumed displacement pattern:

Criterion 1. The assumed pattern should lead to constant strains within the element as the element size is reduced.

Criterion 2. For interelement compatibility the variation of the displacements along an edge should be determined by the displacement of the nodes lying on that edge.

There is a *third* criterion which also requires to be verified; this is one of rigid body modes. The chosen pattern should permit rigid body movement of an element without causing any internal stresses. This also sounds obvious since elements should be free to displace as rigid bodies without getting strained. In many cases the first criterion envelopes this criterion. But in curved elements it needs special attention. Without much loss of generality we may consider that the two criteria mentioned above are necessary and also sufficient in most cases.

c. STRESSES

Consider now the nature of stress distribution implied. These are determined by premultiplying the strains by the elastic constants. Thus

$$\sigma = \mathbf{D}\epsilon = \lfloor \mathbf{DA} \rceil \mathbf{U} \tag{9.17}$$

where matrix **D** is as in Eq. (9.8) or (9.10). It is seen that stresses also vary linearly. These stresses should satisfy the equations of equilibrium (9.1). In this case it will be found that these equations are *not* satisfied. This is the general feature of an assumed strain distribution. Also the stresses may not satisfy the boundary conditions along boundary edges. If we take a point on an interelement edge and compute stresses from the two adjacent elements they will in general not be the same. Also if such a point happens to be along an exterior (boundary) edge the calculated stress may in general not be equal to the applied boundary stresses. Both these issues should not sound new to the reader, as these were pointed out in Chapter 5. Recall the analysis of the beam of Sec. 5.9. Assuming of sinusoidal q^* (Eq. (a_8) of Sec. 5.9) violated the equilibrium within the beam. Also the assumed shape gave zero moments at ends while as it ought to have been equal the moment M_1.

d. EQUIVALENT NODAL FORCES

Since equilibrium cannot be established exactly we approximately ensure it in an overall sense. We define equivalent nodal forces, equivalent statically, to the boundary stresses. This idea is not new but is the same as that introduced in Sec. 5.9. Figure 9.9(a) and (b) show the boundary stresses and the equivalent nodal forces respectively. We choose these forces so as to match with the nodal displacements. The equivalence is established by choosing the criterion that during a deformation, the virtual work of boundary stresses through boundary displacements is equal to the work of the equivalent nodal forces through the nodal displacements. Thus let the displacements be as in (9.15); let $\sigma_b = \{\sigma_n, \sigma_t\}$ be the vector of boundary stresses* with σ_n and σ_t denoting the normal and tangential stresses. Let $\mathbf{u}_b = \{u_n, u_t\}$ with u_n and u_t denoting the normal and tangential displacements of a point on the edge. The equivalent nodal force vector **P** is taken as (8×1) column vector

$$\mathbf{P} = \{\mathbf{P}_1, \mathbf{P}_2, \mathbf{P}_3, \mathbf{P}_4\}; \ \mathbf{P}_i = \{P_i, Q_i\} \tag{a_{10}}$$

which match with **U** (Eq. (a_1)). The definition of **P** implies the equality

$$\mathbf{P}^t\,\mathbf{U} = \int_S \sigma_b{}^t\,\mathbf{u}_b\,dS \tag{9.18}$$

* Strictly, boundary tractions; these are per unit length of the contour curve.

Figure 9.9(b) shows the free body of the element wlth equivalent nodal forces instead of boundary stresses. This is an essential consequence of assuming a strain distribution determining an approximate stress distribution which violates the stress equilibrium conditions.

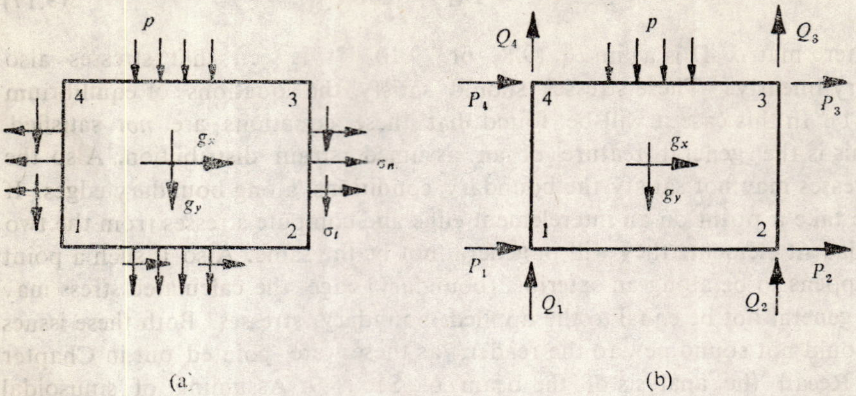

Fig. 9.9

e. Element Stiffness

We derive complete stiffness relations in the usual form

$$\mathbf{P} = \mathbf{KU} + \mathbf{P}_F \equiv \mathbf{KU} - \mathbf{P}_E \tag{9.19}$$

Presently **K** will be a (8×8) symmetric matrix and $\mathbf{P}_F$ is a (8×1) vector of fixing nodal forces. These denote **P** in a condition $\mathbf{U} = \mathbf{0}$. $\mathbf{P}_E = -\mathbf{P}_F$, are equivalent forces.

The derivation proceeds exactly on the lines followed in Sec. 5.9. Let $\mathbf{g} = \{g_x, g_y\}$ be the body forces and $\mathbf{p} = \{p_x, p_y\}$ the x and y components of boundary stresses. Equation of virtual work becomes

$$\int_V \boldsymbol{\sigma}^t \boldsymbol{\epsilon}\, dV = \int_V \mathbf{g}^t \mathbf{u}\, dV + \int_{S_p} \mathbf{p}^t \mathbf{u}\, dS + \mathbf{P}^t \mathbf{U} \tag{9.20}$$

Here the second term denotes integration over those edges where stresses are prescribed. We now substitute relations (9.15) for **u** and (9.16) for $\boldsymbol{\epsilon}$. Then

$$\int_V \boldsymbol{\sigma}^t \mathbf{AU}\, dV = \int_V \mathbf{g}^t \mathbf{N}\,\mathbf{U}\, dV + \int_{S_p} \mathbf{p}^t \mathbf{NU}\, dS + \mathbf{P}^t \mathbf{U}$$

Taking transpose and cancelling $\mathbf{U}^t$ we find,

$$\int_V \mathbf{A}^t \boldsymbol{\sigma}\, dV = \int_V \mathbf{N}^t \mathbf{g}\, dV + \int_{S_p} \mathbf{N}^t \mathbf{p}\, ds + \mathbf{P} \tag{9.21}$$

This is the equation of equilibrium for the element as a whole*. We can express this in terms of displacements by using (9.17) for $\boldsymbol{\sigma}$. The left hand side becomes $\int_V \mathbf{A}^t\mathbf{DAU}\, dV$ which is **KU** with $\mathbf{K} = \int_V \mathbf{A}^t \mathbf{DA}\, dV$.

* Compare with Eq. (5.61) obtained for beams. Also see Eq. (a_{22}) of Sec. (5.10).

This is of the form (9.19) in which

$$\mathbf{K} = \int_v \mathbf{A}^t\mathbf{D}\mathbf{A}\, dV \tag{9.22}$$

$$\mathbf{P}_E = \mathbf{P}_{Eg} + \mathbf{P}_{Ep} = \int_v \mathbf{N}^t\mathbf{g}\, dv + \int_{sp} \mathbf{N}^t\mathbf{p}\, dS \tag{9.23}$$

These are of the same form as of Eqs. (5.72, 73).

Relations (9.19), actually read

$$\begin{Bmatrix} \mathbf{P}_1 \\ \mathbf{P}_2 \\ \mathbf{P}_3 \\ \mathbf{P}_4 \end{Bmatrix} = \begin{bmatrix} \mathbf{K}_{11} & \mathbf{K}_{12} & \mathbf{K}_{13} & \mathbf{K}_{14} \\ \mathbf{K}_{21} & \cdots & \cdots & \cdots \\ \cdots & \text{etc} & & \cdots \\ \cdots & \cdots & \cdots & \mathbf{K}_{44} \end{bmatrix} \begin{Bmatrix} \mathbf{u}_1 \\ \mathbf{u}_2 \\ \mathbf{u}_3 \\ \mathbf{u}_4 \end{Bmatrix} + \begin{Bmatrix} \mathbf{P}_{F1} \\ \mathbf{P}_{F2} \\ \mathbf{P}_{F3} \\ \mathbf{P}_{F4} \end{Bmatrix} \tag{9.24}$$

This is obviously an extension of the general stiffness of the line element with two ends, (Eq. (8.16)), to an element with four 'ends' or nodes.

For the rectangular element in question, we can write the stiffness matrix, using matrix **A** from Eq. (9.16) and **D** from (9.8) or (9.10). The $\mathbf{K}_{ij}$ submatrix can be written as

$$\underset{(2\times 2)}{\mathbf{K}_{ij}} = \int_v \mathbf{A}_i^t\, \mathbf{D}\, \mathbf{A}_j\, dV \tag{9.25}$$

Presently we can integrate these explicitely. However it is best to carry it out within a computer. The explicit integrations are possible only in simple cases such as at hand, the obvious general method being one of numerical integration. The submatrix $\mathbf{K}_{11}$ for a plane stress element of constant thickness becomes

$$\mathbf{K}_{11} = \int_{-b}^{b}\int_{-a}^{a} \mathbf{A}_1^t\mathbf{D}\mathbf{A}_1 t\ dx\, dy$$

$$= abt \int_{-1}^{1}\int_{-1}^{1} (\mathbf{A}_1^t\mathbf{D}\mathbf{A}_1)\, d\xi\, d\eta$$

$$= \frac{Eabt}{1-\nu^2}\int_{-1}^{1}\int_{-1}^{1} \begin{bmatrix} -\frac{1-\eta}{4a} & 0 & -\frac{1-\xi}{4b} \\ 0 & -\frac{1-\xi}{4b} & -\frac{1-\eta}{4a} \end{bmatrix} \begin{bmatrix} 1 & \nu & 0 \\ \nu & 1 & 0 \\ 0 & 0 & \frac{1-\nu}{2} \end{bmatrix} \begin{bmatrix} -\frac{1-\eta}{4a} & 0 \\ 0 & -\frac{1-\xi}{4b} \\ -\frac{1-\xi}{4b} & -\frac{1-\eta}{4a} \end{bmatrix} d\xi\, d\eta \tag{a_{11}}$$

$$= \frac{Eabt}{1-\nu^2} \int_{-1}^{1} \int_{-1}^{1} \begin{bmatrix} \left(\frac{1-\eta}{4a}\right)^2 + \frac{1-\nu}{2}\left(\frac{1-\xi}{4b}\right)^2 & \frac{1+\nu}{2}\frac{1-\xi}{4b}\cdot\frac{1-\eta}{4a} \\ \frac{1+\nu}{2}\frac{1-\xi}{4b}\cdot\frac{1-\eta}{4a} & \left(\frac{1-\xi}{4b}\right)^2 + \frac{1-\nu}{2}\left(\frac{1-\eta}{4a}\right)^2 \end{bmatrix} d\xi \, d\eta \qquad (a_{12})$$

On evaluating these double integrals we obtain

$$\mathbf{K}_{11} = \frac{Et}{3(1-\nu^2)} \begin{bmatrix} \alpha + \frac{1-\nu}{2}\frac{1}{\alpha} & \frac{3}{8}(1+\nu) \\ \frac{3}{8}(1+\nu) & \frac{1}{\alpha} + \frac{1-\nu}{2}\alpha \end{bmatrix} \qquad (a_{13})$$

with $\alpha = a/b$.

Equivalent loads

Let us evaluate the first term of Eq. (9.23) for the simple case of

$$g_x = X_0, \quad g_y = Y_0$$

$$P_{E0} = \int \mathbf{N}^t \mathbf{g} \, dV = \int_{-1}^{1} \int_{-1}^{1} \begin{bmatrix} N_1 & 0 \\ 0 & N_1 \\ \hline N_2 & 0 \\ 0 & N_2 \\ \hline & \text{etc.} \end{bmatrix} \begin{Bmatrix} X_0 \\ Y_0 \end{Bmatrix} t(ab) \, d\xi \, d\eta \qquad (a_{14})$$

$$= \frac{(4abt)}{4} \left\{ X_0, Y_0 \mid X_0, Y_0 \mid X_0, Y_0 \mid X_0, Y_0 \right\} \qquad (a_{15})$$

These show that the total loads ($4abt\ X_0$) and ($4abt\ Y_0$) are equally distributed to all the nodes. These are the equivalent nodal loads.

Let us next consider deriving similar equivalent loads due to edge forces. The integration in Eq. (9.23) has to be carried out along the edges. The displacement variations can be determined from Eq. (9.15) in which we set ξ or $\eta = \pm 1$ approximately for the different edges:

$$\mathbf{u}=\begin{Bmatrix} u_{12} \\ v_{12} \\ \hline u_{43} \\ v_{43} \\ \hline u_{14} \\ v_{14} \\ \hline u_{23} \\ v_{23} \end{Bmatrix}=\left[\begin{array}{cc|cc|cc|cc} \frac{1-\xi}{2} & 0 & \frac{1+\xi}{2} & 0 & & & & \\ 0 & \frac{1-\xi}{2} & 0 & \frac{1+\xi}{2} & & & & \\ \hline & & & & \frac{1+\xi}{2} & 0 & \frac{1-\xi}{2} & 0 \\ & & & & 0 & \frac{1+\xi}{2} & 0 & \frac{1-\xi}{2} \\ \hline \frac{1-\eta}{2} & 0 & & & & & \frac{1+\eta}{2} & 0 \\ 0 & \frac{1-\eta}{2} & & & & & 0 & \frac{1+\eta}{2} \\ \hline & & \frac{1-\eta}{2} & 0 & \frac{1+\eta}{2} & 0 & & \\ & & 0 & \frac{1-\eta}{2} & 0 & \frac{1+\eta}{2} & & \end{array}\right]$$

$$\begin{Bmatrix} u_1 \\ v_1 \\ \hline u_2 \\ v_2 \\ \hline u_3 \\ v_3 \\ \hline u_4 \\ v_4 \end{Bmatrix} = \mathbf{NU} \qquad (a_{16})$$

The equivalent loads are therefore obtained as follows:

$$\left\{\begin{array}{c} P_{Ex1} \\ P_{Ey1} \\ \hline P_{Ex2} \\ P_{Ey2} \\ \hline P_{Ex3} \\ P_{Ey3} \\ \hline P_{Ex4} \\ P_{Ey4} \end{array}\right\} = \int_{-1}^{1} \left[\begin{array}{cc|cc|cc|cc} \frac{1-\xi}{2} & 0 & & & \frac{1-\eta}{2} & 0 & & \\ 0 & \frac{1-\xi}{2} & & & 0 & \frac{1-\eta}{2} & & \\ \hline \frac{1+\xi}{2} & 0 & & & & & \frac{1-\eta}{2} & 0 \\ 0 & \frac{1+\xi}{2} & & & & & 0 & \frac{1-\eta}{2} \\ \hline & & \frac{1+\xi}{2} & 0 & & & \frac{1+\eta}{2} & 0 \\ & & 0 & \frac{1+\xi}{2} & & & 0 & \frac{1+\eta}{2} \\ \hline & & \frac{1-\xi}{2} & 0 & \frac{1+\eta}{2} & 0 & & \\ & & 0 & \frac{1-\xi}{2} & 0 & \frac{1+\eta}{2} & & \end{array}\right]$$

$$\times \lambda \left\{\begin{array}{c} p_{x12} \\ p_{y12} \\ \hline p_{x43} \\ p_{y43} \\ \hline p_{x14} \\ p_{y14} \\ \hline p_{x23} \\ p_{y23} \end{array}\right\} dS \qquad (a_{17})$$

We can evaluate these integrals since the variation of edge loads would be known. For illustration let us take only p_{y12} linearly varying from q_1 to q_2 along edge 12 above, Fig. 9.10. Then,

$$p_{y12} = \frac{1-\xi}{2} q_1 + \frac{1+\xi}{2} q_2 \qquad (a_{18})$$

We therefore see from (a_{17}), that, we obtain

$$P_{Ey1} = \int_{-1}^{1} \frac{1-\xi}{2} p_{y12} \, (a \, d\xi); \quad P_{Ey2} = \int_{-1}^{1} \frac{1+\xi}{2} p_{y12} \; a \, d\xi$$

with all other forces zero. Using (a_{18}) and integrating we obtain

$$P_{Ey1} = \frac{a}{3}(2q_1+q_2);\ P_{Ey2} = \frac{a}{3}(q_1+2q_2) \qquad (a_{19})$$

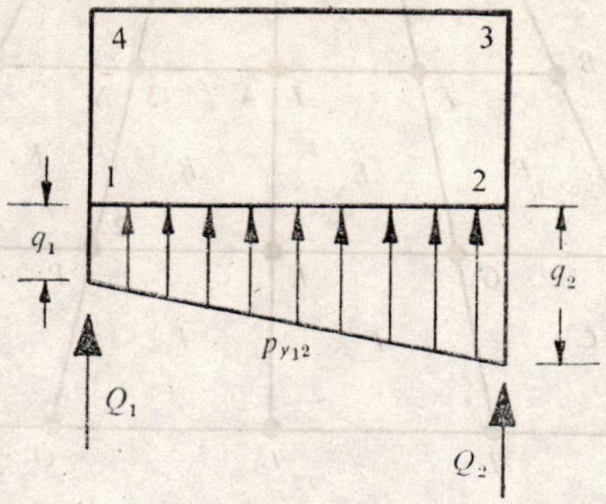

Fig. 9.10.

These loads are the same as the reactions obtained by treating the edge as a simply supported beam. This is consistent with the assumption of linear variation of displacements along the edge. Note the generality of the procedure enabling us to derive the equivalent loads consistent with an assumed general variation of edge displacement. Also note that we can exercise our own judgement and replace distributed loads by statically equivalent nodal loads, without having to follow the process just described.

f. Global Assembly

Having derived the element stiffness and the equivalent nodal loads we proceed to construct the global equations for the assembly of the elements. At each node the conditions of equilibrium are written stating that the sum of the nodal forces of the elements meeting at the node in question be zero. Also we can consider the actual applied concentrated loads in the nodes. However unlike in bar frameworks concentrated loads in the context of a continuum analysis are to be avoided.

The connectivity data would now be an array having as many rows as are elements and, as many columns as are nodes per element, in this case 4; this data for the example shown in Fig. 9.11, where quardrilateral elements are taken rather than rectangular ones (which however does not matter for the present discussion), will be as follows:

Element	Node 1	Node 2	Node 3	Node 4
A	B	F	E	A
B	C	G	F	B
'	'	'	'	'
'	'	'	'	'
R	K	P	N	J

(a_{20})

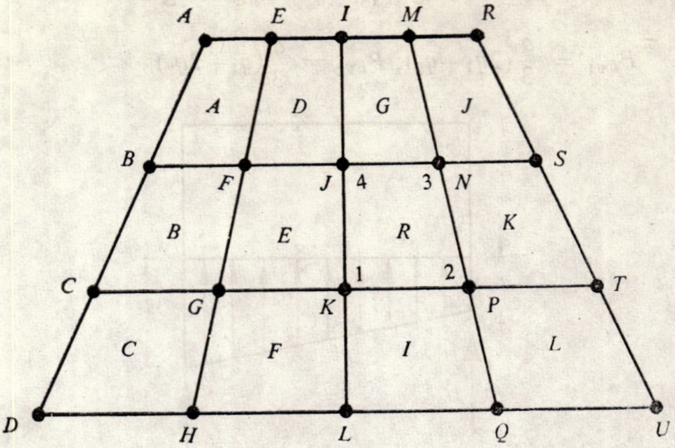

Fig. 9.11

The contribution of an element to the global stiffness matrix follows the same pattern as for a line element dealt with in Chapter 8. The contributions of the typical R^{th} member, to the global stiffness matrix and the load factor are as shown below:

$$\begin{array}{c} \\ J \\ \vdots \\ K \\ \vdots \\ N \\ \vdots \\ P \\ \\ \end{array} \begin{array}{c} \quad J \qquad\quad K \qquad\quad N \qquad\quad P \\ \begin{bmatrix} & \vdots & & \vdots & & \vdots & & \vdots & \\ \cdots & \mathbf{K}_{44} & \cdots & \mathbf{K}_{41} & \cdots & \mathbf{K}_{43} & & \mathbf{K}_{42} & \cdots \\ & \vdots & & \vdots & & \vdots & & \vdots & \\ \cdots & \mathbf{K}_{14} & \cdots & \mathbf{K}_{11} & \cdots & \mathbf{K}_{13} & & \mathbf{K}_{12} & \cdots \\ & \vdots & & \vdots & & \vdots & & \vdots & \\ \cdots & \mathbf{K}_{34} & \cdots & \mathbf{K}_{31} & \cdots & \mathbf{K}_{33} & & \mathbf{K}_{32} & \cdots \\ & \vdots & & \vdots & & \vdots & & \vdots & \\ \cdots & \mathbf{K}_{24} & \cdots & \mathbf{K}_{21} & \cdots & \mathbf{K}_{23} & & \mathbf{K}_{22} & \cdots \\ & \vdots & & \vdots & & \vdots & & \vdots & \end{bmatrix} \end{array} \begin{Bmatrix} \vdots \\ \mathbf{U}_J \\ \vdots \\ \mathbf{U}_K \\ \vdots \\ \mathbf{U}_N \\ \vdots \\ \mathbf{U}_P \\ \vdots \end{Bmatrix} = \begin{Bmatrix} \vdots \\ \mathbf{P}_{E4} \\ \vdots \\ \mathbf{P}_{E1} \\ \vdots \\ \mathbf{P}_{E3} \\ \vdots \\ \mathbf{P}_{E2} \\ \vdots \end{Bmatrix} \qquad (9.26$$

It is seen that the connectivity data gives the desired information to locate the global node address of each node of the element. Thus when $\mathbf{K}_{ij}$ is calculated it is placed in the global stiffness at location with row number equal to the joint number corresponding i^{th} node and column equal to the joint number corresponding to the jth node of element. Similarly $\mathbf{P}_E$ vector is placed in the global load vector at location with row number equal to i^{th} node of element.

Having assembled the stiffness matrix and the load vector, the boundary conditions are effected. This consists in modifying the rows and columns either before calling a standard solution routine, or during the triangularization.

9.3 PARAMETRIC COORDINATE REPRESENTATIONS

The use of intrinsic or parametric coordinates proves very convenient in dealing with triangular, quadrilateral, and curvilinear elements. Observe that we can conveniently define element geometry and deformation shapes in coordinates local to the element, since only the definite integrals taken over the element are involved.

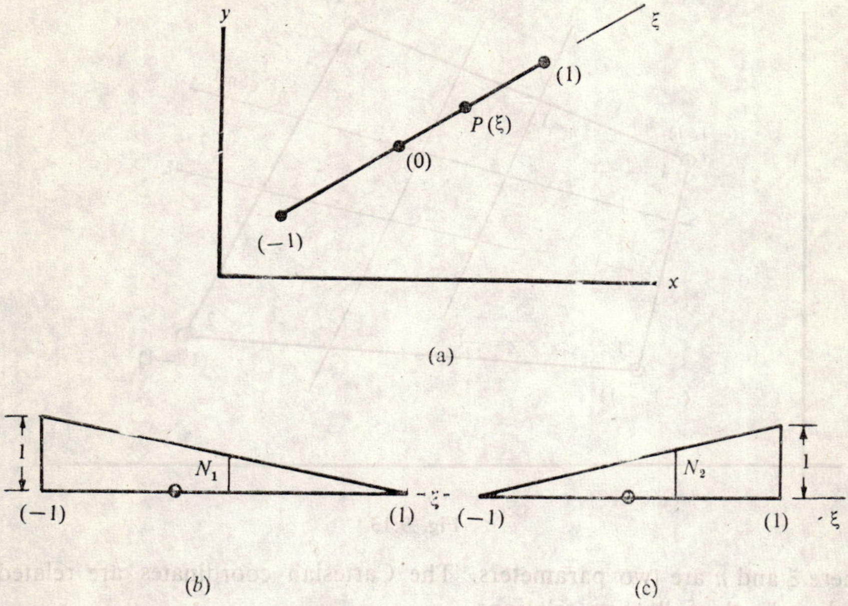

Fig. 9.12

Consider the straight line shown in Fig 9.12. The relationship between the coordinates (x, y) of any point P on the line depends on coordinates (x_1, y_1) and (x_2, y_2) of the two end points. Cartesian equation is not quite convenient. Alternatively we can define a point P in terms of a single local (parametric) coordinate ξ. Both x and y can be expressed as a linear function of ξ. There can be more than one choices for this parameter. One such choice is,

$$x = \frac{1-\xi}{2} x_1 + \frac{1+\xi}{2} x_2 = N_1 x_1 + N_2 x_2$$

$$y = \frac{1-\xi}{2} y_1 + \frac{1+\xi}{2} y_2 = N_1 y_1 + N_2 y_2 \tag{9.27}$$

Here N_1 and N_2 are nondimensional shape functions. These can also be called *interpolation* functions. It is seen that the end points have coordinates (-1) and (1). *A* given value of ξ fixes a pair of values (x, y) of a point P lying on the straight line. $\xi = 0$ corresponds to the midpoint

$(x_1+x_{,2})/2)$, $(y_1+y_2)/2$. Thus we can identify ξ to be an axis along the straight line with origin at midpoint of the given segment. Fig 9.12 (*c,d*) show the shape functions $N_1=(1-\xi)/2$, $N_2=(1+\xi)/2$, having the usual property that $N_i=1$ at end i with $N_j=0$, j being the other end.

The above reasoning admits two and three dimensional generalizations. Figure 9.13 shows a quadrilateral figure. Any point P is defined as P (ξ,η),

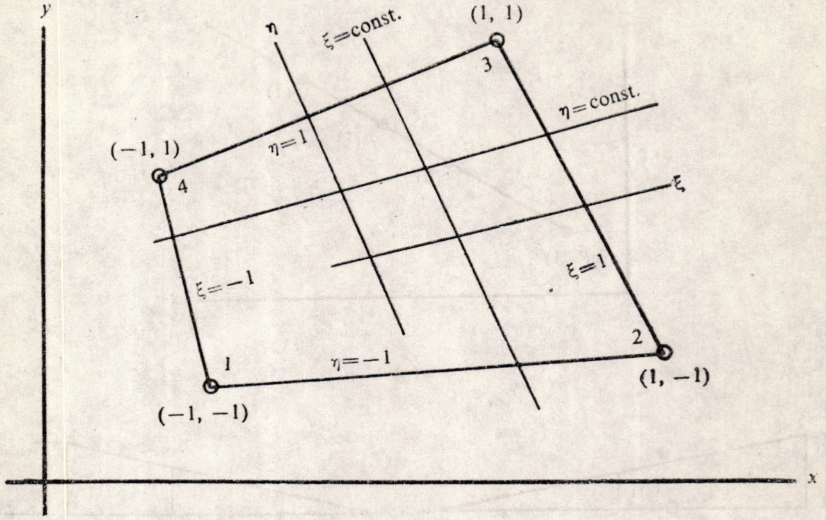

Fig. 9.13

where ξ and n are two parameters. The Cartesian coordinates are related to these by the following relations

$$x=N_1x_1+N_2x_2+N_3x_3+N_4x_4=\sum_1^4 N_ix_i$$

$$y=N_1y_1+N_2y_2+N_3y_3+N_4y_4=\sum_1^4 N_iy_i \tag{9.28}$$

where

$$N_i=\tfrac{1}{4}(1+\xi_i\xi)(1+\eta_i\eta), \quad i=1,2,3,4 \tag{9.29}$$

These shape functions are the *same* as (9.14). The (ξ,η) coordinates of the four corners of the quadrilateral are $(-1,-1)$, $(1,-1)$, $(1,1)$ and $(-1, 1)$. Thus $N_1=\frac{1}{4}(1-\xi)(1-\eta)$, $N_3=\frac{1}{4}(1+\xi)(1+\eta)$, and so on. Observe that ξ=const and η=const. represent straight lines. The origin for (ξ, η) is at a point 0 with Cartesian coordinates $(\frac{1}{4}(x_1+x_2+x_3+x_4)$, $\frac{1}{4}(y_1+y_2+y_3+y_4))$ which is the average point on the figure. Observe that lines ξ=const. and η=constant are nonorthogonal unlike the straight lines x=const. and y=const. Hence they can represent quardrilateral figures. Shape functions (9.29) will be called 'linear' shape functions.

Consider next the 'quadratic' extensions of the above arguments. Figures 9.14 and 9.15 show these generalizations. Any point P on the curve passing through three points can be identified by a coordinate ξ. The relationship between Cartesian coordinates and this can be a quadratic one since three points can define quadratic curve. These are as follows:

$$x = N_1 x_1 + N_2 x_2 + N_3 x_3 = \Sigma N_i x_i$$
$$y = N_1 y_1 + N_2 y_2 + N_3 y_3 = \Sigma N_i y_i \tag{9.30}$$

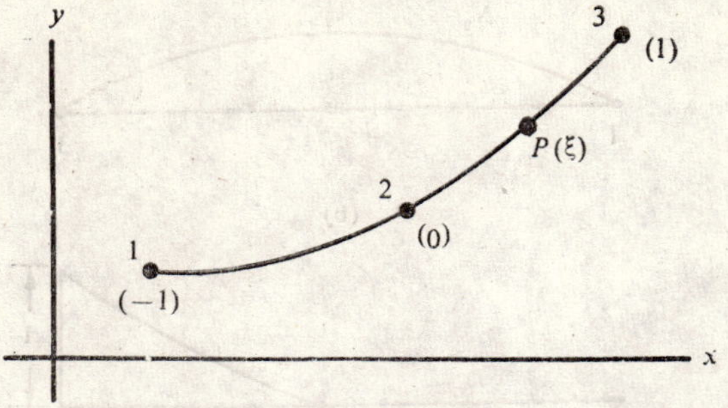

Fig. 9.14

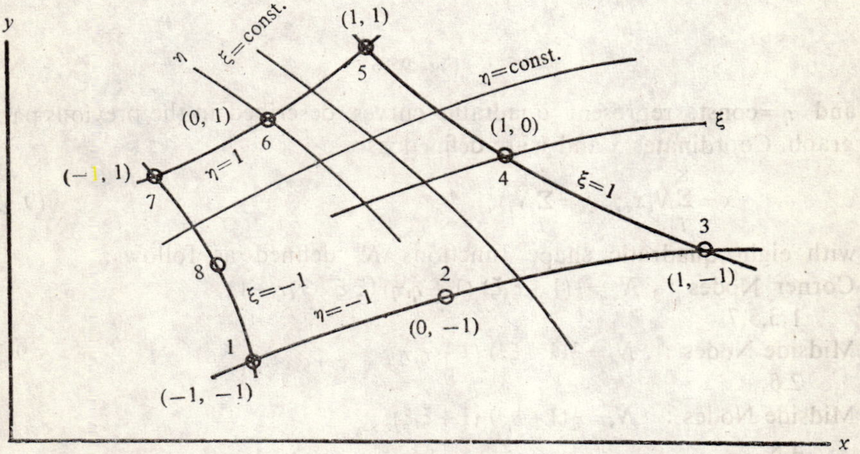

Fig. 9.15

with

$$N_1 = \frac{-\xi(1-\xi)}{2};\ N_2 = 1-\xi^2;\ N_3 = \frac{\xi(1+\xi)}{2} \tag{9.31}$$

These are sketched in Fig. 9.16. Points 1, 2 and 3 have coordinates $\xi = -1, 0$ and 1 respectively.

Figure 9.15 shows the curvilinear quadrilateral, with 8 nodes all lying along the edges as shown. The shape functions are so chosen that ξ=const.

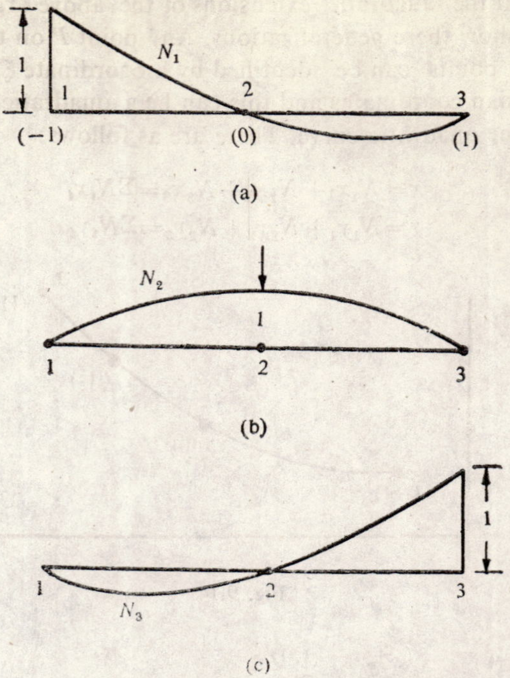

Fig. 9.16

and η=const. represent quadratic curves described in the previous paragraph. Coordinates x and y are defined as

$$x=\sum_{i}^{8} N_i x_i; \quad y=\sum_{i}^{8} N_i y_i \tag{9.32}$$

with eight quadratic shape functions N_i defined as follows:

Corner Nodes : $N_i=\frac{1}{4}(1+\xi_i\xi)(1+\eta_i\eta)(\xi_i\xi+\eta_i\eta-1)$
1,3,5,7

Midside Nodes : $N_i=\frac{1}{2}(1-\xi^2)(1+\eta_i\eta)$ 9.33
2,6

Midside Nodes : $N_i=\frac{1}{2}(1-\eta^2)(1+\xi_i\xi)$
4,8

The coordinates of nodes are as in Fig 9.15. Observe that, along edge 1−2−3, $\xi=-1$ and we otbain N_1, N_2 and N_3 as in (9.31).

The shape functions described so far belong to the so called 'Serendipity' family [26]. Generalization to represent volume elements is obvious. The details can be obtained from Reference [26]

Triangular Coordinates

For triangular element we adopt similar parametric forms, but with a slight change. Instead of (ξ, η), a set of apparently three different coordinates (ξ_1, ξ_2, ξ_3) is adopted but with an understanding that $\xi_1+\xi_2+\xi_3=1$. For a triangle we represent

$$x=\sum_i^3 N_i x_i \; ; \quad y=\sum_i^3 N_i y_i \tag{a$_1$}$$

where

$$N_i=\xi_i, \qquad i=1, 2, 3 \tag{9.34}$$

In this case we can explicitely obtain ξ_1, ξ_2 and ξ_3 in terms of x and y. Therefore,

$$\begin{bmatrix} 1 & 1 & 1 \\ x_1 & x_2 & x_3 \\ y_1 & y_2 & y_3 \end{bmatrix} \begin{Bmatrix} \xi_1 \\ \xi_2 \\ \xi_3 \end{Bmatrix} = \begin{Bmatrix} 1 \\ x \\ y \end{Bmatrix} \tag{a$_2$}$$

This can be inverted and the meaning of ξ_1, ξ_2 and ξ_3 identified :

$$\begin{Bmatrix} \xi_1 \\ \xi_2 \\ \xi_3 \end{Bmatrix} = \frac{1}{2A} \begin{bmatrix} a_1 & b_1 & c_1 \\ a_2 & b_2 & c_2 \\ a_3 & b_3 & c_3 \end{bmatrix} \begin{Bmatrix} 1 \\ x \\ y \end{Bmatrix} \tag{9.35}$$

where

$$a_1=x_2y_3-x_3y_2, \quad b_1=y_2-y_3, \quad c_1=x_3-x_2 \tag{9.36}$$

with a_2, b_2, c_2, etc., obtained by cyclic permutation of indices. A is the area of the triangle given by

$$2A=\begin{vmatrix} 1 & 1 & 1 \\ x_1 & x_2 & x_3 \\ y_1 & y_2 & y_3 \end{vmatrix} \tag{9.37}$$

We observe that ξ_i are linear functions of x and y. Thus ξ_i=const. denote straight lines and these are parallel to the sides (Fig. 9.17). Coordinates of vertices are (1, 0, 0), (0, 1, 0), and (0, 0, 1), and equations to the sides are $\xi_1=0$, $\xi_2=0$, and $\xi_3=0$. Any point $P(\xi_1, \xi_2, \xi_3)$ inside the triangle makes three triangles P_{23}, P_{13} and P_{12}. It can be shown that these respectively are $A\xi_1$, $A\xi_2$ and $A\xi_3$. Thus ξ_1, ξ_2 and ξ_3 have the meaning that they are the ratios of these linear triangles to that of the given triangle. Hence the alternative name 'area coordinates'. This also shows that $\xi_1+\xi_2+\xi_3=1$.

Curvilinear Triangle

We can define a curvilinear triangle using six nodes as shown in Fig. 9.18

$$x=\sum_i^6 N_i x_i \; ; \quad y=\sum_i^6 N_i y_i \tag{9.38}$$

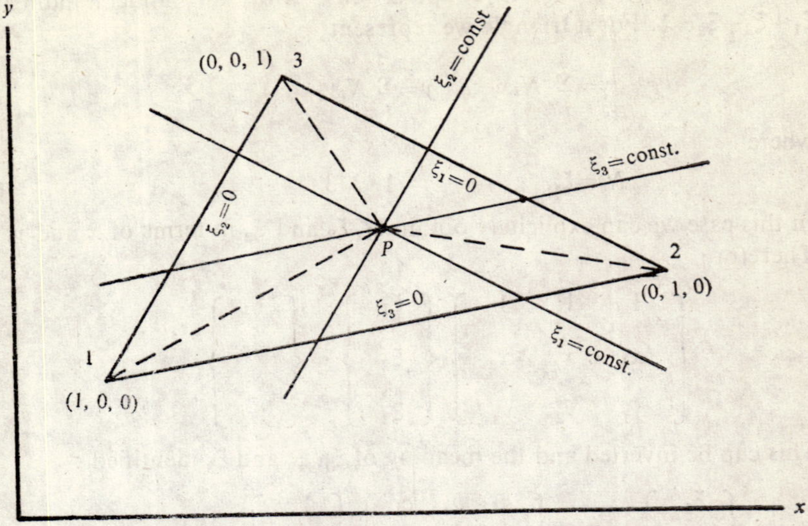

Fig. 9.17

where

$$
\begin{aligned}
N_1 &= \xi_1^2 - \xi_1\xi_2 - \xi_3\xi_1 = \xi_1(2\xi_1 - 1) \\
N_2 &= \xi_2^2 - \xi_1\xi_2 - \xi_2\xi_3 = \xi_2(2\xi_2 - 1) \\
N_3 &= \xi_3^2 - \xi_2\xi_3 - \xi_3\xi_1 = \xi_3(2\xi_3 - 1) \\
N_4 &= 4\xi_1\xi_2\ ;\ N_5 = 4\xi_2\xi_3\ ;\ N_6 = 4\xi_2\xi_1
\end{aligned}
\tag{9.39}
$$

The coordinates for the midside nodes are

$$(\tfrac{1}{2}, \tfrac{1}{2}, 0), (0, \tfrac{1}{2}, \tfrac{1}{2}), \text{ and } (\tfrac{1}{2}, 0, \tfrac{1}{2}).$$

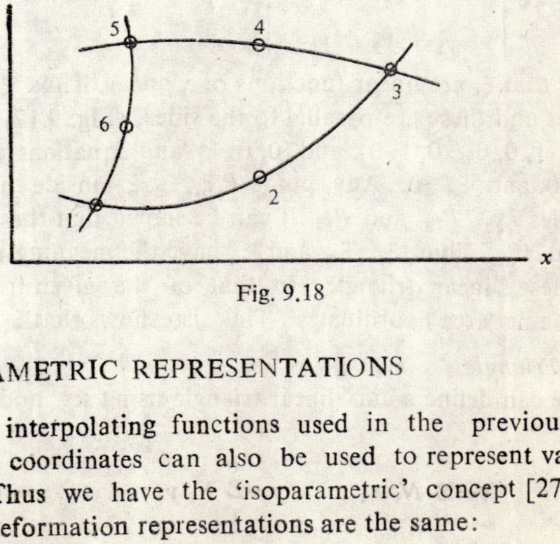

Fig. 9.18

9.4 ISOPARAMETRIC REPRESENTATIONS

The interpolating functions used in the previous section to represent the coordinates can also be used to represent variation of displacements. Thus we have the 'isoparametric' concept [27] in which the element and deformation representations are the same:

$$x = \Sigma N_i x_i \ ; \quad y = \Sigma N_i y_i$$
$$u = \Sigma N_i u_i \ ; \quad v = \Sigma N_i v_i \tag{9.40}$$

We may interpret coordinates (ξ, η) as a common coordinate system in terms of which both the element geometry and deformation patterns are expressed. Thus if we choose four nodes and use shape functions given by (9.29) we have the geometry of the quadrilateral shown in Fig. 9.13. The nodes are characterised by $(\pm 1, \pm 1)$. If we adopt the same shape functions for the deformation pattern we have a geometry of deformation expressed in terms of the displacements occurring at the same nodes. Thus a pair of values (ξ_i, η_i) defines the node i, and u or v evaluated at (ξ_i, η_i) becomes the displacement of the node i. Figure 9.19 shows the variation of N_i. Observe that it is quadratic in ξ and η. It gives linear variations for ξ=const. and η=const.

Let us now consider the two essential criteria to be satisfied by the deformation pattern assumed in terms of parametric coordinates. Consider the *first* criterion of obtaining constant strains as the element size is reduced. This is equivalent to having a linear form for the displacements u and v. Thus let $u = a + bx + cy$. Then

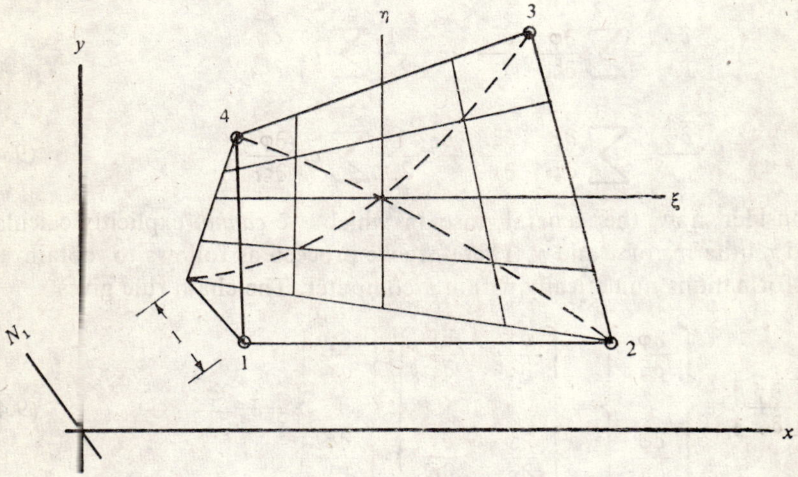

Fig. 9.19

$$u = a + b\,\Sigma N_i x_i + c\,\Sigma N_i y_i \tag{a$_1$}$$

In the limit when (x_i, y_i) tend to a common value (x_0, y_0) we obtain

$$u \rightarrow a + bx_0\,\Sigma N_i + cy_0\,\Sigma N_i \tag{a$_2$}$$

This would be a constant quantity if $\Sigma N_i = 1$. This is readily verified.

And now regarding the *second* criterion of interelement compatibility. Observe that the choice of shape functions for displacements along an edge in terms of only the edge nodes automatically implies satisfaction of this criterion. For instance the displacement variation along edge $1-2(\eta=-1)$ is obtained from shape functions (9.29) as

$$u=\frac{1-\xi}{2}\,u_1+\frac{1+\xi}{2}\,u_2$$

with similar expression for v. These are seen to depend on only the edge nodes. We can similarly verify that the quadratic variation along an edge is solely determined by the displacements of the three edge nodes.

Derivatives of Displacements

Since the components of strains involve derivatives with respect to x and y we need some transformations. Let φ be a typical displacement or some physical variable. Consider first the linear triangular coordinates $(\xi_1\ \xi_2,\ \xi_3)$ where it is relatively easier since ξ_1, ξ_2 and ξ_3 are explicitly known as linear functions of x and y, as stated in Eq. (9.35). Thus

$$\frac{\partial \xi_i}{\partial x}=\frac{1}{2A}\,b_i\,;\qquad \frac{\partial \xi_i}{\partial y}=\frac{1}{2A}\,c_i \tag{9.41}$$

Thus

$$\frac{\partial \varphi}{\partial x}=\sum^{3}\frac{\partial \varphi}{\partial \xi_i}\,\frac{\partial \xi_i}{\partial x}=\frac{1}{2A}\sum^{3} b_i\,\frac{\partial \varphi}{\partial \xi_i}$$

$$\frac{\partial \varphi}{\partial y}=\sum^{3}\frac{\partial \varphi}{\partial \xi_i}\,\frac{\partial \xi_i}{\partial y}=\frac{1}{2A}\sum^{3} c_i\,\frac{\partial \varphi}{\partial \xi_i} \tag{9.42}$$

Consider now the general case in which we *cannot* explicitly calculate ξ and η in terms of x and y. Therefore we proceed as follows to obtain the transformations numerically within a computer. The chain rule gives

$$\left\{\frac{\partial \varphi}{\partial \xi}\right\}=\begin{Bmatrix}\dfrac{\partial \varphi}{\partial \xi}\\[2ex] \dfrac{\partial \varphi}{\partial \eta}\end{Bmatrix}=\begin{bmatrix}\dfrac{\partial x}{\partial \xi} & \dfrac{\partial y}{\partial \xi}\\[2ex] \dfrac{\partial x}{\partial \eta} & \dfrac{\partial y}{\partial \eta}\end{bmatrix}\begin{Bmatrix}\dfrac{\partial \varphi}{\partial x}\\[2ex] \dfrac{\partial \varphi}{\partial y}\end{Bmatrix}=\mathbf{J}\,\frac{\partial \varphi}{\partial \mathbf{x}} \tag{9.43}$$

Here $\mathbf{J}$ is the Jacobian. On inversion

$$\frac{\partial \varphi}{\partial \mathbf{x}}=\begin{Bmatrix}\dfrac{\partial \varphi}{\partial x}\\[2ex] \dfrac{\partial \varphi}{\partial y}\end{Bmatrix}=\left[\ \mathbf{J}^{-1}\ \right]\begin{Bmatrix}\dfrac{\partial \varphi}{\partial \xi}\\[2ex] \dfrac{\partial \varphi}{\partial \eta}\end{Bmatrix}=\mathbf{J}^{-1}\left\{\frac{\partial \varphi}{\partial \xi}\right\} \tag{9.44}$$

Thus we need to compute $\mathbf{J}$ numerically for a given point $(\xi,\ \eta)$ and obtain its inverse. Using Eq. (9.40)

$$\mathbf{J}=\begin{bmatrix} \sum\left(\frac{\partial N_i}{\partial \xi}x_i\right) & \sum\left(\frac{\partial N_i}{\partial \xi}y_i\right) \\ \sum\left(\frac{\partial N_i}{\partial \eta}x_i\right) & \sum\left(\frac{\partial N_i}{\partial \eta}y_i\right) \end{bmatrix} \tag{9.45}$$

The derivatives of the shape functions (9.29) or (9.33) can be evaluated at desired values of ξ and η .*

Since integrations would be performed with respect to (ξ, η) coordinate system we need the expression for the elemental area formed by coordinate lines (ξ, η) and $(\xi+d\xi, \eta+d\eta)$. This can be obtained as

$$dA = dx\,dy = |\,\mathbf{J}\,|\,d\xi\,d\eta \tag{9.46}$$

where $|\,\mathbf{J}\,|$ stands for the determinant of the Jacobian.

The calculation of the Jacobian and differentiation rules for the curvilinear triangular element are similarly derived. See [26].

9.5 NUMERICAL INTEGRATION

The use of more general paramatric representation of element geometry and deformation shapes calls for numerical integration. The Gauss quadrature rule as applied to line elements was demonstrated in Sec. 5.11. We will now give the two dimensional generalizations [28].

The Gauss points and weighting factors given in Table 5.1 are used for both ξ and η directions. The integrations are carried out by a double summation as follows:

$$\begin{aligned} I &= \int_{-1}^{1}\int_{-1}^{1} f(\xi,\eta)\,d\xi\,d\eta = \int_{-1}^{1}\Big[\sum_{j=1}^{m} H_j\, f(\xi_j,\eta)\Big]d\eta \\ &= \sum_{1}^{n} H_i \sum_{1}^{n} H_j\, f(\xi_j,\eta_i) \\ &= \sum_{i=1}^{n}\sum_{j=1}^{n} H_i H_j\, f(\xi_j,\eta_i) \end{aligned} \tag{9.47}$$

It is usual though not necessary to employ same order of integration. The 3×3 scheme for $n=3$ is as shown in Fig. 9.20.

The schemes of integration for a triangular element with one, three and six integrating points are as shown in Fig. 9.20. They are Hammer's rule [26]. An alternative scheme—Radau's—are also available [26]. The form is,

$$I = \int_A (\xi_1\,\xi_2\,\xi_3)\,dA = A\sum_{1}^{n} H_i\, f(\xi_{1i}, \xi_{2i}, \xi_{3i}) \tag{9.48}$$

where A is the area of triangle.

*See subroutine SF2, Sec. 10.4.

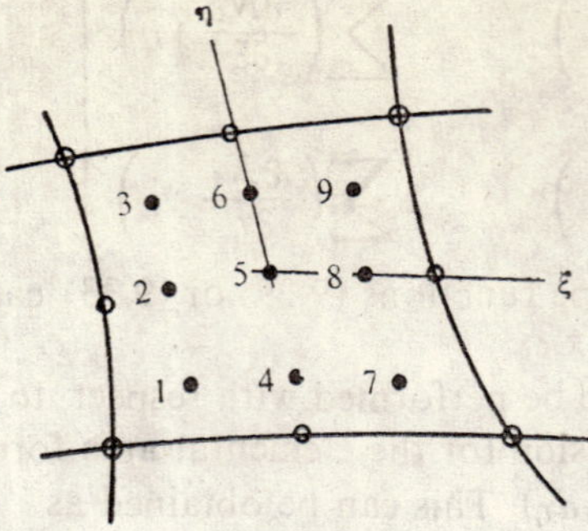

Fig. 9.20

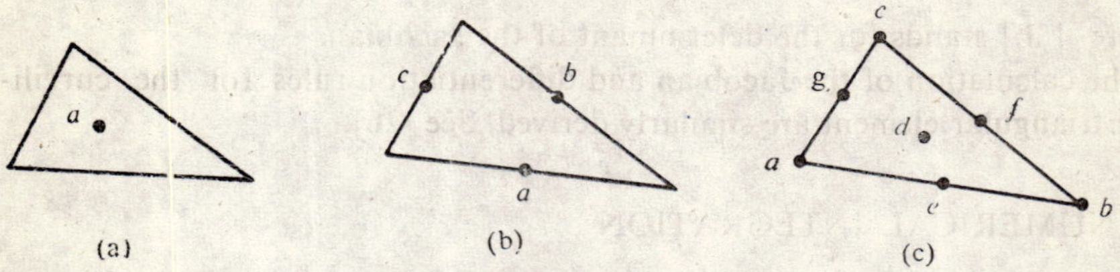

Fig. 9.21

The points and weighting factors are as follows,

Fig.	Point (ξ_1, ξ_2, ξ_3)	Weight factor H_i
9.21 (a)	a(1/3, 1/3, 1/3)	1
9.21 (b)	a(1/2, 1/2, 0)	1/3
	b(0, 1/2, 1/2)	1/3
	c(1/2, 0, 1/2)	1/3
9.21 (c)	a(1, 0, 0)	1/20 (9.49)
	b(0, 1, 0)	1/20
	c(0, 0, 1)	1/20
	d(1/3, 1/3, 1/3)	9/20
	e(1/2, 1/2 0)	2/15
	f(0, 1/2, 1/2)	2/15
	g(1/2, 0, 1/2)	2/15

Some of the integrals have closed form solutions. These are as follows:

$$\int \xi_1^p \, \xi_2^q \, \xi_3^r \, dA = 2A \; \frac{p!\, q!\, r!}{(p+q+r+2)!} \tag{9.50}$$

For instance $\int \xi_1 \xi_2 \xi_3 \, dA = \frac{A}{60}$, etc. This exact value is obtained by the numerical scheme with 7 integrating points:

$$\int \xi_1 \xi_2 \xi_3 \, dA = A \left\{ \frac{1}{20}\,(0+0+0) + \frac{9}{20}\left(\frac{1}{3} + \frac{1}{3} + \frac{1}{3}\right) + \frac{2}{15}(0+0+0) \right\}$$

$$= \frac{A}{60}$$

Considerable number of alternative integration rules are available. We have presented here one commonly adopted for numerical integration in structural and solid mechanics.

9.6 CONSTANT STRAIN TRIANGLE

Here we choose

$$x = \sum^{3} N_i x_i; \qquad y = \sum^{3} N_i y_i \tag{9.51}$$

$$u = \sum^{3} N_i u_i; \qquad v = \sum^{3} N_i v_i$$

with N_i given simply as

$$N_i = \xi_i, \quad i = 1, 2, 3 \tag{9.52}$$

Recall that ξ_i are linear functions of coordinates (Eq. 9.35). We have thus linear variation of displacements over an element. This is equivalent to a constant strain within the element. Hence the abbreviation CST. The strains are

$$\boldsymbol{\epsilon} = \begin{Bmatrix} \epsilon_x \\ \epsilon_y \\ \gamma_{xy} \end{Bmatrix} = \begin{bmatrix} \dfrac{\partial}{\partial x} & 0 \\ 0 & \dfrac{\partial}{\partial y} \\ \dfrac{\partial}{\partial y} & \dfrac{\partial}{\partial x} \end{bmatrix} \begin{Bmatrix} \sum^{3} N_i u_i \\ \sum^{3} N_i v_i \end{Bmatrix}$$

$$= \sum^{3} \begin{bmatrix} \dfrac{\partial N_i}{\partial x} & 0 \\ 0 & \dfrac{\partial N_i}{\partial y} \\ \dfrac{\partial N_i}{\partial y} & \dfrac{\partial N_i}{\partial x} \end{bmatrix} \begin{Bmatrix} u_i \\ v_i \end{Bmatrix} = \sum^{3} \mathbf{A}_i \mathbf{u}_i \tag{9.53}$$

Using Eq. (9.41) we observe that

$$\frac{\partial N_i}{\partial x} = \frac{1}{2A} b_i\, ; \qquad \frac{\partial N_i}{\partial y} = \frac{1}{2A} C_i \tag{9.54}$$

We thus have

$$\boldsymbol{\epsilon} = \sum^{3} \begin{bmatrix} b_i & 0 \\ 0 & c_i \\ c_i & b_i \end{bmatrix} \begin{Bmatrix} u_i \\ v_i \end{Bmatrix} = \sum^{3} \mathbf{A}_i \mathbf{u}_i \tag{9.55}$$

$$= \mathbf{AU}$$

which in full reads,

$$\boldsymbol{\epsilon} = [\, \mathbf{A}_1, \mathbf{A}_2, \mathbf{A}_3 \,] \begin{Bmatrix} \mathbf{u}_1 \\ \mathbf{u}_2 \\ \mathbf{u}_3 \end{Bmatrix} = \mathbf{AU} \tag{9.56}$$

Evaluation of the stiffness matrix $\int \mathbf{A}^t \mathbf{D} \mathbf{A} \, dV$ becomes trivial since $\mathbf{A}$ is constant. The stiffness relations will be of the form

$$\begin{Bmatrix} \mathbf{P}_1 \\ \mathbf{P}_2 \\ \mathbf{P}_3 \end{Bmatrix} = \begin{bmatrix} \mathbf{K}_{11} & \mathbf{K}_{12} & \mathbf{K}_{13} \\ \mathbf{K}_{21} & \mathbf{K}_{22} & \mathbf{K}_{23} \\ \mathbf{K}_{31} & \mathbf{K}_{32} & \mathbf{K}_{33} \end{bmatrix} \begin{Bmatrix} \mathbf{u}_1 \\ \mathbf{u}_2 \\ \mathbf{u}_3 \end{Bmatrix} + \begin{Bmatrix} \mathbf{P}_{F1} \\ \mathbf{P}_{F2} \\ \mathbf{P}_{F3} \end{Bmatrix} \tag{9.57}$$

The typical $\mathbf{K}_{ij}$ (for $i\,;j = 1, 2, 3$) is seen to be

$$\mathbf{K}_{ij} = \int\int \mathbf{A}_i \mathbf{D} \mathbf{A}_j \, (t \, dx \, dy) \tag{9.58}$$

If the thickness is constant then,

$$\mathbf{K}_{ij} = At \, [\mathbf{A}_i \mathbf{D} \mathbf{A}_j] \tag{9.59}$$

where A is the area of the triangle.

Consider now the possibility that the thickness varies linearly over the element. We represent this variation as

$$t = \mathbf{t} = [\,N_1 \quad N_2 \quad N_3\,] \begin{Bmatrix} t_1 \\ t_2 \\ t_3 \end{Bmatrix} = \Sigma N_i \, t_i \tag{a_1}$$

We have then

$$\mathbf{K}_{ij} = \int_A \mathbf{A}_i{}^t \, \mathbf{D} \, \mathbf{A}_j \, (\Sigma N_i t_i) \, dA \tag{a_2}$$

Each term requires to be multiplied by $\int\int(\Sigma \; N_i \; t_i) \, dA$

This multiplying factor is

$$\int_A (\xi_1 \, t_1 + \xi_2 \, t_2 + \xi_3) \, dA \tag{a_3}$$

We can use the formula (9.50) and obtain

$$\int \xi_3 \, dA = \int \xi_2 dA = \int \xi_3 \, dA = A/3$$

Thus the above factor is $(A/3)\,(t_1+t_2+t_3) = \bar{t}$, where $\bar{t}$ is the average thickness. Thus we can adopt the average value for the thickness and evaluate the stiffness as for an element of constant thickness.

Body Forces

If $\mathrm{g} = \{X_0, Y_0\}$ are the body forces then the equivalent loads $\mathbf{P}_{Eg}$ are (See Eq. (9.23))

$$P_{Eg} = \int \mathbf{N}^t \, \mathrm{g} dV = \int_A \begin{bmatrix} \xi_1 & 0 \\ 0 & \xi_1 \\ \hline \xi_2 & 0 \\ 0 & \xi_2 \\ \hline \xi_3 & 0 \\ 0 & \xi_3 \end{bmatrix} \begin{Bmatrix} X_0 \\ Y_0 \end{Bmatrix} t \, dA \tag{a_4}$$

If t, X_0 and Y_0 are constants, we have

$$P_{E_g} = At/3 \{X_0, Y_0 \mid X_0, Y_0 \mid X_0, Y_0\} \tag{9.60}$$

which are equivalent to replacing the loads by equal loads at the nodes.

Similarly we can show that for applied edge loads (**p**) we can treat each edge as a simple beam and replace the loads at nodes by equivalent loads, equal and opposite to the reactions. This is being left as an exercise to the reader.

9.7 CURVILINEAR ELEMENT

The 8 noded curved quadrilateral element is quite an efficient one. It can also be used to represent a straight edged quadrilateral. The coordinates of the nodes are arbitrary, except that the Jacobian **J** is not singular. This may happen when we have very ill conditioned shapes [26]. When the edges are straight the coordinates of midside nodes are interpolated within the computer in terms of coordinates or corner nodes. Consider now some essentials involved in computing element stiffness and equivalent loads.

Consider the isoparametric representations (9.40). As explained in Sec. 9.7 we can evaluate the transformation matrix **J**, (Eq. 9.45) at any desired value of ξ and η. To perform the required numerical integrations we evaluate it at integrating points. The kinematic matrix **A** is obtained as in the following relation:

$$\boldsymbol{\epsilon} = \begin{Bmatrix} \epsilon_x \\ \epsilon_y \\ \gamma_{xy} \end{Bmatrix} = \sum^{8} \begin{bmatrix} \dfrac{\partial N_i}{\partial x} & 0 \\ 0 & \dfrac{\partial N_i}{\partial y} \\ \dfrac{\partial N_i}{\partial y} & \dfrac{\partial N_i}{\partial x} \end{bmatrix} \begin{Bmatrix} u_i \\ v_i \end{Bmatrix} = \sum^{8} \mathbf{A}_i \mathbf{u}_i \tag{9.61}$$

This is formally the same as in (9.53) except the limit of summation over nodes. The derivatives $\partial N_i/\partial x$, $\partial N_i/\partial y$ are obtained by using (9.44):

$$\begin{Bmatrix} \dfrac{\partial N_i}{\partial x} \\ \dfrac{\partial N_i}{\partial y} \end{Bmatrix} = \begin{bmatrix} \mathbf{J}^{-1} \end{bmatrix} \begin{Bmatrix} \dfrac{\partial N_i}{\partial \xi} \\ \dfrac{\partial N_i}{\partial \eta} \end{Bmatrix} \tag{9.62}$$

The kinematic matrix **A** giving $\boldsymbol{\epsilon}$ as **AU** can be written as

$$\mathbf{A} = [\mathbf{A}_1, \mathbf{A}_2, \mathbf{A}_3 \ldots\ldots \mathbf{A}_8] \tag{9.63}$$

with $\mathbf{A}_i$ as in Eq. (9.61). Thus the stiffness matrix is obtained as

$$\mathbf{K}=\int \mathbf{A}^t\mathbf{D}\mathbf{A}dv=\begin{bmatrix} \mathbf{K}_{11}\ldots\mathbf{K}_{12}\ldots\mathbf{K}_{18} \\ \mathbf{K}_{21}\ldots\ldots\ldots\ldots \\ \vdots \\ \mathbf{K}_{81}\ldots\ldots\ldots\mathbf{K}_{88} \end{bmatrix} \tag{9.64}$$

The (2×2) submatrix $\mathbf{K}_{ij}$ is, as before, (See Eq (9.58))

$$\mathbf{K}_{ij}=\int \mathbf{A}_i{}^t\mathbf{D}\mathbf{A}_j\, d_v. \tag{9.65}$$

Here the elemental area is replaced by its equivalent (9.46). Thus the (2×2) submatrix becomes,

$$\mathbf{K}_{ij}=\int_{-1}^{1}\int_{-1}^{1} \mathbf{A}_i{}^t\mathbf{D}\mathbf{A}_j\, t\,|\mathbf{J}|\, d\xi d\eta \tag{9.66}$$

Each element is of the type

$$k=\int_{-1}^{1}\int_{-1}^{1} f(\xi,\eta)d\xi d\eta$$

which is evaluated as discussed in the previous section.

$$k=\sum_{j}^{m}\sum_{i}^{m} f(\xi_i,\eta_j)\,(t\,|\mathbf{J}|\,H_iH_j) \tag{9.67}$$

where m is the order of Gauss integration rule.

The computing details can be grasped by persuing some of the subroutines of program given in see 10.4. Subroutine GAUSS sets up integration constants ξ_i and H_i as tabulated in Sec. 5.11. Another subroutine *SF*2 determines, shape functions N_i, and $\partial N_i/\partial\xi$, $\partial N_i/\partial\eta$ at any given pair (ξ_i, η_i). A third subroutine DERIV computes the Cartesian derivatives $\partial N_i/\partial x$, $\partial N_i/\partial y$ and the inverse of Jacobian **J**, at each of the Gauss points. The subroutine ELSTIF then computes the elements of the stiffness matrix by first composing $\mathbf{A}_i$ matrix of Eq. 9.61; then $\mathbf{A}_i{}^t\,\mathbf{D}\,\mathbf{A}_j$ is evaluated and each term of it is multiplied by the scalar $(t\,|J|\,H_iH_j)$. The double summation is then performed over the number of Gauss points.

Body forces

Evaluation of equivalent nodal loads too involves numerical integration. Consider body forces:

$$\mathbf{P}_E = -\mathbf{P}_F = \int_{-1}^{1}\int_{-1}^{1} \mathbf{N}^t\mathbf{g}t\,|J|\,d\xi d\eta$$

$$=\Sigma\,\Sigma[\mathbf{N}(\xi_i\eta_i)]^t\mathbf{g}\,(\xi_j,\eta_i)t\,|J|\,H_iH_j \tag{9.68}$$

Here, N and g are evaluated at the Gauss points and the double summation carried out. Subroutine BODYF explains the details.

Edge loads

These are evaluated as

$$\mathbf{P}_E = -\mathbf{P}_F = \int \mathbf{N}^t \mathbf{p} ds \tag{9.69}$$

where the integration proceeds over the element edges. Equation to an edge is a function of either ξ or η. The prescribed edge loads p_x and p_y will also be functions of ξ or η. Thus

$$x = \Sigma N'_i x_i;\ y = \Sigma N'_i y_i \qquad (a_1)$$
$$p_x = \Sigma N_i' p_{xi};\ p_y = \Sigma N_i' p_{yi}$$

where N_i' are edge shape functions. Thus along edge 1—2—3, $\eta = -1$ and from Eq. (9.33)

$$N_1' = \frac{-\xi(1-\xi)}{2};\ N_2' = 1-\xi^2;\ N_3' = \frac{\xi(1+\xi)}{2} \qquad (a_2)$$

Along any edge p_{xi} and p_{yi} would be known. These would be the magnitudes of the force intensities p_x and p_y at 1, 2 and 3. Thus we assume that the variable intensities are 'isoparametric', in the sense that variation of edge shape and the edge forces are the same. (See Fig 9.22). These integral of Eq. (9.69) can be written as

$$\mathbf{P}_E = \int \mathbf{N}^t (p_x dy + p_y dx) \qquad (a_3)$$

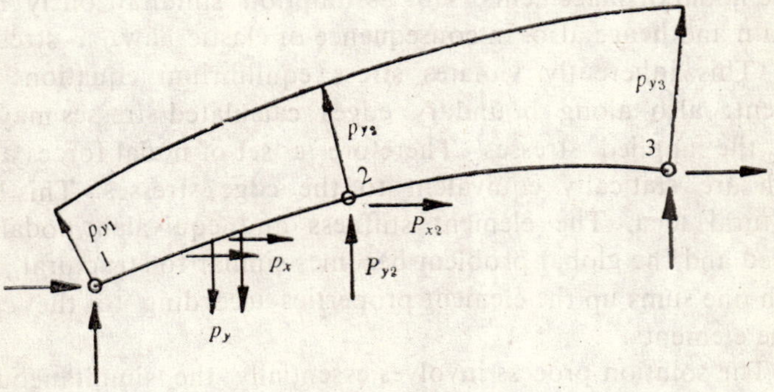

Fig. 9.22

with p_x and p_y components of the edge load intensities. Along any edge, dy and dx can be expressed in terms of $d\xi$ or $d\eta$. Thus along edge 1—2—3,

$$dx = \left(\sum \frac{\partial N_i'}{\partial \xi} x_i\right) d\xi = a d\xi$$
$$dy = \left(\sum \frac{\partial N_i'}{\partial \xi} y_i\right) d\xi = b d\xi \qquad (a_4)$$

where N'_i are as in Eq. (a_2). Thus Eq. (a_3) can be written for an edge k, as

$$\mathbf{P}_E=\int_{-1}^{1}\mathbf{N}_k^t(b_k p_{xk}+a_k p_{yk})\,(d\xi \text{ or } d\eta)$$

$$k=1,2,3,4 \qquad (a_5)$$

to be repeated over the number of edges, namely 4. The integral between -1 and 1 is replaced by its numerical integration form in terms of a summation over Gauss points.
Thus

$$\mathbf{P}_{Ek}=\Sigma_i[\mathbf{N}_k^t(b_k p_{xk}+a_k p_{yk})]_i H_i \qquad (9.70)$$

Subroutine PRESS in the program (See 10.4) explains the above steps. The above contents may be reread in conjunction with it.

9.8 CONCLUDING REMARKS

In retrospect the following features would be recalled:

a. The first step is the one in which we imagine that the given body is subdivided into a number of finite elements. The number of elements depends on its type. If higher order elements are used such as the eight noded curved quadrilateral ones, bigger size elements can be chosen since a higher order variation of stresses and strains are implied. In practice it is usual to repeat the problem by taking different number of elements.

b. The displacement variation within an element is assumed in terms of the nodal displacements. This assumption simultaneously means that a strain and hence also, in consequence of elastic law, a stress distribution. This inherently violates stress equilibrium equations within the element; also along boundary edges calculated stresses may not match with the applied stresses. Therefore a set of nodal forces are assumed which are statically equivalent to the edge stresses. This leads to the structural idea. The element stiffness and equivalent nodal forces are defined and the global problem becomes similar to structural analysis in which one sums up the element properties according to the connectivity of the elements.

c. The solution process involves essentially the simultaneous solution of the nodal displacements with the aid of the computer. The element analysis retraced with calculated nodal displacements determines stresses at any point within the element.

The scope of the method has now been fully exploited and quite a few texts and references are available [26, 34–36]. Generalizations to provide solutions to stress analysis of solids, surface structures like plates and shells are 'obvious,' but the associated details are important without which successful computer programs are not possible.

Methods of *successive elastic solutions* with material properties and loads varied successively each time, have provided solutions to nonlinear problems, and also to time dependent problems. The increasing power of computers has been widening the scope of the method and providing solutions to complex physical problems.

Section 10.4 of the next Chapter provides a simple program. Any student aspiring to use this tool effectively cannot do without grasping the associated programing aspects. The flow of computations is considerably influenced by the size of the problem. Any practical problem worth the name invariably exhausts the main core memory, if all the equations are assembled and triangularized. In Sec 10.4 however a simple extension of the plane frame program (Sec. 10.3) has been given and can be considered as a student program showing the essential aspects of finite element program excluding the complications introduced as a result of block solution of equations.

This Chapter is included not so much with a view to introduce the reader to this subject but to show the common features between the analysis of an assembly of line elements and of finite elements. The direct stiffness method is a general method which does not distinguish between a two noded line element and a multinoded finite element. In fact different types of element can occur together. Such situations are more common in design practice, such as edge beams and panels together constituting a stiffened thin walled element commonly encountered in aircraft structures, columns, beams, wall and floor panels of buildings, etc.

EXERCISES

9.1 Derive the differential equation for the bending of a beam starting from Eq. (9.1) and integrating the stresses across thickness and transforming to moments and shear forces.

9.2 Verify contragredience between Eqs. (9.1) and Eq. (9.5). Can we obtain similar relations using, Eqs. (9.2, 3, 4) instead of Eqs. (9.5)?

9.3 Derive the expressions of Sec. 9.2 using a coordinate system (ξ, η) with $x=2a\xi$ and $y=2b\xi$. This is equivalent to taking the origin at node 1 and (ξ, η) axes along edges 1-2 and 1-4 (Fig. 9.7). Appreciate the advantages of the coordinates chosen in Sec. 9.2.

9.4 Derive shape functions similar to (9.27) by taking the limits for ξ as 0 and 1.

9.5 Derive the quadratic shape functions (9.31) taking the points at $\xi=0$, $\frac{1}{2}$ and 1.

9.6 Derive shape functions (9.29) assuming φ in the form $a+b\xi+c\eta+d\xi\eta$. Why are ξ^2 and η^2 not included?

9.7 Consider the case of triangle with six nodes. The sides of the triangle are straight instead of curvilinear as in Fig. 9.18. We can adopt a 'Sub-parametric' analysis, in which we use only corner nodes for geometry with linear shape functions (9.34), and for displacements we use all six nodes with shape functions (9.39). Derive the expressions for stiffness and equivalent loads. Such a formulation leads to linear strains and the triangle called L.S.T. as opposed to C.S.T.

9.8 Derive Eq. (9.46) using vector algebra.

9.9 Derive Eq. (9.37) using vector algebra.

9.10 Consider a trapezium $ABCD$ with AD parallel to BC. $AB=2a$, $BC=2c$ and $AD=2b$. Take x and y axis along AB and AD. Determine the Jacobian (9.45). Consider the particular case $b=c$ and interpret J.

9.11 Consider C.S.T.; deduce the expressions for equivalent loads due to applied edge stresses as commented in the para below Eq. (9.60).

Chapter 10

Programming Aspects

Undoubtedly, computers are increasingly dominating the field of structural engineering. It is demanding from engineers more and more knowledge of programming and use of program packages and systems. The subject matter of analysis presented so far can be effectively utilized only when the aspects associated in implementing it on computers are also understood. This chapter provides the necessary information to supplement the computer oriented methods of analysis described in the text so far. It is supposed that the student has undergone the basic courses on computer programming. The details associated with this aspect are therefore excluded altogether.

The programs given in this chapter are expected to serve as sample programs to a beginner who encounters the task of programming for the first time. They also serve as examples to those who prefer to remain as program users. To them they serve as typical of utility programs. These programs undoubtedly are small compared to the large packages such as NASTRAN, GENESYS, ASKA etc. The purpose of this chapter is essentially to initiate the beginner to programming structural algorithms.

10.1 SOLUTION OF EQUATIONS

We will not attempt here to discuss the different possible schemes of equation solvers. The oldest and yet popular and efficient one is the Gauss elimination procedure with its possible variants specially designed to handle the problems of displacement method.

Before proceeding with the details we will discuss the Gauss elimination method keeping the following physical issues of the displacement method in view:

(a) convenience of handling multiple right hand sides, and

(b) modifications to handle prescribed displacements and determination of reactions.

The Gauss procedure consists of three distinct stages:

(i) triangularization of coefficient matrix,
(ii) modification of right hand side vector, and
(iii) back substitution.

The first operation is performed only once, while the second and the third operations are performed for each new right hand side (load case). It is therefore necessary to draw separate computer subroutines to achieve the above operations. The triangularized coefficient matrix holds coefficients which are needed in stages (ii) and (iii).

Consider the following small set of *symmetric* equations for illustration. Let us avoid discussion in terms of general sizes and algebraically indexed representations. This in no way would impair generality:

$$\begin{bmatrix} k_{11} & k_{12} & k_{13} & k_{14} \\ k_{12} & k_{22} & k_{23} & k_{24} \\ k_{13} & k_{23} & k_{33} & k_{34} \\ k_{14} & k_{24} & k_{34} & k_{44} \end{bmatrix} \begin{Bmatrix} u_1 \\ u_2 \\ u_3 \\ u_4 \end{Bmatrix} = \begin{Bmatrix} P_1 \\ P_2 \\ P_3 \\ P_4 \end{Bmatrix} \tag{a$_1$}$$

The procedure consists of systematic elimination. The variable u_1 is eliminated from the second, third and fourth equations. Each time, a certain multiple of the first equation is subtracted. Before this all the coefficients are divided by k_{11}:

$$\begin{bmatrix} 1 & \dfrac{k_{12}}{k_{11}} & \dfrac{k_{13}}{k_{11}} & \dfrac{k_{14}}{k_{11}} \\ k_{12} & k_{22} & k_{23} & k_{24} \\ k_{13} & k_{23} & k_{33} & k_{34} \\ k_{14} & k_{24} & k_{34} & k_{44} \end{bmatrix} \begin{Bmatrix} u_1 \\ u_2 \\ u_3 \\ u_4 \end{Bmatrix} = \begin{Bmatrix} \dfrac{P_1}{k_{11}} \\ P_2 \\ P_3 \\ P_4 \end{Bmatrix} \tag{a$_2$}$$

Thus k_{12} times the first equation is subtracted from the second, k_{13} times of again the first is subtracted from the third, and so on. Thus the modified elements are

$$\begin{bmatrix} 1 & \dfrac{k_{12}}{k_{11}} & \dfrac{k_{13}}{k_{11}} & \dfrac{k_{14}}{k_{11}} \\ 0 & k'_{22} & k'_{23} & k'_{24} \\ 0 & k'_{23} & k'_{33} & k'_{34} \\ 0 & k'_{24} & k'_{34} & k'_{44} \end{bmatrix} \begin{Bmatrix} u_1 \\ u_2 \\ u_3 \\ u_4 \end{Bmatrix} = \begin{Bmatrix} \dfrac{P_1}{k_{11}} \\ P'_2 \\ P'_3 \\ P'_4 \end{Bmatrix} \tag{a$_3$}$$

where

$$k'_{22} = k_{22} - \left(\frac{k_{12}}{k_{11}}\right) k_{12}$$

$$k'_{23} = k_{23} - \left(\frac{k_{12}}{k_{11}}\right) k_{13}$$

$$k'_{33} = k_{33} - \left(\frac{k_{13}}{k_{11}}\right) k_{13}, \text{ etc.} \qquad (a_4)$$

and

$$P'_2 = P_2 - \left(\frac{k_{12}}{k_{11}}\right) P_1$$

$$P'_3 = P_3 - \left(\frac{k_{13}}{k_{11}}\right) P_1, \text{ etc.}$$

The second variable is now eliminated by repeating the above process. Thus, finally we reduce them to the following:

$$\begin{bmatrix} 1 & k'_{12} & k'_{13} & k'_{14} \\ & 1 & k'_{23} & k'_{24} \\ & & 1 & k'_{34} \\ & & & 1 \end{bmatrix} = \begin{Bmatrix} u_1 \\ u_2 \\ u_3 \\ u_4 \end{Bmatrix} = \begin{Bmatrix} P'_1/k'_{11} \\ P'_2/k'_{22} \\ P'_3/k'_{33} \\ P'_4/k'_{44} \end{Bmatrix} \qquad (a_5)$$

Here k' and P' are the *successively* modified coefficients and load terms. The general pattern is as in (a_4). The diagonal coefficient k_{ii} is termed the *pivot*, which for structural problems is required to be positive for an admissible solution.

Thus the elimination process reduces the original coefficient matrix to a triangular matrix and hence its alternative name 'triangularization'. Also the right hand side vector is modified by the same multipliers.

The calculation of the unknowns is the process of 'back substitution'. The last equation gives u_4, the third gives u_3 and so on:

$$u_4 = P'_4/k'_{44}$$

$$u_3 = \frac{P'_3}{k'_{33}} - k'_{34}u_4$$

$$u_2 = \frac{P'_2}{k'_{22}} - k'_{24}u_4 - k'_{23}u_3 \qquad (a_6)$$

$$u_1 = \frac{P'_1}{k_{11}} - k'_{14}u_4 - k'_{13}u_3 - k'_{12}u_2$$

We can see that solution of equations for a different right hand side, entails only the modification of the right hand side. The triangularized matrix of coefficients of (a_5) carries the multipliers which are required in modifying the right hand side with an exception of pivotal terms. Therefore instead of storing 1's in the leading diagonal locations the pivots k'_{ii} are stored. Thus the triangularized coefficient matrix of Eq. (a_5) is modified to hold the coefficients as shown in (a_7)

$$\begin{bmatrix} \boxed{k'_{11}} & k'_{12} & k'_{13} & k'_{14} \\ 0 & \boxed{k'_{22}} & k'_{23} & k'_{24} \\ 0 & 0 & \boxed{k'_{33}} & k'_{34} \\ 0 & 0 & 0 & \boxed{k'_{44}} \end{bmatrix} \tag{a_7}$$

Since the new set of right hand side elements are modified using the elements of the triangularized matrix, the computer codes generally have two routines; the first one alters the coefficients matrix and stores the modified elements in the same matrix. The second routine includes the modification of right hand side elements and back substitution. The first routine, which we may term 'elimination' is called only once and the second routine, 'solve' can be called as many times as are right hand sides.

Prescribed variables and reactions

Suppose $u_2 = u_2^*$ is prescribed with a corresponding unknown R_2 to be determined.

$$\begin{bmatrix} k_{11} & k_{12} & k_{13} & k_{14} \\ k_{12} & k_{22} & k_{23} & k_{24} \\ k_{13} & k_{23} & k_{33} & k_{34} \\ k_{14} & k_{24} & k_{34} & k_{44} \end{bmatrix} \begin{Bmatrix} u_1 \\ u_2 \\ u_3 \\ u_4 \end{Bmatrix} = \begin{Bmatrix} P_1 \\ P_2+R_2 \\ P_3 \\ P_4 \end{Bmatrix} \tag{a_8}$$

The equations to be solved are,

$$\begin{bmatrix} k_{11} & 0 & k_{13} & k_{14} \\ 0 & 1 & 0 & 0 \\ k_{13} & 0 & k_{33} & k_{34} \\ k_{14} & 0 & k_{34} & k_{44} \end{bmatrix} \begin{Bmatrix} u_1 \\ u_2 \\ u_3 \\ u_4 \end{Bmatrix} = \begin{Bmatrix} P_1-k_{12}\,u_2^* \\ u_2^* \\ P_3-k_{23}\,u_2^* \\ P_4-k_{24}\,u_2^* \end{Bmatrix} \tag{a_9}$$

and

$$R_2 = \lfloor k_{12} \quad k_{22} \quad k_{23} \quad k_{24} \rfloor \begin{Bmatrix} u_1 \\ u_2 \\ u_3 \\ u_4 \end{Bmatrix} - \{P_2\} \tag{a_{10}}$$

The set of Eq. (a_9) can be solved as usual and after solving we can obtain the reactions using (a_{10}). Observe that the coefficients which were set equal to zero or 1 are the ones which are again required in computing the reaction. Instead of zeros and ones we can continue to store the original coefficients, but modify the elimination and back substitution operations appropriately as governed by Eq. (a_9).

The above issue becomes quite significant in large size problems where it would be uneconomical to separately store the coefficients to compute reactions as required in (a_{10}). It is not very difficult to retain these in the triangularized matrix itself so that they can be reused in computing reactions and in modifying new right hand sides. The triangularized matrix is held on a backing store and can always be called in.

Eliminate the first variable in (a_8) and obtain

$$\boxed{k_{11}} \rightarrow \begin{bmatrix} 1 & k'_{12} & k'_{13} & k'_{14} \\ 0 & k'_{22} & k'_{23} & k'_{24} \\ 0 & k'_{23} & k'_{33} & k'_{34} \\ 0 & k'_{24} & k'_{34} & k'_{44} \end{bmatrix} \begin{Bmatrix} u_1 \\ u_2 \\ u_3 \\ u_4 \end{Bmatrix} = \begin{Bmatrix} P'_1 \\ P'_2 + R_2 \\ P'_3 \\ P'_4 \end{Bmatrix} \tag{a_{11}}$$

The pivot element k_{11} is actually stored in place of 1 as shown.

Since the second variable is a prescribed one we modify the operation of elimination. We do not perform the row operations, but simply place 1 in the diagonal location and zeros on the row positions as required in (a_9). However, while programming we actually store the pivot element in its location, and *change* the signs of the row elements :

$$\boxed{k_{11}} \rightarrow \begin{bmatrix} 1 & k'_{12} & k'_{13} & k'_{14} \\ & & & \boxed{-k'_{24}} \rightarrow \\ 0 & -1 \leftarrow \boxed{k'_{22}} & 0 \leftarrow \boxed{-k'_{23}} & 0 \\ 0 & 0 & k'_{33} & k'_{34} \\ 0 & 0 & k'_{34} & k'_{44} \end{bmatrix} \begin{Bmatrix} u_1 \\ u_2 \\ u_3 \\ u_4 \end{Bmatrix} = \begin{Bmatrix} P'_1 \; (\leftarrow \boxed{P'_2}) \\ -u_2^* \; (\leftarrow \boxed{P'_2 + k_{22}u_2^*}) \\ P_3 - k'_{23}u_2^* \\ P_4 - k'_{24}u_2^* \end{Bmatrix} \tag{a_{12}}$$

The choice of changing the sign is a matter of convenience in the process of back substitution and modification of right hand side. Thus $-k'_{23}$, $-k'_{24}$ are the multipliers to be used while modifying the right hand side. Also $-u_2{}^*$ is required to be placed in the load position since the signs of row elements are changed.

Consider now the modification of the right hand side. The value P'_2 is temporarily stored, and a value $-u_2{}^*$ is placed in the load position. Using k'_{23} and k'_{24} as multipliers the remaining load values P'_3 and P'_4 are obtained. These will give $P_3 - k'_{23}u_2{}^*$ and $P_4 - k'_{24}u_2{}^*$ as can be seen readily from Eq. (a_{11}) when u_2 is eliminated from the third and the fourth equations.

After modification a value of $(-P'_2+k'_{22}u_2{}^*)$ is placed in the load position to suit back substitution. The subsequent modifications of load terms follow the usual pattern.

Consider the back substitution stage: the variables u_4 and u_3 are obtained as usual. The process remains same for the next step also, but the value obtained will be R_2.

Thus

$$\begin{aligned} R_2 &= (-P'_2+k'_{22}u_2{}^*)-(-k'_{24}u_4-k'_{23}u_3) \\ &= -P'_2+k'_{22}u_2{}^*+k'_{23}u_3+k'_{24}u_4 \end{aligned} \tag{a_{13}}$$

which agrees with the value required by the second of Eq. (a_{11}).

After this step the value of $u_2{}^*$ is placed in the location of u_2 and variables 'above' it are calculated; in this case

$$u_1 = P_1/k_{11} - k'_{14}u_4 - k_{13}u_3 - k'_{12}u_2{}^* \tag{a_{14}}$$

The above discussion would be of interest only to programmers. Those who may not be actively concerned with this issue may recall that these are only the modifications for an economic utilization of computer storage and computation time. The equation solvers developed at the University College of Swansea [21] are based on this approach. The code given in the next section is a simpler version in which all the equations are held in main core in the form of a two-dimensional array.

EXAMPLE 10.1

Let us take the following numerical set:

$$\begin{bmatrix} 5 & -4 & 1 & 0 \\ -4 & 6 & -4 & 1 \\ 1 & -4 & 6 & -4 \\ 0 & 1 & -4 & 5 \end{bmatrix} \begin{Bmatrix} u_1 \\ u_2 = 2 \\ u_3 \\ u_4 \end{Bmatrix} = \begin{Bmatrix} -2 \\ 1 \\ -5 \\ 3 \end{Bmatrix} + \begin{Bmatrix} 0 \\ R_2 \\ 0 \\ 0 \end{Bmatrix} \tag{a_1}$$

with $u_2 = u_2{}^* = 2$.

Consider now the triangularization of the coefficient matrix. Eliminate the first variable as usual:

$$\begin{bmatrix} \boxed{5} & -0.8 & 0.2 & 0 \\ 0 & 2.8 & -3.2 & 1.0 \\ 0 & -3.2 & 5.8 & -4.0 \\ 0 & 1.0 & -4.0 & 5.0 \end{bmatrix} \qquad (a_2)$$

Here the pivot element 5 is stored instead of unity. The second variable to be eliminated is a prescribed one; therefore skip the elimination (row modification) operation but change the signs of the elements of this row:

$$\begin{bmatrix} \boxed{5} & -0.8 & 0.2 & 0 \\ 0 & \boxed{2.8} & 3.2 & -1 \\ 0 & 0 & 5.8 & -4 \\ 0 & 0 & -4.0 & 5.0 \end{bmatrix} \qquad (a_3)$$

Here again the pivot 2.8 is stored instead of -1 (see Eq. (a_{12}). The elimination of the third variable is carried out as usual.

$$\begin{bmatrix} \boxed{5} & -0.8 & 0.2 & 0 \\ 0 & \boxed{2.8} & 3.2 & -1 \\ 0 & 0 & \boxed{5.8} & -0.7 \\ 0 & 0 & 0 & \boxed{2.2} \end{bmatrix} \qquad (a_4)$$

This completes elimination or triangularization. Observe that the off-diagonal coefficients are the multipliers required to modify the loads. Consider the load vector modification; corresponding to the first variable, we have

$$\text{RHS} = \begin{Bmatrix} -2 \\ 1-(-0.8)(-2) \\ -5-(0.2)(-2) \\ 3-(0)(-2) \end{Bmatrix} = \begin{Bmatrix} -0.4 \\ \boxed{-0.6} \\ -4.6 \\ 3 \end{Bmatrix} \begin{matrix} \rightarrow \text{store} \\ \leftarrow \text{write } \boxed{-2} \\ \\ \\ \end{matrix} \qquad (a_5)$$

The second variable requires modification as discussed. Assume that (-0.6) is stored temporarily elsewhere and in its place $-u_2^* = -2$ is replaced. The third and the fourth load terms are modified as usual picking up the multipliers from the second row of the coefficient matrix (a_4):

$$\text{RHS} = \begin{Bmatrix} -0.4 \\ \boxed{-2} \\ -4.6-(3.2)(-2) \\ 3-(-1)(-2) \end{Bmatrix} = \begin{Bmatrix} -0.4 \\ \boxed{2} \\ 1.8 \\ 1.0 \end{Bmatrix} \quad \begin{matrix} \rightarrow \text{store} \\ \leftarrow \text{write } \boxed{6.2} \\ \\ \\ \end{matrix} \tag{a_6}$$

Having done this replace the load term corresponding to the prescribed variable by $-P'_2+k'_{22}\,u_2{}^*$, namely $-(-0.6)+2.8\times 2 = 6.2$.
Thus the final load vector will be

$$\{\ 0.4,\ 6.2,\ 1.8,\ 1.0\} \tag{a_7}$$

The modification of the load vector due to the elimination of third variable is carried out normally; thus using the multiplier -0.7 from the third row,

$$\text{RHS} = \begin{Bmatrix} -0.4 \\ 6.2 \\ 1.8/5.8 \\ 1.0-(-0.7)(1.8) \end{Bmatrix} = \begin{Bmatrix} -0.4 \\ 6.2 \\ 0.3 \\ 2.2 \end{Bmatrix} \tag{a_8}$$

When the penultimate variable is eliminated the last load term may or may not be divided by the pivot. To be systematic, on dividing the last element by the pivot $= 2.2$, we obtain,

$$\text{RHS} = \{-0.4,\ 6.2,\ 0.3,\ 1.0\} \tag{a_9}$$

The back substitution yields the variables u_4, u_3 and R_2 :

$$\begin{aligned} u_4 &= 1.0 \\ u_3 &= 0.3-(-0.7)(1) = 1.0 \\ R_2 &= 6.2-(-1)(1)-(3.2)(1) = 4.0 \end{aligned} \tag{a_{10}}$$

Having obtained this we replace the value $u_2 = u_2{}^* = 2$ in place of 6.2 and continue the process:

$$u_1 = -0.4-0.2\times(1)-(-0.8)(2) = 1 \tag{a_{11}}$$

The reader is advised to pursue the above flow of calculations in detail keeping the requirements of storage in view. The essential equations are, Eqs. (8.23) and (8.24). See subroutines ELIM and SOLV given in the next section.

10.2 EQUATION SOLVER

The computer code presented here is based on the method of solution discussed in the previous section. It handles prescribed displacements and provides the corresponding reactive forces. It keeps all the coefficients in the main core as a two-dimensional array.

(*a*) VARIABLES AND ARRAYS

NEQ	Total number of equations
IB	Semi band width
NR	Total number of prescribed variables
LR	List of prescribed variables (NR numbers)
UR	List of the magnitudes of prescribed values of the variables (NR numbers)
SK (NEQMAX, IBMAX)	Coefficient matrix—a rectangular array with NEQMAX number of rows and IBMAX number of columns storing the coefficients as in Eq. (a_6) of Sec. 8.3. NEQMAX, IBMAX are dynamic array bounds.
R (NEQMAX)	Right hand side (load) vector; also to hold later the solution vector.
LR (NRMAX)	Vector listing the prescribed variable numbers. NRMAX is the maximum number of prescribed variables permissible ⩽ NEQMAX.
UR (NRMAX)	Vector listing the magnitudes of the prescribed variables, equal to NR ⩽ NRMAX.
RR (NRMAX)	Vector for the calculated reactive forces corresponding to NR number of prescribed variables listed in LR.

(*b*) ROUTINES ELIM AND SOLV

Subroutine ELIM triangularizes the coefficient array SK and overwrites it. The modified SK array holds the necessary coefficients required for modification of right hand side vector and back substitution. Subroutine SOLV modifies the right hand side vector using the coefficients from the SK array, and then performs back substitution operation. The required modifications for prescribed variables and determination of reactive forces are incorporated at appropriate places as discussed in previous Section. For the solution of different right hand sides (as for more than one loading cases and in nonlinear work) only the subroutine SOLV has to be recalled again and again within a loop statement. The calling segment will be required to generate the coefficient array SK and the load vector R. The subroutine SOLV holds the solution vector in R and the reactive forces in a separate vector RR. These can be picked up by the calling segment for output.

(*c*) COMPUTER CODING

The code shows, on the right hand side the UPDATE numbering of card image held on tape. We have given the extract starting with number 4800 in steps of 100. (100 to 4700 used by Master Segment). See Appendix A.

EXPLANATORY NOTES

CARD NUMBER	COMMENT
5100	IR, counter for prescribed variables processed.
5400	IB1 upper limit for eliminating variables of a row.
5500	Go to release error message if the pivot is less than or equal to zero.
5600	Skip if all the prescribed variables are processed.
5700	Skip if the variable in question is not a prescribed variable.
5900	Change sign of off-diagonal terms.
6200 to 7000	Modify rows as per Gauss Scheme.
7500	Print error message, equation number and the pivot.
8100 to 9500	Modification of right hand side vector.
8600	Copy $(k_{ii}\, u_i^* - P'_i)$ into R. (See Eq. (a_{12}))
8700	Place $-u_i^*$ in the load position.
9900	Compute reaction for the prescribed variable.
10400 to 11000	Back substitution process.
11400	Copy the prescribed displacement into solution vector.

10.3 PLANE FRAME PROGRAM

A simple program for analysing plane frames is presented in this section. It is based on the direct stiffness method discussed in Chapter 8. The purpose here is to introduce to the student of structural analysis a simple program bringing out the essential steps in writing a program. The chief operations to be handled in the direct stiffness method are:

(a) assembling the structure stiffness matrix,
(b) forming the load vector,
(c) incorporating the support conditions,
(d) solving the assembled equations for the displacements, and
(e) computing the internal forces in the members and reactive forces at supports.

We will be discussing in a later section the special ways of handling the above issues as applied to large size problems. We are at this stage discussing a possible arrangement for smaller size problems, avoiding more efficient (and hence complex) ways of handling the above operations. We therefore assume, to begin with, that the problem is of small size and that it is possible to store the upper band of the full stiffness matrix in the main core of the computer during the assembly and solution stages. An in-core solver discussed in the previous section will be used. Handling of a large system calls for simultaneous assembly triangularization, transfer operations between main and backing stores of computer etc., which in effect demand programming skill.

(*a*) FLOW OF COMPUTATIONS

Input data

The first step would be to read in the data. The first card may be assigned to carry the following basic data:

1. Number of joints (NJ)
2. Number of members (NM)
3. Number of restrained joints (NJR)
4. Number of loaded members (NLM)
5. Number of loaded joints (NLJ)

These are the main parameters which control the number of data sets to be read in subsequently. The abbreviations given in the brackets are the possible identifiers which can be used while drawing a program.

The next set of cards would be to read in the joint coordinate data. This consists of NJ sets of data, each containing the coordinates (x, y) of the joints or the projections Lx and Ly if these are found convenient to provide. The instructions in the input routine would read in all NJ sets and store the coordinates (or Lx and Ly) as a two-dimensional array, say CORD (NJ, 2).

The next set of cards would be the ones specifying the member connectivity and also its properties. Thus NM sets of data read. Each data card may contain the following:

<Element No><joint 1><joint 2><Material Property><Cross-sectional parameters<

The member connectivity data can be stored as a two-dimensional array, say ICON (NM, 2). Likewise suitable arrays can be set out to store the cross-section properties required for member analysis.

The next set of data may consist of NJR number of sets of data each set giving the number of the joint, the degrees of freedom prescribed and if nonzero, the values of the prescribed displacements. Thus each card may be assigned to carry the following quantities:

<Joint Number>, <Fixity Code>, $<u^*>$, $<v^*>$, $<\theta^*>$

The choice for the fixity code is arbitrary. One convenient way is to use, for the plane frame problem in question which has three degrees of freedom, a three digit code <*abc*> in which *a*, *b* and *c* are either 1 or 0 denoting fixity or otherwise of the degrees of freedom u, v, θ respectively. Thus 101 denotes the code to convey the information that, the first and the third degrees of freedom, namely u and θ are prescribed. The above information can be stored in appropriate arrays. Thus the fixity codes and the prescribed displacements can be stored in suitable arrays.

The load data is likewise read in NLM number of sets of data are read in to record the number of element, the load type, and load parameters. Similarly NLJ sets of data are read in for the loads acting on the joints, each containing the name of the loaded joint and the magnitudes of the loads applied to the joints. We will not consider here the details. It can be detailed out easily.

Assembling Structure Stiffness Matrix

The structure stiffness matrix say SK (or simply K if declared to be a real element array) is set out as a two-dimensional array storing only the upper band of the symmetric stiffness matrix; see Eq. (a_8) of Sec. 8.3. Thus in terms of submatrices it has NJ rows, ND columns of (3×3) submatrices, where ND is the joint number difference, or in terms of scalars, $NJ \times 3$ number of rows and $(ND+1) \times 3$ number of columns. Here ND is maximum difference in joint numbers determining the band spread, see Eq. (a_5) of Sec. 8.3. Thus the bandwidth has to be calculated and checked if the array size allotted to SK can accommodate this. Calculation of this bandwidth can be done by scanning the connectivity data. (See Subroutine FIX.)

Having verified this aspect the elements of structure stiffness are all set equal to zero, and the contributions from the members are placed memberwise in sequence. Thus an element is considered and the following steps repeated:

(a) forming the element stiffness
(b) placing the coefficients in the structure stiffness matrix, and
(c) storing the element stiffness in the back store (tape or disc).

This is achieved by a Do loop of FORTRAN (or the FOR statement of ALGOL). The stiffness matrix is computed as per Eq. (8.10). From the connectivity data, the coordinates are read and the direction cosines computed. Then using the cross-sectional and material properties of the element, which have already been read in, the stiffness matrices are computed.

The next step is to place the $\mathbf{K}_{ij}$ matrices into the structure stiffness matrix. This is performed according to the rule set out in Eq. (8.20), or if one works with scalar quantities, as in Eq. (8.21). The contributions only at the diagonal and upper off-diagonal locations are placed since that portion alone is being assembled. (See Subroutine FORMSK).

When once this operation is complete, the following information of the element is transferred to a backing store

(a) member connectivity, and properties
(b) member stiffness matrices.

These would be recalled at the time of computing the member forces.

Assembling the Load Vector

Next obvious step is to assemble the load vector, which can conveniently be called R or RHS, which is a single dimension array of size $NJ \times 3$. Observe that we consider all the joints inclusive of the support joints, which would have figured in the member connectivity data and also necessary during the assembly of structure stiffness matrix.

The fixing end forces of loaded members (NLM members) are formed elementwise. With the help of the member connectivity data, those with opposite sign are placed in the appropriate location of the RHS vector as per Eq. (8.20) or (8.21). These are transferred to the back store for reuse. Next the actual applied joint loads are placed. These would have been directly read in as NLJ sets of data.

Support Conditions

Before solving the equations the support conditions have to be incorporated as discussed in Sec. 8.4. The procedure outlined in Sec. 10.1, handles this aspect during elimination and back substitution. If one chooses such a routine, then no separate modifications are necessary. Else if standard solution routines are used, the coefficients of the stiffness matrix and the elements of the load vector are required to be modified as in Eqs. (8.23). Before this is undertaken, the loads and the stiffness coefficients to be modified are to be copied elsewhere, for these would again be required in computing reactions as per Eqs. (8.24). The required modifications either separately or during equation solving, are effected using the data already read in, namely the fixity codes and the prescribed displacement values.

Solution of Equations

The elimination routine is then called to triangularize the coefficient matrix. It is usual to overwrite this on the original stiffness matrix locations. These hold coefficients and pivots required in back substitution and during modification of a fresh right hand side. The solution routine is then called to solve for the displacements. If the reactions corresponding to the prescribed displacements are not obtained one has to take recourse to separate calculations, using the calculated displacements, the prescribed displacements, and the coefficients saved in 'support conditions' above.

Calculation of Member Forces

Calculation of member end forces measured in frame coordinates is then performed memberwise using the complete force displacement relations (8.16). The stiffness coefficients, and the fixing end forces of the members are read in from the backing store, and using the calculated values of the displacements, the member end forces are computed using Eq. (8.16). (See Subroutine FORCES.)

Additional Load Cases

It is usual to have several loading cases. In such situations (and in nonlinear problems too) the load vector is assembled afresh. The resolution routine then only modifies the elements using the elements of the triangularized stiffness matrix and once again using them, it performs the back substitution operation giving a new solution vector.

(*b*) DATA PREPARATION

Every program invariably has to tell the user the exact manner according to which the user has to present the data. Such instruction manuals (or user manuals) have got to be specific, since the user may not have access to the program. When one begins to write a computer program, one needs to keep the data input pattern in mind and form the instructions accordingly. The plane frame program has been written to operate on input data to be presented in the following sequence and format:

Each data set is headed by a data set identifier card (I5), giving the data set number (IDATA). The program has been written to go to STOP when no more data sets are present in the data stream. Thus the data sets should be presented as follows:

CARD 1: DATA SET NUMBER CARD (I5)
Cols. 1-5 Data Set identifier number
Cols. 6-80 Blank.

CARD DECK (*n* cards)	DATA CARDS OF DATA SET 1
⋮	⋮
CARD *n*+1	DATA SET NUMBER CARD
CARD DECK (*m* cards)	DATA CARDS OF DATA SET 2
⋮	⋮ etc. ⋮

Each new data set is therefore a new problem. (In each data set, there can be more than one load cases.) We will now describe each DATA SET.

Description of DATA SET

(A) BASIC DATA CARD (4 I5)
Cols. 1—5 Number of joints (NJ)
Cols. 6—10 Number of members (NM)
Cols. 11—15 Number of joints restrained (NJR)
Cols. 16—20 Number of load cases (NLD).

(B) CONNECTIVITY CARDS NM number of cards; can be in any order.
Each card (3 I5) containing connectivity as follows:
Cols. 1—5 Number of member (I)

Cols. 6—10 Number of the joint at end 1(NJ1(I))
Cols. 11—15 Number of the joint at end 2(NJ2(I)).

(C) JOINT COORDINATE CARDS: NJ number of cards; can be in any order.
Each card (I5, 2F 10.4) giving coordinates:
Cols. 1—5 Number of the joint (I)
Cols. 6—15 x-coordinate (X(I))
Cols. 16—25 y-coordinate (Y(I)).

(D) SUPPORT CONDITIONS: NJR number of cards; to be in ascending order.
Each card (2I5, 3E15.5) specifying support conditions:
Cols. 1—5 Number of support joint (I)
Cols. 6—10 Fixity code (LFIX(I))
Cols. 11—25 Prescribed value u (UR (I, 1))
Cols. 26—40 Prescribed value of v (UR (I, 2))
Cols. 41—55 Prescribed value of θ (UR (I, 3))

The fixity code is a three digit number 111 or 110, or 001, or 010, etc; 1 denotes fixity and zero freedom. The first digit stands for u, second for v and the third for θ respectively. 011 means, fixed along v and θ directions only, and so on. The prescribed values (cols. 11 to 25) to be left blank for supports fully restrained.

(E) MEMBER PROPERTIES: NM number of cards; can be in any order.
Each card (I5, 3E15.5), giving members properties
Cols. 1—5 Number of the member
Cols. 6—20 Magnitude of (AE)—extensional rigidity
Cols. 21—35 Magnitude of (EI)—flexural rigidity
Cols. 36—40 Magnitude of (AG)—shear rigidity.

The above set completes the basic data exclusive of loading data. The program determines the bandwidth, number of equations and compares with the allocated array dimensions. The program goes to STOP if errors are detected before reading the loading data.

(F) LOAD DATA CARDS
F.1 LOADING CONTROL CARD (2I5):
Cols. 1—5 Number of loaded members (NML)
Cols. 6--10 Number of loaded joints (NJL).
F.2 NML sets of member loading data sets; each data set consists of the following:
F2.1 MEMBER LOADING HEADER CARD (5I5), giving
Cols. 1—5 Number of the loaded member
Cols. 6—10 Loading type code LCODE (1)—(Point loads)
Cols. 11—15 Loading type code LCODE (2)—(Uniform loads)
Cols. 16—20 Loading type code LCODE (3)
Cols. 21—25 Loading type code LCODE (4)

Each of LCODE can be 1 or 0, zero indicating the absence of that load type. Subroutines would be called only if LCODE (I) is 1. Presently LCODE (3), LCODE (4) are to be left blank, since these are intended for other loading types to be introduced later by the user. Corresponding subroutines have been left dummy for the present.

F2.2 LOAD DESCRIPTION SETS

(a) IF LCODE (1) = 1, (concentrated loads present) following are necessary:

(i) CARD (I5)—Number (NCONC) giving the number of concentrated loads.

(ii) NCONC number of cards as follows, each card (4F15.5), giving

Cols. 1—15 Magnitude of x-component of load
Cols. 16—30 Magnitude of y-component of load
Cols. 31—45 Magnitude of distance x from end 1
Cols. 46—60 Magnitude of distance y from end 2.

(b) The above set should be followed by the following card if only LCODE (2) = 1, else to be skipped. CARD (2F15.5)

Cols. 1—15 x-component of UDL (total)
Cols. 16—30 y-component of UDL (total)

(c) These should be followed by the sets of cards corresponding to LCODE (3) and LCODE (4), whenever the program is extended to handle these; for the present not required.

F.3 JOINT LOAD CARDS. NJL number of cards.

Each card (I5, 3F15.5) giving

Cols. 1—5 Number of the loaded joint
Cols. 6—20 x-component of load, (P)
Cols. 21—35 y-component of load, Q)
Cols. 36—50 θ-component (M).

(c) VARIABLES AND ARRAY DEFINITIONS

NDF	Degrees of freedom of joint, preset to 3 in Master Segment
NM	Number of members
NJ	Number of joints
NJR	Number of joints restrained
NML	Number of members loaded
NJL	Number of joints loaded
NEQ	Number of equations (NJ × NDF)
IB	Semi bandwidth
NRD	Number of restrained displacement degrees of freedom
NEQMAX	Dynamic array bound for SK and R
IBMAX	Dynamic array bound for SK

NRDMAX	Dynamic array bound for LRD, URD, RRD
NFILE 1	Name of scratch tape (disc)
NLD	Number of load cases
ICON	Connectivity array of size (NM, 2)
LJR	Array of the numbers of the joints restrained of size (NJR)
LFIX	Array of the joint fixity codes of size (NJR)
UR	Array of prescribed displacements of size (NJR, 3)
PROP	Array holding the member properties of size (NM, 5)
PFM	Array holding the fixing end forces of size (NM, 6)
SK	Stiffness array, rectangular of size (NEQMAX, IBMAX)
LRD	Array listing the numbers of the equations with prescribed degrees of freedom; of size (NJR, 3)
URD	Array holding the magnitudes of the prescribed degrees of freedom of size (NJR, 3)
RRD	Array to hold the calculated magnitudes of the reactions, of size (NJR, 3)
R	Array to hold the load terms before elimination and displacements after back substitution, of size (NEQMAX)
ESK	Element stiffness array, of size (6, 6)
P	Element load vector of size (6)
CORD	Array holding joint coordinates (NJ, 2)
DIS	Array to hold the calculated values of reactions of joint displacements of size (NJ, 3)
REA	Array to hold calculated values of reactions of joint (NJ, 3)

(*d*) SUBROUTINES

1. INPUT 1: This is a subroutine which reads in data, excluding the loading data. Reads and prints out

(a) NJ, NM, NJR, NLD
(b) Connectivity data in array ICON
(c) Joint coordinates in array, CORD
(d) Support joints, fixities, and prescribed magnitudes of displacements, in arrays LJR, LFIX and UR respectively, and
(e) Member properties, 3 for each member (AE, EI and AG), in array PROP; also computes two more, (x_2-x_1) and (y_2-y_1) for each member.

2. PRESET: Checks the array sizes allotted to SK with respect to bandwidth and number of equations; returns the error code to the master segment to halt the program if necessary.

3. WRITE1: This is a subroutine which (a) reads the load data (b) computes the fixing end forces for members (c) computes stiffness

matrices of members and (d) writes on to a sequential access scratch disc NFILE 1. Calls subroutines STIFF 1 and LOAD 1, LOAD 2, LOAD 3 and LOAD 4. It also computes the global load vector R.

4. STIFF 1: This routine computes member stiffness matrix, in the array ESK (6, 6).

5. LOAD 1, LOAD 2, LOAD 3, LOAD 4: These are routines to compute fixing end forces for different load cases, and to add on to the load vector R. LOAD 1 and LOAD 2 compute for a set of concentrated loads and uniform loading respectively. LOAD 3 and LOAD 4 are, in the present version set to work as dummy routines.

6. FORMSK: Assembles the upper band of the structure stiffness matrix in rectangular array SK by reading element stiffness matrices and fixing end forces, from NFILE 1.

7. FIX: Assembles the array LRD, and URD from the data held in LJR and UR. Array LRD lists the numbers of the equations whose variables are presented; array URD lists the magnitudes of the prescribed variables. (Note that LJR and UR are jointwise and LRD and URD are in terms of the degress of freedom 3 at each joint.)

8. ELIM: This triangularizes the SK array, and overwrites it. It is a modified routine to change the row operations when the equation number corresponds to a prescribed variable.

9. SOLV: This performs right hand load vector transformations and back substitution. It computes displacements and overwrites the vector R. It also computes the reactions and hold them in vector RRD.

10. DISREA: This merely picks up the calculated values from the vectors R and RRD and prints out displacements and reactions, jointwise.

11. FORCES: Computes and prints the member end forces by reading stiffness matrices and load vectors from NFILE 1, and using the calculated values of the displacements.

Master segment is indeed a mere calling of the various subroutines described above. The two main Do loops are, one over the number of data sets, and within this loop, the other loop over the number of load cases. We will be explaining the flow shortly.

(*e*) Storage Considerations

It is a simple program in which we employ the in-core assembler and solver. A scratch disc has been used to store the stiffness matrix and fixing end forces calculated for the members in succession. Use of common blocks is amply made, and where found convenient variables are returned from one segment to another through argument lists also. ELIM and SOLV routines operate through argument lists so that these routines can be easily extracted out from the program for use in other programs. It may be noted that subroutines depending on common blocks tend to become local to a program and it is relatively difficult to isolate such routines for use elsewhere.

(*f*) Flow of Program

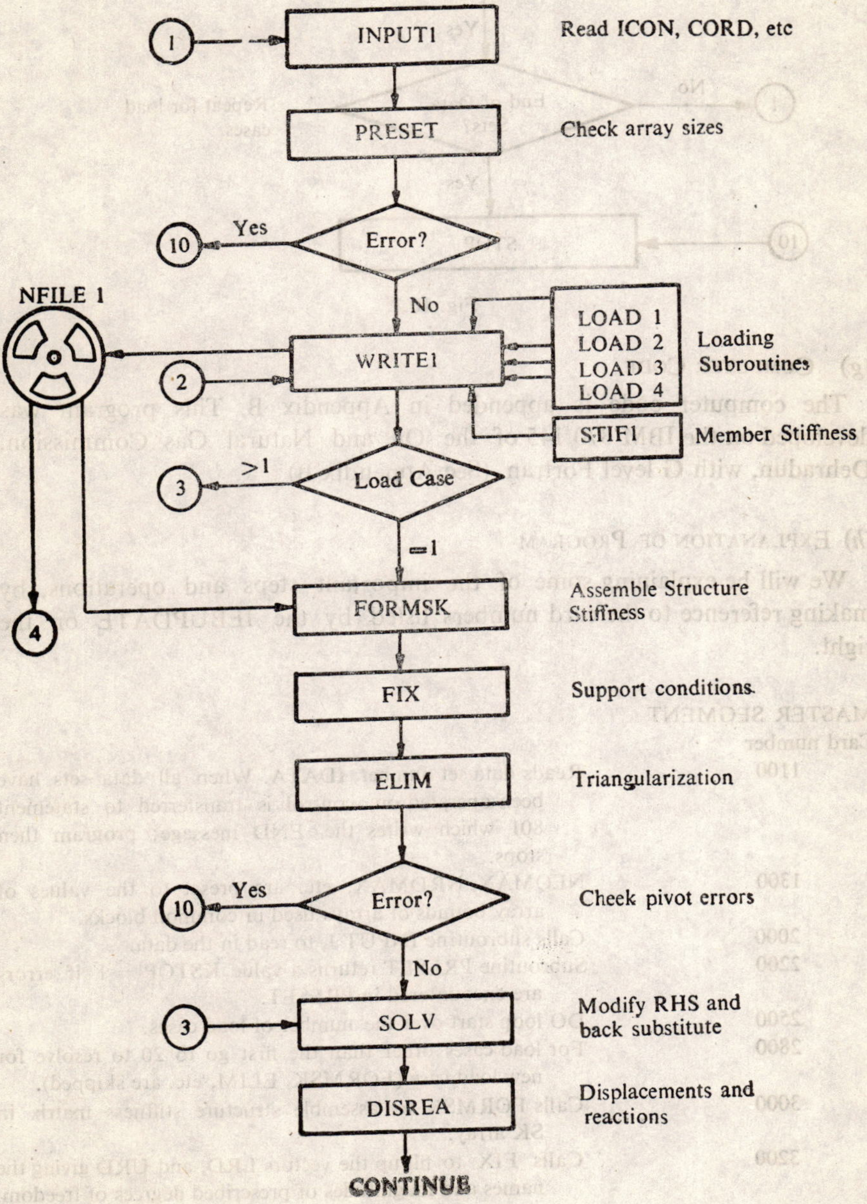

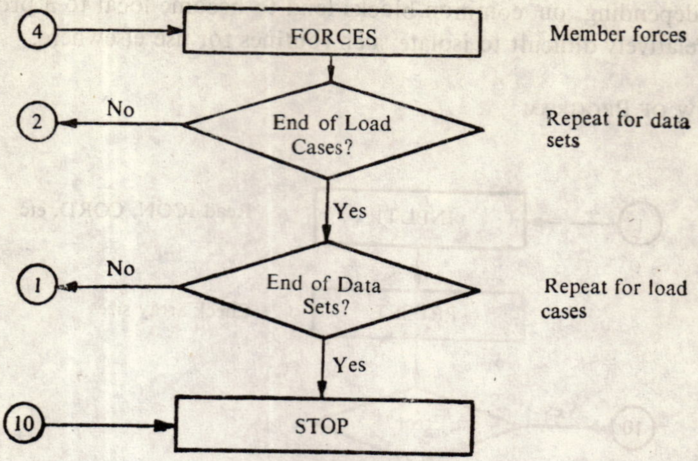

Fig. 10.1

(g) Computer Code

The computer code is appended in Appendix B. This program was developed on the IBM 370/145 of the Oil and Natural Gas Commission, Dehradun, with G level Fortran. (See Appendix B)

(h) Explanation of Program

We will be explaining some of the important steps and operations, by making reference to the card numbers listed by the IEBUPDATE on the right.

MASTER SEGMENT

Card number

1100 Reads data set number, IDATA. When all data sets have been operated on, control is transferred to statement 801 which writes the END message; program then stops.

1300 NEQMAX, NRDMAX, etc. are preset to the values of array bounds of arrays used in common blocks.

2000 Calls subroutine INPUT 1, to read in the data.

2200 Subroutine PRESET returns a value KSTOP = 1 if errors are encountered in PRESET.

2500 DO loop start over the number of load cases.

2800 For load cases other than the first go to 20 to resolve for new load case (FORMSK, ELIM, etc. are skipped).

3000 Calls FORMSK to assemble structure stiffness matrix in SK array.

3200 Calls FIX to fill up the vectors LRD, and URD giving the names and magnitudes of prescribed degrees of freedom.

3500	Calls ELIM to triangularize the SK array and overwrite it.
3700	Calls SOLV to modify the load vector and back substitute. Displacements returned in R and reactions in RRD.
3900	Calls DISREA to compute and print out displacements and reactions in arrays DIS and REA.
4100	Calls FORCES to compute member end forces and print out the same.
4200	CONTINUE, to go back to loop over number of load cases (Card no. 2500).
4300	Go back to 10 to repeat for the new data set.

SUBROUTINE INPUT 1

5800	Temporary array CORD (50, 2) declared to hold coordinates.
6700	Observe that connectivity data can be in any order.
7200	Observe that coordinate data can also be in any order.
7700	Read LJR, LFIX and UR; these are to be in ascending order.
8200	Read PROP data; these can be in any order.
8500, 8600	Compute (x_2-x_1), and (y_2-y_1) and record them as fourth and fifth properties.

SUBROUTINE PRESET

11700	Assume array sizes to be all right.
11900	Do loop over number of members to compute band-width from connectivity data.
12000	LDIF gives the difference between NJ1 and NJ2.
12300	Computes bandwidth (See Eq. (a_5), Sec. 8.3).
12500	If bandwidth is less than the number of columns of arrays SK, and R go to check NEQ, else set error code KSTOP = 1.
12800	Return if actual number of equations is less than number of rows allocated to SK and R, else set error code as 1.

SUBROUTINE WRITE (ILD)

15200	Initialize PFM, to hold fixing end forces of members.
15400	Initialize R, to hold right hand side load vector.
15600, 15700	Counters for loaded members.
15800	If no members are loaded and is not a first load case, go to 10 to consider applied joint loads.
15900	Begin loop over number of members.
16000	Form element stiffness matrix only for the first time, else skip, since these would have already been written on NFILE 1.
16200	Initialize PF to hold member fixing end forces.
16300	If no members are loaded skip the subsequent steps of computing fixing end forces.
16400	Skip reading member data if all members have been considered.
17100	Skip, upto 14 if the member is not a loaded member.
17900	Call Load 1, to return fixing end forces in PF; similarly cards 18000, 18100 and 18200.

18300, 18400	Address locators.
18500	Add on (−PF) to the load vector R. Observe that (IX1+IDF), and (IX2+IDF), for IDF = 1, 2, 3, give the appropriate location in R for the six values of PF. (See Eq. 8.21).
18900	Skip writing on NFILE 1 for load cases other than the first.
19300	End loop over members.
19400	If no joints are loaded skip the subsequent steps upto 50.
19700	Do loop over number of loaded joints.
20000	Address locator.
20200	Add joint loads to load vector R.

SUBROUTINE STIFF 1

21700	This routine is called within a Do loop with IMEM = 1, NM (See cards 15000 and 16000).
22500 to 22900	Read in properties from PROP.
23700 to 25600	Assign values to stiffness coefficients; (See Table 8.1).
26000	Generate symmetric coefficients.

SUBROUTINE LOAD 1

26900	Read the number of total concentrated loads.
27000	Read magnitudes of loads and distances.
27400	Loop over number of concentrated loads.
27600 to 29200	Compute fixing end forces.
29800	Return the vector PF to subroutine WRITE 1.

SUBROUTINE LOAD 2

30600	Read magnitudes of components of loads.
30900 to 31400	Compute fixing end moments and shears.
31800	Return vector PF to WRITE 1.

SUBROUTINE LOAD 3, LOAD 4

32000, 32400	It is seen that these are rendered dummy; appropriate decks can be inserted when required to use for other loading types.

SUBROUTINE FORMSK

32800	Subroutine called by MASTER within a loop over number of members. Generates the elements of the SK array in common block.
33900	Initialize SK array.
34100	Loop over number of members.
34200	Read, from NFILE 1, element stiffness matrix in ESK and the five member properties in P. Note that P have to be read in though not required, so as to keep the sequential access pointer to the next record.
34400	Outer most Do loop, over number of member ends. (Observe that there are 3 nested loops within.)
34500 to 35400	Placing of k_{ij} of element stiffness matrix into the SK array.
35500	Skip operation since only upper band alone is being assembled.
35900	As commented, JCOLB gives the column location in the rectangular SK array. JCOL is the column location in the square stiffness matrix.
36100	End loop over number of members.

[Reader should recall contents of Sec. 8.2, leading to Eq. (8.20).]

SUBROUTINE FIX

37300	Initialize LRD, URD, and RRD.
37700	Set the counter for restrained degrees of freedom to zero.
37800	Do loop over number of joints restrained.
37900	Read the fixity code into JFIX.
38000	Set IFIX = 100.
38100	Read the number of joint restrained into IJT.
38200	Do loop over the three possible degrees of freedom to be examined.
38400	If the degree of freedom is not a prescribed one go to 12, to consider the next degree of freedom.
38700	Copy the prescribed value of the displacement.
38800	Set the code of next degree of freedom to JFIX.
39000	End Do loop over the three degrees of freedom.
39100	End Do loop over the number of restrained joints.

SUBROUTINE DISREA

40700	Set counter for NRD.
40800	Set counter for NJR.
40900	Do loop over number of joints.
41000	Do loop over the three degrees of freedom.
41300	Copy displacements from R into DIS.
41400	Skip if all restrained joints are considered.
41500	Skip if the joint in question is not a restrained joint.
42000	Copy reactions from RRD into REA. Observe that instructions are similar to the ones in subroutine FIX.
42600	End Do loop over number of joints.

SUBROUTINE ELIM and SOLV

These have been explained in Sec. 10.2.

SUBROUTINE FORCES

52300	Do loop over number of members.
52400	Read from NFILE 1, member stiffness matrix and properties, in ESK and PRP.
53600	Copy member displacements from R. Observe that we do this with the help of connectivity data.
53800	Initialize member forces.
53900	Do loop over number of member end displacements.
54100	Compute member forces (Eq. 8.16).
54300 to 55000	Obtain member forces in member coordinates by coordinate transformation.

(*i*) Illustrative Example

Consider the frames shown in Fig. 10.2 and 10.3. Frame of Fig. 10.2 is the the same as the one in Example 8.2 and the one of Fig. 10.3 is adopted from Reference [38]. The analysis of these three frames will be made for different loading cases as follows:

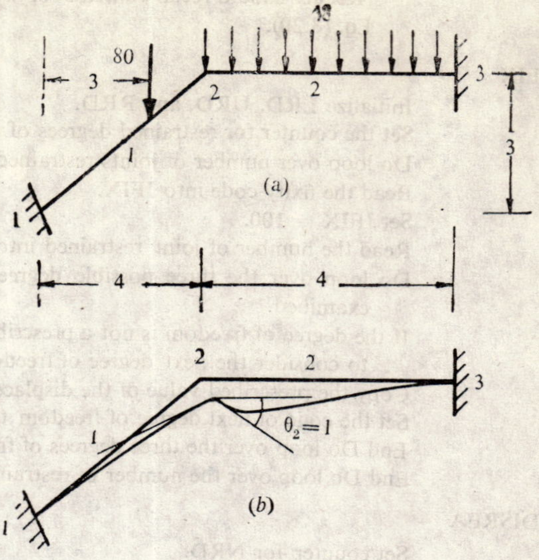

Fig. 10.2

Three data sets are to be presented, each with an identifier number say 1, 2 and 3. Let the loading on frame of Fig. 10.2(a) be the same as in Example 8.2 so that the answers can be checked; the second data set is for the same frame as shown in Fig. 10.2(b) with $\theta_2 = 1$. The frame of Fig. 10.3 will be analysed for the separate load cases shown in (b) and (c).

INPUT DECK

CARD NUMBER	FORMAT	NUMBERS				
1	I5	1				
2	4I5	3	2	2	1	
3	3I5	1	1	2		
4	,,	2	2	3		
5	I5, 2F 10.4	1	0.0	0.0		
6	,,	2	4.0	−3.0		
7	,,	3	8.0	−3.0		
8	2I5, 3E 15.5	1	111	0	0	0
9	,,	3	111	0	0	0
10	I5, 3E 15.5	1	2.0E02	1.25E01	0	
11	,,	2	2.0E02	8.0E00	0	
12	2I5	2	0			
13	5I5	1	1	0	0	0
14	I5	1				
15	4F 15.5	0	80.0	3.0	−2.25	
16	5I5	2	0	1	0	0
17	2F 15.5	0.0	48.0			
18	I5	2				
19	4I5	3	2	3	1	

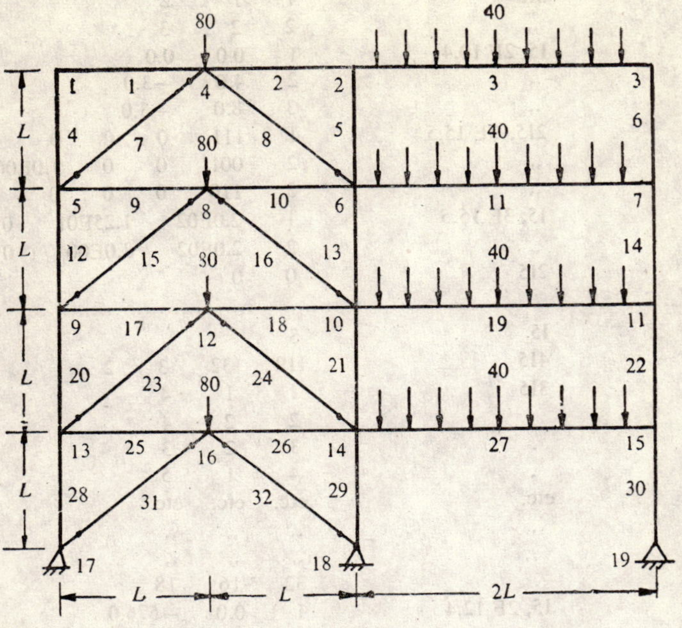
80
40
1
1
4
2
2
3
3
L
4
7
8
5
6
40
80
5
9
8
10
6
11
7
L
12
15
80
16
13
40
14
9
17
18
10
19
11
12
L
20
23
80
24
21
40
22
13
25
16
26
14
27
15
L
28
31
32
29
30
17
18
19
L
L
2L

(a)

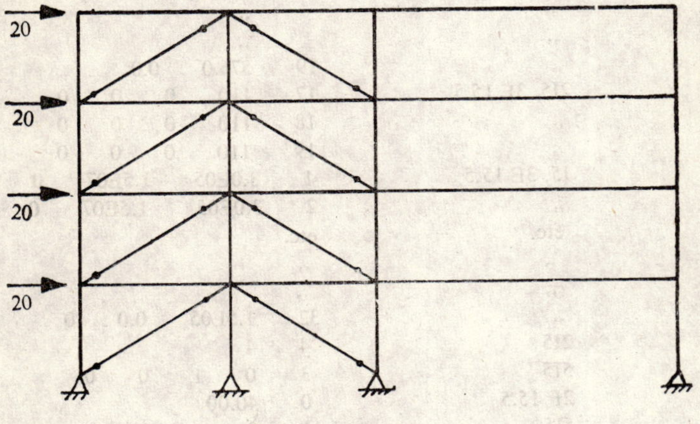
20
20
20
20

(b)

Fig. 10.3

20	3I5	1	1	2		
21	,,	2	2	3		
22	I5, 2F 10.4	1	0.0	0.0		
23	,,	2	4.0	−3.0		
24	,,	3	8.0	−3.0		
25	2I5, 3E 15.5	1	111	0	0	0
26	,,	2	001	0	0	1.0E00
27	,,	3	110	0	0	0
28	I5, 3E 15.5	1	2.0E02	1.25E01	0	
29	,,	2	2.0E02	8.0E00	0	
30	2I5	0	0			

....

31	I5	3				
32	4I5	19	32	3	2	
33	3I5	1	1	4		
34	.	2	2	4		
35	.	3	2	3		
36	.	4	1	5		
etc.	etc.	etc.	etc.	etc.		
,,	,,	,,	,,	,,		
,,	,,	,,	,,	,,		
64	.	32	16	18		
65	I5, 2F 12.4	1	0.0	−576.0		
66	,,	2	288.0	−576.0		
etc.	etc.		etc.			
,,	,,		,,			
,,	,,		,,			
83	,,	19	576.0	0.0		
84	2I5, 3E 15.5	17	110	0	0	0
85	,,	18	110	0	0	0
86	,,	19	110	0	0	0
87	I5, 3E 15.5	1	3.0E05	1.5E07	0	
88	,,	2	3.0E05	1.5E07	0	
etc.	etc.	etc.				
,,	,,	,,				
,,	,,	,,				
118	,,	32	1.5E05	0.0	0	
119	2I5	4	4			
120	5I5	3	0	1	0	0
121	2F 15.5	0	40.00			
122	5I5	11	0	1	0	0
123	2F 15.5	0	40.00			
124	5I5	19	0	1	0	0
125	2F 15.5	0	40.0			
126	5I5	27	0	1	0	0
127	2F 15.5	0	40.0			
128	15, 3F 15.5	4	0	8.0E01	0	
129	,,	8	0	8.0E01	0	
130	,,	12	0	8.0E01	0	
131	,,	16	0	8.0E01	0	
132	2I5	0	4			

133	I5, 3F 15.5	1	2.0E01	0	0
134	,,	5	2.0E01	0	0
135	,,	9	2.0E01	0	0
136	,,	13	2.0E01	0	0

(*j*) OUTPUT (See Appendix B)

10.4 A FINITE ELEMENT PROGRAM

A simple program to illustrate the subject matter of Chapter 9 is presented here. It is for two-dimensional problems of plane stress or strain. It employs 8 noded (parabolic) isoparametric elements and numerical integration. An in-core solver is again used, which limits the utility to small size problems. However, to reemphasize, the purpose here again is to illustrate the physical ideas of the finite element method introduced in Chapter 9. It also serves to bring out the features common to this and the plane frame program. Such a comparison would instil confidence in a programmer to modify a given program to meet slightly different situations and requirements or, for instance, to combine two or three programs into one.

(*a*) INPUT DATA

This program, unlike the one in the previous section, works only on one set of data to be presented in the following formats.

(A) BASIC DATA CARD (9I5)

Cols. 1—5 Number of elements (NEL)
Cols. 6—10 Number of nodes (NODES)
Cols. 11—15 Number of nodes restrained (NODR)
Cols. 16—20 Number of nodes whose coordinates are to be read (NNOD); midside nodes would be interpolated within the program
Cols. 21—25 Gauss integration order, (NG = 2, or 3)
Cols. 26—30 Code for isotropy, (ISO); ISO = 0 for isotropy and 1 for anisotropy
Cols. 31—36 Code for type of problem, (NP), NP = 0 for plane stress and 1 for plane strain
Cols. 36—40 Number of materials, (NMAT); ⩽ 5
Cols. 41—45 Number of load cases (NLD).

(B) ELEMENT CARDS: NEL number of cards each card (10I5) as follows; can be in any order.

Cols. 1—5 Element number
Cols. 6—10 Joint number of node 1
Cols. 11—15 Joint number of node 2
.. ... etc.
.. ... ,,
Cols. 41—45 Joint number of node 8
Cols. 46—50 Material number

(C) NODAL CARDS. NNOD number of cards, with each card (I5, 2F 15.5, I5) as follows; can be in any order. Midside nodes need not be given, except where there is restraint.

Cols. 1—5 Node number

Cols. 6—20 x-coordinate of node

Cols. 21—35 y-coordinate of node

Cols. 36—40 Fixity code (11, 10 or 01); leave blank for unrestrained nodes.

(D) MATERIAL CARDS. NMAT number of sets; For isotropy (ISO = 0), NMAT cards, each card of (2E 15.5) format as follows:

Cols. 1—5 Value of E

Cols. 6—10 Value of Poisson ratio

Cols. 11—80 Blank

If anisotropic, (ISO = 1), NMAT sets to be presented each set containing two cards, each card of 3E 15.5 format; First card (3E 15.5), D_{11}, D_{12} and D_{22} of modulus matrix; Second card (3E 15.5), D_{13}, D_{23}, D_{33} of modulus matrix.

(E) LOAD DATA SETS: NLD sets of loading data will have to be presented. Each set containing sets described in E1 to E6 below:

E1. LOAD PARAMETERS CARD. One card (4I5)

Cols. 1—5 Number of elements loaded NELD

Cols. 6—10 Number of nodes loaded, NODLD

Cols. 11—15 Code for prescribed nonzero displacements—KPRED = 1; zero means no nonzero prescribed displacements

Cols. 16—20 Number of nodes (NPRED) with nonzero prescribed displacements; leave blank if KPRED = 0.

E2. NONZERO PRESCRIBED DISPLACEMENT CARDS. Only if KPRED = 1; NPRED number of cards, each (I5, 2E 15.5) as follows:

Cols. 1—5 Number of the node

Cols. 6—20 x-component of prescribed displacement

Cols. 21—35 y-component of prescribed displacement

E3. LOADED ELEMENTS LISTING CARDS. List of the numbers of elements loaded (LELD(I)) to be given on as many cards as may be necessary, each of 16I5 format.

E4. UNIFORM BODY FORCE CARD, One card (I5, 2F 15.5)

Cols. 1—5 KBODY, body force code; 0 for uniform body forces; 1 if otherwise.

Cols. 6—20 x-component of body force, blank if KBODY = 1.

Cols. 21—35 y-component of body force, blank if KBODY = 1.

E5. ELEMENT LOAD SETS: NELD sets of data to be given as follows:

E5.1. LOADING HEADER CARD (7I5)

Cols. 1—5 Number of the element

Cols. 6—10 LCODE (1), loading code for body force (0 or 1)
Cols. 11—15 LCODE (2), loading code for pressure loading (0 or 1)
Cols. 16—20 LCODE (3), loading code for equivalent nodal loads (0 or 1)
Cols. 21—25 KEDGE (1), code for edge 1 loaded or no. (1 or 0)
Cols. 26—30 KEDGE (2), code for edge 2 loaded for no. (1 or 0)
Cols. 31—35 KEDGE (3), code for edge 3 loaded or no. (1 or 0)
Cols. 36—40 KEDGE (4), code for edge 4 loaded or no. (1 or 0)

LCODE (1), LCODE (2), etc. are 0 if the corresponding loading is absent; KEDGE (1), KEDGE (2), etc. = 0 if free of edge loads, else 1. All these to be left blank if LCODE (2) is itself zero. Load subroutines would be called only if the loading code is not zero.

E5.2 LOAD DESCRIPTION SETS, (upto three sets): The above header card should be followed by sets of cards (upto three), depending on LCODE (1), LCODE (2) and LCODE (3). If any of these codes is zero, the corresponding sets of cards are not to be included.

(a) IF LCODE (1) is 1, and KBODY = 1 then the following card (I5, 2F 15.5) should be given:
Cols. 1—5 Number of the element loaded
Cols. 6—20 x-component of body force
Cols. 21—35 y-component of body force
This set not to be included if KBODY = 0. (See Card E 4)

(b) If LCODE (2) = 1, one card for each of the edges loaded to be provided in (I5, 6F 12.4) format:
Cols. 1—5 Loaded edge number
Cols. 6—17 x-component of pressure at node 1
Cols. 18—29 x-component of pressure at node 2
Cols. 30—41 x-component of pressure at node 3
Cols. 42—53 y-component of pressure at node 1
Cols. 54—65 y-component of pressure at node 2
Cols. 66—77 y-component of pressure at node 3

Note that there is no provision to print the number of the element, for identification in case of trouble with deck.

(c) If LCODE (3) = 1, 4 cards giving the equivalent nodal loads.
First card: (I5, 5F 15.5)
Cols. 1—5 Element number
Cols. 6—20 x-component at node 1
Cols. 21—35 x-component at node 2
. . .
. . .

Cols. 66—80 x-component of node 5
Second card (5F 15.5)
Cols. 1—5 x-component of node 6
Cols. 16—30 x-component of node 7
Cols. 31—45 x-component of node 8
Cols. 46—60 y-component of node 1
Cols. 61—75 y-component of node 2
Third and fourth cards, each (5F 15.5) to bear the remaining values of y-component at other nodes.
Four cards always necessary though blank.

E.6 NODAL LOAD CARDS (only if NODLD≠0): NODLD number of cards each card (I5, 2F 15.5) giving in
Cols. 1—5 Loaded node number
Cols. 6—20 x-component of applied load
Cols. 21—35 y-component of applied load

Recall that cards as detailed in E1 to E6 form a load data set, and that NLD number of such load data sets are to be prescribed keeping one set immediately after the previous one. There are no title cards to describe load data sets. Program, however, prints the serial number of the load case.

(*b*) VARIABLES AND ARRAYS

NDF	Degrees of freedom per node (to be set as 2)
NEL	Number of elements
NODES	Number of nodes
NODR	Number of nodes restrained
NODEL	Number of nodes per element (to be set as 8)
NDFEL	Number of degrees of freedom per element = NODEL *NDF (= 16)
NG	Order of Gauss integration value (= 2 or 3)
NUMGP	Number of Gauss integration points (NG * NG)
KSF	Order of shape function, presently KSF = 2 since only parabolic elements are incorporated
ISO	Code for isotropy (= 0), 1 for anisotropy
NP	Nature of problem, = 0 for plane stress and 1 for plane strain
NMAT	Number of materials (⩽ 5)
NEQ	Number of equations (NODES * NDF)
NB	Semi bandwidth (IB used in previous program)
NRD	Number of restrained displacement degrees of freedom
NEQMAX, NBMAX, NRDMAX	Dynamic array bounds for SK, R, URD and RRD
PROP (5, 6)	Array (5, 6) holding the 6 material properties for the possible 5 sets of materials
SFN (9, 8)	Array holding the 8 shape functions. $N_1, N_2, \ldots, N_8$, evaluated at maximum possible number of Gauss points

DSF (9, 2, 8)	Array holding (ξ, η) derivatives of shape functions evaluated at 9 (or 4) Gauss points. First array index for Gauss point, second for $\frac{\partial}{\partial \xi}$ and $\frac{\partial}{\partial \eta}$, and the third for $N_1, N_2, \ldots, N_8$.
DETJ	Determinant of the Jacobian (Eq. 9.45)
NFILE 1	Scratch disc number
NNOD	Number of nodes whose coordinates are to be read
NELD	Number of elements loaded
NODLD	Number of nodes loaded
NLD	Number of load cases
IELNO	Element number (on NFILE 1 records)
ICONE (8)	Array for element connectivity
ELX (8)	Array for x-coordinates of element nodes
ELY (8)	Array for y-coordinates of element nodes
DXY (9, 2, 8)	Array holding derivatives of shape functions with respect to x and y, stored as in DSF
DV (9)	Array holding (det $\mid J \mid * H_i * H_j$) at each of the nine (or four) Guass points (See Eq. 9.67)
D1 (3×3)	Modulus matrix (Eq. 9.8, 10)
ELSK (16, 16)	Element stiffness matrix
COMN	An array (2000), declared to be shared by different arrays through EQUIVALENCE.

The other arrays, such as SK, R, LRD, URD, LFIX, ICON, CORD, etc. have same meaning as in previous program.

(*c*) Subroutines

(1) INPUT 1: This is similar to the routine of plane frame program of Appendix B, with minor modifications. The nodal cards only include fixity codes, without the magnitudes of the prescribed displacements. These are left to be read in by the load data routine.

(2) PRESET: Some scope as in previous program; checks array sizes.

(3) GAUSS: Routine to set up constants required for Gauss integration, for a given order of integration. Arrays F and H hold (ξ_i) and (H_i) (Table 5.1) respectively.

(4) SF2: Shape Function (second order) routine, to evaluate the shape functions and their derivatives with respect to ξ and η at a given Gauss point (ξ_i, η_i). Shape functions are kept in SFN, and their derivatives in DSF. This routine is called in a Do loop over number of Gauss points, nested within another over the number of elements.

(5) MOD: Routine to compute the elements of modulus matrix and place them in array D (3×3), for plane stress or plane strain appropriately.

(6) DERIV: Computes for an element the Cartesian derivatives of shape functions, at a Gauss point. Computes the terms of the Jacobian and sums over the nodes as required by Eq. (9.86); called within a Do loop over number of Gauss points nested within another over number of elements.

(7) ELSTIF: Routine to compute element stiffness matrix using numerical integration; (Eq. 9.66); called within a Do loop over number of elements.

(8) FORMSK: Assembles the global stiffness matrix in SK, by reading the element stiffness matrices from NFILE1. Very similar to the one of previous program. A comparison will bring out the common essence of the direct stiffness method, which does not distinguish between the types of structural elements.

(9) RHS: Reads load data, computes element equivalent loads, and assembles the global load vector R. Calls subroutines BODYF, PRESS, EQLOAD and PTLOAD.

(10) BODYF: Reads body force data, computes equivalent nodal loads and adds them to the load vector R appropriately.

(11) PRESS: Reads data of applied edge loads, computes nodal loads and adds to the load vector.

(12) EQLOAD: Routine to read the equivalent nodal loads to the global load vector appropriately.

(13) PTLOAD: Routine to read applied nodal forces and add to the global load vector.

(14) FIX, ELIM, SOLV, DISREA: Routines of same scope as in previous program.

(15) STRESS: Routine (similar in scope to FORCES of the previous program) to compute and print out stresses at Gauss points.

(*d*) STORAGE CONSIDERATIONS

A common block/COM/COMN (2000) has been introduced, and various DIMENSIONEd arrays are EQUIVALENCEd. It serves the purpose of scratch in-core storage area. The use of this is explained in the explanatory notes at appropriate places.

(*e*) FLOW DIAGRAM

See Fig. 10.4.

(*f*) THE PROGRAM

The program is developed on IBM 370/145 with G level Fortran Compiler. It has been written keeping NEL=100, NODES=50, NMAT=5. (See Appendix C)

(*g*) EXPLANATORY NOTES

These are given with respect to the card numbers printed at the right by the IEBUPDATE. These notes are less elaborate than in the previous program. Comments on similar features are excluded.

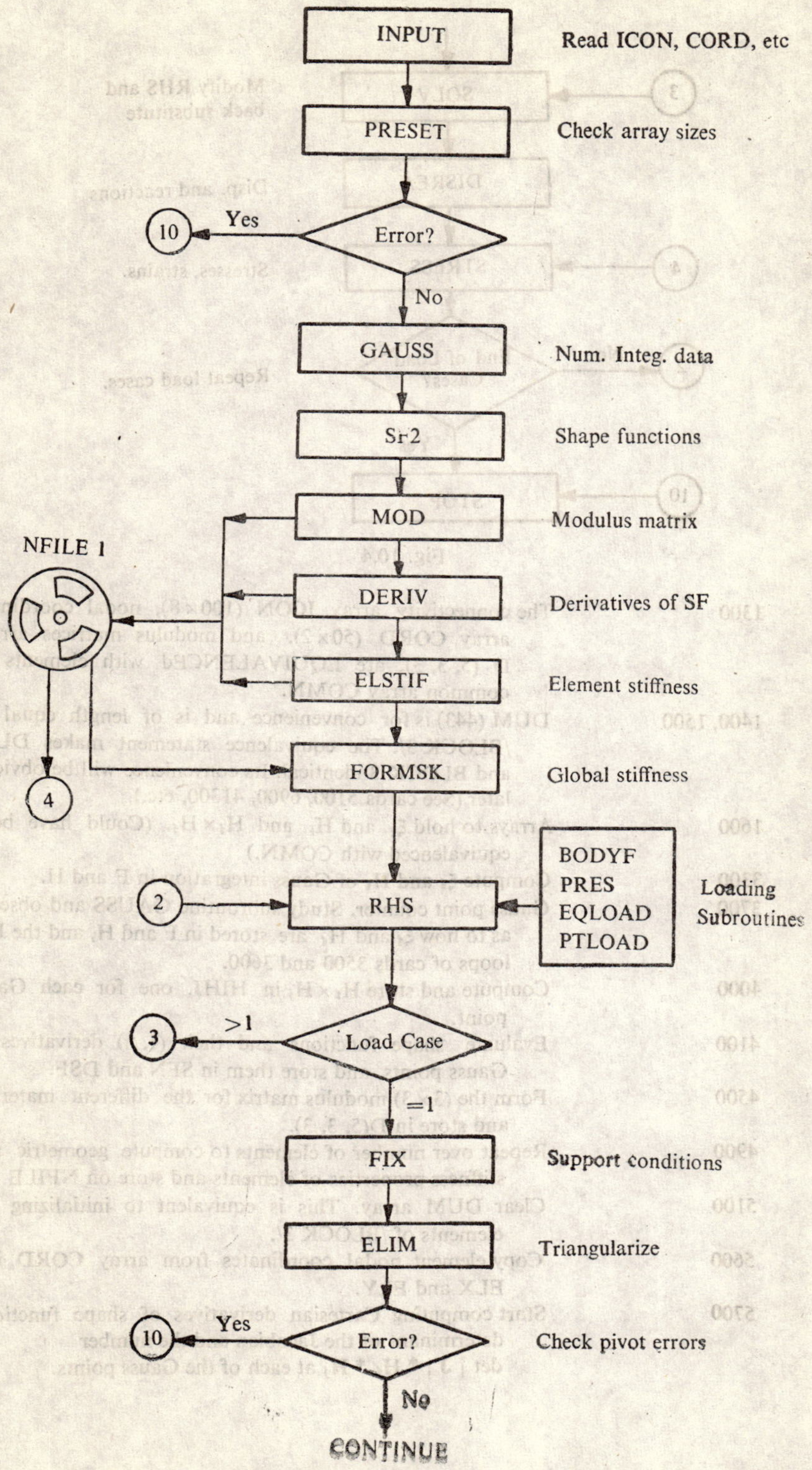
INPUT
Read ICON, CORD, etc
PRESET
Check array sizes
Error?
Yes
10
No
GAUSS
Num. Integ. data
Sr2
Shape functions
MOD
Modulus matrix
NFILE 1
DERIV
Derivatives of SF
ELSTIF
Element stiffness
FORMSK
Global stiffness
4
BODYF
PRES
EQLOAD
PTLOAD
Loading Subroutines
2
RHS
3
>1
Load Case
=1
FIX
Support conditions
ELIM
Triangularize
Yes
10
Error?
Check pivot errors
No
CONTINUE

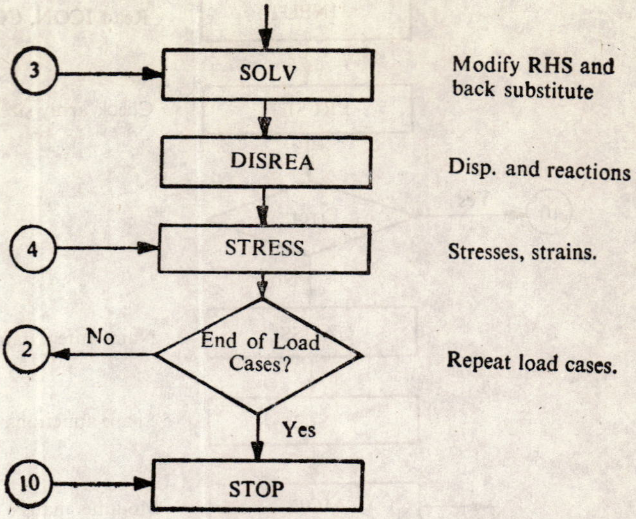

Fig. 10.4

1300 The connectivity array ICON (100×8), nodal coordinate array CORD (50×2), and modulus matrices array, D (5, 3, 3), are EQUIVALENCEd with elements of common array COMN.

1400, 1500 DUM (443) is for convenience and is of length equal to /BLOCK 3/. The equivalence statement makes DUM and BLOCK 3 identical. Its convenience will be obvious later (See cards 5100, 6900, 41300, etc.).

1600 Arrays to hold ξ_i and H_i and $H_i \times H_j$. (Could have been equivalenced with COMN.)

3300 Compute ξ_i and H_i of Gauss integration in F and H.

3700 Gauss point counter. Study subroutine GAUSS and observe as to how ξ_i and H_i are stored in F and H, and the DO loops of cards 3500 and 3600.

4000 Compute and store $H_i \times H_j$ in HIHJ, one for each Gauss point.

4100 Evaluate shape functions and their (ξ, η) derivatives at Gauss points, and store them in SFN and DSF.

4500 Form the (3 × 3) modulus matrix for the different materials and store in D(5, 3, 3).

4900 Repeat over number of elements to compute geometric and stiffness properties of elements and store on NFILE 1.

5100 Clear DUM array. This is equivalent to initializing the elements of /BLOCK 3/.

5600 Copy element nodal coordinates from array CORD into ELX and ELY.

5700 Start computing Cartesian derivatives of shape functions, determinant of the Jacobian and the number det | J | * H_i * H_j at each of the Gauss points.

5800 DERIV subroutine computes the determinant of the Jacobian, Cartesian derivatives of the shape functions, and returns the values in DETJ, and DSF, in common storage block, /BLOCK 1/.

5900 Compute the scalar multiplier giving elemental area multiplied by weight factors.

6100 Copy element number in IELNO to be transferred to NFILE 1.

6300 Pick up element connectivity data from ICON into ICONE (8) to be transferred to NFILE 1.

6600 Copy modulus matrix from D into D1, to be transferred to NFILE 1.

6700 Evaluate element stiffness matrix from routine ELSTIF which computes the same in common block array ELSK (16×16).

6800 Write 443 values of DUM (443).

6900 By virtue of EQUIVALENCE statement (card 1300), transfer of DUM array is equivalent to transferring the values of the variables and arrays of common block, /BLOCK 3/, element number, connectivity, derivatives of shape functions, etc.

7000 Go back to statement of card 5000 to repeat for the next element.

7300 Do loop over number of load cases.

7700 For the first load case go to subroutine FIX, to incorporate boundary conditions.

7800 If there are no nonzero prescribed displacements, skip the next instruction of calling subroutine FIX, (for boundary conditions).

8000 Skip the triangularization routine for subsequent loading cases, and proceed to SOLV.

SUBROUTINE INPUT 1

10600 to 12000 Steps very similar to those in previous program.

13200 Determine coordinates of mid side nodes as an average of corner nodes.

14300 Read E and Poisson ratio for isotropic case.

14700 Read elements of modulus matrix for anisotropic case.

SUBROUTINE PRESET

This subroutine is similar to the one of the previous program.

SUBROUTINE GAUSS

21800 Computes ξ_i and H_i in F and H.

23700, 23800 Compute ξ_i and H_i using symmetry.

SUBROUTINE SF2

24600, 24700 AI and BI hold the values of ξ_i and η_i of the node i, $i = 1$ to 8.

25800 Shape functions at nodes 1, 3, 5, 7 (Eq. 9.33).

25900 $\dfrac{\partial N_i}{\partial \xi}$ at $i = 1, 3, 5, 7$.

26000 $\dfrac{\partial N_i}{\partial \eta}$ at $i = 1, 3, 5, 7$.

26800 to 27000 Similar expression for nodes 2 and 6.
27200 to 27400 Similar expression for nodes 4 and 8.

SUBROUTINE MOD

28500 Go to case of anisotropy, plane stress/strain.
28600, 28700 *M* would be material number.
28800 Go to plane strain isotropic case.
28900 to 29400 Compute elements of modulus matrix (see Eq. 9.8).
29600 to 30100 Copy the element of modulus matrix directly read into PROP earlier.
30400 to 30800 Plane strain case (Eq. 9.10).

SUBROUTINE DERIV

32500 Outer Do loop over ξ and η.
32800 Inner summation loop over the element nodes.
33100, 33200 Elements of Jacobian matrix (Eq. 9.45).
33600 Go to error if determinant of Jacobian is non-positive.
34500 Do loop over x and y.
34600 Summation over nodes of the element.
35000 Elements of array holding $\dfrac{\partial N_i}{\partial x}$, $\dfrac{\partial N_i}{\partial y}$.

SUBROUTINE ELSTIF

36300 GPSTIFF holds the summation value of stiffness matrix evaluated at a Gauss point, and *A* holds the elements of the kinematic matrix (Eq. 9.61).
36400 DUM, a dummy array for ease of initializing. (See card 38200).
37000 Outer summation loop over number of Gauss points.
37200 to 38000 Elements of kinematic matrix (Eq. 9.61), $\dfrac{\partial N_i}{\partial x}$, $\dfrac{\partial N_i}{\partial y}$ etc.
38400 to 38800 The summations to perform $\mathbf{A}^t\mathbf{DA}$. Note that NDFEL = 16. A good instance showing the convenience of matrix notations (See Eqs. 9.64, 9.67).
39200 Multiply GPSTIF by $dV = \det |J| H_i H_j$.
39300 End summation over number of Gauss points.

SUBROUTINE FORMSK

Very similar to the one used in plane frame program given in the previous section.

SUBROUTINE RHS

44900 Observe the use of common block area /COM/COMN.
45900 Assign zero prescribed displacements in array UR.
46100 Skip if there are no nodes with nonzero prescribed displacements.
46900 Skip if there are no loaded elements.
47500 Read uniform body pressure code, and uniform body force components. The latter are zero of KBODY = 1, which are subsequently read in (See card 52600).
47700 Counter for loaded elements processed.
48000 Read element properties into Common/BLOCK 3/ from NFILE 1.
48400 Read loading header card (Eq. 5.1, card, Sec. D1).

48600 to 48800	Call subroutines only if required. Therefore note the care required in presenting the data (E 5.2, Sec. D1).
49100	Consider if required, point loads directly applied at nodes.

SUBROUTINE BODYF

52600	Read body force components (which can be different for different element).
53000, 53100	Double summations over number of Gauss points and nodes of the element.
53400, 53500	Equivalent nodal forces as per Eq. (9.68).
54300	Add equivalent nodal forces to the global load vector. As in previous program; IX is the address locater, etc.

SUBROUTINE PRESS

56800	Assign values of η along edges to AMULT.
57400	Read components of applied edge pressures in PRES (4, 6), with first index for edge, the first three of a row for x-component and the remaining three for y-component.
58200	Do loop for numerical integration over the order of Gauss rule.
58300	Form ξ_i and η_i for the different edges.
59000	FORM shape functions along edges, and their derivatives with respect to ξ and η.

SUBROUTINE EQLOAD

64100	Read directly the given equivalent nodal forces.
64800	Add to the global load vector appropriately.

SUBROUTINE PTLOAD

66600	Read x- and y-components of nodal loads.
66900	Add to the global load vector appropriately.

SUBROUTINE FIX, ELIM, SOLV, DISREA

These are same as in the previous program.

SUBROUTINE STRESS

83600	Read element properties from NFILE 1.
83800	Numerical integration loop.
84100	Inner loop for (isoparametric) summation over number of element nodes.
84200, 84300	Compute coordinates of Gauss point.
84700, 84800	Pick up nodal displacements from array R.
84900 to 85400	Assemble kinematic matrix A (Eq. 9.61).
86400	Compute strains (ϵ_x, ϵ_y, γ_{xy}) at the Gauss point.
86800	Compute stresses (σ_x, σ_y, τ_{xy}) at the Gauss point.
86900	If plane stress go to 51 to compute σ_z else compute $\sigma_z = \nu(\sigma_x+\sigma_y)$.
87700	Print Gauss point coordinates and stresses in a convenient format.

(*h*) Illustrative Example

We will explain here the use of the above program by considering the assembly of elements shown in Fig. 10.5 for which an appropriate analysis is possible. If we neglect the two-dimensional action ($\nu = 0$) we have a strip loaded by a load p_0 as centre. By symmetry, we have in the elements of left hand half, $\sigma_x = \frac{p_0}{2}$ and in the right hand half $\sigma_x = -\frac{p_0}{2}$, with other stresses equal to zero. The longitudinal displacement is

$$\frac{p_0}{2} \times \frac{1}{E} \times \frac{L}{2} = \frac{p_0 L}{4E},$$

with L the total length. With $p_0 = 40{,}000$ and $L = 2.0$ and $E = 2.0 \times 10^5$ this would be 0.1. This serves as a rough check.

(*i*) Output (See Appendix C)

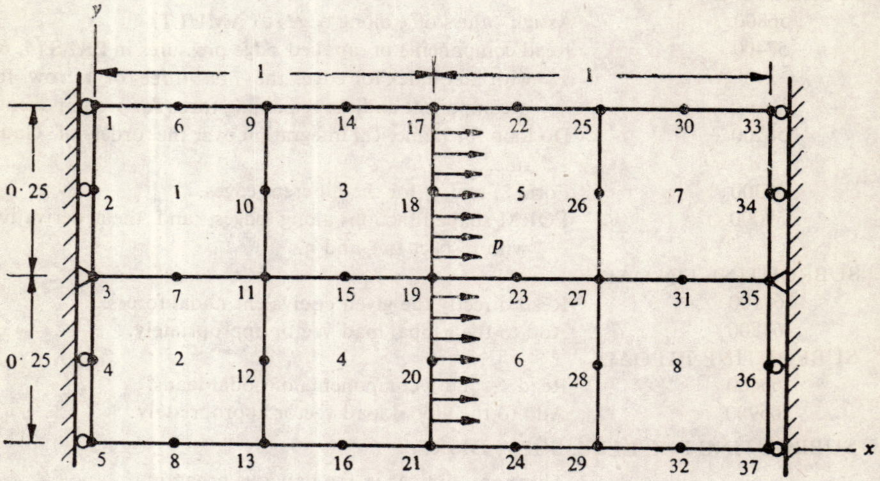

Fig. 10.5

10.5 SOLUTION OF LARGE SYSTEM OF EQUATIONS

The computing procedure involving assembling of the stiffness matrix, incorporation of support conditions and solution of equations as applied to practical problems has to be modified since it is almost impossible to keep all the equations in the main core of the computer. Obviously therefore any solution procedure has to use a backing store of a computer during assembly of equations, triangularization of the coefficient matrix and back substitution.

We discussed in Sec. 10.1 the simple procedure in which all the equations were assumed to be in the main core of the computer. The essential feature of any procedure of solving a large set is to consider the minimum number of equations required to eliminate a variable and as soon as it is eliminated the elements of the corresponding row are transferred to a backing store. This minimum number of equations is controlled by the semi bandwidth. Therefore if we adopt such a procedure, what really matters is not the total number of equations to be solved but the minimum number of equations that are required to be handled simultaneously.

Consider the banded set shown below in (a_1) below:

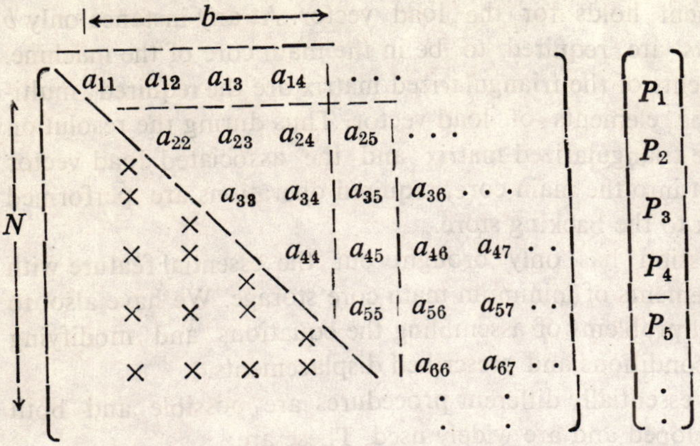

(a_1)

The semi bandwidth is 4 and the total number N can be arbitrary. It is seen that to eliminate the first variable we need only four equations. Thus only $b\,(b+1)/2$ terms are to be stored; in this case $4\,(4+1)/2=10$ terms. Observe that this number increases as the square of the bandwidth. Thus when the bandwidth is say 100, this number is 5050 and about 20000 for a bandwidth of 200. The row operations are performed on these elements as required in the Gauss elimination procedure. After these operations, the elements of the first row divided by the pivot, are transferred to a backing store. All the remaining modified coefficients in the triangular back are shifted up as shown in (a_2).

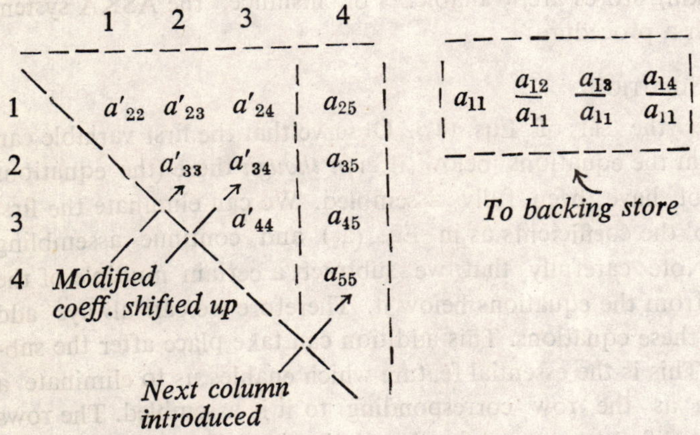

(a_2)

The new set of coefficients, a_{25}, a_{35}, a_{45} and a_{55} are entered in to fill the remaining space. This 'shifting' up of coefficients so as to occupy the $b\,(b+1)/2$ locations is simple, but entails computer time.

The elimination process is repeated. In this manner the equations are pulled up into the common area of $b\,(b+1)/2$ locations, modified and stored in a backing store. Since this kind of store is virtually unlimited, the total number of equations does not pose any problem with respect to storage.

A similar argument holds for the load vector. At any instance only b number of elements are required to be in the main core of the machine. Recall that the elements of the triangularized matrix are the required multipliers to modify the elements of load vector. Thus during the resolution routine, a row of the triangularized matrix and the associated load vector elements are brought into the main core, required operations are performed and transferred back to the backing store.

The above discussion has only brought out the essential feature with respect to the requirements of minimum main core storage. We have also to consider the essential problems of assembling the equations and modifying them for boundary conditions and prescribed displacements.

In this regard two essentially different procedures are possible and both these have been developed and are widely used. These are:

(i) a procedure in which the assembly of equations, incorporation of boundary conditions and solution process are handled simultaneously.

(ii) a procedure in which the stiffness equations and load vectors are first assembled and modified for boundary conditions. The solution process is carried out next in an uninterrupted manner.

The first procedure is an extension of the procedure already discussed in Sec. 10.1 but modified to operate using backing store of a computer. This is quite efficient and suitable where only the sequentially accessible tape storage units are available. The second procedure enjoys the advantage of providing a modular program structure, and becomes a practical proposal where direct access backing stores are available. For instance, the ASKA system [29] employs such a procedure.

(*a*) The Band Solution

Consider the set of Eqs. (a_1). Observe that the first variable can be eliminated from the equations 'below' it *even though* these (the equations below) might not have been fully assembled. We can eliminate the first variable, shift 'up' the coefficients as in Eq. (a_2) and continue assembling the equations. Note carefully that we subtract a certain multiple of the pivotal equation from the equations below it. Therefore we can always add contributions to these equations. This addition can take place after the subtraction process. This is the essential feature which enables us to eliminate a variable as soon as the row corresponding to it is assembled. The rows below it can be modified to account for the elimination, even though these lower rows are not fully assembled.

Consider the assembly shown in Fig. 10.6. Here the maximum difference in node numbers is 5. Thus the semi bandwidth is $(5+1)\times 2=12$. The maximum locations required at the time of assembly (and elimination, which is simultaneous) are $12\times(12+1)/2=78$ locations. The procedure adopts

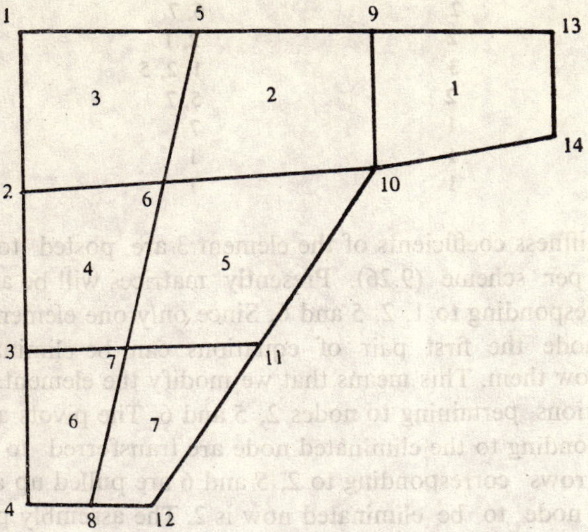

Fig. 10.6

elimination of nodes in a serial order. Suppose that we have assembled the element stiffness matrices and have stored them sequentially on a disc or a tape. Since elimination of nodes is required to proceed in a sequential order we see that we require the stiffness of elements whose order may not be sequential. For the assembly of Fig. 10.6 we see that we require the data of element in the order 3, 4, 6, 2, 5, 7 and 1. Therefore a modified sequence is adopted for storage of element data. This reordering of element storages is generally achieved by a subroutine which scans through the connectivity data and arrives at a desirable sequence of numbers so that the element stiffnesses are always read in a forward fashion. Alternatively the element stiffnesses can be grouped into blocks so that at the time of assembling the equations, the scanning through the element data to find out if it has any contribution to make to the node in question, can be minimised. This problem is altogether avoided if direct access backing store is used. By a suitable procedure of scanning through element connectivity data we arrive at the total number of elements connected to a given node, and the corresponding elements. Thus for the problem of Fig. 10.6 this information would be as follows:

Node	*No. of elements*	*Element numbers*
1	1	3
2	2	3, 4
3	2	4, 6
4	1	6
5	2	3, 2
6	4	2. 3, 4, 5

7	4	4, 5, 6, 7	
8	2	6, 7	(a_3)
9	2	2, 1	
10	3	1, 2, 5	
11	2	5, 7	
12	1	7	
13	1	1	
14	1	1	

Thus the stiffness coefficients of the element 3 are posted to appropriate locations as per scheme (9.26). Presently matrices will be at row-column locations corresponding to 1, 2, 5 and 6. Since only one element is connected to this node the first pair of equations can be eliminated from the equations below them. This means that we modify the elements of the rows of the equations pertaining to nodes 2, 5 and 6. The pivots and the multipliers corresponding to the eliminated node are transferred to the backing store.* The rows corresponding to 2, 5 and 6 are pulled up as in Eq. (a_2), so that the node to be eliminated now is 2. The assembly process is now continued. Since two elements are connected to this node, and one of which namely 3 is already considered, the contributions of the remaining one, namely 4 are taken up and are placed accordingly. Observe that there would be contributions at nodes 2, 3, 6 and 7. Further observe that the coefficients corresponding to node 7 occupy the positions vacated by those occupied previously which corresponded to node 6. In other words, we have to keep relative addressing between the node number and the location number in the triangular main core area. This is a matter of detailing and depends upon how the locations of this triangular area are addressed. The node 2 is eliminated in a similar manner; the rows corresponding to this node are transferred to the back store, and the remaining rows shifted up the triangular block.

The modifications required for the boundary conditions do not pose any special problem. It is handled in the same way as was indicated in Sec. 10.1. The elimination process is modified accordingly. When all the equations are modified the backing store tape is rewound to the start to help the subsequent steps.

Consider now resolution and back substitution process. To modify the load vector multipliers and pivots are read in from the back store in a forward sense and the required modification effected. The back substitution is performed by reading the triangulated matrix in a backward manner and operating on the load vector.

Another powerful variant of the band solution is a procedure in which at any time only very few equations are kept in the main core. Very large band-

*If convenient they can be collected within the core in a small buffer area and transferred in bulk to the backing store.

widths can be accommodated. See Reference [31]. Evidently more transfer operations between main and backing store are involved.

(*b*) THE FRONT SOLUTION SCHEME

This technique [32] is a very useful variant of the procedure discussed so far. In the previous procedure called 'band' solution, the nodes are eliminated in a sequential order. After the rows corresponding to a node are eliminated the remaining equations are shifted up the triangular block as shown in Eqs. (a_2). This operation, though simple, entails computer time. The frontal solution technique avoids this time by lifting the requirement of having to eliminate the nodes in a sequential manner. The sequence of the nodes to be eliminated are determined by the *element numbering.* Thus here, element numbering becomes important. This will determine a parameter 'front width', analogous to 'bandwidth'.

Consider again the assembly of Fig. 10.6. Let us suppose that we have adopted first the element numbering as shown. We then arbitrarily number the nodes. Let the same numbers be considered now. The front solution processes the stiffness *element-wise sequentially* and investigates at each stage if any nodes can be eliminated. Thus when we post the contributions of element 1 we can conclude that at nodes 13 and 14 there would not be any more contributions from other elements. At 9 and 10, however, contributions of elements 2 and 5 would be required. Thus nodes 13 and 14 can be eliminated immediately. The locations occupied by them are thus free to be given to the new nodes. Thus when the contribution of element 2 is considered next, these locations are allowed to the 'incoming' nodes 5 or 6. It is seen that node 9 can now be eliminated. When the contribution of element 3 is considered, the locations occupied by coefficients corresponding to node 9 can be allocated to those of node 1 and fresh locations to those of node 2. In this manner we can economically utilize the locations without having to shift up the coefficients as in band solution. This procedure has evidently a good physical appeal but involves relatively complicated book keeping and relative addressing. It has been shown that in finite element assemblies with midside nodes the front width is generally smaller than the bandwidth and thus more efficient. But the flow of the program becomes very different. For details the reader may consult Reference [32].

(*c*) CHOLESKI SCHEME

The Choleski scheme has also been successfully used for positive definite systems. The hypermatrix scheme [33] is a procedure which belongs to the type (b) mentioned before, wherein the equations are assembled first and modified for boundary effects, followed by a solution process. This procedure employs the scheme of submatrices and what is important the direct access backing store devices. The matrix algebra has been programmed in terms of sub-matrices directly using matrices of addresses giving the addresses of variables kept in direct access backing store. Very large systems have been solved with no limitation on bandwidths. [See References 29, 33.]

10.6 SUBSTRUCTURE TECHNIQUES

In the previous section methods of handling systems of equations with large bandwidth were discussed. The technique of substructuring is also employed to handle a large system [29]. The essential ideas of substructuring were introduced in Sec. 8.8 and an application demonstrated in which an arched member was analysed as an assembly of straight beam segments. The idea can be generalized to a situation where a structure is imagined to be an assembly of a small number of substructures. Each of these are *handled separately* and results of analysis *stored into the backing store* of the computer and *later assembled* according to the *connectivity of the substructures.* We will consider some details of this approach.

(*a*) SERIES SUBDIVISION

Consider the frame* shown in Fig. 10.7(a). This can be imagined to be an assembly of two substructures shown in 10.7(b). The force vectors of the joints 5 and 6 are the mutual forces between these two imaginary substructures. These are obviously equal and opposite as any set of internal forces. In the terminology introduced in Sec. 8.6, we define respectively for substructures A and B

$$\mathbf{U}_I{}^A = \{\mathbf{U}_1, \mathbf{U}_2, \mathbf{U}_3, \mathbf{U}_4\}$$
$$\mathbf{U}_I{}^B = \{\mathbf{U}_7, \mathbf{U}_8, \mathbf{U}_9, \ldots, \mathbf{U}_{13}\} \qquad (a_1)$$

as the set of 'internal' joints. The set of joints

$$\mathbf{U}_E{}^{AB} = \{\mathbf{U}_5, \mathbf{U}_6\} \qquad (a_2)$$

will be called 'external' joints for the interface between substructures *A* and *B*. (We could alternatively term $\mathbf{U}_E$ as vector of *interface joints* if we so please to term it.) So we can analyse substructure *A* and determine the forces $\mathbf{R}_5{}^A$ and $\mathbf{R}_6{}^A$ at 5 and 6, to balance the forces acting at the joints 1, 2, ..., 6. Similarly we can analyse the substructure *B* and determine $\mathbf{R}_5{}^B$ and $\mathbf{R}_6{}^B$ to equilibrate the loads acting on the joints 5 to 13. Since the forces **R** at joints 5 and 6 are mutual internal actions, we have the conditions

$$\mathbf{R}^A + \mathbf{R}^B = \begin{Bmatrix} \mathbf{R}_5{}^A \\ \mathbf{R}_6{}^B \end{Bmatrix} + \begin{Bmatrix} \mathbf{R}_5{}^B \\ \mathbf{R}_6{}^B \end{Bmatrix} = \begin{Bmatrix} \mathbf{0} \\ \mathbf{0} \end{Bmatrix} = \mathbf{0} \qquad (a_3)$$

Then, the mutual forces **R** will be obtained in terms of $\mathbf{U}_E{}^{AB}$ and these equations will be sufficient to determine these interface displacements. Let us now consider the details of the above procedure.

*Obviously, we can consider an assembly of finite elements also.

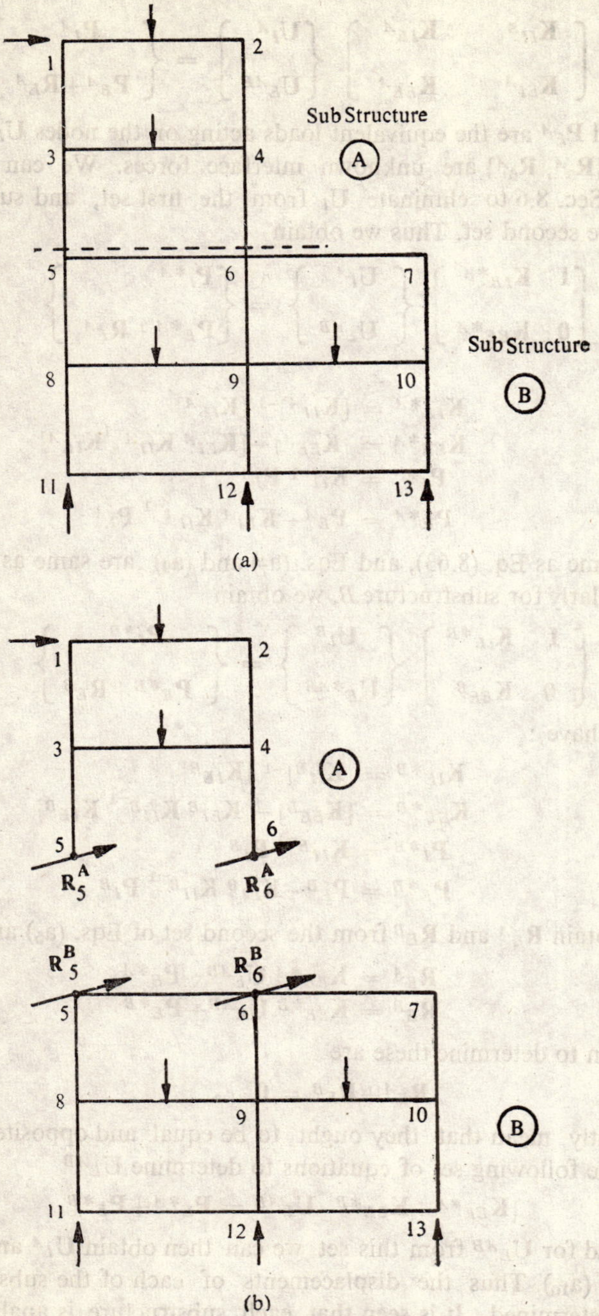
1
2
Sub Structure
3
4
A
5
6
7
Sub Structure
8
9
10
B
11
12
13
1
2
3
4
A
5
6
R5A
R6A
R5B
R6B
5
6
7
8
9
10
B
11
12
13

(a)

(b)

Fig. 10.7

Assemble the stiffness relations for the substructure A

$$\begin{bmatrix} \mathbf{K}_{II}{}^{A} & \mathbf{K}_{IE}{}^{A} \\ \mathbf{K}_{EI}{}^{A} & \mathbf{K}_{EE}{}^{A} \end{bmatrix} \begin{Bmatrix} \mathbf{U}_{I}{}^{A} \\ \mathbf{U}_{E}{}^{AB} \end{Bmatrix} = \begin{Bmatrix} \mathbf{P}_{I}{}^{A} \\ \mathbf{P}_{E}{}^{A}+\mathbf{R}_{E}{}^{A} \end{Bmatrix} \qquad (a_4)$$

Here $\mathbf{P}_I{}^A$ and $\mathbf{P}_E{}^A$ are the equivalent loads acting on the nodes $\mathbf{U}_I$ and $\mathbf{U}_E{}^{AB}$, and $\mathbf{R}_E{}^A \equiv \{\mathbf{R}_5{}^A, \mathbf{R}_6{}^A\}$ are unknown interface forces. We can proceed as discussed in Sec. 8.6 to eliminate $\mathbf{U}_I$ from the first set, and substitute the result into the second set. Thus we obtain

$$\begin{bmatrix} \mathbf{I} & \mathbf{K}_{IE}{}^{*A} \\ \mathbf{0} & \mathbf{K}_{EE}{}^{*A} \end{bmatrix} \begin{Bmatrix} \mathbf{U}_{I}{}^{A} \\ \mathbf{U}_{E}{}^{AB} \end{Bmatrix} = \begin{Bmatrix} \mathbf{P}_{I}{}^{*A} \\ \mathbf{P}_{E}{}^{*A}+\mathbf{R}_{E}{}^{A} \end{Bmatrix} \qquad (a_5)$$

where

$$\mathbf{K}_{IE}{}^{*A} = [\mathbf{K}_{II}{}^{A}]^{-1}\,[\mathbf{K}_{IE}{}^{A}] \qquad (a_6)$$

$$\mathbf{K}_{EE}{}^{*A} = [\mathbf{K}_{EE}{}^{A}]-[\mathbf{K}_{EI}{}^{A}\,\mathbf{K}_{II}{}^{A^{-1}}\mathbf{K}_{IE}{}^{A}] \qquad (a_7)$$

$$\mathbf{P}_{I}{}^{*A} = \mathbf{K}_{II}{}^{-1}\,\mathbf{P}_{I}{}^{A} \qquad (a_8)$$

$$\mathbf{P}_{E}{}^{*A} = \mathbf{P}_{E}{}^{A}-\mathbf{K}_{EI}{}^{A}\,\mathbf{K}_{II}{}^{A^{-1}}\,\mathbf{P}_{I}{}^{A} \qquad (a_9)$$

Eq. (a_5) is same as Eq. (8.63), and Eqs. (a_7) and (a_9) are same as (8.59) and (8.60). Similarly for substructure B, we obtain

$$\begin{bmatrix} \mathbf{I} & \mathbf{K}_{IE}{}^{*B} \\ \mathbf{0} & \mathbf{K}_{EE}{}^{B} \end{bmatrix} \begin{Bmatrix} \mathbf{U}_{I}{}^{B} \\ \mathbf{U}_{E}{}^{*AB} \end{Bmatrix} = \begin{Bmatrix} \mathbf{P}_{I}{}^{*B} \\ \mathbf{P}_{E}{}^{*B}+\mathbf{R}_{E}{}^{B} \end{Bmatrix} \qquad (a_{10})$$

Similarly we have :

$$\mathbf{K}_{IE}{}^{*B} = [\mathbf{K}_{II}{}^{B}]^{-1}\,[\mathbf{K}_{IE}{}^{B}] \qquad (a_{11})$$

$$\mathbf{K}_{EE}{}^{*B} = [\mathbf{K}_{EE}{}^{B}]-[\mathbf{K}_{EI}{}^{B}\,\mathbf{K}_{II}{}^{B^{-1}}\,\mathbf{K}_{IE}{}^{B}] \qquad (a_{12})$$

$$\mathbf{P}_{I}{}^{*B} = \mathbf{K}_{II}{}^{B^{-1}}\,\mathbf{P}_{I}{}^{B} \qquad (a_{13})$$

$$\mathbf{P}_{E}{}^{*B} = \mathbf{P}_{E}{}^{B}-\mathbf{K}_{EI}{}^{B}\,\mathbf{K}_{II}{}^{B^{-1}}\,\mathbf{P}_{I}{}^{B} \qquad (a_{14})$$

We can obtain $\mathbf{R}_E{}^A$ and $\mathbf{R}_E{}^B$ from the second set of Eqs. (a_5) and (a_{10}):

$$\mathbf{R}_{E}{}^{A} = \mathbf{K}_{EE}{}^{*A}\,\mathbf{U}_{E}{}^{AB}-\mathbf{P}_{E}{}^{*A}$$

$$\mathbf{R}_{E}{}^{B} = \mathbf{K}_{EE}{}^{*B}\,\mathbf{U}_{E}{}^{AB}-\mathbf{P}_{E}{}^{*B} \qquad (a_{15})$$

The condition to determine these are

$$\mathbf{R}_{E}{}^{A}+\mathbf{R}_{E}{}^{B} = \mathbf{0} \qquad (a_{16})$$

which presently mean that they ought to be equal and opposite. Therefore we obtain the following set of equations to determine $\mathbf{U}_E{}^{AB}$

$$[\mathbf{K}_{EE}{}^{*A}+\mathbf{K}_{EE}{}^{*B}]\,\mathbf{U}_{E}{}^{AB} = \mathbf{P}_{E}{}^{*A}+\mathbf{P}_{E}{}^{*B} \qquad (a_{17})$$

Having solved for $\mathbf{U}_E{}^{AB}$ from this set we can then obtain $\mathbf{U}_I{}^A$ and $\mathbf{U}_I{}^B$ from Eqs. (a_5) and (a_{10}). Thus the displacements of each of the substructure are completely determined. It is seen that each substructure is analysed partly; the internal joints $\mathbf{U}_I$ are eliminated and substructure stiffness relations (Eq. (a_{15}) are obtained. These are treated as units and the assembly process involved in Eq. (a_{17}) is then carried out to determine the interface displacements.

(b) Parallel Subdivision

Consider now a relatively more general (parallel) subdivision of a structure as shown in Fig. 10.8(a), (b). Here the structure is divided into three substructures *A*, *B* and *C*. Let the common (interface) nodes be labelled as shown in (a). Presently we have seven such joints. Let the numbering of these joints in each substructure be as shown in (b). The relationships between these two systems are:

$$\begin{array}{lllll} \mathbf{U}_1^A = \mathbf{U}_7; & \mathbf{U}_2^A = \mathbf{U}_6; & \mathbf{U}_3^A = \mathbf{U}_2; & \mathbf{U}_4^A = \mathbf{U}_1 & \\ \mathbf{U}_1^B = \mathbf{U}_5; & \mathbf{U}_2^B = \mathbf{U}_4; & \mathbf{U}_3^B = \mathbf{U}_3; & \mathbf{U}_4^B = \mathbf{U}_6; & \mathbf{U}_5^B = \mathbf{U}_7 \\ \mathbf{U}_1^C = \mathbf{U}_1; & \mathbf{U}_2^C = \mathbf{U}_2; & \mathbf{U}_3^C = \mathbf{U}_3; & \mathbf{U}_4^C = \mathbf{U}_4; & \mathbf{U}_5^C = \mathbf{U}_5 \end{array} \qquad (a_{18})$$

This is *analogous to the connectivity data* in which we relate element node address to the global node address.

Each substructure can be analysed, and the substructure stiffness relations of type (a_5) or (a_{10}) obtained. The second set of each give $\mathbf{R}_E$ for each substructure. Thus

$$\begin{aligned} \mathbf{R}_E^A &= \mathbf{K}_{EE}^{*A}\,\mathbf{U}_E^A - \mathbf{P}_E^{*A} \\ \mathbf{R}_E^B &= \mathbf{K}_{EE}^{*B}\,\mathbf{U}_E^B - \mathbf{P}_E^{*B} \\ \mathbf{R}_E^C &= \mathbf{K}_{EE}^{*C}\,\mathbf{U}_E^C - \mathbf{P}_E^{*C} \end{aligned} \qquad (a_{19})$$

where we have

$$\begin{array}{ll} \mathbf{U}_E^A = \{\mathbf{U}_1^A, \mathbf{U}_2^A, \mathbf{U}_3^A, \mathbf{U}_4^A\}; & \mathbf{R}_E^A = \{\mathbf{R}_1^A, \mathbf{R}_2^A\, \mathbf{R}_3^A, \mathbf{R}_4^A\} \\ \mathbf{U}_E^B = \{\mathbf{U}_1^B, \mathbf{U}_2^B, \ldots, \mathbf{U}_5^B\}; & \mathbf{R}_E^B = \{\mathbf{R}_1^B, \mathbf{R}_2^B, \ldots, \mathbf{R}_5^B\} \\ \mathbf{U}_E^C = \{\mathbf{U}_1^C, \mathbf{U}_2^C, \ldots, \mathbf{U}_5^C\}; & \mathbf{R}_E^C = \{\mathbf{R}_1^C, \mathbf{R}_2^C \ldots, \mathbf{R}_5^C\} \end{array} \qquad (a_{20})$$

The matrices $\mathbf{K}_{EE}^{*A}$; $\mathbf{K}_{EE}^{*B}$, etc. are appropriate modified stiffness matrices modified according to Eq. (a_7). The vectors $\mathbf{P}_E^{*A}$, $\mathbf{P}_E^{*B}$ and $\mathbf{P}_E^{*C}$ are the modified vectors according to (a_9). Recall that $-\mathbf{P}_E^{*A}$, $-\mathbf{P}_E^{*B}$, etc. are the fixing end forces, and $\mathbf{P}_E^{*A}$, $\mathbf{P}_E^{*B}$, etc. are the equivalent loads.

We have to write now the *conditions of equilibrium* to determine the internal forces $\mathbf{R}_E^A$, $\mathbf{R}_E^B$ and $\mathbf{R}_E^C$. The equations at nodes

$$\begin{array}{lll} \mathbf{R}_4^A + \mathbf{R}_1^C = 0; & \mathbf{R}_3^A + \mathbf{R}_2^C = 0; & \mathbf{R}_2^B + \mathbf{R}_4^C = 0 \\ \mathbf{R}_1^B + \mathbf{R}_5^C = 0; & \mathbf{R}_2^A + \mathbf{R}_4^B = 0; & \mathbf{R}_1^A + \mathbf{R}_5^B = 0 \end{array} \qquad (a_{21})$$

These simply state that they are mutual actions.

The next step would be to express these in terms of displacements of global nodes, $\mathbf{U}_1, \mathbf{U}_2, \ldots, \mathbf{U}_7$, using relations ($a_{19}$) and the connectivity ($a_{18}$). This problem is the same as encountered in the displacement analysis.

Let us write the first set of Eqs. (a_{19}) in full as

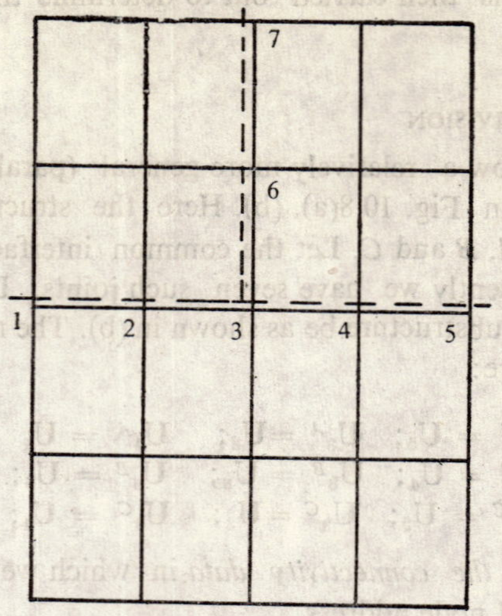

(a)

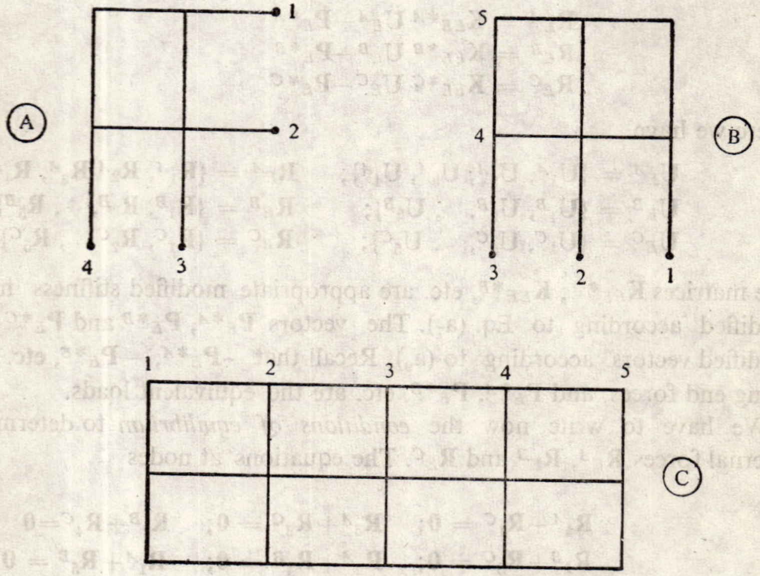

(b)

Fig. 10.8

$$\begin{Bmatrix} \mathbf{R}_1{}^A \\ \mathbf{R}_2{}^A \\ \mathbf{R}_3{}^A \\ \mathbf{R}_4{}^A \end{Bmatrix} = \begin{bmatrix} \mathbf{K}_{11}{}^{*A} & \mathbf{K}_{12}{}^{*A} & \mathbf{K}_{13}{}^{*A} & \mathbf{K}_{14}{}^{*A} \\ \cdots & \cdots & \cdots & \cdots \\ \cdots & \text{etc.} & \cdots & \cdots \\ \mathbf{K}_{41}{}^{*A} & \cdots & \cdots & \mathbf{K}_{44}{}^{*A} \end{bmatrix} \begin{Bmatrix} \mathbf{U}_1{}^A \\ \mathbf{U}_2{}^A \\ \mathbf{U}_3{}^A \\ \mathbf{U}_4{}^A \end{Bmatrix} + \begin{Bmatrix} \mathbf{P}_1{}^{*A} \\ \mathbf{P}_2{}^{*A} \\ \mathbf{P}_3{}^{*A} \\ \mathbf{P}_4{}^{*A} \end{Bmatrix} \tag{a$_{22}$}$$

Here we can take into account (a_{18}). Thus in terms of global node addresses

$$\begin{Bmatrix} \mathbf{R}_1{}^A \\ \mathbf{R}_2{}^A \\ \mathbf{R}_3{}^A \\ \mathbf{R}_4{}^A \end{Bmatrix} = \begin{bmatrix} \\ \textit{Same as} \\ \textit{in Eq. } (a_{22}) \\ \\ \end{bmatrix} \begin{Bmatrix} \mathbf{U}_7 \\ \mathbf{U}_6 \\ \mathbf{U}_2 \\ \mathbf{U}_1 \end{Bmatrix} + \begin{Bmatrix} \mathbf{P}_7{}^{*A} \\ \mathbf{P}_6{}^{*A} \\ \mathbf{P}_2{}^{*A} \\ \mathbf{P}_1{}^{*A} \end{Bmatrix} \tag{a$_{23}$}$$

In this manner we can write the other two sets of Eq. (a_{19}) in terms of global node numbering. With the help of these we can convert the equilibrium relations (a_{21}) in terms of global nodes:

$$\begin{bmatrix} \mathbf{K}_{11} & \mathbf{K}_{12} \cdots & \mathbf{K}_{17} \\ \vdots & \vdots & \vdots \\ \vdots & \vdots & \vdots \\ \mathbf{K}_{71} & \mathbf{K}_{72} \cdots & \mathbf{K}_{77} \end{bmatrix} \begin{Bmatrix} \mathbf{U}_1 \\ \mathbf{U}_2 \\ \vdots \\ \mathbf{U}_7 \end{Bmatrix} = \begin{Bmatrix} \mathbf{P}_1 \\ \mathbf{P}_2 \\ \vdots \\ \mathbf{P}_7 \end{Bmatrix} \tag{a$_{24}$}$$

Here $\mathbf{K}_{ij}$ are the appropriate sums of $\mathbf{K}_{ij}{}^*$ of different substructures, and $\mathbf{P}^*$ are the sums of equivalent loads of different substructures as in the usual direct stiffness method.

Tridiagonal Partitioning

Program structure to handle the more general interconnection, such as the one in Fig. 10.8(a) is quite involved and we will not be discussing its details. However, the serially connected substructure analysis is relatively simple. The example of Fig. 10.7 belongs to this type. Fig. 10.9 shows a more general case, in which the frame has been imagined to be subdivided into four substructures serially connected to one another. In such a connection any substructure is connected to the previous and the next substructure. The assembled stiffness matrix would, in terms of submatrices, be a tridiagonal one. Let us denote

$$\begin{aligned} \mathbf{U}_1 &= \{\mathbf{u}_1, \mathbf{u}_2, \mathbf{u}_3, \mathbf{u}_4, \mathbf{u}_5\} \\ \mathbf{U}_2 &= \{\mathbf{u}_6, \mathbf{u}_7, \dots, \mathbf{u}_{10}\} \\ \mathbf{U}_3 &= \{\mathbf{u}_{11}, \mathbf{u}_{12}, \dots, \mathbf{u}_{15}\} \\ \mathbf{U}_4 &= \{\mathbf{u}_{16}, \mathbf{u}_{17}, \dots, \mathbf{u}_{20}\} \\ \mathbf{U}_5 &= \{\mathbf{u}_{21}, \mathbf{u}_{22}, \dots, \mathbf{u}_{25}\} \end{aligned} \tag{a$_{25}$}$$

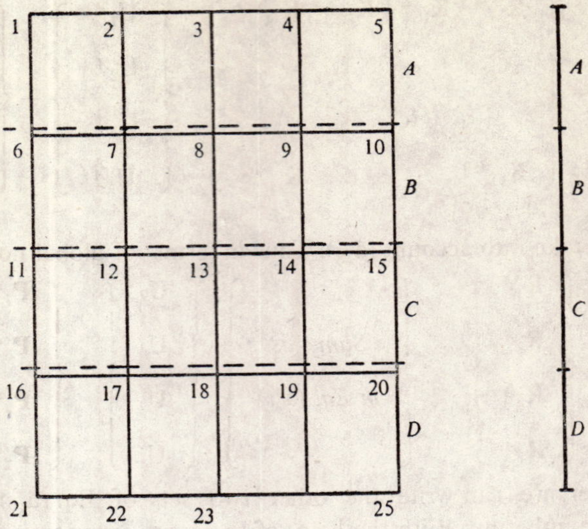

Fig. 10.9

The structure stiffness relations will be of the following form

$$\begin{bmatrix} \mathbf{K}_{11} & \mathbf{K}_{12} & \cdots & \cdots & \cdots \\ \mathbf{K}_{12}{}^t & \mathbf{K}_{22} & \mathbf{K}_{23} & \cdots & \cdots \\ \cdots & \mathbf{K}_{23}{}^t & \mathbf{K}_{33} & \mathbf{K}_{34} & \cdots \\ \cdots & \cdots & \mathbf{K}_{34}{}^t & \mathbf{K}_{44} & \mathbf{K}_{45} \\ \cdots & \cdots & \cdots & \mathbf{K}_{45}{}^t & \mathbf{K}_{55} \end{bmatrix} \begin{Bmatrix} \mathbf{U}_1 \\ \mathbf{U}_2 \\ \mathbf{U}_3 \\ \mathbf{U}_4 \\ \mathbf{U}_5 \end{Bmatrix} = \begin{Bmatrix} \mathbf{P}_1 \\ \mathbf{P}_2 \\ \mathbf{P}_3 \\ \mathbf{P}_4 \\ \mathbf{P}_5 \end{Bmatrix} \qquad (a_{26})$$

Here $\mathbf{K}_{11}$, etc. are submatrices of appropriate sizes. If the frame is a plane frame, then $\mathbf{U}_1$ is of size (15×1) and $\mathbf{K}_{11}$, $\mathbf{K}_{12}$, etc. are of size (15×15).

In problems of this type we can assemble and triangularize blocks of equations, shift them to backing store of computer, and in the core space vacated by them, assemble the next block of equations. This is similar to the procedure outlined in the previous section in which *each degree of freedom* is handled and a *single row* is transferred to the backing core.

Consider that the stiffness and the load vectors for joints $\mathbf{U}_1$ and $\mathbf{U}_2$ are assembled in which contributions of the members of the substructure A alone are considered:

$$\begin{bmatrix} \mathbf{K}_{11}{}^A & \mathbf{K}_{12}{}^A \\ \mathbf{K}_{12}{}^{At} & \mathbf{K}_{22}{}^A \end{bmatrix} \begin{Bmatrix} \mathbf{U}_1 \\ \mathbf{U}_2 \end{Bmatrix} = \begin{Bmatrix} \mathbf{P}_1{}^A \\ \mathbf{P}_2{}^A \end{Bmatrix} \qquad (a_{27})$$

Since the rows corresponding to $\mathbf{U}_2$ are *not fully* assembled we cannot eliminate the equations fully. We, therefore, eliminate only upto $\mathbf{U}_1$; this process consists of the usual row operations:

$$\begin{bmatrix} \mathbf{I} & \mathbf{K}_{12}^* \\ \mathbf{0} & \mathbf{K}_{22}^{*A} \end{bmatrix} \begin{Bmatrix} \mathbf{U} \\ \mathbf{U}_2 \end{Bmatrix} = \begin{Bmatrix} \mathbf{P}_1^* \\ \mathbf{P}_2^{*A} \end{Bmatrix} \tag{a_{28}}$$

After these operations, the modified coefficients of $\mathbf{K}_{12}^*$ and $\mathbf{P}_1^*$ are shifted from main core to the backing store. In the space vacated by them we shift up the coefficients of $\mathbf{K}_{22}^{*A}$ and $\mathbf{P}_2^{*A}$. Using the space vacated by these we assemble the contributions of members of substructure B:

$$\begin{bmatrix} \mathbf{K}_{22}^{*A} + \mathbf{K}_{22}^B & \mathbf{K}_{23}^B \\ \mathbf{K}_{23}^{B^t} & \mathbf{K}_{33}^B \end{bmatrix} \begin{Bmatrix} \mathbf{U}_2 \\ \mathbf{U}_3 \end{Bmatrix} = \begin{Bmatrix} \mathbf{P}_2^{*A} + \mathbf{P}_2^B \\ \mathbf{P}_3^B \end{Bmatrix} \tag{a_{29}}$$

Now the row operations are performed on variables up to $\mathbf{U}_2$ alone, since the rows corresponding to $\mathbf{U}_3$ are not fully assembled. Thus we obtain

$$\begin{bmatrix} \mathbf{I} & \mathbf{K}_{23}^* \\ \mathbf{0} & \mathbf{K}_{33}^{*B} \end{bmatrix} \begin{Bmatrix} \mathbf{U}_2 \\ \mathbf{U}_3 \end{Bmatrix} = \begin{Bmatrix} \mathbf{P}_2^* \\ \mathbf{P}_3^{*B} \end{Bmatrix} \tag{a_{30}}$$

We now shift $\mathbf{K}_{23}^*$ and $\mathbf{P}_2^*$ to backing store and in their place shift up $\mathbf{K}_{33}^{*B}$ and $\mathbf{P}_3^{*B}$. Using the space vacated by these, we assemble the coefficients due to the contributions from the members of substructure C. In this manner we traverse onwards.

The reader may like to see as to how the above process includes the physical idea of the substructuring. Treating $\mathbf{U}_2$ as *external* nodes to substructure A, we have

$$\begin{bmatrix} \mathbf{K}_{11}^A & \mathbf{K}_{12}^A \\ \mathbf{K}_{12}^{*A^t} & \mathbf{K}_{22}^A \end{bmatrix} \begin{Bmatrix} \mathbf{U}_1 \\ \mathbf{U}_2 \end{Bmatrix} = \begin{Bmatrix} \mathbf{P}_1^A \\ \mathbf{P}_2^A + \mathbf{R}^{AB} \end{Bmatrix} \tag{a_{31}}$$

Similarly for substructure B, $\mathbf{U}_2$ are external and therefore

$$\begin{bmatrix} \mathbf{K}_{22}^B & \mathbf{K}_{23}^B \\ \mathbf{K}_{23}^{B^t} & \mathbf{K}_{33}^B \end{bmatrix} \begin{Bmatrix} \mathbf{U}_2 \\ \mathbf{U}_3 \end{Bmatrix} = \begin{Bmatrix} \mathbf{P}_2^B - \mathbf{R}^{AB} \\ \mathbf{P}_3^B \end{Bmatrix} \tag{a_{32}}$$

Here, in the first set, we have introduced $-\mathbf{R}^{AB}$, since these forces are required to be opposite to those acting on structure A at $\mathbf{U}_2$. We can now eliminate $\mathbf{R}^{AB}$ from these two equations to obtain

$$\left[\begin{bmatrix} \mathbf{K}_{11}^A & \mathbf{K}_{12}^A & \\ \mathbf{K}_{12}^{A^t} & \mathbf{K}_{22}^A & \\ & & \mathbf{0} \end{bmatrix} + \begin{bmatrix} \mathbf{0} & & \\ & \mathbf{K}_{22}^B & \mathbf{K}_{23}^B \\ & \mathbf{K}_{23}^{B^t} & \mathbf{K}_{23}^B \end{bmatrix} \right] \begin{Bmatrix} \mathbf{U}_1 \\ \mathbf{U}_2 \\ \mathbf{U}_3 \end{Bmatrix} = \begin{Bmatrix} \mathbf{P}_1^A \\ \mathbf{P}_2^A + \mathbf{P}_2^B \\ \mathbf{P}_3^B \end{Bmatrix} \tag{a_{33}}$$

One can now see that the process described earlier handles these equations but with the considerations of assembling partly, triangularizing, shifting of coefficients to back store of computer, and shifting up of coefficients within the main core area.

It is seen that the above procedure is possible only for *serially* partitioned substructures. For a more general (parallel) connection as in Fig. 10.8(a) the above process is not possible. The programming style will have to be very different.

```
      SUBROUTINE ELIM(SK,R,IB,NEQ,NEQMAX,IBMAX,LR,UR,RR,NR,NRMAX)
      DIMENSION SK(NEQMAX,IBMAX),R(NEQMAX),LR(NRMAX),UR(NRMAX),RR(NRMAX)
      NEQ1=NEQ-1
      IR=1
      DO 106 N=1,NEQ1
      M=N-1
      IB1=MINO(NEQ-M,IB)
      IF(SK(N,1).LE.0) GO TO 101
      IF(IR.GT.NR) GO TO 108
      IF(N.NE.LR(IR)) GO TO 108
      DO 105 L=2,IB1
  105 SK(N,L)=-SK(N,L)
      IR=IR+1
      GO TO 106
  108 DO 100 L=2,IB1
      RATIO=SK(N,L)/SK(N,1)
      I=M+L
      J=0
      DO 102 K=L,IB1
      J=J+1
  102 SK(I,J)=SK(I,J)-RATIO*SK(N,K)
  100 SK(N,L)=RATIO
  106 CONTINUE
      GO TO 103
  101 WRITE(6,200) N,SK(N,1)
      STOP
  103 RETURN
  200 FORMAT(//,'PIVOT ERROR',/,5X,'EQ.NUM.=',5X,'PIVOT=',1PE20.10)
      END
```

```
      SUBROUTINE SOLV(SK,R,IB,NEQ,NEQMAX,IBMAX,LR,UR,RR,NR,NRMAX)
      DIMENSION SK(NEQMAX,IBMAX),R(NEQMAX),LR(NRMAX),UR(NRMAX),RR(NRMAX)
      NEQ1=NEQ-1
      IR=1
      DO 203 N=1,NEQ1
      M=N-1
      IB1=MINO(NEQ-M,IB)
      IF(IR.GT.NR) GO TO 202
      IF(N.NE.LR(IR)) GO TO 202
      R(N)=SK(N,1)*UR(IR)-R(N)
      P=-UR(IR)
      IR=IR+1
      GO TO 204
  202 P=R(N)
      R(N)=P/SK(N,1)
  204 DO 200 L=2,IB1
      I=M+L
  200 R(I)=R(I)-SK(N,L)*P
  203 CONTINUE
C     BACK SUBSTITUTION
      IR=NR
      IF(NEQ.NE.LR(IR)) GO TO 206
      RR(IR)=SK(NEQ,1)*UR(IR)-R(NEQ)
      R(NEQ)=UR(IR)
      IR=IR-1
      GO TO 205
  206 R(NEQ)=R(NEQ)/SK(NEQ,1)
  205 DO 208 I=1,NEQ1
      N=NEQ-I
      M=N-1
      IB1=MINO(NEQ-M,IB)
      DO 207 K=2,IB1
      L=M+K
  207 R(N)=R(N)-SK(N,K)*R(L)
      IF(IR.EQ.0) GO TO 208
      IF(N.NE.LR(IR)) GO TO 208
      RR(IR)=R(N)
      R(N)=UR(IR)
      IR=IR-1
  208 CONTINUE
      RETURN
      END
```

```
C       MASTER PROGRAM FOR PLANE FRAME
C
        COMMON/BLOCK1/NDF,NM,NJ,NJR,NML,NJL,NEQ,IB,NRD,NEQMAX,IBMAX,
       1NRDMAX,NFILE1,NLD
        COMMON/BLOCK2/ICON(200,2),LJR(25),LFIX(25),UR(25,3),PROP(200,5),
       1PFM(200,6)
        COMMON/BLOCK3/XL,YL,AL,SIN,COS,NJ1,NJ2,ESK(6,6),P(6)
        COMMON/BLOCK4/SK(150,30),LRD(75),URD(75),RRD(75),R(150)
C
        WRITE(6,901)
10      READ(5,101,END=801)IDATA
        WRITE(6,201) IDATA
C       ASSIGN PRESCRIBED VALUES TO PARAMETERS
        NFILE1=1
        NDF=3
        NEQMAX=150
        IBMAX=30
        NRDMAX=75
C       READ DATA
        CALL INPUT1
C       CHECK FOR BAND WIDTH
        CALL PRESET(KSTOP)
        IF(KSTOP.EQ.1)GO TO 999
C       WRITE MEMBER STIFFNESS,MAT PROP,FIXING END FORCES ON NFILE1
        DO 100 ILD=1,NLD
        WRITE(6,902)ILD
        CALL WRITE1(ILD)
        IF(ILD.NE.1)GO TO 20
C       ASSEMBLE GLOBAL STIFF MAT
        CALL FORMSK
C       DETERMINE PRESCRIBED DISP VARIABLES AND THEIR MAGNITUDES
        CALL FIX
C       TRIANGULARIZE SK MATRIX AND MAKE MODIFICATION
C       FOR PRESCRIBED VARIABLES
        CALL ELIM(SK,R,IB,NEQ,NEQMAX,IBMAX,LRD,URD,RRD,NRD,NRDMAX)
C       FORM RHS AND MODIFY FOR PRESC VARIABLES AND FOR NEW LOADS
20      CALL SOLV(SK,R,IB,NEQ,NEQMAX,IBMAX,LRD,URD,RRD,NRD,NRDMAX)
C       WRITE DISPLACEMENTS AND REACTIONS
        CALL DISREA
C       COMPUTE MEMBER FORCES
        CALL FORCES
100     CONTINUE
        GO TO 10
801     WRITE(6,903)
101     FORMAT(I5)
201     FORMAT(///,5X,'PROBLEM NO=',I5)
901     FORMAT(/////,45X,'PLANE FRAME PROGRAM')
902     FORMAT(////,5X,'LOAD CASE=',I5,////)
903     FORMAT(///,5X,'END OF DATA SETS')
999     STOP
        END
```

```
      SUBROUTINE INPUT1
C
      COMMON/ BLOCK1/NDF,NM,NJ,NJR,NML,NJL,NEQ,IB,NRD,NEQMAX,IBMAX,
     1NRDMAX,NFILE1,NLD
      COMMON/BLOCK2/ICON(200,2),LJR(25),LFIX(25),UR(25,3),PROP(200,5)
      COMMON/BLOCK3/XL,YL,AL,SIN,COS,NJ1,NJ2,ESK(6,6),P(6)
      DIMENSION CORD(50,2)
C
C     READ DATA
      WRITE(6,901)
      READ(5,801)NJ,NM,NJR,NLD
      WRITE(6,902)NJ,NM,NJR,NLD
C     CONNECTIVITY DATA
      WRITE(6,903)
      DO 101 I=1,NM
      READ(5,801)L,(ICON(L,J),J=1,2)
  101 WRITE(6,904)L,(ICON(L,J),J=1,2)
C     READ JOINT COORDINATES
      WRITE(6,905)
      DO 102 I=1,NJ
      READ(5,802) L,(CORD(L,J),J=1,2)
  102 WRITE(6,906)L,(CORD(L,J),J=1,2)
C      BOUNDARY NODES AND FIXITY CODES
      WRITE(6,907)
      DO 103 I=1,NJR
      READ(5,804) LJR(I),LFIX(I),(UR(I,J),J=1,3)
  103 WRITE(6,908) LJR(I),LFIX(I),(UR(I,J),J=1,3)
C     READ MATERIAL PROPERTIES
      WRITE(6,909)
      DO 104 I=1,NM
      READ(5,803) L,(PROP(L,J),J=1,3)
      NJ1=ICON(L,1)
      NJ2=ICON(L,2)
      PROP(L,4)=CORD(NJ2,1)-CORD(NJ1,1)
      PROP(L,5)=CORD(NJ2,2)-CORD(NJ1,2)
  104 WRITE(6,910)L,(PROP(L,J),J=1,5)
  801 FORMAT(16I5)
  802 FORMAT(I5,2F10.4)
  803 FORMAT(I5,5E15.5)
  804 FORMAT(2I5,3E15.5)
  901 FORMAT(5X,'BASIC DATA',//)
  902 FORMAT(5X,'NUMBER OF JOINTS',11X,I5,/,
     15X,'NUMBER OF MEMBERS',10X,I5,/,
     15X,'NUMBER OF RESTRAINED JOINTS',I5,/,
     15X,'NUMBER OF LOAD CASE',8X,I5,//)
  903 FORMAT(//,5X,'MEMBER CONNECTIVITY',//,5X,'MEMBER',6X,
     1'END1',6X,'END2',//)
  904 FORMAT(3(5X,I5))
  905 FORMAT(//,5X,'JOINT COORDINATES',//,5X,'JOINT',10X,'CORDX',
     110X,'CORDY',//)
  906 FORMAT(5X,I5,7X,F10.4,5X,F10.4)
```

```
907   FORMAT(//,5X,'SUPPORT CCNDITICNS',//,10X,'JCINT',10X,'FIXITY',
     112X,'UX',20X,'UY',20X,'UM',//)
908   FORMAT(2(10X,I5),3(7X,E15.5))
909   FORMAT(//,5X,'MEMBER PROPERTIES',//,9X,'MEMBER',3X,'PROP1(AE)',
     17X,'PROP2(EI)',7X,'PROP3(AG)',12X,'X2-X1',12X,'Y2-Y1',//)
910   FORMAT(8X,I5,5(5X,1PE12.5),//)
      RETURN
      END


      SUBROUTINE PRESET(KSTCP)
C
      COMMON/BLOCK1/NDF,NM,NJ,NJR,NML,NJL,NEQ,IB,NRD,NEQMAX,IBMAX,
     1NRDMAX,NFILE1,NLD
      COMMON/BLOCK2/ICON(200,2),LJR(25),LFIX(25),UR(25,3),PROP(200,5)
C
      KSTOP=0
      MAX=0
      DO 10 IM=1,NM
      LDIF=IABS(ICON(IM,2)-ICCN(IM,1))
      IF(LDIF.GT.MAX) MAX=LDIF
   10 CONTINUE
      IB=NDF*(MAX+1)
      NEQ=NDF*NJ
      IF(IB.LE.IBMAX)GC TC 15
      KSTCP=1
      WRITE(6,101) IB
15    IF(NEQ.LE.NEQMAX)RETURN
      KSTOP=1
      WRITE(6,102)NEQ
101   FORMAT(//,'ERRCR---','BAND WIDTH EXCEEDED',5X,'IB=',I5)
102   FORMAT(//,'ERROR---','NEQMAX SIZE EXCEEDEC',5X,'NEQ=',I5)
      RETURN
      END


      SUBROUTINE WRITE1(ILD)
C
      COMMON/BLOCK1/NDF,NM,NJ,NJR,NML,NJL,NEQ,IB,NRC,NEQMAX,IBMAX,
     1NRDMAX,NFILE1,NLD
      COMMCN/BLCCK2/ICCN(200,2),LJR(25),LFIX(25),UR(25,3),PROP(200,5),
     1PFM(200,6)
      COMMON/BLOCK3/XL,YL,AL,SIN,COS,NJ1,NJ2,ESK(6,6),P(6)
      COMMON/BLCCK4/SK(150,30),LRD(75),URD(75),RRC(75),R(150)
      DIMENSICN LCODE(4),PF(6)
C
      READ(5,101)NML,NJL
      WRITE(6,1001)NML,NJL
C     SUBROUTINE TO FORM MEM STIFFNESS,TO READ LOAD DATA AND TO WRITE
C     MEMBER ANALYSIS CN NFILE1 MEMBERWISE
      NDF2=NDF*2
      DO 39 IM=1,NM
      DO 39 I=1,NDF2
```

```
39    PFM(IM,I)=0.
      DO 40 I=1,150
40     R(I)=0.
      REWIND NFILE1
      IC1=1
      IC2=1
      IF((NML.EQ.0).AND.(ILD.NE.1))GO TO 10
      DO 10 IM=1,NM
      IF(ILD.EQ.1) CALL STIF1(IM,ESK)
      DO 12 I=1,NDF2
12    PF(I)=0.
      IF(NML.EQ.0) GO TC 14
      IF(IC1.GT.NML) GO TO 14
      IF(IC1.NE.IC2) GO TO 15
      WRITE(6,2001)
C     LCODE STANDS FOR CODE FOR DIFFERENT TYPES OF LOADING
      READ(5,102) LML,(LCCDE(J),J=1,4)
      WRITE(6,2002) LML,(LCODE(J),J=1,4)
      IC2=IC2+1
15    IF(IM.NE.LML) GC TO 14
      NJ1=ICON(IM,1)
      NJ2=ICON(IM,2)
      XL=PROP(IM,4)
      YL=PROP(IM,5)
      AL=SQRT(XL*XL+YL*YL)
      COS=XL/AL
      SIN=YL/AL
      IF(LCODE(1).NE.0) CALL LOAD1(PF)
      IF(LCODE(2).NE.0) CALL LCAD2(PF)
      IF(LCODE(3).NE.0) CALL LOAD3(PF)
      IF(LCODE(4).NE.0) CALL LOAD4(PF)
      IX1=NDF*(NJ1-1)
      IX2=NDF*(NJ2-1)
      DO 13 I=1,NDF
      R(IX1+I)=R(IX1+I)-PF(I)
13    R(IX2+I)=R(IX2+I)-PF(I+NDF)
      IC1=IC1+1
14    IF(ILD.NE.1)GC TC 17
      WRITE(NFILE1)IM,((ESK(I,J),I=1,6),J=1,6),(PROP(IM,J),J=1,5)
17    DO 49 I=1,NDF2
49    PFM(IM,I)=PF(I)
10     CONTINUE
      IF(NJL.EQ.0) GO TC 50
C     ADD APPLIED JCINT LCADS
      WRITE(6,2004)
      DO 46 IJ=1,NJL
      READ(5,103) LJL,(P(IDF),IDF=1,NDF)
      WRITE(6,2005) LJL,(P(IDF),IDF=1,NDF)
      IX=(LJL-1)*NDF
      DO 46 IDF=1,NDF
46    R(IX+IDF)=R(IX+IDF)+P(IDF)
50    CONTINUE
```

```
1001  FORMAT(5X,'NUMBER OF LOADED MEMBERS',3X,I5,/,
     15X,'NUMBER OF LOADED JOINTS',4X,I5,//)
101   FORMAT(16I5)
 102  FORMAT(5I5)
 103  FORMAT(I5,3F15.5)
2001  FORMAT(/,5X,'MEMBER LOADED',5X,'LOAD TYPE1',5X,'LOAD TYPE2',
     15X,'LOAD TYPE3',5X,'LOAD TYPE4')
2002  FORMAT(//,5X,I5,13X,4(I5,10X))
2004  FORMAT(/,5X,'APPLIED JOINT LOADS',//,
     15X,'JOINT',10X,'PX',12X,'QY',12X,' M',/)
2005  FORMAT(4X,I5,9X,3(1PE12.5,2X))
      RETURN
      END

      SUBROUTINE STIF1(IMEM,A)
C
      COMMON/BLOCK1/NDF,NM,NJ,NJR,NML,NJL,NEQ,IB,NRD,NEQMAX,IBMAX,
     1NRDMAX,NFILE1,NLD
      COMMON/BLOCK2/ICON(200,2),LJR(25),LFIX(25),UR(25,3),PROP(200,5)
      COMMON/BLOCK3/XL,YL,AL,SIN,COS,NJ1,NJ2,ESK(6,6),P(6)
      DIMENSION A(6,6)
C
      AE=PROP(IMEM,1)
      EI=PROP(IMEM,2)
      AG=PROP(IMEM,3)
      XL=PROP(IMEM,4)
      YL=PROP(IMEM,5)
      AL=SQRT(XL*XL+YL*YL)
      C=XL/AL
      S=YL/AL
      A1=AE/AL
      A2=12.*EI/(AL**3)
      A3=.5*AL*A2
      A4=4.*EI/AL
      A(1,1)=C*C*A1+S*S*A2
      A(4,4)=A(1,1)
      A(5,5)=S*S*A1+C*C*A2
      A(2,2)=A(5,5)
      A(1,2)=C*S*(A1-A2)
      A(4,5)=A(1,2)
      A(3,4)=S*A3
      A(4,6)=A(3,4)
      A(2,3)=C*A3
      A(2,6)=A(2,3)
      A(3,3)=A4
      A(6,6)=A4
      A(3,6)=A4/2.
      A(1,3)=-A(4,6)
      A(1,6)=A(1,3)
```

```
      A(3,5)=-A(2,3)
      A(5,6)=A(3,5)
      A(1,5)=-A(1,2)
      A(2,4)=A(1,5)
      A(1,4)=-A(1,1)
      A(2,5)=-A(2,2)
      DO 10 I=1,6
      DO 10 J=I,6
      IF(J.NE.I) A(J,I)=A(I,J)
   10 CONTINUE
      RETURN
      END

      SUBROUTINE LOAD1(PF)
C
      COMMON/BLOCK3/XL,YL,AL,SIN,COS,NJ1,NJ2,ESK(6,6),P(6)
      DIMENSION WX(10),WY(10),AX(10),AY(10),PF(6)
C
      READ(5,101) NCONC
      READ(5,105) ((WX(I),WY(I),AX(I),AY(I)),I=1,NCCNC)
      WRITE (6,102) NCONC
      WRITE (6,103)
      WRITE(6,104)((WX(I),WY(I),AX(I),AY(I)),I=1,NCCNC)
      DO 11 I=1,NCONC
      A=SQRT(AX(I)*AX(I)+AY(I)*AY(I))
      WXBAR=WX(I)*COS+WY(I)*SIN
      WYBAR=WY(I)*CCS-WX(I)*SIN
      B1=AL-A
      B2=B1*B1*(3.0*A+B1)/(AL**3)
      B3=A*B1*B1/(AL*AL)
      PBAR=-WXBAR*B1/AL
      QBAR=-WYBAR*B2
      PF(1)=PF(1)+PBAR*COS-QBAR*SIN
      PF(2)=PF(2)+PBAR*SIN+QBAR*COS
      PF(3)=PF(3)-WYBAR*B3
      B2=A*A*(3.0*B1+A)/(AL**3)
      B3=A*A*B1/(AL*AL)
      PBAR=-WXBAR*A/AL
      QBAR=-WYBAR*B2
      PF(4)=PF(4)+PBAR*COS-QBAR*SIN
      PF(5)=PF(5)+PBAR*SIN+QBAR*COS
   11 PF(6)=PF(6)+WYBAR*B3
  101 FORMAT(I5)
  102 FORMAT(//,5X,'NUMBER OF CONC LOADS',I5,//)
  103 FORMAT(//,13X,'WX',20X,'WY',20X,'AX',20X,'AY')
  104 FORMAT(5X,4(F15.5,5X),//)
  105 FORMAT(4F15.5)
      RETURN
      END
```

```
      SUBROUTINE LOAD2(PF)
C
      COMMON/BLOCK3/XL,YL,AL,SIN,COS,NJ1,NJ2,ESK(6,6),P(6)
      DIMENSION PF(6)
C
C     WX AND WY REPRESENT TOTAL  U.D.L.
      READ(5,101) WX,WY
      WRITE(6,102)
      WRITE(6,103) WX,WY
      PF(1)=PF(1)-WX/2.
      PF(4)=PF(1)
      PF(2)=PF(2)-WY/2.
      PF(5)=PF(2)
      PF(3)=PF(3)-(WX*YL+WY*XL)/12.
      PF(6)=-PF(3)
101   FORMAT(2F15.5)
 102  FORMAT(//,5X,'UNIFORMLY DIST     TOTAL LOAD',//,15X,'WX',20X,'WY')
103   FORMAT(5X,2(F15.5,5X))
      RETURN
      END

      SUBROUTINE LOAD3(PF)
C     DUMMY SUBROUTINE TO BE USED FOR OTHER LOAD CASES
      RETURN
      END

      SUBROUTINE LOAD4(PF)
C     DUMMY SUBROUTINE TO BE USED FOR OTHER LOAD CASES
      RETURN
      END

      SUBROUTINE FORMSK
C
      COMMON/BLOCK1/NDF,NM,NJ,NJR,NML,NJL,NEQ,IB,NRD,NEQMAX,IBMAX,
     1NRDMAX,NFILE1,NLD
      COMMON/BLOCK2/ICON(200,2),LJR(25),LFIX(25),UR(25,3),PROP(200,5)
      COMMON/BLOCK3/XL,YL,AL,SIN,COS,NJ1,NJ2,ESK(6,6),P(6)
      COMMON/BLOCK4/SK(150,30),LRD(75),URD(75),RRC(75),R(150)
      DIMENSION PF(6)
C
      DO 10 I=1,NEQ
      DO 10 J=1,IB
   10 SK(I,J)=0.
      REWIND NFILE1
      DO 11 JMEM=1,NM
      READ(NFILE1) IMEM,((ESK(I,J),I=1,6),J=1,6),(P(I),I=1,5)
C     II AND JJ ARE STARTING ROW AND COLUMN LOCATIONS IN STIFF MATRIX
      DO 100 I=1,2
      II=(ICON(IMEM,I)-1)*NDF
      DO 100 J=1,2
      JJ=(ICON(IMEM,J)-1)*NDF
```

```
C     IROW AND JCCL ARE LCCATICN IN SQUARE STIFF MAT
C     IERCW AND JECOL  ARE LOCATICN IN MEM STIFF MAT
      DO 100 IDF=1,NDF
      IROW=II+IDF
      IEROW=(I-1)*NDF+IDF
      DO 100 JDF=1,NDF
      JCOL=JJ+JDF
      IF(IRCW.GT.JCCL) GO TC 100
      JECOL=(J-1)*NDF+JCF
      JCOLB=JCOL+1-IROW
C     JCOLB GIVES THE CCLUMN LCCATICN IN THE EQUIVALENT SK ARRAY
      SK(IROW,JCOLB)=SK(IROW,JCOLB)+ESK(IERCW,JECCL)
  100 CONTINUE
   11 CONTINUE
      RETURN
      END

      SUBROUTINE FIX
C
      COMMON/BLOCK1/NDF,NM,NJ,NJR,NML,NJL,NEC,IB,NRC,NEQMAX,IBMAX,
     1NRDMAX,NFILE1,NLD
      COMMON/BLOCK2/ICON(200,2),LJR(25),LFIX(25),UR(25,3),PROP(200,5)
      COMMON/BLOCK3/XL,YL,AL,SIN,COS,NJ1,NJ2,ESK(6,6),P(6)
      COMMON/BLOCK4/SK(150,30),LRD(75),URC(75),RRC(75),R(150)
C
      NJR3=NJR*NDF
      DO 1 I=1,NJR3
      LRD(I)=0
      URD(I)=0.
    1 RRD(I)=0.
      NRD=0
      DO 13 JR=1,NJR
      JFIX=LFIX(JR)
      IFIX=10**(NDF-1)
      IJT=LJR(JR)
      DO 11 IDF=1,NDF
      IX=(IJT-1)*NDF+IDF
      IF(JFIX.LT.IFIX) GO TO 12
      NRD=NRD+1
      LRD(NRD)=IX
      URD(NRD)=UR(JR,IDF)
      JFIX=JFIX-IFIX
   12 IFIX=IFIX/10
   11 CONTINUE
   13 CONTINUE
      RETURN
      END
```

```
      SUBROUTINE DISREA
C
      COMMON/BLOCK1/NDF,NM,NJ,NJR,NML,NJL,NEQ,IB,NRD,NEQMAX,IBMAX,
     1NRDMAX,NFILE1,NLD
      COMMON/BLOCK2/ICON(200,2),LJR(25),LFIX(25),UR(25,3),PROP(200,5)
      COMMON/BLOCK3/XL,YL,AL,SIN,COS,NJ1,NJ2,ESK(6,6),P(6)
      COMMON/BLOCK4/SK(150,30),LRD(75),URD(75),RRD(75),R(150)
      DIMENSION DIS(50,3),REA(50,3)
C
      DO 1 J=1,NDF
      DO 1 I=1,50
      DIS(I,J)=0.
    1 REA(I,J)=0.
      IRD=1
      IJR=1
      DO 10 IJ=1,NJ
      DO 12 IDF=1,NDF
      REA(IJ,IDF)=0.
      IX=(IJ-1)*NDF+IDF
   12 DIS(IJ,IDF)=R(IX)
      IF(IJR.GT.NJR)GO TO 10
      IF(IJ.NE.LJR(IJR))GO TO 10
      JFIX=LFIX(IJR)
      IFIX=10**(NDF-1)
      DO 14 IDF=1,NDF
      IF(JFIX.LT.IFIX) GO TO 15
      REA(IJ,IDF)=RRD(IRD)
      IRD=IRD+1
      JFIX=JFIX-IFIX
   15 IFIX=IFIX/10
   14 CONTINUE
      IJR=IJR+1
   10 CONTINUE
      WRITE(6,101)
      DO 16 I=1,NJ
   16 WRITE(6,102) I,((DIS(I,J),J=1,NDF),(REA(I,J),J=1,NDF))
  101 FORMAT(/,5X,'DISPLACEMENTS AND REACTIONS',///,
     15X,'JOINT',7X,'UX',14X,'UY',14X,'UM',20X,'RX',14X,'RY',
     114X,'RM',//)
  102 FORMAT(5X,I5,6(5X,1PE12.5),/)
      RETURN
      END
```

```
      SUBROUTINE ELIM(SK,R,IB,NEQ,NEQMAX,IBMAX,LR,UR,RR,NR,NRMAX)
C
      DIMENSICN SK(NEQMAX,IBMAX),R(NEQMAX),LR(NRMAX),UR(NRMAX),RR(NRMAX)
C
      NEQ1=NEQ-1
      IR=1
      DO 106 N=1,NEQ1
      M=N-1
      IB1=MINO(NEC-M,IB)
      IF(SK(N,1).LE.0) GO TO 101
      IF(IR.GT.NR) GO TO 108
      IF(N.NE.LR(IR)) GO TO 108
      DO 105 L=2,IB1
  105 SK(N,L)=-SK(N,L)
      IR=IR+1
      GO TO 106
  108 DO 100 L=2,IB1
      RATIO=SK(N,L)/SK(N,1)
      I=M+L
      J=0
      DO 102 K=L,IB1
      J=J+1
  102 SK(I,J)=SK(I,J)-RATIC*SK(N,K)
  100 SK(N,L)=RATIO
  106 CONTINUE
      GO TO 103
  101 WRITE(6,200) N,SK(N,1)
      STOP
  103 RETURN
  200 FORMAT(//,'PIVCT ERROR',/,'EQ NUM =',I5,5X,'PIVOT=',1PE20.10)
      END
```

```
      SUBROUTINE SOLV(SK,R,IB,NEQ,NEQMAX,IBMAX,LR,UR,RR,NR,NRMAX)
C
      DIMENSION SK(NEQMAX,IBMAX),R(NEQMAX),LR(NRMAX),UR(NRMAX),RR(NRMAX)
C
      NEQ1=NEQ-1
      IR=1
      DO 203 N=1,NEQ1
      M=N-1
      IB1=MIN0(NEQ-M,IB)
      IF(IR.GT.NR) GO TO 202
      IF(N.NE.LR(IR)) GO TO 202
      R(N)=SK(N,1)*UR(IR)-R(N)
      P=-UR(IR)
      IR=IR+1
      GO TO 204
  202 P=R(N)
      R(N)=P/SK(N,1)
  204 DO 200 L=2,IB1
      I=M+L
  200 R(I)=R(I)-SK(N,L)*P
  203 CONTINUE
C     BACK SUBSTITUTION
      IR=NR
      IF(NEQ.NE.LR(IR)) GO TO 206
      RR(IR)=SK(NEQ,1)*UR(IR)-R(NEQ)
      R(NEQ)=UR(IR)
      IR=IR-1
      GO TO 205
  206 R(NEQ)=R(NEQ)/SK(NEQ,1)
  205 DO 208 I=1,NEQ1
      N=NEQ-I
      M=N-1
      IB1=MIN0(NEQ-M,IB)
      DO 207 K=2,IB1
      L=M+K
  207 R(N)=R(N)-SK(N,K)*R(L)
      IF(IR.EQ.0) GO TO 208
      IF(N.NE.LR(IR)) GO TO 208
      RR(IR)=R(N)
      R(N)=UR(IR)
      IR=IR-1
  208 CONTINUE
      RETURN
      END
```

```
      SUBROUTINE FORCES
C
      COMMON/BLOCK1/NDF,NM,NJ,NJR,NML,NJL,NEQ,IB,NRD,NEQMAX,IBMAX,
     1NRDMAX,NFILE1,NLD
      COMMON/BLOCK2/ICON(200,2),LJR(25),LFIX(25),LR(25,3),PROP(200,5),
     1PFM(200,6)
      COMMON/BLOCK3/XL,YL,AL,SIN,COS,NJ1,NJ2,ESK(6,6),P(6)
      COMMON/BLOCK4/SK(150,30),LRD(75),URD(75),RRD(75),R(150)
      DIMENSION PRP(5),DISMEM(6),FORCE(6)
C
      REWIND NFILE1
      WRITE(6,1001)
      DO 15 IM=1,NM
      READ(NFILE1)IMEM,((ESK(I,J),I=1,6),J=1,6),(PRP(J),J=1,5)
      NJ1=ICON(IMEM,1)
      NJ2=ICON(IMEM,2)
      IX1=(NJ1-1)*NDF
      IX2=(NJ2-1)*NDF
      XL=PRP(4)
      YL=PRP(5)
      AL=SQRT(XL*XL+YL*YL)
      COS=XL/AL
      SIN=YL/AL
      DO 20 IDF=1,NDF
      DISMEM(IDF)=R(IX1+IDF)
   20 DISMEM(IDF+NDF)=R(IX2+IDF)
      DO 30 K=1,6
   30 FORCE(K)=0.
      DO 25 K=1,6
      FORCE(K)=FORCE(K)+PFM(IM,K)
      DO 25 L=1,6
   25 FORCE(K)=FORCE(K)+ESK(K,L)*DISMEM(L)
      P1BAR=FORCE(1)*COS+FORCE(2)*SIN
      Q1BAR=-FORCE(1)*SIN+FORCE(2)*COS
      FORCE(1)=P1BAR
      FORCE(2)=Q1BAR
      P2BAR=FORCE(4)*COS+FORCE(5)*SIN
      Q2BAR=-FORCE(4)*SIN+FORCE(5)*COS
      FORCE(4)=P2BAR
      FORCE(5)=Q2BAR
      WRITE(6,1002)IMEM,(FORCE(K),K=1,6)
   15 CONTINUE
 1001 FORMAT(//,5X,'INTERNAL FORCES',////,
     15X,'MEMBER',8X,'P1BAR',12X,'Q1BAR',
     112X,'M1    ',12X,'P2BAR',12X,'Q2BAR',12X,'M2    ')
 1002 FORMAT(//,5X,I5,6(5X,1PE12.5))
      RETURN
      END
```

OUTPUT PLANE FRAME PROGRAM

```
PROBLEM NO=      1
BASIC DATA

NUMBER OF JOINTS                 3
NUMBER OF MEMBERS                2
NUMBER OF RESTRAINED JOINTS      2
NUMBER OF LOAD CASE              1
```

MEMBER CONNECTIVITY

MEMBER	END1	END2
1	1	2
2	2	3

JOINT COORDINATES

JOINT	CORDX	CORDY
1	0.0	0.0
2	4.0000	-3.0000
3	8.0000	-3.0000

SUPPORT CONDITIONS

JOINT	FIXITY	UX	UY	UM
1	111	0.0	0.0	0.0
3	111	0.0	0.0	0.0

MEMBER PROPERTIES

MEMBER	PROP1(AE)	PROP2(EI)	PROP3(AG)	X2-X1	Y2-Y1
1	2.00000E 02	1.25000E 01	0.0	4.00000E 00	-3.00000E 00
2	2.00000E 02	8.00000E 00	0.0	4.00000E 00	0.0

LOAD CASE= 1

NUMBER OF LOADED MEMBERS 2
NUMBER OF LOADED JOINTS 0

MEMBER LOADED	LOAD TYPE1	LOAD TYPE2	LOAD TYPE3	LOAD TYPE4
1	1	0	0	0

NUMBER OF CONC LOADS 1

WX	WY	AX	AY
0.0	80.00000	3.00000	-2.25000

MEMBER LOADED	LOAD TYPE1	LOAD TYPE2	LOAD TYPE3	LOAD TYPE4
2	0	1	0	0

UNIFORMLY DIST TOTAL LOAD

WX	WY
0.0	48.00000

DISPLACEMENTS AND REACTIONS

JOINT	UX	UY	UM	RX	RY	RM
1	0.0	0.0	0.0	9.13826E 01	-9.78830E 01	-4.45019E 01
2	1.82765E 00	7.43003E 00	-1.67601E 00	0.0	0.0	0.0
3	0.0	0.0	0.0	-9.13824E 01	-3.01170E 01	3.15860E 01

INTERNAL FORCES

MEMBER	P1BAR	Q1BAR	M1	P2BAR	Q2BAR	M2
1	1.31836E 02	-2.34769E 01	-4.45019E 01	-8.38359E 01	-4.05231E 01	7.11803E 00
2	9.13825E 01	-1.78830E 01	-7.11800E 00	-9.13825E 01	-3.01170E 01	3.15860E 01

PROBLEM NO= 2
BASIC DATA

NUMBER OF JOINTS	3
NUMBER OF MEMBERS	2
NUMBER OF RESTRAINED JOINTS	3
NUMBER OF LOAD CASE	1

MEMBER CONNECTIVITY

MEMBER	END1	END2
1	1	2
2	2	3

JOINT COORDINATES

JOINT	CORDX	CORDY
1	0.0	0.0
2	4.0000	-3.0000
3	8.0000	-3.0000

SUPPORT CONDITIONS

JOINT	FIXITY	UX	UY	UM
1	111	0.0	0.0	0.0
2	1	0.0	0.0	0.10000E 01
3	111	0.0	0.0	0.0

MEMBER PROPERTIES

MEMBER	PROP1(AE)	PROP2(EI)	PROP3(AG)	X2-X1	Y2-Y1
1	2.00000E 02	1.25000E 01	0.0	4.00000E 00	-3.00000E 00
2	2.00000E 02	8.00000E 00	0.0	4.00000E 00	0.0

LOAD CASE= 1

NUMBER OF LOADED MEMBERS 0
NUMBER OF LOADED JOINTS 0

DISPLACEMENTS AND REACTIONS

JOINT	UX	UY	UM	RX	RY	RM
1	0.0	0.0	0.0	1.02276E 00	2.98029E 00	4.99472E 00
2	2.04552E-02	-1.31416E-02	1.00000E 00	0.0	0.0	1.79553E 01
3	0.0	0.0	0.0	-1.02276E 00	-2.98029E 00	3.96058E 00

INTERNAL FORCES

MEMBER	P1BAR	Q1BAR	M1	P2BAR	Q2BAR	M2
1	-9.69964E-01	2.99789E 00	4.99472E 00	9.69964E-01	-2.99789E 00	9.99472E 00
2	1.02276E 00	2.98029E 00	7.96058E 00	-1.02276E 00	-2.98029E 00	3.96058E 00

PROBLEM NO= 3
BASIC DATA

NUMBER OF JOINTS 19
NUMBER OF MEMBERS 32
NUMBER OF RESTRAINED JOINTS 3
NUMBER OF LOAD CASE 2

MEMBER CONNECTIVITY

MEMBER	END1	END2
1	1	4
2	2	4
3	2	3
4	1	5
5	2	6
6	3	7
7	4	5
8	4	6
9	5	8
10	6	8
11	6	7
12	5	9
13	6	10
14	7	11
15	8	9
16	8	10
17	9	12
18	10	12
19	10	11
20	9	13
21	1J	14
22	11	15
23	12	13
24	12	14
25	13	16
26	14	16
27	14	15
28	13	17
29	14	18
30	15	19
31	16	17
32	16	18

JOINT COORDINATES

JOINT	CORDX	CORDY
1	0.0	-576.000,
2	288.0000	-576.0000
3	576.0000	-576.0000
4	144.0000	-576.0000
5	0.0	-432.0000
6	288.0000	-432.0000
7	576.0000	-432.0000
8	144.0000	-432.0000
9	0.0	-288.0000
10	288.0000	-288.0000
11	576.0000	-288.0000
12	144.0000	-288.0000
13	0.0	-144.0000
14	288.0000	-144.0000
15	576.0000	-144.0000
16	144.0000	-144.0000
17	0.0	0.0
18	288.0000	0.0
19	576.0000	0.0

SUPPORT CONDITIONS

JOINT	FIXITY	UX	UY	UM
17	110	0.0	0.0	0.0
18	110	0.0	0.0	0.0
19	110	0.0	0.0	0.0

MEMBER PROPERTIES

MEMBER	PROP1(AE)	PROP2(EI)	PROP3(AG)	X2-X1	Y2-Y1
1	3.00000E 05	1.50000E 07	0.0	1.44000E 02	0.0
2	3.00000E 05	1.50000E 07	0.0	-1.44000E 02	0.0
3	3.00000E 05	1.50000E 07	0.0	2.88000E 02	0.0
4	4.50000E 05	7.50000E 06	0.0	0.0	1.44000E 02
5	4.50000E 05	7.50000E 06	0.0	0.0	1.44000E 02
6	4.50000E 05	7.50000E 06	0.0	0.0	1.44000E 02
7	1.50000E 05	0.0	0.0	-1.44000E 02	1.44000E 02
8	1.50000E 05	0.0	0.0	1.44000E 02	1.44000E 02
9	3.00000E 05	1.50000E 07	0.0	1.44000E 02	0.0
10	3.00000E 05	1.50000E 07	0.0	-1.44000E 02	0.0
11	3.00000E 05	1.50000E 07	0.0	2.88000E 02	0.0
12	4.50000E 05	7.50000E 06	0.0	0.0	1.44000E 02
13	4.50000E 05	7.50000E 06	0.0	0.0	1.44000E 02
14	4.50000E 05	7.50000E 06	0.0	0.0	1.44000E 02

15	1.50000E 05	0.0	0.0	-1.44000E 02	1.44000E 02
16	1.50000E 05	0.0	0.0	1.44000E 02	1.44000E 02
17	3.00000E 05	1.50000E 07	0.0	1.44000E 02	0.0
18	3.00000E 05	1.50000E 07	0.0	-1.44000E 02	0.0
19	3.00000E 05	1.50000E 07	0.0	2.88000E 02	0.0
20	4.50000E 05	7.50000E 06	0.0	0.0	1.44000E 02
21	4.50000E 05	7.50000E 06	0.0	0.0	1.44000E 02
22	4.50000E 05	7.50000E 06	0.0	0.0	1.44000E 02
23	1.50000E 05	0.0	0.0	-1.44000E 02	1.44000E 02
24	1.50000E 05	0.0	0.0	1.44000E 02	1.44000E 02
25	3.00000E 05	1.50000E 07	0.0	1.44000E 02	0.0
26	3.00000E 05	1.50000E 07	0.0	-1.44000E 02	0.0
27	3.00000E 05	1.50000E 07	0.0	2.88000E 02	0.0
28	4.50000E 05	7.50000E 06	0.0	0.0	1.44000E 02
29	4.50000E 05	7.50000E 06	0.0	0.0	1.44000E 02
30	4.50000E 05	7.50000E 06	0.0	0.0	1.44000E 02
31	1.50000E 05	0.0	0.0	-1.44000E 02	1.44000E 02
32	1.50000E 05	0.0	0.0	1.44000E 02	1.44000E 02

```
LOAD CASE=     1

NUMBER OF LOADED MEMBERS        4
NUMBER OF LOADED JOINTS         4

MEMBER LOADED     LOAD TYPE1     LOAD TYPE2     LOAD TYPE3     LOAD TYPE4

   3                  0              1              0              0

UNIFORMLY DIST    TOTAL LOAD

       WX                 WY
      0.0            40.00000
MEMBER LOADED     LOAD TYPE1     LOAD TYPE2     LOAD TYPE3     LOAD TYPE4

  11                  0              1              0              0

UNIFORMLY DIST    TOTAL LOAD

       WX                 WY
      0.0            40.00000
MEMBER LOADED     LOAD TYPE1     LOAD TYPE2     LOAD TYPE3     LOAD TYPE4

  19                  0              1              0              0

UNIFORMLY DIST    TOTAL LOAD

       WX                 WY
      0.0            40.00000
MEMBER LOADED     LOAD TYPE1     LOAD TYPE2     LOAD TYPE3     LOAD TYPE4

  27                  0              1              0              0

UNIFORMLY DIST    TOTAL LOAD

       WX                 WY
      0.0            40.00000
```

APPLIED JOINT LOADS

JOINT	PX	QY	M
4	0.0	8.00000E 01	0.0
8	0.0	8.00000E 01	0.0
12	0.0	8.00000E 01	0.0
16	0.0	8.00000E 01	0.0

DISPLACEMENTS AND REACTIONS

JOINT	UX	UY	UM	RX	RY	RM
1	5.93369E-02	8.14314E-02	1.06721E-03	0.0	0.0	0.0
2	5.60325E-02	1.50649E-01	9.79108E-04	0.0	0.0	0.0
3	4.85702E-02	6.26382E-02	-2.51205E-03	0.0	0.0	0.0
4	5.82569E-02	2.19593E-01	-1.51071E-04	0.0	0.0	0.0
5	1.41199E-02	8.00388E-02	5.97601E-04	0.0	0.0	0.0
6	4.74555E-02	1.41449E-01	5.86594E-04	0.0	0.0	0.0
7	4.94589E-02	5.65583E-02	-1.08219E-03	0.0	0.0	0.0
8	3.12261E-02	2.00474E-01	2.37951E-05	0.0	0.0	0.0
9	-6.83869E-03	6.59410E-02	5.01167E-04	0.0	0.0	0.0
10	2.67776E-02	1.13413E-01	6.71385E-04	0.0	0.0	0.0
11	2.61202E-02	4.40458E-02	-1.21394E-03	0.0	0.0	C.C
12	1.03775E-02	1.65496E-01	-4.58883E-05	0.0	0.0	0.0
13	-1.84260E-02	3.91861E-02	3.11003E-04	0.0	0.0	0.0
14	1.66858E-02	6.62978E-02	9.69593E-04	0.0	0.0	0.0
15	2.09321E-02	2.51544E-02	-1.64983E-03	0.0	0.0	C.0
16	-7.39381E-04	9.74181E-02	-1.78942E-04	0.0	0.0	0.0
17	0.0	0.0	-3.47439E-04	3.66261E 01	-1.58606E 02	0.0
18	0.0	0.0	-3.10985E-04	-3.46789E 01	-2.42786E 02	0.0
19	0.0	0.0	1.04296E-03	-1.94791E 00	-7.86075E 01	0.0

INTERNAL FORCES

MEMBER	P1BAR	Q1BAR	M1	P2BAR	Q2BAR	M2
1	2.24998E 00	-4.35228E 00	-1.86459E 02	-2.24998E 00	4.35228E 00	-4.40268E 02
2	4.63399E 00	7.74994E 00	6.75722E 02	-4.63399E 00	-7.74994E 00	4.40269E 02
3	7.77324E 00	-2.10001E 01	-9.22193E 02	-7.77324E 00	-1.89998E 01	6.34144E 02
4	4.35207E 00	2.25000E 00	1.86459E 02	-4.35207E 00	-2.25000E 00	1.37541E 02
5	2.87498E 01	3.13927E 00	2.46471E 02	-2.87498E 01	-3.13927E 00	2.05584E 02
6	1.89996E 01	-7.77322E 00	-6.34144E 02	-1.89996E 01	7.77322E 00	-4.85200E 02
7	4.96963E 01	0.0	0.0	-4.96963E 01	0.0	0.0
8	4.63254E 01	0.0	0.0	-4.63254E 01	0.0	0.0
9	-3.56378E 01	-4.56299E 00	-2.68764E 02	3.56378E 01	4.56299E 00	-3.88307E 02
10	-3.38112E 01	6.20737E 00	5.05555E 02	3.38112E 01	-6.20737E 00	3.88306E 02
11	-2.08684E 00	-1.98981E 01	-8.58409E 02	2.08684E 00	-2.01019E 01	8.87758E 02
12	4.40557E 01	1.75277E 00	1.31222E 02	-4.40557E 01	-1.75277E 00	1.21177E 02
13	8.76128E 01	2.10674E 00	1.47269E 02	-8.76128E 01	-2.10674E 00	1.56102E 02
14	3.91017E 01	-5.68636E 00	-4.02556E 02	-3.91017E 01	5.68636E 00	-4.16280E 02
15	5.02437E 01	0.0	0.0	-5.02437E 01	0.0	0.0

16	4.76610E 01	0.0	0.0	-4.76610E 01	0.0	0.0
17	-3.58671E 01	-4.02530E 00	-2.32837E 02	3.58671E 01	4.02530E 00	-3.46807E 02
18	-3.41669E 01	5.85448E 00	4.96239E 02	3.41669E 01	-5.85448E 00	3.46806E 02
19	6.84784E-01	-2.00660E 01	-8.71311E 02	-6.84784E-01	-1.99340E 01	8.52301E 02
20	8.36090E 01	1.41327E 00	1.11660E 02	-8.36090E 01	-1.41327E 00	9.18511E 01
21	1.47235E 02	3.25697E 00	2.18970E 02	-1.47235E 02	-3.25697E 00	2.50034E 02
22	5.90356E 01	-6.37115E 00	-4.36020E 02	-5.90356E 01	6.37115E 00	-4.81426E 02
23	5.07845E 01	0.0	0.0	-5.07845E 01	0.0	0.0
24	4.83802E 01	0.0	0.0	-4.83802E 01	0.0	0.0
25	-3.68471E 01	-2.93714E 00	-1.60438E 02	3.68471E 01	2.93714E 00	-2.62510E 02
26	-3.63025E 01	5.30763E 00	5.01788E 02	3.63025E 01	-5.30763E 00	2.62510E 02
27	-4.42326E 00	-2.04281E 01	-8.85216E 02	4.42326E 00	-1.95719E 01	7.61927E 02
28	1.22457E 02	4.76303E-01	6.85876E 01	-1.22457E 02	-4.76303E-01	0.0
29	2.07181E 02	9.26345E-01	1.33394E 02	-2.07181E 02	-9.26345E-01	1.22070E-04
30	7.86074E 01	-1.94791E 00	-2.80499E 02	-7.86074E 01	1.94791E 00	1.83105E-04
31	5.11237E 01	0.0	0.0	-5.11237E 01	0.0	0.0
32	5.03535E 01	0.0	0.0	-5.03535E 01	0.0	0.0

LOAD CASE= 2

NUMBER OF LOADED MEMBERS 0
NUMBER OF LOADED JOINTS 4

APPLIED JOINT LOADS

JOINT	PX	QY	M
1	2.00000E 01	0.0	0.0
5	2.00000E 01	0.0	0.0
9	2.00000E 01	0.0	0.0
13	2.00000E 01	0.0	0.0

DISPLACEMENTS AND REACTIONS

JOINT	UX	UY	UM	RX	RY	RM
1	3.82359E-01	-3.09625E-02	2.53452E-04	0.0	0.0	0.0
2	3.72220E-01	2.75442E-02	1.90596E-04	0.0	0.0	0.0
3	3.71441E-01	2.46823E-03	1.06618E-04	0.0	0.0	0.0
4	3.72944E-01	-5.26818E-03	1.93710E-04	0.0	0.0	0.0
5	3.23737E-01	-3.08371E-02	3.83749E-04	0.0	0.0	0.0
6	3.14325E-01	2.78071E-02	2.93307E-04	0.0	0.0	0.0
7	3.14170E-01	2.30457E-03	3.14871E-04	0.0	0.0	0.0
8	3.10097E-01	-4.81817E-03	1.36174E-04	0.0	0.0	0.0
9	2.37737E-01	-2.77057E-02	5.01300E-04	0.0	0.0	0.0
10	2.28293E-01	2.53651E-02	3.81328E-04	0.0	0.0	0.0
11	2.28028E-01	1.86824E-03	4.32977E-04	0.0	0.0	0.0
12	2.19556E-01	-3.79158E-03	5.57529E-05	0.0	0.0	0.0
13	1.29911E-01	-1.84315E-02	5.28804E-04	0.0	0.0	0.0
14	1.21353E-01	1.71344E-02	3.39295E-04	0.0	0.0	0.0
15	1.22150E-01	1.09250E-03	4.61602E-04	0.0	0.0	0.0
16	1.06760E-01	8.68400E-04	-3.17860E-05	0.0	0.0	0.0
17	0.0	0.0	1.08884E-03	-3.94032E 01	9.65967E 01	0.0

18	0.0	0.0	1.09445E-03	-4.01840E 01	-9.31828E 01	0.0
19	0.0	0.0	1.04160E-03	-4.19552E-01	-3.41408E 00	0.0

INTERNAL FORCES

MEMBER	P1BAR	Q1BAR	M1	P2BAR	Q2BAR	M2
1	1.96157E 01	3.91911E-01	3.44408E 01	-1.96157E 01	-3.91911E-01	2.19944E 01
2	1.50684E 00	-3.09987E-01	-2.26435E 01	-1.50684E 00	3.09987E-01	-2.19948E 01
3	8.11523E-01	5.11451E-01	7.80228E 01	-8.11523E-01	-5.11451E-01	6.92751E 01
4	-3.91907E-01	-3.84092E-01	-3.44407E 01	3.91907E-01	3.84092E-01	-2.08682E 01
5	-8.21472E-01	-6.94861E-01	-5.53795E 01	8.21472E-01	6.94861E-01	-4.46804E 01
6	5.11449E-01	-8.11512E-01	-6.92752E 01	-5.11449E-01	8.11512E-01	-4.75823E 01
7	-1.23111E 01	0.0	0.0	1.23111E 01	0.0	0.0
8	1.33037E 01	0.0	0.0	-1.33037E 01	0.0	0.0
9	2.84180E 01	6.88144E-01	7.53355E 01	-2.84180E 01	-6.88144E-01	2.37572E 01
10	-8.80957E 00	-1.02636E-01	8.97832E 00	8.80957E 00	1.02636E-01	-2.37579E 01
11	1.61865E-01	8.52082E-01	1.21577E 02	-1.61865E-01	-8.52082E-01	1.23823E 02
12	-9.78554E 00	-6.71457E-01	-5.44675E 01	9.78554E 00	6.71457E-01	-4.22226E 01
13	7.63110E 00	-1.12903E 00	-8.58748E 01	-7.63110E 00	1.12903E 00	-7.67058E 01
14	1.36353E 00	-9.73463E-01	-7.62406E 01	-1.36353E 00	9.73463E-01	-6.39379E 01
15	-2.57670E 01	0.0	0.0	2.57670E 01		0.0

16	2.68856E 01	0.0	0.0	-2.68856E 01	0.0	0.0
17	3.78770E 01	9.76180E-01	1.16696E 02	-3.78770E 01	-9.76180E-01	2.38738E 01
18	-1.82031E 01	1.39440E-01	4.39538E 01	1.82031E 01	-1.39440E-01	-2.38744E 01
19	2.76413E-01	1.06063E 00	1.50041E 02	-2.76413E-01	-1.06063E 00	1.55421E 02
20	-2.89818E 01	-1.01448E 00	-7.44744E 01	2.89818E 01	1.01448E 00	-7.16097E 01
21	2.57210E 01	-1.65941E 00	-1.17288E 02	-2.57210E 01	1.65941E 00	-1.21666E 02
22	2.42416E 00	-1.24988E 00	-9.14819E 01	-2.42416E 00	1.24988E 00	-8.85002E 01
23	-3.90647E 01	0.0	0.0	3.90647E 01	0.0	0.0
24	4.02481E 01	0.0	0.0	-4.02481E 01	0.0	0.0
25	4.82320E 01	9.93761E-01	1.29946E 02	-4.82320E 01	-9.93761E-01	1.31561E 01
26	-3.04026E 01	3.54132E-01	6.41516E 01	3.04026E 01	-3.54132E-01	-1.31568E 01
27	-8.30307E-01	9.89907E-01	1.36176E 02	8.30307E-01	-9.89907E-01	1.48917E 02
28	-5.75986E 01	-4.05111E-01	-5.83365E 01	5.75986E 01	4.05111E-01	9.30786E-04
29	5.35450E 01	-5.46251E-01	-7.86611E 01	-5.35450E 01	5.46251E-01	1.00708E-03
30	3.41407E 00	-4.19546E-01	-6.04155E 01	-3.41407E 00	4.19546E-01	1.09863E-03
31	-5.51517E 01	0.0	0.0	5.51517E 01	0.0	0.0
32	5.60563E 01	0.0	0.0	-5.60563E 01	0.0	0.0

END OF DATA SETS

```
      FINITE ELEMENT PROGRAMME FOR PLANE STRESS/STRAIN

      COMMON/BLOCK1/NDF,NEL,NODES,NODR,NODEL,NDFEL,NG,NUMGP,KSF,ISO,NP,
     1NMAT,NEQ,NB,NRD,NEQMAX,NBMAX,NRDMAX,PROP(5,6),SFN(9,8),DSF(9,2,8),
     1DETJ,NFILE1
      COMMON/BLOCK1/KBODY,BODYFX,BODYFY,KPRED,NELD,NODLD,NLD
      COMMON/BLOCK2/LFIX(50),LMAT(100)
      COMMON/BLOCK3/IELNO,ICONE(8),ELX(8),ELY(8),DXY(9,2,8),DV(9),
     1D1(3,3),ELSK(16,16)
      COMMON/BLOCK4/SK(100,50),LRD(100),URD(100),RRD(100),R(100)
      COMMON/COM/COMN(2000)
      DIMENSION ICON(100,8),CORD(50,2),D(5,3,3)
      EQUIVALENCE (ICON,COMN),(CORD,COMN(801)),(D,COMN(901))
      DIMENSION DUM(443)
      EQUIVALENCE (DUM,IELNO)
      DIMENSION F(3),H(3),HIHJ(9)
C
      WRITE(6,301)
C     ASSIGN VALUES TO PARAMETERS
      NDF=2
      NODEL=8
      KSF=2
      NEQMAX=100
      NBMAX=50
      NRDMAX=100
      NFILE1=1
      NDFEL=NODEL*NDF
      CALL INPUT1
C     CHECK BAND WIDTH
      CALL PRESET(KSTOP)
      IF(KSTOP.NE.0)GO TO 999
C     READ GAUSS PTS AND SHAPE FUNCTIONS
      CALL GAUSS(NG,F,H)
      NUMGP=0
      DO 10 IG=1,NG
      DO 10 JG=1,NG
      NUMGP=NUMGP+1
      XSI=F(IG)
      ETA=F(JG)
      HIHJ(NUMGP)=H(IG)*H(JG)
      CALL SF2(XSI,ETA,NUMGP,SFN,DSF)
10    CONTINUE
C     FORM MODULUS MATRIX
      DO 25 IMAT=1,NMAT
      CALL MOD(IMAT,D)
25    CONTINUE
C     COMPUTE CARTESIAN DERIVATIVES OF SHAPE FUN,ELEM STIFF AND
C     STORE ON NFILE1
      DO 30 IEL=1,NEL
      DO 29 I=1,443
```

```
29    DUM(I)=0.
      MAT=LMAT(IEL)
      DO 35 I=1,NODEL
      NJ=ICON(IEL,I)
      ELX(I)=CORD(NJ,1)
35    ELY(I)=CORD(NJ,2)
      DO 40 IGP=1,NUMGP
      CALL DERIV(IEL,IGP)
      DV(IGP)=HIHJ(IGP)*DETJ
40    CONTINUE
      IELNO=IEL
      DO 42 I=1,NODEL
42    ICONE(I)=ICON(IEL,I)
      DO 43 I=1,3
      DO 43 J=1,3
43    D1(I,J)=D(MAT,I,J)
      CALL ELSTIF
      WRITE(NFILE1)(DUM(I),I=1,443)
C     QUANTITIES STORED ON  NFILE1 AS IN COMMON BLOCK3
30    CONTINUE
C     ASSEMBLE GLOBAL STIFF MAT
      CALL FORMSK
      DO 39 ILD=1,NLD
      WRITE(6,209)ILD
C     FORM RHS FOR VARIOUS LOADS
      CALL RHS
      IF(ILD.EQ.1)GO TO 92
      IF(KPRED.EQ.0)GO TO 93
92    CALL FIX
93    IF(ILD.NE.1)GO TO 94
      CALL ELIM(SK,R,NB,NEQ,NEQMAX,NBMAX,LRD,URD,RRD,NRD,NRDMAX)
94    CALL SOLV(SK,R,NB,NEQ,NEQMAX,NBMAX,LRD,URD,RRD,NRD,NRDMAX)
C     COMPUTE DISPLACEMENTS AND REACTIONS
      CALL DISREA
C     COMPUTE STRAINS AND STRESSES
      CALL STRESS
39    CONTINUE
      WRITE(6,302)
209   FORMAT(//,10X,'LOAD CASE',I5,//)
301   FORMAT(/////,49X,'FEM PROGRAM',/,
     143X,'PLANE STRESS PLANE STRAIN',//)
302   FORMAT(//,10X,'END')
999   STOP
      END
```

```
      SUBROUTINE INPUT1
C
      COMMON/BLOCK1/NDF,NEL,NODES,NODR,NODEL,NDFEL,NG,NUMGP,KSF,ISO,NP,
     1NMAT,NEQ,NB,NRD,NEQMAX,NBMAX,NRDMAX,PROP(5,6),SFN(9,8),DSF(9,2,8),
     1DETJ,NFILE1
      COMMON/BLOCK1/KBODY,BODYFX,BODYFY,KPRED,NELD,NODLD,NLD
      COMMON/BLOCK2/LFIX(50),LMAT(100)
      COMMON/COM/COMN(2000)
      DIMENSION ICON(100,8),CCRD(50,2),D(5,3,3)
      EQUIVALENCE (ICON,COMN),(CORD,COMN(801)),(D,COMN(901))
C
C     READ-IN BASIC DATA
      WRITE(6,301)
      READ(5,101)NEL,NODES,NODR,NNOD,NG,ISO,NP,NMAT,NLD
      WRITE(6,201)NDF,NEL,NCDES,NCDR,NCDEL,NDFEL,NG,KSF,ISO,NP,
     1NMAT,NEQMAX,NBMAX
C     READ-IN CONNECTIVITY AND MAT USED
      WRITE(6,302)
      DO 20 IEL=1,NEL
      READ(5,102)L,(ICON(L,J),J=1,NODEL),LMAT(L)
20    WRITE(6,202)L,(ICON(L,J),J=1,NCDEL),LMAT(L)
C     READ-IN NODAL COORDS AND FIXITY
      DO 29 I=1,NODES
29    LFIX(I)=0
      DO 30 I=1,NNOD
30    READ(5,103)N,(CORD(N,J),J=1,2),LFIX(N)
      IF(NNOD.EQ.NODES)GO TO 40
C     GENERATE COORDS FOR INTERMEDIATE NODES
      DO 40 IEL=1,NEL
      DO 40 NCORN1=1,NODEL,KSF
      ITEMP1=ICON(IEL,NCORN1)
      NCORN2=NCORN1+KSF
      IF(NCORN2.GT.NODEL)NCORN2=1
      ITEMP2=ICON(IEL,NCORN2)
      NMIDLE=NCORN1+1
      ITEMP3=ICON(IEL,NMIDLE)
      DO 50 K=1,2
50    CORD(ITEMP3,K)=(CCRD(ITEMP1,K)+CORD(ITEMP2,K))/2.
40    CONTINUE
C     WRITE COORDS AND FIXITIES
      WRITE(6,303)
      DO 55 I=1,NODES
55    WRITE(6,203)I,(CORD(I,J),J=1,2),LFIX(I)
C     READ-IN MAT PROP
      WRITE(6,304)
      DO 65 MAT=1,NMAT
      IF(ISO.NE.0)GO TO 70
      WRITE(6,305)
      READ(5,104)(PROP(MAT,J),J=1,2)
      WRITE(6,204)MAT,(PRCP(MAT,J),J=1,2)
      GO TO 65
```

```
70    WRITE(6,306)
      READ(5,105)(PROP(MAT,J),J=1,6)
      WRITE(6,205)MAT,(PROP(MAT,J),J=1,6)
65    CONTINUE
101   FORMAT(16I5)
102   FORMAT(16I5)
103   FORMAT(I5,2F15.5,I5)
104   FORMAT(2E15.5)
105   FORMAT(3E15.5)
201   FORMAT(////,10X,I5,4X,'DEGREES OF FREEDOM',/,
     110X,I5,4X,'NUMBER OF ELEMENTS',/,
     110X,I5,4X,'NUMBER OF NODES',/,
     110X,I5,4X,'NUMBER OF NODES WITH PRESCRIBED DISP',/,
     110X,I5,4X,'NUMBER OF NODES PER ELEMENT',/,
     110X,I5,4X,'NUMBER OF DEGREES OF FREEDOM PER ELEMENT',/,
     110X,I5,4X,'ORDER OF GAUSS PT RULE',/,
     110X,I5,4X,'ORDER OF SHAPE FUNCTION',/,
     110X,I5,4X,'CODE FOR ISOTROPY',/,
     110X,I5,4X,'TYPE OF PROBLEM',/,
     110X,I5,4X,'NUMBER OF MATERIALS USED',/,
     110X,I5,4X,'SIZE OF NEQMAX ALLOWED',/,
     110X,I5,4X,'SIZE OF NBMAX ALLOWED',//)
202   FORMAT(10X,10(I5,5X))
203   FORMAT(9X,I5,17X,1PE12.5,13X,1PE12.5,17X,I5)
204   FORMAT(10X,I5,13X,1PE12.5,10X,1PE12.5)
205   FORMAT(10X,I5,11X,1PE12.5,5(6X,1PE12.5))
301   FORMAT(////,10X,'BASIC DATA',//)
302   FORMAT(//,10X,'ELEMENT CONNECTIVITY',//,
     110X,'ELEM  NUM ',80X,'MAT NO')
303   FORMAT(//,10X,'COORDINATES AND FIXITIES',//,
     110X,'NODE',20X,'CORDX',20X,'CORDY',20X,'FIXITY',//)
304   FORMAT(//,10X,'MATERIAL PROPERTIES',//)
305   FORMAT(//,10X,'MAT USED',16X,'E',19X,'MUE',//)
306   FORMAT(//,10X,'MAT USED',11X,'PROP(1)',11X,'PROP(2)',
     C11X,'PROP(3)',11X,'PROP(4)',11X,'PROP(5)',11X,'PROP(6)')
      RETURN
      END

      SUBROUTINE PRESET(KSTCP)
C
      COMMON/BLOCK1/NDF,NEL,NODES,NODR,NODEL,NDFEL,NG,NUMGP,KSF,ISO,NP,
     1NMAT,NEQ,NB,NRD,NEQMAX,NBMAX,NRDMAX,PROP(5,6),SFN(9,8),DSF(9,2,8),
     1DETJ,NFILE1
      COMMON/COM/COMN(2000)
      DIMENSION ICON(100,8),CORD(50,2),D(5,3,3)
      EQUIVALENCE (ICON,COMN),(CORD,COMN(801)),(D,COMN(901))
C
C     SET AND CHECK BAND WIDTH AND NUMBER OF EQUATIONS
      KSTCP=0
      NB=0
```

```
      DO 10 IEL=1,NEL
      MAX=ICON(IEL,1)
      MIN=ICON(IEL,1)
      DO 20 IDL=1,NCDEL
      ND=ICON(IEL,IDL)
      IF(MAX.LT.ND)MAX=ND
      IF(MIN.LE.ND)GO TO 20
      MIN=ND
20    CONTINUE
      ITEMP=(MAX-MIN+1)*NDF
      IF(NB.LT.ITEMP)NB=ITEMP
10    CONTINUE
      NEQ=NODES*NDF
      IF(NB.LE.NBMAX)GO TC 15
      KSTOP=1
      WRITE(6,201)NB
15    IF(NEQ.LE.NEQMAX)RETURN
      KSTCP=1
      WRITE(6,202)
      RETURN
201   FORMAT(//,10X,'ERROR---',5X,'BAND WIDTH EXCEEDED','NB=',I5)
202   FORMAT(//,10X,'ERROR---',5X,'NUM OF EQU EXCEEDED','NEQ=',I5)
      END

      SUBROUTINE GAUSS(NG,F,H)
C
      DIMENSICN F(3),H(3)
C
      GO TO(1,2,3),NG
1     F(1)=0.
      H(1)=1.
      RETURN
2     F(1)=-0.577350269189626
      H(1)=1.
      GO TO 100
3     F(1)=-0.774596669241483
      F(2)=0.
      H(1)=0.555555555555555
      H(2)=0.888888888888888
100   N=NG/2
      NNG=NG+1
      DO 200 J=1,N
      ITEMP=NNG-J
      F(ITEMP)=-F(J)
200   H(ITEMP)=H(J)
      RETURN
      END
```

```
      SUBROUTINE SF2(XSI,ETA,NUMGP,SFN,DSF)
C
      DIMENSION SFN(9,8),DSF(9,2,8)
      DIMENSICN AI(8),BI(8)
C
      DATA AI/-1.,0.,1.,1.,1.,0.,-1.,-1./
      DATA BI/-1.,-1.,-1.,0.,1.,1.,1.,0./
C
C     PARABOLIC ELEMENT SHAPE FUNCTION
C
      XSISQ=XSI*XSI
      ETASQ=ETA*ETA
      DO 200 I=1,7,2
      XI=AI(I)
      YI=BI(I)
      X=1.+XSI*XI
      Y=1.+ETA*YI
      SFN(NUMGP,I)=0.25*X*Y*(X+Y-3.)
      DSF(NUMGP,1,I)=0.25*Y*XI*(Y+2.*X-3.)
      DSF(NUMGP,2,I)=0.25*X*YI*(X+2.*Y-3.)
200   CONTINUE
      DO 201 I=2,8,2
      XI=AI(I)
      YI=BI(I)
      X=1.+XSI*XI
      Y=1.+ETA*YI
      IF(XI)203,202,203
202   SFN(NUMGP,I)=0.5*(1.-XSISC)*Y
      DSF(NUMGP,1,I)=-XSI*Y
      DSF(NUMGP,2,I)=0.5*YI*(1.-XSISQ)
      GO TO 2C1
203   SFN(NUMGP,I)=0.5*(1.-ETASQ)*X
      DSF(NUMGP,1,I)=0.5*XI*(1.-ETASQ)
      DSF(NUMGP,2,I)=-ETA*X
201   CONTINUE
      RETURN
      END

      SUBROUTINE MOD(M,D)
C
      COMMON/BLOCK1/NDF,NEL,NODES,NODR,NODEL,NDFEL,NG,NUMGP,KSF,ISO,NP,
     1NMAT,NEQ,NB,NRD,NEQMAX,NBMAX,NRDMAX,PROP(5,6),SFN(9,8),DSF(9,2,8),
     1DETJ,NFILE1
      DIMENSION D(5,3,3)
C
      IF(ISO.NE.0)GC TO 20
      E=PROP(M,1)
      AMU=PROP(M,2)
      IF(NP.NE.0)GO TC 10
      D(M,1,1)=E/(1.-AMU*AMU)
```

```
      D(M,1,2)=E*AMU/(1.-AMU*AMU)
      D(M,2,2)=D(M,1,1)
      D(M,1,3)=0.
      D(M,2,3)=0.
      D(M,3,3)=E*(1.-AMU)/(2.*(1.-AMU*AMU))
      GO TO 30
20    D(M,1,1)=PROP(M,1)
      D(M,1,2)=PROP(M,2)
      D(M,2,2)=PROP(M,3)
      D(M,1,3)=PROP(M,4)
      D(M,2,3)=PROP(M,5)
      D(M,3,3)=PROP(M,6)
      GO TO 30
C     PLANE STRAIN
10    D(M,1,1)=E*(1.-AMU)/((1.+AMU)*(1.-2.*AMU))
      D(M,1,2)=E*AMU*(1.-AMU)*(1.-AMU)/((1.+AMU)*(1.-2.*AMU))
      D(M,2,2)=D(M,1,1)
      D(M,1,3)=0.
      D(M,2,3)=0.
      D(M,3,3)=E/(2.*(1.+AMU))
30    DO 60 I=1,3
      DO 60 J=1,3
      IF(J.NE.I)D(M,J,I)=D(M,I,J)
60    CONTINUE
      RETURN
      END

      SUBROUTINE DERIV(IEL,IGP)
C
      COMMON/BLOCK1/NDF,NEL,NODES,NODR,NODEL,NDFEL,NG,NUMGP,KSF,ISO,NP,
     1NMAT,NEQ,NB,NRD,NEQMAX,NBMAX,NRDMAX,PROP(5,6),SFN(9,8),DSF(9,2,8),
     1DETJ,NFILE1
      COMMON/BLOCK3/IELNO,ICONE(8),ELX(8),ELY(8),DXY(9,2,8),DV(9),
     1D1(3,3),ELSK(16,16)
      DIMENSION COJ(2,2),COJIN(2,2)
C
      DO 10 I=1,2
      TEMP1=0.
      TEMP2=0.
      DO 20 IDL=1,NODEL
      TEMP1=TEMP1+ELX(IDL)*DSF(IGP,I,IDL)
      TEMP2=TEMP2+ELY(IDL)*DSF(IGP,I,IDL)
      COJ(I,1)=TEMP1
20    COJ(I,2)=TEMP2
10    CONTINUE
C     COMPUTE DETJ AND INVERSE OF COJ
      DETJ=COJ(1,1)*COJ(2,2)-COJ(1,2)*COJ(2,1)
      IF(DETJ)100,100,101
100   WRITE(6,301)IEL,DETJ
      STOP
```

```
101   CONTINUE
      COJIN(1,1)=COJ(2,2)/DETJ
      COJIN(2,2)=COJ(1,1)/DETJ
      COJIN(1,2)=-COJ(1,2)/DETJ
      COJIN(2,1)=-COJ(2,1)/DETJ
C     COMPUTE CARTESIAN DERIVATIVES
      DO 25 I=1,2
      DO 25 IDL=1,NODEL
      TEMP=0.
      DO 30 J=1,2
30    TEMP=TEMP+COJIN(I,J)*DSF(IGP,J,IDL)
      DXY(IGP,I,IDL)=TEMP
25    CONTINUE
301   FORMAT(//,5X,'ERROR---',5X,'DETJ NONPOSITIVE',//,5X,
     1'ELEMENT NO.',5X,I5,//,5X,'DET IS',5X,1PE12.5)
      RETURN
      END

      SUBROUTINE ELSTIF
C
      COMMON/BLOCK1/NDF,NEL,NODES,NODR,NODEL,NDFEL,NG,NUMGP,KSF,ISO,NP,
     1NMAT,NEQ,NB,NRD,NEQMAX,NBMAX,NRDMAX,PROP(5,6),SFN(9,8),DSF(9,2,8),
     1DETJ,NFILE1
      COMMON/BLOCK3/IELNO,ICONE(8),ELX(8),ELY(8),DXY(9,2,8),DV(9),
     1D1(3,3),ELSK(16,16)
      DIMENSION GPSTIF(16,16),A(3,16)
      DIMENSION DUM(256)
      EQUIVALENCE(DUM,GPSTIF)
C
      DO 10 I=1,NDFEL
      DO 10 J=1,NDFEL
10    ELSK(I,J)=0.
      DO 20 IGP=1,NUMGP
C     FORM A MATRIX AT EACH GAUSS PT
      DO 30 I=2,NDFEL,2
      ID=I-1
      JD=I/2
      A(1,ID)=DXY(IGP,1,JD)
      A(2,ID)=0.
      A(3,ID)=DXY(IGP,2,JD)
      A(1,I)=0.
      A(2,I)=DXY(IGP,2,JD)
30    A(3,I)=DXY(IGP,1,JD)
      DO 40 I=1,256
40    DUM(I)=0.
C     FORM ATDA
      DO 50 I=1,NDFEL
      DO 50 J=1,NDFEL
      DO 50 K=1,3
      DO 50 L=1,3
      GPSTIF(I,J)=GPSTIF(I,J)+A(K,I)*D1(K,L)*A(L,J)
```

```
50    CONTINUE
      DO 60 I=1,NDFEL
      DO 60 J=1,NDFEL
60    ELSK(I,J)=ELSK(I,J)+GPSTIF(I,J)*DV(IGP)
20    CONTINUE
39    FORMAT(///,5X,8E15.6,/,5X,8E15.6)
      RETURN
      END

      SUBROUTINE FORMSK
C
      COMMON/BLOCK1/NDF,NEL,NODES,NCDR,NODEL,NDFEL,NG,NUMGP,KSF,ISO,NP,
     1NMAT,NEQ,NB,NRD,NEQMAX,NBMAX,NRDMAX,PROP(5,6),SFN(9,8),DSF(9,2,8),
     1DETJ,NFILE1
      COMMON/BLOCK3/IELNO,ICONE(8),ELX(8),ELY(8),DXY(9,2,8),DV(9),
     1D1(3,3),ELSK(16,16)
      COMMON/BLOCK4/SK(100,50),LRD(100),URD(100),RRD(100),R(100)
      DIMENSION DUM(443)
      EQUIVALENCE (DUM,IELNO)
C
      DO 10 J=1,NB
      DO 10 I=1,NEQ
10    SK(I,J)=0.
      REWIND NFILE1
      DO 11 IEL=1,NEL
      READ(NFILE1)(DUM(I),I=1,443)
C     II AND JJ ARE STARTING ROW AND CCLUMN LCCATICN IN STIFF MAT
      DO 100 I=1,NODEL
      II=(ICONE(I)-1)*NDF
      DO 100 J=1,NODEL
      JJ=(ICONE(J)-1)*NDF
C     IROW AND JCOL ARE LOCATION IN SQUARE STIFF MAT
C     IEROW AND JECOL ARE LOCATION IN ELEP STIFF MAT
      DO 100 IDF=1,NDF
      IROW=II+IDF
      IEROW=(I-1)*NDF+IDF
      DO 100 JDF=1,NDF
      JCOL=JJ+JDF
      IF(IROW.GT.JCOL)GC TC 100
      JECOL=(J-1)*NDF+JDF
      JCOLB=JCOL+1-IROW
C     LCOLB GIVES THE CCLUMN LOCATION IN EQUIVALENT SK ARRAY
      SK(IROW,JCOLB)=SK(IROW,JCOLB)+ELSK(IEROW,JECOL)
100   CONTINUE
11    CONTINUE
      RETURN
      END

      SUBROUTINE RHS
C
      COMMON/BLOCK1/NDF,NEL,NODES,NODR,NODEL,NDFEL,NG,NUMGP,KSF,ISO,NP,
     1NMAT,NEQ,NB,NRD,NEQMAX,NBMAX,NRDMAX,PROP(5,6),SFN(9,8),DSF(9,2,8),
```

```
     1DETJ,NFILE1
      COMMCN/BLCCK1/KBOCY,BODYFX,BODYFY,KPRED,NELD,NODLD,NLD
      COMMCN/BLCCK2/LFIX(50),LMAT(100)
      COMMON/BLOCK3/IELNO,ICONE(8),ELX(8),ELY(8),CXY(9,2,8),DV(9),
     1D1(3,3),ELSK(16,16)
      COMMON/BLOCK4/SK(100,50),LRD(100),URD(100),RRC(100),R(100)
      COMMON/COM/COMN(2000)
      DIMENSICN DUM(443)
      EQUIVALENCE (DUM,IELNC)
      DIMENSION UR(50,2),LELD(100),LCODE(3),KEDGE(4)
      EQUIVALENCE(UR,COMN),(LELD,COMN(101)),(LCCDE,COMN(201)),
     1(KEDGE,COMN(204))
C
      READ(5,101)NELD,NODLD,KPRED,NPRED
      WRITE(6,201)NELD,NODLD,KPRED,NPRED
      DO 10 I=1,100
   10 R(I)=0.
      DO 11 I=1,NODES
      IF(LFIX(I).EQ.0) GO TO 11
      DO 11 IDF=1,NCF
      UR(I,IDF)=0.
   11 CONTINUE
      IF(KPREC.EQ.0)GO TO 25
C     READ NONZERO PRESCRIBED DISP
      WRITE(6,301)
      DO 20 I=1,NPRED
      IF(LFIX(I).EQ.0)GO TO 20
      READ(5,102)INCD,(UR(INOC,J),J=1,NDF)
      WRITE(6,202)INOD,LFIX(I),(UR(INCD,J),J=1,NDF)
   20 CONTINUE
   25 IF(NELD.EQ.0)GO TC 45
C     READ LIST OF ELEMENTS LCADED
      READ(5,103)(LELD(I),I=1,NELD)
      WRITE(6,302)
      DO 30 I=1,NELD
   30 WRITE(6,203)I,LELD(I)
      READ(5,104)KBODY,BODYFX,BODYFY
      WRITE(6,204)KBODY,BCDYFX,BCDYFY
      IC=1
      REWIND NFILE1
      DO 40 IEL=1,NEL
      READ(NFILE1) (DUM(I),I=1,443)
      IF(IC.GT.NELD)GC TO 45
      IF(IEL.NE.LELD(IC))GC TC 40
      WRITE(6,303)
      READ(5,105)INEL,(LCCDE(I),I=1,3),(KEDGE(I),I=1,4)
      WRITE(6,205)INEL,(LCODE(I),I=1,3),(KEDGE(I),I=1,4)
      IF(LCODE(1).EQ.1)CALL BODYF
      IF(LCODE(2).EQ.1)CALL PRESS
      IF(LCODE(3).EC.1)CALL ECLCAD
      IC=IC+1
```

```
40    CONTINUE
45    IF(NODLD.NE.0)CALL PTLOAD
      RETURN
101   FORMAT(16I5)
102   FORMAT(I5,2E15.5)
103   FORMAT(16I5)
104   FORMAT(I5,2F15.5)
105   FORMAT(16I5)
201   FORMAT(//,10X,I5,4X,'NUMBER OF ELEMENTS LCADED',/,
     110X,I5,4X,'NUMBER OF NODES LCADED',/,
     110X,I5,4X,'CODE FCR PRESCRIBED DISP',/,
     110X,I5,4X,'NUMBER CF NODES WITH NON ZERO DISP',//)
202   FORMAT(/,8X,I5,10X,I5,2(8X,E15.5),//)
203   FORMAT(/,13X,I5,22X,I5)
204   FORMAT(//,10X,'CODE FCR UNIFORM BODY FORCES',I5,//,
     110X,'BODYFX',15X,'BCDYFY',//,
     14X,F15.5,5X,F15.5,/)
205   FORMAT(/,10X,I5,12X,I5,16X,I5,17X,I5,22X,4I2,//)
301   FORMAT(//,10X,'NODE',10X,'FIXITY',10X,'UX',24X,'UY',//)
302   FORMAT(//,10X,'ELEMENT COUNT',10X,'LIST OF ELEMENTS LCADED',//)
303   FORMAT(//,10X,'ELEMENT',10X,'BODY FCRCE',10X,'EDGE LOADS',12X,
     1'EQ NODAL LOACS',
     110X,'LOADED ECGE CODE',//)
      END

      SUBROUTINE BODYF
C
      COMMON/BLOCK1/NDF,NEL,NCDES,NODR,NODEL,NDFEL,NG,NUMGP,KSF,ISO,NP,
     1NMAT,NEQ,NB,NRD,NEQMAX,NBMAX,NRDMAX,PROP(5,6),SFN(9,8),DSF(9,2,8),
     1DETJ,NFILE1
      COMMON/BLOCK1/KBODY,BODYFX,BODYFY,KPRED,NELC,NODLD,NLD
      COMMON/BLCCK3/IELNO,ICONE(8),ELX(8),ELY(8),DXY(9,2,8),DV(9),
     1DI(3,3),ELSK(16,16)
      COMMON/BLOCK4/SK(10C,50),LRD(100),URD(100),RRD(100),R(100)
      DIMENSICN EGNF(16)
C
      IF(KBODY.EQ.0)GO TO 1
      READ(5,101) INEL,BCDYFX,BCDYFY
      WRITE(6,201)INEL,BCCYFX,BCDYFY
1     DO 5 I=1,NDFEL
5     EQNF(I)=0.
      DO 10 IGP=1,NUMGP
      DO 10 IDL=1,NCDEL
       IOCD=2*IDL-1
      IEVEN=2*IDL
      EQNF(IODD)=EQNF(ICDD)+SFN(IGP,IDL)*BOCYFX*DV(IGP)
      EQNF(IEVEN)=EGNF(IEVEN)+SFN(IGP,IDL)*BODYFY*DV(IGP)
```

```
10    CONTINUE
C     ADD EQ NODAL LOADS TO RHS AT APPROPRIATE LOCATION
      DO 50 IDL=1,NODEL
      I=IDL*NDF-2
      NN=ICONE(IDL)
      IX=(NN-1)*NDF
      DO 50 IDF=1,NDF
50    R(IX+IDF)=R(IX+IDF)+EQNF(I+IDF)
101   FORMAT(I5,2F15.5)
201   FORMAT(//,10X,'ELEMENT BODY FORCE',/,
     110X,I5,4X,'ELEMENT NUMBER',/,
     110X,'BODYFX',15X,'BODYFY',//,
     14X,F15.5,5X,F15.5,/)
C
C
      RETURN
      END

      SUBROUTINE PRESS
C
      COMMON/BLOCK1/NDF,NEL,NODES,NCDR,NODEL,NDFEL,NG,NUMGP,KSF,ISO,NP,
     1NMAT,NEQ,NB,NRD,NEQMAX,NBMAX,NRDMAX,PROP(5,6),SFN(9,8),DSF(9,2,8),
     1DETJ,NFILE1
      COMMON/BLOCK3/IELNO,ICONE(8),ELX(8),ELY(8),DXY(9,2,8),DV(9),
     1D1(3,3),ELSK(16,16)
      COMMON/BLOCK4/SK(100,50),LRD(100),URD(100),RRD(100),R(100)
      COMMON/COM/COMN(2000)
      DIMENSION UR(50,2),LELD(100),LCODE(3),KEDGE(4)
      EQUIVALENCE(UR,COMN),(LELD,COMN(101)),(LCODE,COMN(201)),
     1(KEDGE,COMN(204))
      DIMENSION EQNF(16),F(3),H(3),ESFN(9,8),EDSF(9,2,8),AMULT(4),
     CPRES(4,6)
C
      DATA AMULT/-1.,1.,1.,-1./
C
      WRITE(6,201)IELNO
      WRITE(6,301)
      DO 5 N=1,4
      IF(KEDGE(N).EQ.0)GO TO 5
      READ(5,101)ISIDE,(PRES(ISIDE,J),J=1,6)
      WRITE(6,202)ISIDE,(PRES(ISIDE,J),J=1,6)
5     CONTINUE
      CALL GAUSS(NG,F,H)
      DO 10 I=1,NDFEL
10    EQNF(I)=0.
      DO 20 ISIDE=1,4
      IF(KEDGE(ISIDE).EQ.0)GO TO 20
      DO 20 IG=1,NG
      GO TO (11,12,11,12),ISIDE
```

```
11    XSI=F(IG)
      ETA=AMULT(ISIDE)
      GO TO 13
12    XSI=AMULT(ISIDE)
      ETA=F(IG)
13    CALL SF2(XSI,ETA,IG,ESFN,EDSF)
      ID=2*ISIDE-2
      PRESX=0.
      PRESY=0.
      DERX=0.
      DERY=0.
      DO 25 IP=1,3
      INDEX1=ID+IP
      INDEX2=IP+3
      IF(INDEX1.GT.NODEL)INDEX1=1
      PRESX=PRESX+ESFN(IG,INDEX1)*PRES(ISIDE,IP)
      PRESY=PRESY+ESFN(IG,INDEX1)*PRES(ISIDE,INDEX2)
      GO TO (26,27,26,27),ISIDE
26    DERX=DERX+EDSF(IG,1,INDEX1)*ELX(INDEX1)
      DERY=DERY+EDSF(IG,1,INDEX1)*ELY(INDEX1)
      GO TO 25
27    DERX=DERX+EDSF(IG,2,INDEX1)*ELX(INDEX1)
      DERY=DERY+EDSF(IG,2,INDEX1)*ELY(INDEX1)
25    CONTINUE
      DO 30 IP=1,3
      INDEX1=ID+IP
      IF(INDEX1.GT.NODEL)INDEX1=1
      IODD=2*INDEX1-1
      IEVEN=2*INDEX1
      EQNF(IODD)=EQNF(IODD)+ESFN(IG,INDEX1)*PRESX*DERY*H(IG)
      EQNF(IEVEN)=EQNF(IEVEN)+ESFN(IG,INDEX1)*PRESY*DERX*H(IG)
30    CONTINUE
20    CONTINUE
      DO 50 IDL=1,NODEL
      I=IDL*NDF-2
      NN=ICONE(IDL)
      IX=(NN-1)*NDF
      DO 50 IDF=1,NDF
50    R(IX+IDF)=R(IX+IDF)+EQNF(I+IDF)
101   FORMAT(I5,6F12.4)
201   FORMAT(10X,'ELEMENT',I5,//)
202   FORMAT(8X,I5,8X,6(F15.5,3X))
301   FORMAT(8X,'SIDE',15X,'PX1',15X,'PX2',15X,'PX3',15X,'PY1',15X,
     1'PY2',15X,'PY3',/)
      RETURN
      END

      SUBROUTINE EQLOAD
C
      COMMON/BLOCK1/NDF,NEL,NCDES,NCDR,NODEL,NDFEL,NG,NUMGP,KSF,ISO,NP,
     1NMAT,NEQ,NB,NRD,NEQMAX,NBMAX,NRDMAX,PROP(5,6),SFN(9,8),DSF(9,2,8),
```

```
     1DETJ,NFILE1
      COMMON/BLOCK3/IELNO,ICONE(8),ELX(8),ELY(8),DXY(9,2,8),DV(9),
     1D1(3,3),ELSK(16,16)
      COMMON/BLOCK4/SK(100,50),LRD(100),URD(100),RRD(100),R(100)
      DIMENSION EQLD(8,2)
C
      WRITE(6,301) IELNO
      READ(5,101)INEL,((EQLD(I,J),I=1,NODEL),J=1,NDF)
      DO 30 I=1,NODEL
30    WRITE(6,201) I,(EQLD(I,J),J=1,NDF)
      DO 20 I=1,NODEL
      NN=ICONE(I)
      DO 20 IDF=1,NDF
      IX=(NN-1)*NDF+IDF
20    R(IX)=R(IX)+EQLD(I,IDF)
      RETURN
101   FORMAT(I5,5F15.5/5F15.5/5F15.5/F15.5)
201   FORMAT(//,15X,I5,10X,F15.5,7X,F15.5)
301   FORMAT(//,10X,'ELEMENT NO',I5,//,
     110X,'ELEMENT LOCAL NODE',10X,'EQLD-X',15X,'EQLD-Y',//)
      END

      SUBROUTINE PTLOAD
C
      COMMON/BLOCK1/NDF,NEL,NODES,NODR,NODEL,NDFEL,NG,NUMGP,KSF,ISO,NP,
     1NMAT,NEQ,NB,NRD,NEQMAX,NBMAX,NRDMAX,PROP(5,6),SFN(9,8),DSF(9,2,8),
     1DETJ,NFILE1
      COMMON/BLOCK1/KBODY,BODYFX,BODYFY,KPRED,NELD,NODLD,NLD
      COMMON/BLOCK4/SK(100,50),LRD(100),URD(100),RRD(100),R(100)
      DIMENSION PTLD(2)
C
      WRITE(6,301)
      DO 10 I=1,NODLD
      READ(5,101)NODE,(PTLD(J),J=1,NDF)
      DO 20 IDF=1,NDF
      IX=(NODE-1)*NDF+IDF
20    R(IX)=R(IX)+PTLD(IDF)
10    WRITE(6,201)NODE,(PTLD(J),J=1,NDF)
101   FORMAT(I5,2F15.5)
201   FORMAT(/,8X,I5,7X,F15.5,6X,F15.5)
301   FORMAT(//,10X,'NODE',10X,'X-COMPONENT',10X,'Y-COMPONENT',/)
      RETURN
      END

      SUBROUTINE FIX
C
      COMMON/BLOCK1/NDF,NEL,NODES,NODR,NODEL,NDFEL,NG,NUMGP,KSF,ISO,NP,
     1NMAT,NEQ,NB,NRD,NEQMAX,NBMAX,NRDMAX,PROP(5,6),SFN(9,8),DSF(9,2,8),
```

```
     1DETJ,NFILE1
      COMMON/BLOCK2/LFIX(50),LMAT(100)
      COMMON/BLOCK4/SK(100,50),LRD(100),URD(100),RRD(100),R(100)
      COMMON/COM/COMN(2000)
      DIMENSICN UR(50,2)
      EQUIVALENCE (COMN,UR)
C
      DO 1 I=1,100
      LRD(I)=0
      URD(I)=0.
1     RRD(I)=0.
      NRD=0
      DO 13 I=1,NCDES
      IF(LFIX(I).EQ.0)GO TO 13
      JFIX=LFIX(I)
      IFIX=10**(NDF-1)
      DO 11 IDF=1,NDF
      IX=(I-1)*NDF+IDF
      IF(JFIX.LT.IFIX)GO TO 12
      NRD=NRD+1
      LRD(NRD)=IX
      URD(NRD)=UR(I,IDF)
      JFIX=JFIX-IFIX
12    IFIX=IFIX/10
11    CONTINUE
13    CONTINUE
      RETURN
      END
      SUBROUTINE ELIM(SK,R,IB,NEQ,NEQMAX,IBMAX,LR,UR,RR,NR,NRMAX)
      DIMENSICN SK(NEQMAX,IBMAX),R(NEQMAX),LR(NRMAX),UR(NRMAX),RR(NRMAX)
      NEQ1=NEQ-1
      IR=1
      DO 106 N=1,NEQ1
      M=N-1
      IB1=MINO(NEQ-M,IB)
      IF(SK(N,1).LE.0) GO TO 101
      IF(IR.GT.NR) GO TO 108
      IF(N.NE.LR(IR)) GO TO 108
      DO 105 L=2,IB1
105   SK(N,L)=-SK(N,L)
      IR=IR+1
      GO TO 106
108   DO 100 L=2,IB1
      RATIO=SK(N,L)/SK(N,1)
      I=M+L
      J=0
      DO 102 K=L,IB1
      J=J+1
102   SK(I,J)=SK(I,J)-RATIO*SK(N,K)
```

```
100   SK(N,L)=RATIO
106   CONTINUE
      GO TO 103
101   WRITE(6,200) N,SK(N,1)
      STOP
103   RETURN
200   FORMAT(//,'PIVOT ERROR---',/,5X,'EQ NUM=',I5,5X,'PIVOT=',1PE20.10)
      END

      SUBROUTINE SOLV(SK,R,IB,NEQ,NEQMAX,IBMAX,LR,UR,RR,NR,NRMAX)
      DIMENSION SK(NEQMAX,IBMAX),R(NEQMAX),LR(NRMAX),UR(NRMAX),RR(NRMAX)
      NEQ1=NEQ-1
      IR=1
      DO 203 N=1,NEQ1
      M=N-1
      IB1=MINO(NEQ-M,IB)
      IF(IR.GT.NR) GO TO 202
      IF(N.NE.LR(IR)) GO TO 202
      R(N)=SK(N,1)*UR(IR)-R(N)
      P=-UR(IR)
      IR=IR+1
      GO TO 204
202   P=R(N)
      R(N)=P/SK(N,1)
204   DO 200 L=2,IB1
      I=M+L
200   R(I)=R(I)-SK(N,L)*P
203   CONTINUE
C     BACK SUBSTITUTION
      IR=NR
      IF(NEQ.NE.LR(IR)) GO TO 206
      RR(IR)=SK(NEQ,1)*UR(IR)-R(NEQ)
      R(NEQ)=UR(IR)
      IR=IR-1
      GO TO 205
206   R(NEQ)=R(NEQ)/SK(NEQ,1)
205   DO 208 I=1,NEQ1
      N=NEQ-I
      M=N-1
      IB1=MINO(NEQ-M,IB)
      DO 207 K=2,IB1
      L=M+K
207   R(N)=R(N)-SK(N,K)*R(L)
      IF(IR.EQ.0) GO TO 208
      IF(N.NE.LR(IR)) GO TO 208
      RR(IR)=R(N)
      R(N)=UR(IR)
      IR=IR-1
208   CONTINUE
      RETURN
      END
```

```
      SUBROUTINE DISREA
C
      COMMON/BLOCK1/NDF,NEL,NCDES,NODR,NODEL,NDFEL,NG,NUMGP,KSF,ISC,NP,
     1NMAT,NEQ,NB,NRD,NEQMAX,NBMAX,NRDMAX,PROP(5,6),SFN(9,8),DSF(9,2,8),
     1DETJ,NFILE1
      COMMON/BLOCK2/LFIX(50),LMAT(100)
      COMMON/BLCCK4/SK(100,50),LRD(100),URD(100),RRD(100),R(100)
      DIMENSICN DIS(50,2),REA(50,2)
C
      DO 1 J=1,NDF
      DO 1 I=1,50
      DIS(I,J)=0.
1     REA(I,J)=0.
      IRD=1
      DO 10 I=1,NODES
      DO 12 IDF=1,NDF
      IX=(I-1)*NDF+IDF
12    DIS(I,IDF)=R(IX)
      IF(LFIX(I).EQ.0)GC TO 10
      JFIX=LFIX(I)
      IFIX=10**(NDF-1)
      DO 14 IDF=1,NCF
      IF(JFIX.LT.IFIX)GO TO 15
      REA(I,IDF)=RRD(IRC)
      IRD=IRD+1
      JFIX=JFIX-IFIX
15    IFIX=IFIX/10
14    CONTINUE
10    CONTINUE
      WRITE(6,301)
      DO 16 I=1,NODES
16    WRITE(6,201) I,(DIS(I,J),J=1,NDF),(REA(I,J),J=1,NDF)
201   FORMAT(/,8X,I5,11X,1PE12.5,4X,1PE12.5,14X,1PE12.5,4X,1PE12.5)
301   FORMAT(/,10X,'DISPLACEMENTS AND REACTIONS',///,
     110X,'NODE',15X,'UX',15X,'UY',25X,'RX',15X,'RY',//)
      RETURN
      END
      SUBROUTINE STRESS
C
      COMMON/BLOCK1/NDF,NEL,NCDES,NODR,NODEL,NDFEL,NG,NUMGP,KSF,ISO,NP,
     1NMAT,NEQ,NB,NRD,NEQMAX,NBMAX,NRDMAX,PROP(5,6),SFN(9,8),DSF(9,2,8),
     1DETJ,NFILE1
      COMMON/BLOCK2/LFIX(50),LMAT(100)
      COMMON/BLOCK3/IELNO,ICONE(8),ELX(8),ELY(8),DXY(9,2,8),DV(9),
     1D1(3,3),ELSK(16,16)
      COMMON/BLOCK4/SK(10C,50),LRD(100),URD(100),RRD(100),R(100)
      COMMCN/COM/COMN(2000)
      DIMENSICN STRAIN(9,4),STRS(9,4)
```

```
      EQUIVALENCE(STRAIN,COMN),(STRS,COMN(37))
      DIMENSION A(3,16),DISP(16)
      DIMENSION GPX(9),GPY(9)
      DIMENSION DUM(443)
      EQUIVALENCE (DUM,IELNO)
C
      WRITE(6,302)
      REWIND NFILE1
      DO 49 IEL=1,NEL
      READ(NFILE1) (DUM(I),I=1,443)
      WRITE(6,201)IELNO
      DO 5 IGP=1,NUMGP
      TEMP1=0.
      TEMP2=0.
      DO 10 IDL=1,NODEL
      TEMP1=TEMP1+SFN(IGP,IDL)*ELX(IDL)
      TEMP2=TEMP2+SFN(IGP,IDL)*ELY(IDL)
      IX=(ICONE(IDL)-1)*NDF
      IODD=2*IDL-1
      IEVEN=2*IDL
      DISP(IODD)=R(IX+1)
      DISP(IEVEN)=R(IX+2)
      A(1,IODD)=DXY(IGP,1,IDL)
      A(2,IODD)=0.
      A(3,IODD)=DXY(IGP,2,IDL)
      A(1,IEVEN)=0.
      A(2,IEVEN)=DXY(IGP,2,IDL)
10    A(3,IEVEN)=DXY(IGP,1,IDL)
      GPX(IGP)=TEMP1
      GPY(IGP)=TEMP2
      DO 20 J=1,3
20    STRAIN(IGP,J)=0.
      DO 25 J=1,3
25    STRS(IGP,J)=0.
C     COMPUTE STRAINS
      DO 30 J=1,3
      DO 30 K=1,NDFEL
30    STRAIN(IGP,J)=STRAIN(IGP,J)+A(J,K)*DISP(K)
C     COMPUTE STRESSES
      DO 35 J=1,3
      DO 35 K=1,3
35    STRS(IGP,J)=STRS(IGP,J)+D1(J,K)*STRAIN(IGP,K)
      IF(NP.EQ.0)GO TO 51
      MAT=LMAT(IELNO)
      AMU=PROP(MAT,2)
      STRS(IGP,4)=(STRS(IGP,1)+STRS(IGP,2))*AMU
      GO TO 5
51    STRS(IGP,4)=0.
5     CONTINUE
      DO 55 IGP=1,NUMGP
55    WRITE(6,202)IGP,GPX(IGP),GPY(IGP),(STRS(IGP,J),J=1,4)
```

```
49     CONTINUE                                                          00087800
201    FORMAT(//,10X,'ELEMENT NO',I5,//)                                 00087900
202    FORMAT(7X,I5,11X,1PE12.5,3X,1PE12.5,4(6X,1PE12.5))                00088000
302    FORMAT(//,10X,'STRESSES',//,10X,'GAUSS PT',10X,'GPX',12X,'GPY',   00088100
      112X,'SIGMA-X',12X,'SIGMA-Y',12X,'TAU-XY',12X,'SIGMA-Z',//)        00088200
       RETURN                                                            00088300
       END                                                               00088400
```

OUTPUT

```
                    FEM PROGRAM
             PLANE STRESS PLANE STRAIN

BASIC DATA

   2     DEGREES OF FREEDOM
   8     NUMBER OF ELEMENTS
  37     NUMBER OF NODES
  10     NUMBER OF NODES WITH PRESCRIBED DISP
   8     NUMBER OF NODES PER ELEMENT
  16     NUMBER OF DEGREES OF FREEDOM PER ELEMENT
   3     ORDER OF GAUSS PT RULE
   2     ORDER OF SHAPE FUNCTION
   0     CODE FOR ISOTROPY
   0     TYPE OF PROBLEM
   1     NUMBER OF MATERIALS USED
 100     SIZE OF NEQMAX ALLOWED
  50     SIZE OF NBMAX ALLOWED


ELEMENT CONNECTIVITY

ELEM  NUM                                                                MAT NO
  1       3       7      11      10       9       6       1       2        1
  2       5       8      13      12      11       7       3       4        1
  3      11      15      19      18      17      14       9      10        1
  4      13      16      21      20      19      15      11      12        1
  5      19      23      27      26      25      22      17      18        1
  6      21      24      29      28      27      23      19      20        1
  7      27      31      35      34      33      30      25      26        1
  8      29      32      37      36      35      31      27      28        1
```

COORDINATES AND FIXITIES

NODE	CORDX	CORDY	FIXITY
1	0.0	5.00000E-01	10
2	0.0	3.75000E-01	10
3	0.0	2.50000E-01	11
4	0.0	1.25000E-01	10
5	0.0	0.0	10
6	2.50000E-01	5.00000E-01	0
7	2.50000E-01	2.50000E-01	0
8	2.50000E-01	0.0	0
9	5.00000E-01	5.00000E-01	0
10	5.00000E-01	3.75000E-01	0
11	5.00000E-01	2.50000E-01	0
12	5.00000E-01	1.25000E-01	0
13	5.00000E-01	0.0	0
14	7.50000E-01	5.00000E-01	0
15	7.50000E-01	2.50000E-01	0
16	7.50000E-01	0.0	0
17	1.00000E 00	5.00000E-01	0
18	1.00000E 00	3.75000E-01	0
19	1.00000E 00	2.50000E-01	0
20	1.00000E 00	1.25000E-01	0
21	1.00000E 00	0.0	0
22	1.25000E 00	5.00000E-01	0
23	1.25000E 00	2.50000E-01	0
24	1.25000E 00	0.0	0
25	1.50000E 00	5.00000E-01	0
26	1.50000E 00	3.75000E-01	0
27	1.50000E 00	2.50000E-01	0
28	1.50000E 00	1.25000E-01	0
29	1.50000E 00	0.0	0
30	1.75000E 00	5.00000E-01	0
31	1.75000E 00	2.50000E-01	0
32	1.75000E 00	0.0	0
33	2.00000E 00	5.00000E-01	10
34	2.00000E 00	3.75000E-01	10
35	2.00000E 00	2.50000E-01	11
36	2.00000E 00	1.25000E-01	10
37	2.00000E 00	0.0	10

MATERIAL PROPERTIES

MAT USED	E	MUE
1	2.00000E 05	2.00000E-01

```
LOAD CASE    1

  2     NUMBER OF ELEMENTS LOADED
  0     NUMBER OF NODES LOADED
  0     CODE FOR PRESCRIBED DISP
  0     NUMBER OF NODES WITH NON ZERO DISP

ELEMENT COUNT          LIST OF ELEMENTS LOADED

      1                      3
      2                      4

CODE FOR UNIFORM BODY FORCES   0
BODYFX             BODYFY
 0.0                0.0

ELEMENT           BODY FORCE        EDGE LOADS        EQ NODAL LOADS        LOADED EDGE CODE

   3                  0                 1                   0                  0 1 0 0

ELEMENT    3

SIDE           PX1               PX2               PX3               PY1               PY2               PY3
  2        40000.00000       40000.00000       40000.00000           0.0               0.0               0.0

ELEMENT           BODY FORCE        EDGE LOADS        EQ NODAL LOADS        LOADED EDGE CODE

   4                  0                 1                   0                  0 1 0 0
```

ELEMENT

SIDE	PX1	PX2	PX3	PY1	PY2	PY3
2	40000.00000	40000.00000	40000.00000	0.0	0.0	0.0

DISPLACEMENTS AND REACTIONS

NODE	UX	UY	RX	RY
1	0.0	-7.47957E-03	-8.21276E 02	0.0
2	0.0	-3.69273E-03	-3.35996E 03	0.0
3	0.0	0.0	-1.63709E 03	1.63574E-02
4	0.0	3.69268E-03	-3.35997E 03	0.0
5	0.0	7.47952E-03	-8.21284E 02	0.0
6	2.50841E-02	-7.44546E-03	0.0	0.0
7	2.50234E-02	-5.05242E-08	0.0	0.0
8	2.50843E-02	7.44541E-03	0.0	0.0
9	5.02262E-02	-7.33387E-03	0.0	0.0
10	4.98940E-02	-3.74132E-03	0.0	0.0
11	4.99452E-02	9.59777E-08	0.0	0.0
12	4.98942E-02	3.74153E-03	0.0	0.0
13	5.02266E-02	7.33414E-03	0.0	0.0
14	7.38821E-02	-7.83829E-03	0.0	0.0
15	7.57942E-02	4.71076E-07	0.0	0.0
16	7.38829E-02	7.83928E-03	0.0	0.0
17	9.63218E-02	6.71494E-07	0.0	0.0
18	9.97438E-02	7.64154E-07	0.0	0.0
19	1.00659E-01	8.19564E-07	0.0	0.0
20	9.97443E-02	8.50996E-07	[illegible]	[illegible]

22	7.38829E-02	7.83910E-03	0.0	0.0
23	7.57946E-02	7.90460E-07	0.0	0.0
24	7.38829E-02	-7.83822E-03	0.0	0.0
25	5.02273E-02	7.33452E-03	0.0	0.0
26	4.98945E-02	3.74196E-03	0.0	0.0
27	4.99457E-02	5.37606E-07	0.0	0.0
28	4.98942E-02	-3.74091E-03	0.0	0.0
29	5.02269E-02	-7.33352E-03	0.0	0.0
30	2.50846E-02	7.44563E-03	0.0	0.0
31	2.50236E-02	1.62749E-07	0.0	0.0
32	2.50844E-02	-7.44528E-03	0.0	0.0
33	0.0	7.47961E-03	-8.21292E 02	0.0
34	0.0	3.69272E-03	-3.36001E 03	0.0
35	0.0	0.0	-1.63711E 03	-3.66211E-03
36	0.0	-3.69272E-03	-3.36000E 03	0.0
37	0.0	-7.47959E-03	-8.21278E 02	0.0

STRESSES

ELEMENT NO 1

GAUSS PT	GPX	GPY	SIGMA-X	SIGMA-Y	TAU-XY	SIGMA-Z
1	5.63508E-02	2.78175E-01	2.00459E 04	1.13582E 02	-2.77253E 01	0.0
2	5.63508E-02	3.75000E-01	1.99739E 04	9.27734E 00	-1.68042E 01	0.0
3	5.63508E-02	4.71824E-01	2.00029E 04	-6.47227E 01	1.65545E 01	0.0
4	2.50000E-01	2.78175E-01	1.99593E 04	-2.47266E 00	-7.76075E 01	0.0
5	2.50000E-01	3.75000E-01	1.99678E 04	3.40078E 01	1.12166E 01	0.0
6	2.50000E-01	4.71824E-01	2.00774E 04	1.00813E 02	1.22488E 02	0.0
7	4.43649E-01	2.78175E-01	1.98849E 04	-7.77539E 01	-9.80621E 01	0.0
8	4.43649E-01	3.75000E-01	1.99739E 04	9.95039E 01	6.86843E 01	0.0
9	4.43649E-01	4.71824E-01	2.01640E 04	3.07102E 02	2.57860E 02	0.0

1	5.63508E-02	2.81754E-02	2.00031E 04	-6.46836E 01	-1.65936E 01	0.0
2	5.63508E-02	1.25000E-01	1.99740E 04	9.32031E 00	1.67466E 01	0.0
3	5.63508E-02	2.21824E-01	2.00460E 04	1.13668E 02	2.76550E 01	0.0
4	2.50000E-01	2.81754E-02	2.00775E 04	1.00746E 02	-1.22480E 02	0.0
5	2.50000E-01	1.25000E-01	1.99679E 04	3.39883E 01	-1.12406E 01	0.0
6	2.50000E-01	2.21824E-01	1.99593E 04	-2.44141E 00	7.75853E 01	0.0
7	4.43649E-01	2.81754E-02	2.01642E 04	3.07043E 02	-2.57847E 02	0.0
8	4.43649E-01	1.25000E-01	1.99740E 04	9.94688E 01	-6.86840E 01	0.0
9	4.43649E-01	2.21824E-01	1.98850E 04	-7.77578E 01	9.80486E 01	0.0

ELEMENT NO 3

1	5.56350E-01	2.78175E-01	2.07946E 04	-4.66203E 02	-1.79127E 02	0.0
2	5.56351E-01	3.75000E-01	2.05047E 04	-3.89469E 02	-4.77950E 02	0.0
3	5.56350E-01	4.71824E-01	1.94527E 04	-5.41398E 02	-7.90516E 02	0.0
4	7.50000E-01	2.78175E-01	2.02194E 04	-2.97426E 02	4.95326E 01	0.0
5	7.50000E-01	3.75000E-01	1.98446E 04	-3.17621E 02	-1.26178E 01	0.0
6	7.50000E-01	4.71824E-01	1.87077E 04	-5.66473E 02	-8.85117E 01	0.0
7	9.43648E-01	2.78175E-01	2.09644E 04	4.27227E 03	2.35229E 02	0.0
8	9.43649E-01	3.75000E-01	2.05048E 04	4.15513E 03	4.09824E 02	0.0
9	9.43648E-01	4.71824E-01	1.92830E 04	3.80934E 03	5.70607E 02	0.0

ELEMENT NO 4

1	5.56350E-01	2.81754E-02	1.94530E 04	-5.41383E 02	7.90591E 02	0.0
2	5.56351E-01	1.25000E-01	2.05050E 04	-3.89469E 02	4.78026E 02	0.0
3	5.56350E-01	2.21824E-01	2.07947E 04	-4.66211E 02	1.79173E 02	0.0
4	7.50000E-01	2.81754E-02	1.87078E 04	-5.66523E 02	8.85311E 01	0.0
5	7.50000E-01	1.25000E-01	1.98447E 04	-3.17637E 02	1.25814E 01	0.0
6	7.50000E-01	2.21824E-01	2.02194E 04	-2.97426E 02	-4.96281E 01	0.0
7	9.43648E-01	2.81754E-02	1.92828E 04	3.80936E 03	-5.70586E 02	0.0
8	9.43649E-01	1.25000E-01	2.05048E 04	4.15514E 03	-4.09846E 02	0.0
9	9.43648E-01	2.21824E-01	2.09645E 04	4.27232E 03	-2.35417E 02	0.0

ELEMENT NO 5

1	1.05635E 00	2.78175E-01	-2.09639E 04	-4.27215E 03	2.35510E 02	0.0
2	1.05635E 00	3.75000E-01	-2.05043E 04	-4.15510E 03	4.09873E 02	0.0
3	1.05635E 00	4.71824E-01	-1.92822E 04	-3.80923E 03	5.70616E 02	0.0
4	1.25000E 00	2.78175E-01	-2.02191E 04	2.97703E 02	4.94707E 01	0.0
5	1.25000E 00	3.75000E-01	-1.98444E 04	3.17867E 02	-1.26254E 01	0.0
6	1.25000E 00	4.71824E-01	-1.87073E 04	5.66742E 02	-8.83948E 01	0.0
7	1.44365E 00	2.78175E-01	-2.07948E 04	4.66254E 02	-1.79287E 02	0.0
8	1.44365E 00	3.75000E-01	-2.05048E 04	3.89535E 02	-4.78040E 02	0.0
9	1.44365E 00	4.71824E-01	-1.94528E 04	5.41441E 02	-7.90500E 02	0.0

ELEMENT NO 6

1	1.05635E 00	2.81754E-02	-1.92828E 04	-3.80927E 03	-5.70496E 02	0.0
2	1.05635E 00	1.25000E-01	-2.05047E 04	-4.15519E 03	-4.09964E 02	0.0
3	1.05635E 00	2.21824E-01	-2.09641E 04	-4.27216E 03	-2.35610E 02	0.0
4	1.25000E 00	2.81754E-02	-1.87077E 04	5.66684E 02	8.84238E 01	0.0
5	1.25000E 00	1.25000E-01	-1.98447E 04	3.17813E 02	1.25419E 01	0.0
6	1.25000E 00	2.21824E-01	-2.02192E 04	2.97695E 02	-4.95288E 01	0.0
7	1.44365E 00	2.81754E-02	-1.94530E 04	5.41426E 02	7.90452E 02	0.0
8	1.44365E 00	1.25000E-01	-2.05050E 04	3.89535E 02	4.77982E 02	0.0
9	1.44365E 00	2.21824E-01	-2.07949E 04	4.66254E 02	1.79252E 02	0.0

ELEMENT NO 7

1	1.55635E 00	2.78175E-01	-1.98850E 04	7.77305E 01	-9.80449E 01	0.0
2	1.55635E 00	3.75000E-01	-1.99742E 04	-9.95703E 01	6.86737E 01	0.0
3	1.55635E 00	4.71824E-01	-2.01644E 04	-3.07164E 02	2.57952E 02	0.0
4	1.75000E 00	2.78175E-01	-1.99595E 04	2.42969E 00	-7.77234E 01	0.0
5	1.75000E 00	3.75000E-01	-1.99680E 04	-3.40156E 01	1.11951E 01	0.0
6	1.75000E 00	4.71824E-01	-2.00778E 04	-1.00848E 02	1.22649E 02	0.0
7	1.94364E 00	2.78175E-01	-2.00462E 04	-1.13676E 02	-2.76149E 01	0.0
8	1.94365E 00	3.75000E-01	-1.99741E 04	-9.30078E 00	-1.68011E 01	0.0
9	1.94364E 00	4.71824E-01	-2.00034E 04	6.46992E 01	1.65182E 01	0.0

ELEMENT NO 8

1	1.55635E 00	2.81754E-02	-2.01642E 04	-3.07082E 02	-2.57902E 02	0.0
2	1.55635E 00	1.25000E-01	-1.99741E 04	-9.95000E 01	-6.87780E 01	0.0
3	1.55635E 00	2.21824E-01	-1.98850E 04	7.77617E 01	9.80727E 01	0.0
4	1.75000E 00	2.81754E-02	-2.00776E 04	-1.00805E 02	-1.22636E 02	0.0
5	1.75000E 00	1.25000E-01	-1.99679E 04	-3.40117E 01	-1.12879E 01	0.0
6	1.75000E 00	2.21824E-01	-1.99595E 04	2.39453E 00	7.77375E 01	0.0
7	1.94364E 00	2.81754E-02	-2.00032E 04	6.46836E 01	-1.65110E 01	0.0
8	1.94365E 00	1.25000E-01	-1.99740E 04	-9.29688E 00	1.67411E 01	0.0
9	1.94364E 00	2.21824E-01	-2.00462E 04	-1.13688E 02	2.76344E 01	0.0

END

References

1. NEAL B.G., *Structural Theorems and their Applications*, Pergamon Press, 1964.
2. CHARLTON T.M., *Energy Principles in Theory of Structures*, Oxford Univ. Press London, 1973.
3. MATHESON J.A.L., *Hyperstatic Structure*, Vol. I, Butterworths, London, 1959.
4. PRZEMIENIECKI J.S., *Theory of Matrix Structural Analysis*, McGraw-Hill, 1968.
5. COOK R.D., *Concepts and Applications of Finite Element Analysis*, Wiley, 1975.
6. LIVESLEY R.K., *Matrix Methods of Structural Analysis*, Pergamon Press, 1964; (Revised ed., 1975).
7. ARGYRIS J.H. and S. KELSEY, *Energy Theorems and Structural Analysis*, Butterworths, London, 1960.
8. TIMOSHENKO S.P. and D.H. YOUNG, *Theory of Structures*, 2 ed. McGraw-Hill 1965.
9. HORNE M.R. and W. MERCHANT, *Elastic Stability of Frames*, Pergamon Press, 1965.
10. TIMOSHENKO S.P. and J.M. GERE, *Theory of Elastic Stability*, McGraw-Hill, 2nd ed., 1961.
11. TIMOSHENKO S.P., *Elements of Strength of Materials* Part II, Van Nostrand, 3rd ed., 1956.
12. ABROMOVITZ M. and I.A. STEGUN (Eds.) *Handbook of Mathematical Functions with Formulas, Graphs and Tables*, Dover, 1965.
13. TOTTENHAM H., *Matrix Progression Method in Structural Analysis—Structural Problems in Nuclear Reactor Vessels*, Pergamon Press, 1962.
14. TOTTENHAM H. and M.B. KANCHI, "Foundations of the Generalised Matrix progression method and its applications to the problems of Structural Mechanics", *Jour. of Inst. of Engrs*, India, **49**, Sept. 1969, 115-135.
15. PESTEL E.C. and F.A. LECKIE, *Matrix Methods in Elastomechanics*, McGraw-Hill, 1963.
16. TOTTENHAM H., "Matrix Progression Techniques in Structural Analysis." *Symposium on the Use of Computers in Civil Engineering*, Lisbon, 1962.
17. ARGYRIS J.H., "Continua and Discontinua", *Proc. Conf. Matrix Methods in Struct. Mech.* Wright Patterson, Air Force Base, Ohio, 1965.
18. DENKE P.H., *A Computerised Static and Dynamic Aircraft Structural Ananlysis System*, Soc. Automotive Engrs. S.P. 264, Dec., 1964.
19. ROBINSON J., *Structural Matrix Analysis for the Engineer*, Wiley, 1966.
20. TURNER N.J., R. CLOUGH, H.C. MARTIN and L.J. TOPP, "Stiffness and Deflection Analysis of Complex Structures", *J. Aero. Sci.* **23**, 805-23, 1956.
21. *Finite Element Solution System*, Part 1, Univ. College of Swansea, CNME Report CR 32, Feb., 1970.
22. RITZ W., "Uber eine neue Methode zur Losung gewisser Variationsprobleme der mathematischen Physik", *J. reine u angew Math*, **135** 1909, pp. 1-61.

23. GALERKIN B.G., Series Solution of some problems of Elastic Equilibrium of Rods and Plates (In Russian), *Vestnik Inzhenerov*, **1** 1915 pp. 897-908.

24. COURANT R., Variational methods for solution of problems of equilibrium and vibration, *Bull. Amer. Math. Soc.*, **49**, 1943, pp. 1-23.

25. TONG P., New Displacement Hybrid Finite Element Model for Solid Continua, *Int. Jour. Num. Method in Engrg.*, **2**, No. 1, 1970, pp. 73-84.

26. ZIENKIEWICZ O.C., *The Finite Element Method in Engineering Science*, McGraw Hill, 1973.

27. ERGATOUDIS J.G., B.M. IRONS and O.C. ZIENKIEWICZ, "Curved, Isoparametric Quadrilateral Elements for Finite Element Analysis", *Int. J. Solids and Struct.*, **4**, 31-42, 1968.

28. IRONS B.M., "Engineering Applications for Numerical Integration in Stiffness Method, *J. AIAA*, **4**, 2035-7, 1966.

29. SCHREM E. and J.R. ROY, "An Automatic System for Kinematic Analysis", *IUTAM Symp. on High Speed Computing of Elastic Structures*, Univ. Liege, Belgium, 1970.

30. MAJID, K.I., *Nonlinear Structures*, Butterworths, London, 1972.

31. WILSON E.L., K.J. BATHE and W.P. DOHERTY, "Direct Solution of large system of linear equations", *Computers and Structures*, **4**, 1974, pp. 363-372.

32. IRONS B.M., A Frontal solution program for Finite Element Analysis, *Int. Jour. Num. Math. in Engg.*, **2**, No. 1, 1970.

33. FUCHS VON G., J.R. ROY and E. SCHREM, "Hypermatrix Solution of Large Sets of Symmetric Positive Definite Linear Equations", *Comp. Method in App. Mech. and Engg.*, **1**, 1972, pp. 197-216.

34. DESAI C.S. and J.F. ABEL, *Introduction to the Finite Element Method*, Van Nostrand Reinhold, 1972.

35. GALLAGHER R.H., *Finite Element Analysis*, Prentice-Hall, 1975

36. BREBBIA C.A. and J.J. CONNOR, *Fundamentals of Finite Element Techniques*, Butterworths, London, 1973.

37. MEEK J.L., *Matrix Structural Analysis*, McGraw-Hill, 1971.

38. WEAVER, *Computer Programs for Structural Analysis*, Van Nostrand, 1969.

Index

Errata

Page	Position	For	Read
195	RHS of eq. (a_{11})	$-\begin{Bmatrix} -127.9920 \\ -172.7530 \\ -61.6 \end{Bmatrix}$	$-\begin{Bmatrix} -127.9920 \\ -172.7530 \\ -61.6 \end{Bmatrix} \frac{1}{EI_0}$
208	LHS of eq. (a_2)	$\mathbf{A}^t S$	$\mathbf{A}^t \mathbf{S}$
212	3rd expression on RHS of eq. (8.14)	$\begin{Bmatrix} u_1 \\ \hline v_1 \\ \hline u_2 \\ \hline v_2 \end{Bmatrix}$	$\begin{Bmatrix} u_1 \\ v_1 \\ u_2 \\ v_2 \end{Bmatrix}$
213	Table 8.1 1st row 2nd expression	(k_5-k_3cs)	$(k_5-k_3)cs$
	1st row 6th expression	$-ks_4$	$-k_4s$
216	eq. between (a_1) and (a_2) 2nd brackets	$\begin{bmatrix} \mathbf{K}_{11A} & \mathbf{K}_{12A} \\ \mathbf{K}_{21A} & \mathbf{K}_{12A} \end{bmatrix}$	$\begin{bmatrix} \mathbf{K}_{11A} & \mathbf{K}_{12A} \\ \mathbf{K}_{21A} & \mathbf{K}_{22A} \end{bmatrix}$
217	Table a column headings	K	k
218	In eq. (a_8)	$\begin{Bmatrix} u_B \\ v_B \\ \theta_B \end{Bmatrix} \begin{Bmatrix} 3.6 \\ 88.3 \\ -29.0 \end{Bmatrix}$	$\begin{Bmatrix} u_B \\ v_B \\ \theta_B \end{Bmatrix} = \begin{Bmatrix} 3.6 \\ 88.8 \\ -29.0 \end{Bmatrix}$
226	Sec. 8.3, 1st line	srtuctural	structural
228	8th line	moments	movements
234	1st of eq. (a_1)	$\mathbf{P}_{FA}$	$\overline{\mathbf{P}}_{FA}$

eq. (a_2) Read as : $\mathbf{P}_{FA} = \{0.8, -0.6, 0, -0.8, 0.6, 0\}\,(AE\epsilon^{\circ})_A$

238 eq. (8.30) Read as :

$$\begin{Bmatrix} \bar{M}_{x1} \\ \bar{M}_{y1} \\ \overline{Q_1} \\ \bar{M}_{x2} \\ \bar{M}_{y2} \\ \overline{Q_2} \end{Bmatrix} = \left[\begin{array}{ccc|ccc} k_5 & 0 & 0 & -k_5 & 0 & 0 \\ 0 & k_1 & -k_4 & 0 & k_2 & k_4 \\ 0 & -k_4 & k_3 & 0 & -k_4 & -k_3 \\ \hline -k_5 & 0 & 0 & k_5 & 0 & 0 \\ 0 & k_2 & -k_4 & 0 & k_1 & k_4 \\ 0 & k_4 & -k_3 & 0 & k_4 & k_3 \end{array}\right] \begin{bmatrix} \overline{\theta_{x1}} \\ \overline{\theta_{y1}} \\ \overline{w_1} \\ \overline{\theta_{x2}} \\ \overline{\theta_{y2}} \\ \overline{w_2} \end{bmatrix}$$